D1497883

MICROWAVE
MOBILE
COMMUNICATIONS

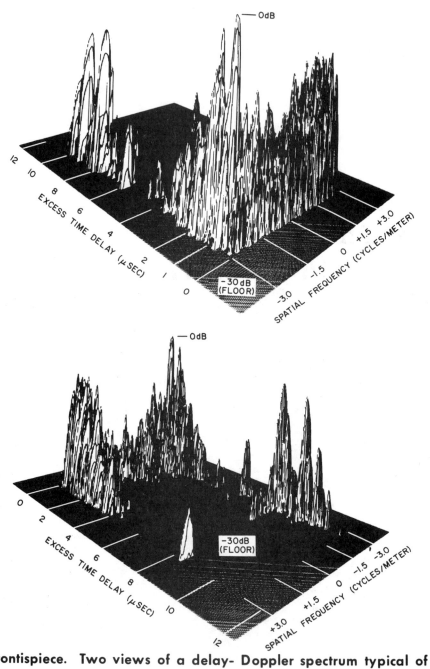

Frontispiece. Two views of a delay- Doppler spectrum typical of mobile radio transmission in New York City. (See Section 1.5)

MICROWAVE
MOBILE
COMMUNICATIONS

Edited by William C. Jakes

FORMERLY DIRECTOR, RADIO
TRANSMISSION LABORATORY
BELL TELEPHONE LABORATORIES
NORTH ANDOVER, MASSACHUSETTS

IEEE
PRESS

IEEE COMMUNICATIONS SOCIETY, *SPONSOR*

THE INSTITUTE OF ELECTRICAL AND ELECTRONICS ENGINEERS, INC., NEW YORK

IEEE Press
445 Hoes Lane
Piscataway, NJ 08855-1331

Reissued in cooperation with the
IEEE Communications Society

IEEE Communications Society Liaison to IEEE PRESS
Jack M. Holtzman

contents

preface to the IEEE edition   ix

preface to the first edition   xi

foreword   xiii

introduction   1
    Wm. C. Jakes

PART I MOBILE RADIO PROPAGATION

chapter 1 multipath interference   11
    Wm. C. Jakes

    Synopsis of Chapter   11
1.1   Spatial Distribution of the Field   13
1.2   Power Spectra of the Fading Signal   19
1.3   Power Spectrum and Other Properties of the Signal Envelope   24
1.4   Random Frequency Modulation   39
1.5   Coherence Bandwidth   45
1.6   Spatial Correlations at the Base Station   60
1.7   Laboratory Simulation of Multipath Interference   65

chapter 2 large-scale variations of the average signal   79
    D. O. Reudink

    Synopsis of Chapter   79
2.1   Factors Affecting Transmission   80
2.2   Observed Attenuation on Mobile Radio Paths over Smooth Terrain   90
2.3   Effects of Irregular Terrain   112
2.4   Statistical Distribution of the Local Mean Signal   119
2.5   Prediction of Field Strength   123

chapter 3 antennas and polarization effects   133
Y. S. Yeh

Synopsis of Chapter   133
3.1 Mobile Antennas   134
3.2 Base Station Antennas   150
3.3 Polarization Effects   152

PART II MOBILE RADIO SYSTEMS

chapter 4 modulation, noise, and
interference   161
M. J. Gans and Y. S. Yeh

Synopsis of Chapter   161
4.1 Frequency Modulation   162
4.2 Digital Modulation   218
4.3 Channel Multiplexing   240
4.4 Man-Made Noise   295

chapter 5 fundamentals of diversity systems   309
Wm. C. Jakes, Y. S. Yeh, M. J. Gans,
and D. O. Reudink

Synopsis of Chapter   309
5.1 Basic Diversity Classifications   310
5.2 Combining Methods   313
5.3 Antenna Arrays for Space Diversity   329
5.4 Effect of Diversity on FM Noise and Interference 341
5.5 Diversity Against Shadowing   377

chapter 6 diversity techniques   389
D. O. Reudink, Y. S. Yeh, and
Wm. C. Jakes

Synopsis of Chapter   389
6.1 Postdetection Diversity   390

6.2 Switched Diversity   399
6.3 Coherent Combining Using Carrier Recovery   423
6.4 Coherent Combining Using a Separate Pilot   464
6.5 Retransmission Diversity   489
6.6 Multicarrier AM Diversity   512
6.7 Digital Modulation-Diversity Systems   517
6.8 Comparison of Diversity Systems   531

chapter 7 layout and control of high-capacity systems   545

D. C. Cox and D. O. Reudink

     Synopsis of Chapter   545
7.1 Large Radio Coverage Area Systems   546
7.2 Coverage Layout of Small Cell Systems   562
7.3 Base Station Assignment in Small Cell Systems   568
7.4 Channel Assignment in Small Cell Systems   572

appendix a computation of the spectra of phase-modulated waves by means of Poisson's sum formula   623

M. J. Gans

appendix b click rate for a nonsymmetrical noise spectrum   627

M. J. Gans

appendix c median values of transmission coefficient variations   631

M. J. Gans

index   635

preface to the IEEE edition

Since the original publication of this book in 1974 mobile radio (now called *cellular*) has experienced dramatic growth. In addition, the new field of "wireless communications" is beginning to appear, which experiences some of the same types of problems as mobile radio.

High interest in both of these disciplines has markedly increased the demand for the present work; thus, we are extremely pleased that the IEEE Press has decided to republish our book and make it more readily available. This also provides an opportunity to include a number of corrections made known to us over the years.

We hope that this book will continue to help those involved in these fast-growing fields.

January 1994

Wm. C. Jakes
Editor

preface to the first edition

Radio communication with moving vehicles is almost as old as radio itself. Only recently, however, has technology advanced to the point where realization of the dream of instant communication with the mobile public appears possible. A more difficult set of constraints for the system designer is hard to imagine: The transmission medium is very lossy and dispersive, suffers extreme random fades, and the location of one of the terminals is unknown! Provision of a high-capacity service also requires use of a significant portion of a natural resource—the frequency spectrum. In 1970 the Federal Communications Commission took a giant step by proposing authorization of a band of frequencies 75 MHz wide around 840 MHz for mobile telephony (Docket 18262). This made it possible, for the first time, to seriously consider the introduction of high-capacity systems capable of serving hundreds of thousands of subscribers in metropolitan areas.

In the past the investigation of radio propagation and systems particularly suited to large-scale mobile telephony has proceeded at a relatively slow pace. Without an allocated frequency band the necessary incentive was lacking; without demonstrably viable systems there was little motivation to allocate such a band. Over the years, however, knowledge has slowly been acquired, and in the recent past the efforts have increased markedly. In view of the current expansion of interest in mobile telephony it seems timely to collect in one place a substantial portion of the existing information, some of it heretofore unpublished, for easy access and use by all workers in the field.

The emphasis of the book is on mobile communication in the microwave range, from about 450 MHz up to 10 or 20 GHz. It is in this frequency range that the very high-capacity systems will come into being. The first three chapters, making up Part I, treat basic propagation phenomena over the mobile radio transmission path and include a discussion of mobile radio antennas. Part II contains four chapters devoted to various factors entering into the design of mobile radio systems, such as modulation, space diversity fundamentals and techniques, and coverage and organization considerations. The authors have drawn freely on published literature that is available and germane to the subject. The bulk of the material presented is based on work done by our colleagues in the Bell Telephone Laboratories, where research directed specifically toward high-capacity systems has been underway for the past decade. Many of these individuals, in addition to those directly involved with preparation of material for the book, have given generously of their time and advice. Although there are too many to recognize individually, their participation is gratefully acknowledged.

April 1974

Wm. C. Jakes
Editor

foreword

It is a personal pleasure to write a foreword to the IEEE reprinting of this classic book, which I will take the liberty of calling "Jakes" as everyone else does. In 1990, I joined the Wireless Information Network Laboratory (WINLAB) at Rutgers University and changed technical fields to wireless communications. Wireless communications technology is unusually multi-faceted and it is easy to get overwhelmed while on the steep part of the learning curve. I searched for good first reading sources. Everyone I asked had different lists of sources but Jakes was common to all. Then, when I had learned enough to start asking semi-intelligent questions, I would often be referred back to Jakes. A problem with such references to the book was locating a copy. Many people now in wireless communications, like me, were new and did not have a copy and were unable to locate one. It was a genuine problem for all to get access to the book. I am, naturally, delighted to see a reprinting of the book so I can finally get my own copy!

We normally think of literary classics as permanent, but, technical classics are constantly being superseded (how many of us have ever read Newton or Darwin?). However, in this case, Jakes remains eminently worth reading. Wireless communications is a hot topic and a number of books are now appearing. As valuable as any of these books may be, there is nothing that compares with reading this classic and going back to the "source." This book provides a fundamental understanding of many issues that are currently being investigated further. In addition, the book gives a sense of how many of these fundamental ideas were conceived.

An example of a topic that is superbly covered is Rayleigh fading. To really understand Rayleigh fading, from both the physical and mathematical viewpoints, I can think of no better prescription than first reading the classic paper by R. H. Clarke, "A Statistical Theory of Mobile-Radio Reception," *BSTJ*, July–August 1968, then reading Jakes, and then entering the more recent books.

I hope that you gain as much as I did from this book.

Jack M. Holtzman
WINLAB
Department of Electrical and Computer Engineering
Rutgers University
Piscataway, New Jersey 08855-0909

introduction

W. C. Jakes

HISTORICAL

The history of mobile communications begins with the first experiments of the radio pioneers. The startling demonstrations of Hertz in the 1880s inspired the entrepreneur Marconi to seek a market for this marvelous new commodity. Among his early feats was a transmission to a tugboat over an 18-mile path from Needles on the Isle of Wight in 1897. After limited use of radio communications in World War I, more as a curiosity than anything else, the first land mobile radiotelephone system was installed in 1921 by the Detroit Police Department for police car dispatch. The New York City Police Department followed suit in 1932.[1]

These first systems operated in the 2-MHz frequency band. As technology and needs increased during the next decade, the trend was to higher frequencies. In 1933 the Federal Communications Commission (FCC) authorized four channels in the 30–40-MHz band on an experimental basis, and ruled for regular service in 1938. World War II imposed a temporary hiatus on installation of commercial systems, but the technological advances made during the war made it possible to exploit ever-higher frequencies. Experimental work at 150 MHz directed specifically toward mobile systems was started in 1945 at Bell Telephone Laboratories. With the authorization by the FCC of a few channels in the vicinity of 35 and 150 MHz, commercial service in these two bands was initiated in 1946 by the Bell system with the installation of a 35-MHz system in Green Bay, Wisconsin and a 150-MHz system in St. Louis, Missouri. Operation was simplex (push-to-talk) and call placement was handled by a mobile telephone operator. The mobile customer also had to search manually for an idle channel before placing a call. In 1956 the same type of service was introduced on newly authorized channels in the 450-MHz band.

Work continued at Bell Laboratories and other places with the twofold objective of improving existing service and pushing on to higher frequencies. In 1964 a new 150-MHz Bell Telephone system was made available which provided full duplex operation, automatic channel search, and dialing to and from the mobile station.[2] This was followed in 1969 with the

1

introduction of the same kind of improved operation at 450 MHz.[3]

During the years following World War II many other kinds of mobile telephone systems were introduced. These generally operated at frequencies below 460 MHz and provided service to specialized groups falling into three main classes. The Public Safety category includes police cars, ambulances, civil defense units, fire trucks, rescue squad vehicles, and other units associated with local government activities. A large variety of users are authorized in the Industrial and Land Transportation class, such as taxicabs, cement trucks, oil distributors, forest products, manufacturers, railroads, businesses, newspapers, motion picture producers, and power utilities. Private individuals were also given the opportunity to use mobile radio for personal purposes with the establishment of a number of Citizens Bands at 460 MHz and below. The extent of interest in mobile communications may be judged by the fact that in the Citizens Bands alone over 800,000 authorizations existed in 1970.

The development and expansion of land mobile systems described above has been paralleled, although to a much lesser degree, by introduction of mobile telephone service to boats and aircraft. The basic needs have been somewhat different, characterized more by emergency or vehicle control considerations than business or casual conversations. Eventually, however, the traveling public will expect accessibility to this convenience regardless of their mode of travel.

Thus since the beginning days of mobile radio telephone in the early 1920s, the picture has been one of steady growth characterized by advancing technology and increasing demand for service that always exceeded available system capabilities.

EVOLUTIONARY FORCES

The brief history of mobile radiotelephone summarized in the preceding section can only be regarded as a prelude. The ultimate objective of mobile communications is to enable anyone on the move to communicate instantly, easily, and effectively with anyone else. The fundamental problem has already surfaced: lack of frequency bandwidth to handle the service demand in regions of the frequency spectrum where technology can provide reasonably economical hardware and systems. A measure of the potential demand for service is indicated by the fact that in 1970 there were 100 million moving vehicles on U. S. highways.

In the first place, then, advances are needed in radio devices and techniques to conquer the more severe transmission problems faced at higher frequencies. In turn, large portions of the spectrum need to be set aside for mobile telephone service. The FCC acts as custodian of this natural resource—the frequency spectrum—and must balance the sometimes conflicting needs of the different services required by society.

FREQUENCY ALLOCATIONS

The majority of potential users of mobile telephones falls into the category denoted Domestic Public Land Mobile Service, which is primarily made up of ordinary, private citizens. The FCC has responded to the needs for this type of service over the years by allocations in three bands at 40, 150, and 450 MHz. As time went on the number of designated channels in the upper two bands was increased by the simple expedient of reducing the frequency separation between channels. Thus the first allocation at 150 MHz in 1946 was for three channels spaced 120 kHz apart. The spacing was shortly reduced to 60 kHz, and then to 30 kHz in the early 1960s. Similarly, at 450 MHz the channel spacing was initially 100 kHz, reduced to 50 kHz in the early 1960s, and to 25 kHz in the late 1960s. The channel allocation scheme existing in 1970 is shown in the accompanying table, and we see that the total bandwidth available at this time for Domestic Land Mobile was 1.95 MHz.

Domestic Public Land Mobile Frequencies in 1970

Base Transmit (MHz)	Mobile Transmit (MHz)	Number of Channels	Channel Spacing (kHz)	Total Bandwidth (MHz)
35.26–35.66	43.26–43.66	10	40	0.8
152.51–152.81	157.77–158.07	11	30	0.6
454.375–454.15	459.375–459.65	12	25	0.55

Increasing the number of channels by "channel splitting" is soon limited, of course, and does not even begin to approach the number of channels required for a truly high-capacity system. There are also technical problems of frequency stability and adjacent channel interference that limit the usefulness of this technique. In May 1970, the FCC (Docket 18262) recognized the situation and proposed to allocate a significant portion of the UHF band, from 806 to 881 MHz, for Domestic Public Land Mobile Service. This was an increase in bandwidth of almost 40 times the previous total of 1.9 MHz. At present writing, plans and proposals are being considered by industry to make efficient use of this new band for FCC approval.

THE USE OF HIGHER FREQUENCIES

Suppose wide frequency bands are available at microwave frequencies—what then? Transmission problems at these frequencies reach staggering proportions, and brute-force techniques used at the lower frequencies are entirely inappropriate. The signal received by a mobile* from a base station as it moves about is very weak, on the average, since it is generally obstructed by buildings and natural obstacles. In addition, the signal undergoes violent and rapid variations in amplitude and phase; these variations introduce intolerable amounts of noise and distortion into a voice-band signal. Simply increasing transmitter power does not overcome these effects; thus new techniques are needed to provide acceptable speech transmission.[4] These new techniques can only result from detailed and intensive study of the basic propagation phenomena.

In order to provide mobile telephone service to hundreds of thousands of users in a metropolitan area, it is quite clear that the available radio channels must be re-used a number of times at different locations within the overall area, in order to use the assigned spectrum most efficiently. Thus there is a potential problem of mobiles simultaneously using the same channel in different locations interfering with each other. The severity of this interference depends on relative signal strengths, which in turn depend on the various factors governing radio transmission at these high frequencies.[5,6]

In a nutshell, then, the natural evolution of demand for mobile telephone service has brought us face-to-face with the ultimate objective of providing it for everyone, and to the realization that high frequencies must be used with new and unusual transmission techniques.

SCOPE OF THE BOOK

Many previous books devoted to mobile communications have been largely equipment-oriented. The intent of the present work is to expose in detail what is now known about mobile radio transmission at frequencies

---

*The word "mobile" will be used to denote a vehicle equipped with a two-way radio telephone.

from VHF to the upper microwave range, and to explore the fundamental design principles of modulation and signal-processing techniques that promise to provide acceptable voice transmission in the mobile radio environment. The reader should have a background in radio communication systems and some familiarity with statistical communication theory.

REVIEW OF CHAPTER CONTENTS

Each chapter begins with a detailed synopsis of the chapter, and the reader may find this helpful. A brief overall summary is given here for preliminary orientation.

Chapter 1 treats the most serious problem of all, the rapid and extreme amplitude and phase modulation introduced by movement of the mobile. The cause of this modulation is attributed to multipath propagation, and it is shown how statistical communication theory can be applied to predict many of the observed properties of transmission. The Rayleigh fading distribution of the signal envelope is an immediate result; then the Doppler shifts introduced by vehicle motion are shown to cause both a broadening of the RF spectrum and random frequency modulation. The coherence bandwidth of the transmission depends on the transmission delays of the various transmission paths, and phase and envelope correlation properties for different transmission frequencies are thereby derived. The effect of asymmetries in the mobile-base path on spatial correlations at the mobile and base station is explored, and the chapter then concludes with a discussion of devices that can faithfully simulate virtually all of the properties of the mobile transmission path, thus making it possible to test new systems in the laboratory before actual field trials.

Chapter 2 is a study of the variations in the "local mean" signal strength caused by terrain, environment, and antenna sites. The term "local mean" denotes the average signal level, averaged over several tens of wavelengths so that the rapid, multipath effects are essentially removed. After a review of basic transmission factors, such as line-of-sight path loss, diffraction, and rain or foliage attenuation, the measured dependence of path loss in flat urban areas on frequency, range, and antenna heights is studied. Then the effects of different environments are considered, such as suburban instead of urban areas, rolling hilly terrain, mountains, and land–sea boundaries. Propagation through tunnels is also discussed.

The local mean signal is generally observed to obey a log-normal distribution as a function of receiver location, and the variance of the distribution is shown to depend on the general physical parameters of the environment. Examples of predicted path loss for various circumstances are given at the end of the chapter.

Chapter 3 is a discussion of antennas suitable for mobile radio transmission in the microwave range, with the emphasis on principles rather than specific hardware designs. At these short wavelengths there is greater freedom to take advantage of the improvements afforded by more directive radiation patterns, and this is discussed from the different points of view of antennas at the base station and mobile. It is shown, for instance, how a fixed directive antenna on the mobile can cause a reduction in the fading rate, which has important implications for system design. The effects of the environment on the apparent polarization of the received field at the mobile are also explored.

Chapter 4 begins Part II, Mobile Radio Systems, with a treatment of various kinds of modulation, some traditional and others as yet untried for mobile radio. Frequency modulation is discussed first, starting with a review of threshold considerations in conventional applications and then developing the effect of the mobile radio multipath fading on signal-to-noise ratios and audio distortion. A comparison is made with other types of analog modulation in this environment. The possibility of using some form of digital modulation is then discussed, and trade-offs are established between the design parameters in the fading situation. Various forms of channel multiplexing that might be considered for high-capacity systems are compared. The last section is a brief discussion of man-made noise.

Chapter 5 introduces the concept of diversity as the most promising scheme for elimination of the effects of multipath fading and begins with a discussion of the basic diversity classifications. A review of the various methods of combining the diversity signals then follows. Space diversity appears to be the most promising type for mobile use, and the factors involved in providing this with an array of antennas at the mobile or base station are explored. A treatment of the effect of diversity on baseband signal-to-noise ratio is given, and the chapter concludes with a discussion of the use of diversity on a large scale to reduce terrain shadowing effects.

Chapter 6 continues the treatment of space diversity systems, specializing in the details of techniques that would be particularly applicable to UHF and microwave mobile communications. Combining the diversity signals after detection is discussed first, and then methods of switching antennas to select the best diversity branch. Several types of coherent predetection combining methods are treated, and it is shown how they are able to reduce or completely eliminate the random FM caused by multipath interference, in addition to reducing the amplitude fading. Special systems that provide diversity advantage for both directions of the two-way base-mobile communication path, while keeping the diversity antenna array and processing equipment at only one end, are then discussed.

A special type of frequency diversity using AM is then described, followed by a study of diversity systems using digital modulation. The overall performance of several diversity systems is then compared with respect to their operational parameters.

Chapter 7 is addressed to the general problem of the layout and organization of a high-capacity mobile telephone system. As mentioned earlier the most appropriate way to cover a large service area is to divide it into small contiguous coverage cells, each served by, at most, only a few base stations. For reference the traffic in the conventional single large coverage area is studied first, then the methods of laying out multiple small cells to cover a large area are discussed. The considerations affecting base station assignments are studied next.

A variety of ways to allocate the radio channels of the assigned frequency band to the geographical cells has been proposed in the past. Two general categories of these methods, using either a fixed or variable number of channels assignable in a given cell, are treated from the standpoint of layout options and traffic handling capabilities. The chapter concludes with a discussion of ways to reassign channels to calls in progress, so that the most effective use is made of the total number of available radio channels.

REFERENCES

1. E. M. Webster, "Utilization and Expansion of Vehicular Radio Communications," *IEEE Trans. Veh. Comm.* 1, February, 1952, p. 66.

2. V. A. Douglas, "The MJ Mobile Radio Telephone System," *Bell Lab. Record*, December, 1964, pp. 383–389.

3. R. J. Cormier and H. M. Owendoff, "The MK Mobile Radio Telephone System," *IEEE Veh. Tech. Conf.*, December 4–5, 1969, Columbus, Ohio.

4. W. C. Jakes, Jr., "New Techniques for Mobile Radio," *Bell Labs Record*, December, 1970, pp. 326–330.

5. W. D. Lewis, "Coordinated Broadband Mobile Telephone System," *IRE Trans. Veh. Comm.*, May, 1960, pp. 43–48.

6. H. J. Schulte, Jr. and W. A. Cornell, "Multi-Area Mobile Radiotelephone System," *IRE Trans. Veh. Comm.*, May, 1960, pp. 49–53.

part I

mobile radio
propagation

chapter 1

multipath interference

W. C. Jakes

SYNOPSIS OF CHAPTER

Nature is seldom kind. One of the most appealing uses for radio-telephone systems—communication with people on the move—must over-come radio transmission problems so difficult they challenge the imagination. A microwave radio signal transmitted between a fixed base station and a moving vehicle in a typical urban environment exhibits extreme variations in both amplitude and apparent frequency. Fades of 40 dB or more below the mean level are common, with successive minima occurring about every half wavelength (every few inches) of the carrier transmission frequency. A vehicle driving through this fading pattern at speeds up to 60 mi/hr can experience random signal fluctuations occurring at rates of 100–1000 Hz, thus distorting speech when transmitted by conventional methods. These effects are due to the random distribution of the field in space, and arise directly from the motion of the vehicle. If the vehicle is stationary the fluctuation rates are orders of magnitude less severe.

These observations seem to defy any attempt at a systematic interpretation or quantitative analysis. However, starting from a model based on multipath wave interference, arising from multiple scattering of the waves by the buildings and other structures in the vicinity of the mobile unit, we shall see that a great many of the observable properties of the transmission can be successfully predicted by using the powerful techniques of statistical communication theory. The fundamental relationships are established in Section 1.1, where the fields are expressed as a linear superposition of plane waves of random phase. This leads directly to expressions giving the probability that the signal envelope can be expected to be found within a narrow range around a given level (probability density), and the percent of time it lies below a given level (cumulative distribution).

Besides a random phase, each component plane wave has associated with it a Doppler shift depending on the mobile speed, the carrier frequency, and the angle its propagation vector makes with the mobile velocity vector. This implies that the apparent power spectrum of each of the three received field components is broadened to occupy a narrow band about the carrier frequency. General expressions for the shapes of these power spectra are derived in Section 1.2 and show their dependence on the assumed density of the arrival angles and the mobile antenna directivity pattern. As a by-product of these derivations it is also shown that, with relatively loose restrictions on the distribution of arrival angles, the three field components are statistically uncorrelated if they are observed simultaneously.

The radio frequency characteristics of signals are usually somewhat difficult to observe; the signal envelope is more directly accessible. A number of properties of the fading signal envelopes associated with the mobile transmission path are derived in Section 1.3. Some general correlations and statistical moments of the in-phase and quadrature signal components are presented, preparatory to deriving the power spectrum of the envelope from the Fourier transform of its autocorrelation. The inclusion of a small steady signal from the base station is shown to explain some observed fine structure in the power spectra. The rate at which the envelope crosses a specified signal level is frequently of interest; expressions are presented for this property, along with expressions for the average length of time the envelope spends below a specified level. Finally, the auto- and cross-covariance functions for the envelopes of the three electromagnetic field components are derived, and it is shown how they may be interpreted in terms of spatial instead of time coordinates.

The instantaneous frequency of the signal received at the mobile undergoes rapid, random variations due to the variations of the in-phase and quadrature signal components. This random frequency modulation is explored in Section 1.4, where its probability density and cumulative distribution are derived first. The larger frequency deviations are shown to coincide with the deeper fades and are many times larger than the Doppler shift of any of the constituent plane waves. The power spectrum of the random frequency modulation is also derived; beyond a frequency equal to twice the maximum possible Doppler shift it drops as $1/f$.

If two carriers are transmitted from the base station at slightly different frequencies, then their statistical properties, as observed at the mobile antenna, are independent if the frequency separation is large enough. The frequency separation for which the signals are still strongly correlated is called the coherence bandwidth, and is studied in Section 1.5. The basic mechanism responsible for this property is shown to be the difference in propagation time delays associated with the various scattered waves making up the total signal. The relationship between the coherence bandwidth and the standard deviation of the time delay distribution is developed and compared with measurements for several models of the delay distribution. Expressions are derived for the correlation of the carrier amplitudes and phases as a function of the frequency separation.

The auto-covariance of any field component envelope measured at two separate points on the mobile is shown in Section 1.3 to decrease rapidly as the spatial separation increases. When the mobile transmits to the base station, however, the analogous auto-covariance measured at the base decreases much more slowly with spatial separation. The reasons for this effect are studied in Section 1.6. They are shown to be related to the geometrical asymmetry in the mobile-base transmission path, which arises because the base antenna is usually located well above any nearby scattering objects. Theoretical expressions for the auto-covariance are derived and compared with the relatively few measurements available.

The final acceptability of mobile radio systems is usually established by tests in the field. Preliminary comparisons between alternative system designs would be considerably expedited if such tests could be carried out in the laboratory, using signals that provide the same characteristics of the fading signals observed in the field. A simulation scheme is described in Section 1.7 that duplicates the envelope fading statistics, correlation function, power spectrum, and random frequency modulation of the mobile radio signal. The method is patterned after the basic multipath interference model described in Section 1.1. The validity of the technique is established by appropriate statistical measurements.

1.1 SPATIAL DISTRIBUTION OF THE FIELD

1.1.1 Envelope Measurements and Mathematical Model

One readily accessible property of the signal transmitted over a mobile radio propagation path is the amplitude variation of its envelope as the position of the mobile terminal is moved. This information is generally presented in the form of time recordings of the signal level; with uniform vehicle motion there is, of course, a 1 : 1 correspondence between distance measured on the recording and distance traveled in the street. A typical recording[1] is shown in Figure 1.1-1 for a run made at 836 MHz in a suburban environment. The occasional deep fades and quasiperiodic occurrence of minima are clearly evident in the expanded section of the record.

Recordings such as these made by many workers in the field over the frequency range from 50 to 11,200 MHz have shown that the envelope of the mobile radio signal is Rayleigh distributed[2-5] when measured over distances of a few tens of wavelengths where the mean signal is sensibly constant. This suggests the assumption,[6] reasonable on physical grounds, that at any point the received field is made up of a number of horizontally traveling plane waves with random amplitudes and angles of arrival for different locations. The phases of the waves are uniformly distributed from 0 to 2π. The amplitudes and phases are assumed to be statistically independent. Other models have also been proposed,[7] but they lead to comparable statistical properties of the field for large numbers of constituent waves.

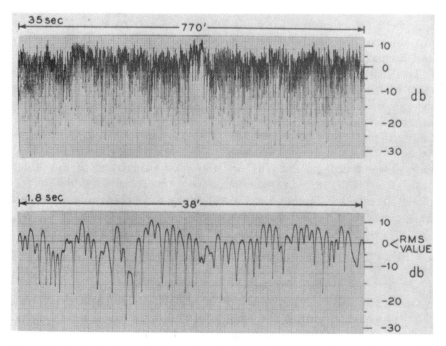

Figure 1.1-1 Typical received signal variations at 836 MHz measured at a mobile speed of 15 miles/hr. Records taken on the same street with different recording speeds.

A diagram of this simple model is shown in Figure 1.1-2 with plane waves from stationary scatterers incident on a mobile traveling in the x-direction with velicity v. The x-y plane is assumed to be horizontal. The vehicle motion introduces a Doppler shift in every wave:

$$\omega_n = \beta v \cos \alpha_n, \qquad (1.1\text{-}1)$$

where $\beta = 2\pi/\lambda$, λ being the wavelength of the transmitted carrier frequency.

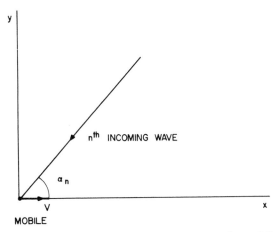

Figure 1.1-2   A typical component wave incident on the mobile receiver.

If the transmitted signal is vertically polarized, the field components seen at the mobile can thus be written

$$E_z = E_0 \sum_{n=1}^{N} C_n \cos(\omega_c t + \theta_n), \qquad (1.1\text{-}2)$$

$$H_x = -\frac{E_0}{\eta} \sum_{n=1}^{N} C_n \sin\alpha_n \cos(\omega_c t + \theta_n), \qquad (1.1\text{-}3)$$

$$H_y = \frac{E_0}{\eta} \sum_{n=1}^{N} C_n \cos\alpha_n \cos(\omega_c t + \theta_n), \qquad (1.1\text{-}4)$$

where

$$\theta_n = \omega_n t + \phi_n, \qquad (1.1\text{-}5)$$

and ω_c is the carrier frequency of the transmitted signal, η is the free-space wave impedance, $E_0 C_n$ is the (real) amplitude of the nth wave in the E_z field. The ϕ_n are random phase angles uniformly distributed from 0 to 2π. Furthermore, the C_n are normalized so that the ensemble average $\langle \sum_{n=1}^{N} C_n^2 \rangle = 1$.

We note from Eq. (1.1-1) that the Doppler shift is bounded by the values $\pm \beta v$ which, in general, will be very much less than the carrier frequency. For example, for $f_c = \omega_c/2\pi = 1000$ MHz, $v = 60$ mi/hr:

$$\frac{1}{2\pi}\beta v = \frac{v}{\lambda} \doteq 90 \text{ Hz.} \tag{1.1-6}$$

The three field components may thus be described as narrow-band random processes. Furthermore, as a consequence of the central limit theorem, for large values of N they are approximately Gaussian random processes, and the considerable body of literature devoted to such processes may be utilized. It must be kept in mind that this is still an approximation; for example, Eq. (1.1-2) implies that the mean signal power is constant with time, whereas it actually undergoes slow variations as the mobile moves distances of hundreds of feet. Nevertheless, the Gaussian model is successful in predicting the measured statistics of the signal to good accuracy in most cases over the ranges of interest for the variables involved; thus its use is justified.

Following Rice[8] we can express E_z as

$$E_z = T_c(t)\cos\omega_c t - T_s(t)\sin\omega_c t, \tag{1.1-7}$$

where

$$T_c(t) = E_0 \sum_{n=1}^{N} C_n \cos(\omega_n t + \phi_n), \tag{1.1-8}$$

$$T_s(t) = E_0 \sum_{n=1}^{N} C_n \sin(\omega_n t + \phi_n), \tag{1.1-9}$$

are Gaussian random processes, corresponding to the in-phase and quadrature components of E_z, respectively. We denote by T_c and T_s the random variables corresponding to $T_c(t)$ and $T_s(t)$ for fixed t. They have zero mean and equal variance:

$$\langle T_c^2 \rangle = \langle T_s^2 \rangle = \frac{E_0^2}{2} = \langle |E_z|^2 \rangle. \tag{1.1-10}$$

The brackets indicate an ensemble average over the α_n, ϕ_n, and C_n. T_c and T_s are also uncorrelated (and therefore independent):

$$\langle T_c T_s \rangle = 0. \tag{1.1-11}$$

1.1.2 Probability Distributions

Since T_c and T_s are Gaussian, they have probability densities of the form

$$p(x) = \frac{1}{\sqrt{2\pi b}} e^{-x^2/2b} \qquad (1.1\text{-}12)$$

where $b = E_0^2/2$ is the mean power, and $x = T_c$ or T_s.
The envelope of E_Z is given by

$$r = (T_c^2 + T_s^2)^{1/2}, \qquad (1.1\text{-}13)$$

and Rice[8] has shown that the probability density of r is

$$p(r) = \begin{cases} \dfrac{r}{b} e^{-r^2/2b}, & r \geqslant 0 \\ 0, & r < 0 \end{cases} \qquad (1.1\text{-}14)$$

which is the Rayleigh density formula. The Gaussian and Rayleigh densities are shown in Figure 1.1-3 for illustration.

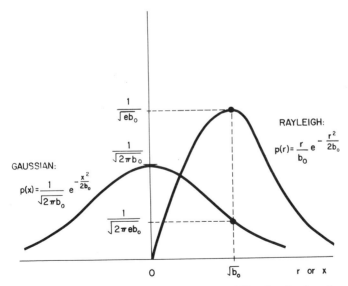

Figure 1.1-3 Gaussian and Rayleigh probability density functions.

The cumulative distribution functions of T_c (or T_s) are also of interest:

$$P[x \leqslant X] = \int_{-\infty}^{X} p(x)\,dx = \frac{1}{2}\left[1 + \text{erf}\left(\frac{X}{\sqrt{2b}}\right)\right], \qquad (1.1\text{-}15)$$

where the error function is defined by

$$\text{erf}(y) = \frac{2}{\sqrt{\pi}} \int_{0}^{y} e^{-t^2}\,dt. \qquad (1.1\text{-}16)$$

Similarly for the envelope,

$$P[r \leqslant R] = \int_{-\infty}^{R} p(r)\,dr = 1 - e^{-R^2/2b}. \qquad (1.1\text{-}17)$$

These distribution functions are illustrated in Figure 1.1-4.

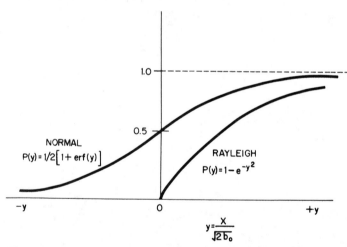

Figure 1.1-4   Normal and Rayleigh cumulative distributions.

Thus the simple model has predicted the widely observed Rayleigh nature of the fading. Some of the measurements are shown in Figure 1.1-5 for tests made at 836 and 11,200 MHz in a suburban area.[1] The coordinates in Figure 1.1-5 are scaled so that the Rayleigh cumulative distribution appears as a straight line.

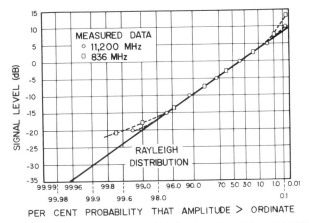

Figure 1.1-5 Cumulative probability distributions for 836 and 11,200 MHz.

The random processes T_c and T_s defined by Eqs. (1.1-8) and (1.1-9) will form the basis for much of the statistical analysis to follow in succeding sections of this chapter. Arguments were advanced earlier to justify the assumption that they are Gaussian random processes. In addition, for times that are short compared to the slow variations it will be assumed that they are wide-sense stationary; since they are assumed to be Gaussian this implies that they are stationary in the strict sense.[9] Such processes (e.g., the wide-sense stationary uncorrelated scattering channel of Bello[10, 11]) have been extensively studied, and expressions have been obtained for many statistical properties that will be freely applied in this work. The final justification of these assumptions, of course, is the accuracy with which the analytical results agree with measurements, but in most cases the agreement is good enough to lend credence to the model.

1.2 POWER SPECTRA OF THE FADING SIGNAL

From the viewpoint of an observer on the mobile unit, the signal received from a CW transmission as the mobile moves with constant velocity may be represented as a carrier whose phase and amplitude are randomly varying, with an effective bandwidth corresponding to twice the maximum Doppler shift of βv. Many of the statistical properties of this random process can be determined from its moments, which, in turn, are most easily obtained from the power spectrum.

1.2.1   RF Spectra of the Field Components

We assume that the field may be represented by the sum of N waves, as in Eq. (1.1-2). As $N \to \infty$ we would expect to find that the incident power included in an angle between α and $\alpha + d\alpha$ would approach a continuous, instead of discrete, distribution. Let us denote by $p(\alpha)d\alpha$ the fraction of the total incoming power within $d\alpha$ of the angle α, and also assume that the receiving antenna is directive in the horizontal plane with power gain pattern $G(\alpha)$. The differential variation of received power with angle is then* $bG(\alpha)p(\alpha)d\alpha$; we equate this to the differential variation of received power with frequency by noting the relationship between frequency and angle of Eq. (1.1-1):

$$f(\alpha) = f_m \cos \alpha + f_c, \tag{1.2-1}$$

where $f_m = \beta v/2\pi = v/\lambda$, the maximum Doppler shift. Since $f(\alpha) = f(-\alpha)$, the differential variation of power with frequency may be expressed as

$$S(f)|df| = b[p(\alpha)G(\alpha) + p(-\alpha)G(-\alpha)]|d\alpha|. \tag{1.2-2}$$

But

$$|df| = f_m|-\sin \alpha \, d\alpha| = \sqrt{f_m^2 - (f - f_c)^2}\ |d\alpha|;$$

thus

$$S(f) = \frac{b}{\sqrt{f_m^2 - (f - f_c)^2}} [p(\alpha)G(\alpha) + p(-\alpha)G(-\alpha)], \tag{1.2-3}$$

where

$$\alpha = \cos^{-1}\left(\frac{f - f_c}{f_m}\right) \quad \text{and} \quad S(f) = 0 \quad \text{if} \quad |f - f_c| > f_m. \tag{1.2-4}$$

Equation (1.2-3) gives the power spectrum of the output of a receiving antenna. In general this power spectrum depends on the antenna gain pattern and differs from the power spectrum of the field components. However, within the assumptions of the present model, there are antennas that respond to the field components directly. For example, we will assume the transmitted signal is vertically polarized. The electric field will then be in the z-direction and may be sensed by a vertical whip antenna on the mobile, with $G(\alpha) = 1.5$. Substituting in Eq. (1.2-3), the power spectrum of

---

*b is the average power that would be received by an isotropic antenna, that is, $G(\alpha) = 1$.

the electric field is

$$S_{E_z}(f) = \frac{1.5b}{\sqrt{f_m^2 - (f - f_c)^2}} [p(\alpha) + p(-\alpha)]. \qquad (1.2\text{-}5)$$

Small loops may likewise be used to sense the magnetic field, a loop along the x-axis for H_y and one along the y-axis for H_x. The assumed antenna patterns are then

$$H_x: \qquad G(\alpha) = \tfrac{3}{2}\sin^2\alpha, \qquad\qquad (1.2\text{-}6)$$

$$H_y: \qquad G(\alpha) = \tfrac{3}{2}\cos^2\alpha. \qquad\qquad (1.2\text{-}7)$$

Substituting these in Eq. (1.2-3),

$$S_{H_x}(f) = \frac{1.5b}{f_m^2}\sqrt{f_m^2 - (f - f_c)^2}\,[p(\alpha) + p(-\alpha)], \qquad (1.2\text{-}8)$$

$$S_{H_y}(f) = \frac{1.5b(f - f_c)^2}{f_m^2\sqrt{f_m^2 - (f - f_c)^2}}[p(\alpha) + p(-\alpha)]. \qquad (1.2\text{-}9)$$

The simplest assumption for the distribution of power with arrival angle α is a uniform distribution:

$$p(\alpha) = \tfrac{1}{2\pi}, \qquad -\pi \leqslant \alpha \leqslant \pi. \qquad (1.2\text{-}10)$$

The three power spectra become

$$S_{E_z}(f) = \frac{3b}{\omega_m}\left[1 - \left(\frac{f - f_c}{f_m}\right)^2\right]^{-1/2}, \qquad (1.2\text{-}11)$$

$$S_{H_x}(f) = \frac{3b}{\omega_m}\left[1 - \left(\frac{f - f_c}{f_m}\right)^2\right]^{1/2} \qquad (1.2\text{-}12)$$

$$S_{H_y}(f) = \frac{3b}{\omega_m}\left(\frac{f - f_c}{f_m}\right)^2\left[1 - \left(\frac{f - f_c}{f_m}\right)^2\right]^{-1/2}. \qquad (1.2\text{-}13)$$

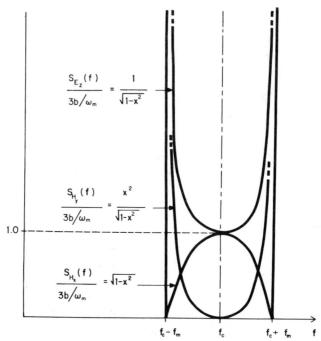

$$\frac{S_{E_z}(f)}{3b/\omega_m} = \frac{1}{\sqrt{1-x^2}}$$

$$\frac{S_{H_y}(f)}{3b/\omega_m} = \frac{x^2}{\sqrt{1-x^2}}$$

1.0

$$\frac{S_{H_x}(f)}{3b/\omega_m} = \sqrt{1-x^2}$$

$f_c - f_m$ f_c $f_c + f_m$ f

Figure 1.2-1   Power spectra of the three field components for uniformly distributed arrival angles. $[x = (f - f_c)/f_m.]$

These spectra are shown in Figure 1.2-1.

Measurement of the RF spectrum is generally very difficult in practice, because of its very small fractional bandwidth of $2v/c$, where c is the velocity of light. Some measurements[12] were made at 910 MHz using oscillators with high frequency stability and yielded the frequency spectogram of the electric field shown in Figure 1.2-2. Spectral density is shown as dark intensity on the figure, frequency is plotted on the ordinate, and time on the abscissa. The mobile was first stationary, as shown by the narrow line to the left. As the mobile speed increased the trace broadened, and the spectrogram width corresponds very closely to the predicted value of $2v/\lambda$.

1.2.2 Correlations and Cross Spectra of the Field Components

Additional spectra that occasionally are of interest correspond to the cross correlations between the three electromagnetic field components, E_z,

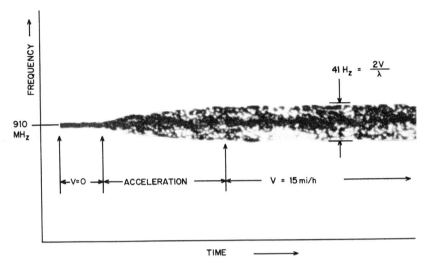

Figure 1.2-2 Frequency spectogram of RF signal at 910 MHz.

H_x, and H_y. Consider E_z and H_x:

$$R_{E_z H_x}(\tau) = \langle E_z(t) H_x(t+\tau) \rangle$$

$$= -\frac{E_0^2}{\eta} \sum_{n,m} C_n C_m \sin \alpha_m \langle \cos[(\omega_c + \omega_n)t + \phi_n] \cos[(\omega_c + \omega_m)(t+\tau) + \phi_m] \rangle.$$

$$(1.2\text{-}14)$$

The product of the cosines can be expanded into cosines of the sums and differences of their angles, which includes terms of the form $\phi_m \pm \phi_n$. The ensemble average over these angles is zero except for the terms $\phi_m - \phi_n$ with $m = n$, in which case

$$R_{E_z H_x}(\tau) = -\frac{b}{\eta} \sum_{n=1}^{N} C_n^2 \sin \alpha_n \cos(\omega_c + \omega_n)\tau. \qquad (1.2\text{-}15)$$

Since $C_n^2 = p(\alpha_n)d\alpha$, in the limit as $N \to \infty$ we can write

$$R_{E_z H_x}(\tau) = -\frac{b}{\eta} \int_{-\pi}^{\pi} p(\alpha) \sin \alpha \cos(\omega_c \tau + \omega_m \tau \cos \alpha) d\alpha. \qquad (1.2\text{-}16)$$

The integrand is an odd function of α if $p(-\alpha) = p(\alpha)$; thus $R_{E_z H_x}(\tau) = 0$ for any τ if $p(\alpha)$ is even. It can easily be shown also that $R_{H_x H_y}(\tau) = 0$ for any τ if $p(\alpha)$ is even; thus H_x is uncorrelated with both E_z and H_y in this case, and the corresponding cross spectra are also zero. Comparing the expressions for E_z and H_y, we see that H_y can be regarded as the output from a filter with frequency transfer function

$$H(\omega) = \frac{\cos \alpha}{\eta} = \frac{\omega - \omega_c}{\eta \omega_m}, \tag{1.2-17}$$

whose input is E_z. The cross spectrum in this case is simply given by[9]

$$S_{E\,H_y}(f) = S_E(f)H^*(f) = \frac{f - f_c}{\eta f_m} S_E(f). \tag{1.2-18}$$

If $p(\alpha) = 1/2\pi$, then

$$S_{E_z H_y}(f) = \frac{3b}{\eta \omega_m} \left(\frac{f - f_c}{f_m} \right) \left[1 - \left(\frac{f - f_c}{f_m} \right)^2 \right]^{-1/2} \tag{1.2-19}$$

Note that this spectrum is an odd function about the center frequency f_c. The cross correlation of E_z and H_y is

$$R_{E_z H_y}(\tau) = \langle E_z(t) H_y(t + \tau) \rangle$$

$$= -\frac{b}{\eta} \int_{-\pi}^{\pi} p(\alpha) \cos \alpha \cos (\omega_c \tau + \omega_m \tau \cos \alpha) \, d\alpha. \tag{1.2-20}$$

For $p(\alpha) = 1/2\pi$,

$$R_{E_z H_y}(\tau) = \frac{b}{\eta} \sin \omega_c \tau J_1(\beta v \tau). \tag{1.2-21}$$

Thus we have the important result that all three field components are uncorrelated at $\tau = 0$.

1.3 POWER SPECTRUM AND OTHER PROPERTIES OF THE SIGNAL ENVELOPE

1.3.1 In-Phase and Quadrature Moments

From the expressions for the power spectral densities given in the preceding section we can derive a number of interesting properties of the envelopes corresponding to the three field components, again with the assumption that the incoming power is uniformly distributed in angle. First we need certain correlations and moments of the in-phase and quadrature components of the signal, T_c and T_s. The subscripts 1 and 2 refer to the times t and $t+\tau$, respectively. Following Rice,[8]

$$\langle T_{c_1} T_{c_2} \rangle = \langle T_{s_1} T_{s_2} \rangle = g(\tau),$$

$$\langle T_{c_1} T_{s_2} \rangle = -\langle T_{s_1} T_{c_2} \rangle = h(\tau),$$

$$\langle T_{c_1} T'_{c_2} \rangle = \langle T_{s_1} T'_{s_2} \rangle = -\langle T'_{c_1} T_{c_2} \rangle = -\langle T'_{s_1} T_{s_2} \rangle = g'(\tau),$$

$$\langle T_{c_1} T'_{s_2} \rangle = \langle T'_{s_1} T_{c_2} \rangle = -\langle T'_{c_1} T_{s_2} \rangle = -\langle T_{s_1} T'_{c_2} \rangle = h'(\tau), \tag{1.3-1}$$

$$\langle T'_{c_1} T'_{c_2} \rangle = \langle T'_{s_1} T'_{s_2} \rangle = -g''(\tau),$$

$$\langle T'_{c_1} T'_{s_2} \rangle = -\langle T'_{s_1} T'_{c_2} \rangle = -h''(\tau),$$

where

$$g(\tau) = \int_{f_c - f_m}^{f_c + f_m} S_i(f) \cos 2\pi (f - f_c) \tau \, df, \tag{1.3-2}$$

$$h(\tau) = \int_{f_c - f_m}^{f_c + f_m} S_i(f) \sin 2\pi (f - f_c) \tau \, df, \tag{1.3-3}$$

and $S_i(f)$ is the input spectrum defined for $f_c - f_m \leqslant f \leqslant f_c + f_m$. (Primes denote differentiation with respect to time.)

From the above correlations evaluated at $\tau = 0$ the moments can also be obtained:

$$b_n = (2\pi)^n \int_{f_c - f_m}^{f_c + f_m} S_i(f)(f - f_c)^n \, df. \tag{1.3-4}$$

Thus

$$\langle T_c^2 \rangle = \langle T_s^2 \rangle = g(0) = b_0,$$

$$\langle T_c T_s \rangle = h(0) = 0,$$

$$\langle T_c T'_c \rangle = \langle T_s T'_s \rangle = g'(0) = 0,$$

$$\langle T_c T'_s \rangle = -\langle T'_c T_s \rangle = h'(0) = b_1, \tag{1.3-5}$$

$$\langle T_c'^2 \rangle = \langle T_s'^2 \rangle = -g''(0) = b_2,$$

$$\langle T'_c T'_s \rangle = -h''(0) = 0.$$

Substituting the expressions for the three spectra given in Eqs. (1.2-11) to (1.2-13) into Eqs. (1.3-2) and (1.3-3):

$$h(\tau)=0 \quad \text{for all three field components.}$$

(This is a consequence of the symmetry of the spectra about f_c.) All b_n likewise equal zero for n odd.

Electric field:

$$b_n = b_0\omega_m^n \frac{1\cdot3\cdot5\cdots(n-1)}{2\cdot4\cdot6\cdots n}, \qquad b_0=1.5b=\frac{3E_0^2}{4} \tag{1.3-6}$$

$$g(\tau)=b_0 J_0(\omega_m\tau). \tag{1.3-7}$$

Magnetic field, x-component:

$$b_n = b_{0H}\omega_m^n \frac{2}{n+2}\frac{1\cdot3\cdot5\cdots(n-1)}{2\cdot4\cdot6\cdots n}, \qquad b_{0H}=\frac{3E_0^2}{8} \tag{1.3-8}$$

$$g(\tau)=b_{0H}[J_0(\omega_m\tau)+J_2(\omega_m\tau)]. \tag{1.3-9}$$

Magnetic field, y-component:

$$b_n = 2b_{0H}\omega_m^n \frac{1\cdot3\cdot5\cdots(n+1)}{2\cdot4\cdot6\cdots(n+2)}, \tag{1.3-10}$$

$$g(\tau)=b_{0H}[J_0(\omega_m\tau)-J_2(\omega_m\tau)]. \tag{1.3-11}$$

Comparison of b_0 and b_{0H} indicates that the average output power of the loop antenna is 3 dB weaker than that from the vertical dipole.*

1.3.2   General Expression for Envelope Autocorrelation and Spectrum

The autocorrelation function of the envelope, r, of a narrow-band Gaussian process can be expressed in terms of a hypergeometric function (Ref. 13, p. 170):

$$R_r(\tau)=\frac{\pi}{2}b_0 F[-\tfrac{1}{2}, -\tfrac{1}{2};1;\rho^2(\tau)], \tag{1.3-12}$$

where

$$\rho^2(\tau)=\frac{1}{b_0^2}[g^2(\tau)+h^2(\tau)] \tag{1.3-13}$$

*Since the vertical dipole will be used often as an example, we do not include the subscript identification, E (i.e., b_{0E} as in b_{0H}) in order to simplify notation.

and $g(\tau)$, $h(\tau)$ are the correlations defined above. At $\tau=0$, $\rho^2(0)=1$; thus

$$R_r(0) = \langle r^2 \rangle = \frac{\pi}{2} b_0 F(-\tfrac{1}{2}, -\tfrac{1}{2}; 1; 1) = 2b_0. \qquad (1.3\text{-}14)$$

(Note that $\langle r^2 \rangle = \int_0^\infty r^2 p(r)\,dr$; using Eq. (1.1-14) for $p(r)$, we also get $\langle r^2 \rangle = 2b_0$.)

The hypergeometric function may be expanded in an infinite series:

$$R_r(\tau) = \frac{\pi}{2} b_0 [1 + \tfrac{1}{4}\rho^2(\tau) + \tfrac{1}{64}\rho^4(\tau) + \cdots]. \qquad (1.3\text{-}15)$$

Dropping terms beyond second degree,

$$R_r(\tau) \doteq \frac{\pi}{2} b_0 [1 + \tfrac{1}{4}\rho^2(\tau)]. \qquad (1.3\text{-}16)$$

At $\tau=0$ this expression gives

$$R_r(0) = \frac{5\pi}{8} b_0 = 1.964 b_0, \qquad (1.3\text{-}17)$$

which differs from the true value of $2b_0$ by only 1.8%; thus Eq. (1.3-16) serves as a good approximation to the exact $R_r(\tau)$.

The power spectral density of the envelope can now be expressed as the Fourier transform of $R_r(\tau)$, using Eq. (1.3-16):

$$S_e(f) = \frac{\pi}{2} b_0 \int_{-\infty}^{\infty} [1 + \tfrac{1}{4}\rho^2(\tau)] e^{-i\omega\tau}\,d\tau. \qquad (1.3\text{-}18)$$

$$= \frac{\pi}{2} b_0 \delta(f) + \frac{\pi}{8b_0} \int_{-\infty}^{\infty} [g^2(\tau) + h^2(\tau)] e^{-i\omega\tau}\,d\tau. \qquad (1.3\text{-}19)$$

To evaluate this integral we first combine $g(\tau)$ and $h(\tau)$ into a complex quantity:

$$\phi(\tau) = g(\tau) + ih(\tau) = \frac{1}{2\pi} \int_{-\infty}^{\infty} F(\omega) e^{i\omega\tau}\,d\omega, \qquad (1.3\text{-}20)$$

where $F(\omega)$ is a two-sided, real, usually even spectrum:

$$F(2\pi f) = S_i(f + f_c). \qquad (1.3\text{-}21)$$

Then $g^2(\tau) + h^2(\tau) = \phi(\tau)\phi^*(\tau)$. Equation (1.3-20) indicates that the Fourier transform of $\phi(\tau)$ is $F(\omega)$; to get the transform of $\phi^*(\tau)$ we note that

$$\phi^*(\tau) = \frac{1}{2\pi} \int_{-\infty}^{\infty} F^*(\omega) e^{-i\omega\tau} d\omega$$

$$= \frac{-1}{2\pi} \int_{\infty}^{-\infty} F(-\omega) e^{i\omega\tau} d\omega$$

$$= \frac{1}{2\pi} \int_{-\infty}^{\infty} F(\omega) e^{i\omega\tau} d\omega, \qquad (1.3\text{-}22)$$

Thus $\phi^*(\tau)$ has the same transform as $\phi(\tau)$. Applying the convolution theorem,

$$\int_{-\infty}^{\infty} \phi(\tau)\phi^*(\tau) e^{-i\omega\tau} d\tau = \frac{1}{2\pi} \int_{-\infty}^{\infty} F(y) F(y-\omega) dy,$$

or

$$\int_{-\infty}^{\infty} [g^2(\tau) + h^2(\tau)] e^{-i\omega\tau} d\tau = \frac{1}{2\pi} \int_{-\infty}^{\infty} F(y) F(y-\omega) dy. \quad (1.3\text{-}23)$$

The spectrum of the envelope is then

$$S_e(f) = \frac{\pi}{2} b_0 \delta(f) + \frac{1}{16 b_0} \int_{-\infty}^{\infty} F(y) F(y-\omega) dy. \qquad (1.3\text{-}24)$$

$S_e(f)$ is always positive, real, and even. The first term in Eq. (1.3-24) is just a dc component. The second term contains one band of width $4f_m$, centered at $f = 0$. This band represents the continuous spectral content of the varying envelope. For positive frequencies it may be expressed as

$$S_0(f) = \frac{\pi}{8 b_0} \int_{f_c - f_m}^{f_c + f_m - f} S_i(x) S_i(x+f) dx, \qquad 0 \leqslant f \leqslant 2f_m. \quad (1.3\text{-}25)$$

1.3.3   Envelope Spectra of the Field Components

Using Eq. (1.3-25) we can now derive expressions for the baseband envelope spectral densities of the three electromagnetic field components, using Eqs. (1.2-11) to (1.2-13) for $S_i(f)$.

Electric field:

$$S_{0E_z}(f) = \frac{\pi b_0}{2\omega_m^2} \int_{f_c - f_m}^{f_c + f_m - f} \left\{ \left[1 - \left(\frac{x - f_c}{f_m} \right)^2 \right] \left[1 - \left(\frac{x + f - f_c}{f_m} \right)^2 \right] \right\}^{-1/2} dx.$$

(1.3-26)

This can be integrated exactly[14] to give

$$S_{0E_z}(f) = \frac{b_0}{4\omega_m} K \left[\sqrt{1 - \left(\frac{f}{2f_m} \right)^2} \right],$$ (1.3-27)

where $K(x)$ is the complete elliptic integral of the first kind.

Magnetic field, x-component:

$$S_{0H_x}(f) = \frac{2\pi b_{0H}}{\omega_m^2} \int_{f_c - f_m}^{f_c + f_m - f} \left\{ \left[1 - \left(\frac{x - f_c}{f_m} \right)^2 \right] \left[1 - \left(\frac{x + f - f_c}{f_m} \right)^2 \right] \right\}^{1/2} dx$$

$$= \frac{4 b_{0H}}{3\omega_m} \left\{ \left[1 + \left(\frac{f}{2f_m} \right)^2 \right] E \left[\sqrt{1 - \left(\frac{f}{2f_m} \right)^2} \right] \right.$$

$$\left. - 2 \left(\frac{f}{2f_m} \right)^2 K \left[\sqrt{1 - \left(\frac{f}{2f_m} \right)^2} \right] \right\},$$ (1.3-28)

where $E(x)$ is the complete elliptic integral of the second kind.

Magnetic field, y-component:

$$S_{0H_y}(f) = \frac{2\pi b_{0H}}{\omega_m^2} \int_{f_c-f_m}^{f_c+f_m-f} \left\{ \left[1 - \left(\frac{x-f_c}{f_m} \right)^2 \right] \left[1 - \left(\frac{x+f-f_c}{f_m} \right)^2 \right] \right\}^{-1/2}$$

$$\times \left(\frac{x-f_c}{f_m} \right)^2 \left(\frac{x+f-f_c}{f_m} \right)^2 dx$$

$$= \frac{b_{0H}}{\omega_m} \left\{ \left[1 + \frac{1}{3} \left(\frac{f}{f_m} \right)^2 \right] K \left[\sqrt{1 - \left(\frac{f}{2f_m} \right)^2} \right] \right.$$

$$\left. - \frac{1}{3} \left[8 - \left(\frac{f}{f_m} \right)^2 \right] E \left[\sqrt{1 - \left(\frac{f}{2f_m} \right)^2} \right] \right\}. \qquad (1.3\text{-}29)$$

These spectra are shown in Figure 1.3-1. Experimental measurements of the spectrum of the electric field envelope[15] in general show reasonably good agreement with the theoretical expression of Eq. (1.3-27). A modification to the original scattering model helps to explain some of the fine structure observed in the experimental spectra. In many actual cases the environment is such that a constant direct wave from the transmitter could be expected, as shown in Figure 1.3-2(a). This wave, arriving at an angle α_0 with respect to the mobile velocity vector, would experience a Doppler shift of $f_m \cos \alpha_0$. The resultant electric field spectrum density would appear as shown in Figure 1.3-2(b), and can be written

$$S_{E_z}'(f) = S_{E_z}(f) + B\delta(f-f_c-f_a), \qquad (1.3\text{-}30)$$

where B is a weighting factor. The modified baseband output spectrum of the resulting envelope is then, using Eq. (1.3-25),

$$S_{0E_z}'(f) = \frac{b_0}{b_0 + B} S_{0E_z}(f)$$

$$+ \frac{\pi B}{8(b_0 + B)} \left[S_{E_z}(f_c + f_a + f) + S_{E_z}(f_c + f_a - f) \right] + dc. \qquad (1.3\text{-}31)$$

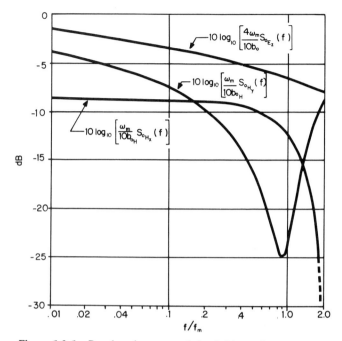

Figure 1.3-1   Baseband spectra of the field envelopes.

The ac part of the new spectrum is simply the original spectrum, $S_{0E_z}(f)$, plus a shifted and folded portion of the input spectrum, $S_{E_z}(f)$. The result will thus contain two new peaks, located at $f_m(1 \pm \cos\alpha_0)$, but will still cut off at $2f_m$, as before. Figure 1.3-2(c) shows an example of the modified spectrum for $\alpha = 60°$. Two experimental spectra are shown in the solid curves of Figure 1.3-3,[7] one with $\alpha_0 = 90°$ and the other with $\alpha_0 = 0°$. The dotted curves are taken from Eq. (1.3-31) with the constant B adjusted arbitrarily for best fit. The modified model gives the basic form of the experimental spectra, but there remain differences in detail. This is probably due to departures of $p(\alpha)$ from the assumed uniform distribution, and also to the fact that the process represented by the received signal is not truly stationary, as assumed, but contains slowly varying terms associated with gross changes in terrain features.

1.3.4 Level Crossing Rates

As illustrated in Figure 1.1-1 the signal envelope experiences very deep fades only occasionally; the shallower the fade the more frequently it is

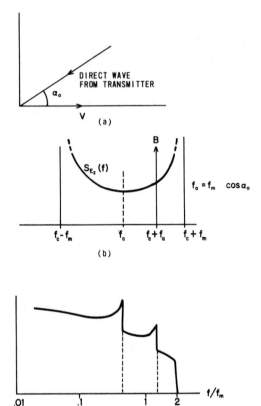

Figure 1.3-2   Effect of the addition of a direct wave from the transmitter on the envelope spectrum. (*a*) Geometry, (*b*), input spectrum, (*c*) spectrum of the envelope.

likely to occur. A quantitative expression of this property is the level crossing rate, N_R, which is defined as the expected rate at which the envelope crosses a specified signal level, R, in the positive direction. In general, it is given by[8]

$$N_R = \int_0^\infty \dot{r} p(R, \dot{r}) \, d\dot{r}, \qquad (1.3\text{-}32)$$

where the dot indicates the time derivative and $p(R, \dot{r})$ is the joint density

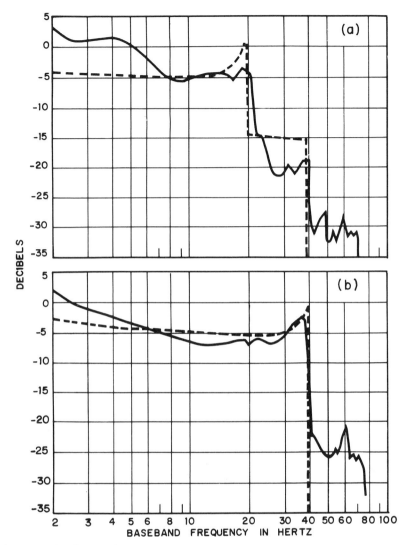

Figure 1.3-3 Comparison of theoretical (dashed) and experimental baseband spectra with direct wave from transmitter. (a) $\alpha_0 = 90°$, (b) $\alpha_0 = 0°$.

function of r and $\dot{r}$ at $r = R$. Rice[8] gives the joint density function in the four random variables r, $\dot{r}$, θ, $\dot{\theta}$ of a Gaussian process for the case $b_1 = 0$:

$$p(r,\dot{r},\theta,\dot{\theta}) = \frac{r^2}{4\pi^2 b_0 b_2} \exp\left[-\frac{1}{2}\left(\frac{r^2}{b_0} + \frac{\dot{r}^2}{b_2} + \frac{r^2\dot{\theta}^2}{b_2} \right) \right], \quad (1.3\text{-}33)$$

where $\tan\theta = -T_s/T_c$. Integrating this expression over θ from 0 to 2π and $\dot{\theta}$ from $-\infty$ to $+\infty$ we get

$$p(r,\dot{r}) = \underbrace{\frac{r}{b_0} e^{-r^2/2b_0}}_{p(r)} \underbrace{\frac{1}{\sqrt{2\pi b_2}} e^{-\dot{r}^2/2b_2}}_{p(\dot{r})}. \quad (1.3\text{-}34)$$

Since $p(r,\dot{r}) = p(r)p(\dot{r})$, r and $\dot{r}$ are independent and uncorrelated. Substituting Eq. (1.3-34) into Eq. (1.3-32) we get the level crossing rate:

$$N_R = \frac{p(R)}{\sqrt{2\pi b_2}} \int_0^\infty \dot{r} e^{-\dot{r}^2/2b_2} d\dot{r}$$

$$= \sqrt{\frac{b_2}{\pi b_0}} \, \rho e^{-\rho^2}, \quad (1.3\text{-}35)$$

where

$$\rho = \frac{R}{\sqrt{\langle r^2\rangle}} = \frac{R}{\sqrt{2b_0}} = \frac{R}{R_{rms}}. \quad (1.3\text{-}36)$$

Substituting the appropriate values of the moments, b_0 and b_2, we get the expressions for the level crossing rates of the three field components*:

$$E_z: \qquad N_R = \sqrt{2\pi}\, f_m \rho e^{-\rho^2}, \quad (1.3\text{-}37)$$

$$H_x: \qquad N_R = \sqrt{\pi}\, f_m \rho e^{-\rho^2}, \quad (1.3\text{-}38)$$

$$H_y: \qquad N_R = \sqrt{3\pi}\, f_m \rho e^{-\rho^2}. \quad (1.3\text{-}39)$$

*In case $b_1 \neq 0$:

$$N_R = \frac{\rho e^{-\rho^2}}{\sqrt{\pi b_0}} \sqrt{b_2 - \frac{b_1^2}{b_0}}\ .$$

These expressions are plotted in Figure 1.3-4 along with some measured values.[1] The rms level of E_z is crossed at a rate of $0.915 f_m$; for example, at $f = 1000$ MHz and $v = 60$ mi/hr, $f_m = 90$ Hz; thus $N_R = 82/$sec at $\rho = 0$ dB. Lower signal levels are crossed less frequently, as shown by the curves. The maximum level crossing rate occurs at $\rho = -3$ dB.

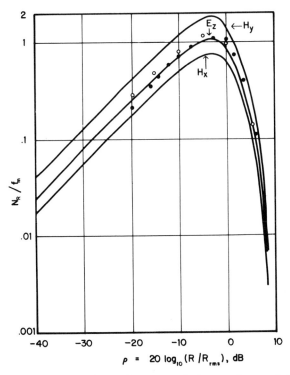

Figure 1.3-4 Normalized level crossing rates of the envelopes of the three field components. Measured values; ○, 11,215 MHz, ●, 836 MHz (E_z).

1.3.5 Duration of Fades

The average duration of fades below $r = R$ is also of interest. Let τ_i be the duration of the ith fade. Then the probability that $r \leqslant R$ for a total time interval of length T is

$$P[r \leqslant R] = \frac{1}{T} \sum \tau_i. \tag{1.3-40}$$

The average fade duration is

$$\bar{\tau} = \frac{1}{TN_R} \sum \tau_i = \frac{1}{N_R} P[r \leqslant R], \tag{1.3-41}$$

$$P[r \leqslant R] = \int_0^R p(r)\,dr = 1 - e^{-\rho^2}, \tag{1.3-42}$$

so that

$$\bar{\tau} = \sqrt{\frac{\pi b_0}{b_2}}\,\frac{1}{\rho}(e^{\rho^2} - 1), \tag{1.3-43}$$

using Eq. (1.3-35) for N_R. Substituting the appropriate values of the moments, we get

$$E_z: \qquad \bar{\tau} = \frac{e^{\rho^2} - 1}{\rho f_m \sqrt{2\pi}}, \tag{1.3-44}$$

$$H_z: \qquad \bar{\tau} = \frac{e^{\rho^2} - 1}{\rho f_m \sqrt{\pi}}, \tag{1.3-45}$$

$$H_y: \qquad \bar{\tau} = \frac{e^{\rho^2} - 1}{\rho f_m \sqrt{3\pi}}. \tag{1.3-46}$$

These expressions are shown in Figure 1.3-5, again with some measured values.[1]

1.3.6 Envelope Autocorrelations and Cross Correlations of the Field Components

From the expressions for $g(\tau)$ and $h(\tau)$ the autocorrelation functions for the envelopes of the three field components may also be derived. In all three cases $h(\tau) = 0$; thus $\rho^2(\tau) = g^2(\tau)/b_0^2$. Using the approximate expression for the envelope autocorrelation of Eq. (1.3-16) we can get the autocovariance function (mean value removed). For a stationary process $r(t)$,

$$L(\tau) \equiv \langle [r(t) - \langle r \rangle][r(t+\tau) - \langle r \rangle] \rangle = R(\tau) - \langle r \rangle^2. \tag{1.3-47}$$

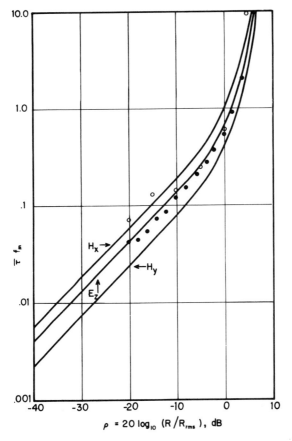

Figure 1.3-5 Normalized durations of fade of the envelopes of the three field components. Measured values: ○, 11,215 MHz, ●, 836 MHz (E_z).

In our case $p(r) = (r/b_0)e^{-r^2/2b_0}$; thus

$$\langle r \rangle = \frac{1}{b_0} \int_0^\infty r^2 e^{-r^2/2b_0} dr = \sqrt{\frac{\pi b_0}{2}} \quad . \tag{1.3-48}$$

Substituting in Eq. (1.3-16),

$$L_e(\tau) \doteq \frac{\pi b_0}{8} \rho^2(\tau) = \frac{\pi}{8b_0} g^2(\tau). \tag{1.3-49}$$

For the three field components,

$$E_z: \qquad L_e(\tau) = \frac{\pi}{8} b_0 J_0^2(\omega_m \tau), \tag{1.3-50}$$

$$H_x: \qquad L_e(\tau) = \frac{\pi}{8} b_{0H} [J_0(\omega_m \tau) + J_2(\omega_m \tau)]^2, \tag{1.3-51}$$

$$H_y: \qquad L_e(\tau) = \frac{\pi}{8} b_{0H} [J_0(\omega_m \tau) - J_2(\omega_m \tau)]^2. \tag{1.3-52}$$

The cross-covariance functions between the envelopes of the three field components are also of interest. We have shown earlier that E_z and H_x are uncorrelated for any value of τ if $p(\alpha)$ is even, and likewise for H_x and H_y. Thus the envelopes of these fields will also be uncorrelated. Using the series expressions for E_z and H_y, it can be shown that the correlation between the envelopes of E_z and H_y is given in terms of a hypergeometric function if $p(\alpha) = 1/2\pi$:

$$R_e(\tau) = \frac{\pi}{2} \sqrt{b_0 b_{0H}} \; F[-\tfrac{1}{2}, -\tfrac{1}{2}; 1; 2J_1^2(\omega_m \tau)], \tag{1.3-53}$$

which can be approximated to better than 1% accuracy for any value of τ by

$$R_e(\tau) \doteq \frac{\pi}{2} \sqrt{b_0 b_{0H}} \; [1 + \tfrac{1}{2} J_1^2(\omega_m \tau)]. \tag{1.3-54}$$

The cross-covariance function is then

$$L_e(\tau) \doteq \frac{\pi}{4} \sqrt{b_0 b_{0H}} \; J_1^2(\omega_m \tau). \tag{1.3-55}$$

These four envelope covariances, Eqs. (1.3-50)–(1.3-52) and (1.3-55), are shown in Figure 1.3-6. (the envelope covariances for E_zH_x and H_xH_y are zero as pointed out earlier, and thus are not shown in Figure 1.3-6.) These would be the functions measured in the mobile as a function of time while it moves with uniform velocity; alternatively, they can be regarded as spatial correlations by setting $\zeta = v\tau$; thus $\omega_m \tau = \beta\zeta/\lambda$. This equivalence between time and spatial correlations is important and should be noted, since it will be used in space diversity calculations and other applications.

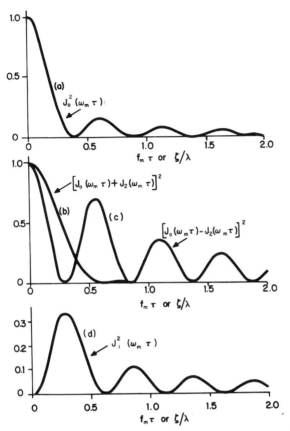

Figure 1.3-6   Covariance functions of various field envelopes. (a) $|E_z|$, (b) $|H_x|$, (c) $|H_y|$, (d) cross-covariance of $|E_z|$ and $|H_y|$.

1.4 RANDOM FREQUENCY MODULATION

1.4.1 Probability Distribution of Random FM

The time-varying nature of the in-phase and quadrature components of the fading signal means that the apparent frequency of the signal varies with time in a random manner; that is, the signal exhibits random frequency modulation. The characteristics of this random FM can be most easily described in terms of its probability distributions and power spectrum. The probability density of $\dot{\theta}$ can be easily obtained from

$p(r,\dot{r},\theta,\dot{\theta})$ by integrating over r, $\dot{r}$, and θ. Using Eq. (1.3-33),

$$p(\dot{\theta}) = \frac{1}{4\pi^2 b_0 b_2} \int_0^\infty dr \int_{-\infty}^\infty d\dot{r} \int_0^{2\pi} d\theta$$

$$\times r^2 \exp\left[-\frac{1}{2}\left(\frac{r^2}{b_0} + \frac{\dot{r}^2}{b_2} + \frac{r^2 \dot{\theta}^2}{b_2} \right) \right]$$

$$= \frac{1}{2}\sqrt{\frac{b_0}{b_2}} \left(1 + \frac{b_0}{b_2}\dot{\theta}^2 \right)^{-3/2}. \tag{1.4-1}$$

From the expression for $p(\dot{\theta})$ we can deduce the somewhat surprising result that the mean square value of the random FM is infinite:

$$\langle \dot{\theta}^2 \rangle = \int_{-\infty}^\infty \dot{\theta}^2 p(\dot{\theta})\, d\dot{\theta}$$

$$= \frac{b_0}{b_2}\left[-1 + \lim_{\dot{\theta}\to\infty} \log\left(2\dot{\theta}\sqrt{\frac{b_0}{b_2}} \right) \right] = \infty.* \tag{1.4-2}$$

In actual FM receivers, of course, the discriminator or audio amplifiers will limit at some value of frequency deviation. With this assumption it then becomes possible to calculate the rms baseband noise due to the random FM, as is shown in Chapter 4.

For the E_z field $b_2/b_0 = \omega_m^2/2$, using Eq. (1.3-6), thus $p(\dot{\theta})$ can be written:

$$E_z{:}p(\dot{\theta}) = \frac{1}{\omega_m\sqrt{2}}\left[1 + 2\left(\frac{\dot{\theta}}{\omega_m} \right)^2 \right]^{-3/2}. \tag{1.4-3}$$

---

*When log appears without subscript, the natural log is assumed.

The cumulative distribution function is

$$P[\dot\theta \le \dot\theta_0] = \frac{1}{2}\sqrt{\frac{b_0}{b_2}} \int_{-\infty}^{\dot\theta_0} \left(1 + \frac{b_0}{b_2}\dot\theta^2\right)^{-3/2} d\dot\theta$$

$$= \frac{1}{2}\left[1 + \sqrt{\frac{b_0}{b_2}}\,\dot\theta_0\left(1 + \frac{b_0}{b_2}\dot\theta_0^2\right)^{-1/2}\right], \qquad (1.4\text{-}4)$$

$$E_z: P[\dot\theta \le \dot\theta_0] = \frac{1}{2}\left[1 + \sqrt{2}\,\frac{\dot\theta_0}{\omega_m}\left(1 + 2\frac{\dot\theta_0^2}{\omega_m^2}\right)^{-1/2}\right]. \qquad (1.4\text{-}5)$$

These probability functions are similar for the other two field components, differing only in scale. Equations (1.4-3) and (1.4-5) are shown in Figure 1.4-1. Note that, in contrast to the sharply defined power spectrum of the signal (Figure 1.2-1), there is a nonzero probability of finding its frequency at any value, although the larger excursions occur only rarely since they are associated with the deeper fades of the signal. This can be seen by examining the probability density function of $\dot\theta$ conditioned on the signal level R:

$$p(\dot\theta|R) = \frac{p(\dot\theta, R)}{p(R)} = \frac{R}{\sqrt{2\pi b_2}}e^{-R^2\dot\theta^2/2b_2}. \qquad (1.4\text{-}6)$$

For fixed R this is a Gaussian distribution with standard deviation $\sqrt{b_2}/R$. Thus as R decreases (deep fades) the frequency deviations of interest increase proportionately. If $R = R_{rms} = \sqrt{2b_0}$ the significant deviations occupy a bandwidth approximately equal to ω_m in the case of E_z; for a 20-dB fade the bandwidth is ten times greater.

1.4.2 Power Spectrum of Random FM

The power spectrum of $\dot\theta$ may be derived by conventional methods. The autocorrelation function of $\dot{\theta}$ is given by Rice[8]:

$$R_{\dot{\theta}}(\tau) = \langle \dot{\theta}(t)\dot{\theta}(t+\tau) \rangle$$

$$= -\frac{1}{2}\left\{ \left[\frac{g'(\tau)}{g(\tau)} \right]^2 - \left[\frac{g''(\tau)}{g(\tau)} \right] \right\} \log\left\{ 1 - \left[\frac{g(\tau)}{g(0)} \right]^2 \right\}. \qquad (1.4\text{-}7)$$

Then the power spectrum is the Fourier transform of $R_{\dot{\theta}}(\tau)$:

$$S_{\dot{\theta}}(f) = \int_{-\infty}^{\infty} R_{\dot{\theta}}(\tau)e^{-i\omega\tau}\,d\tau = 4\int_{0}^{\infty} R_{\dot{\theta}}(\tau)\cos\omega\tau\,d\tau, \qquad (1.4\text{-}8)$$

where a factor of 2 introduced into the second factor since we are only interested in positive frequencies.

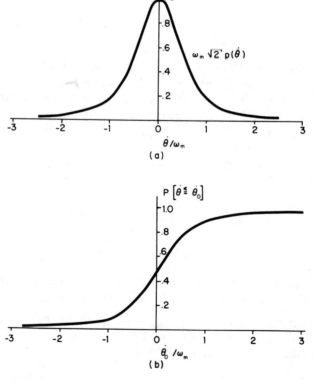

Figure 1.4-1 Probability functions for the random frequency modulation, $\dot{\theta}$, of the electric field. (a) Probability density, (b) cumulative distribution.

Considering now the E_z field for the case $p(\alpha) = 1/2\pi$, the values of $g(\tau)$ and its derivatives can be obtained from Eq. (1.3-7):

$$\frac{g(\tau)}{g(0)} = J_0(u), \qquad u = \omega_m \tau,$$

$$\frac{g'(\tau)}{g(\tau)} = -\omega_m \frac{J_1(u)}{J_0(u)}, \qquad (1.4\text{-}9)$$

$$\frac{g''(\tau)}{g(\tau)} = \omega_m^2 \left[\frac{J_1(u)}{u J_0(u)} - 1 \right],$$

and

$$R_{\dot{\theta}}(\tau) = \frac{\omega_m^2}{2 J_0^2(u)} \left[\frac{J_0(u) J_1(u)}{u} - J_0^2(u) - J_1^2(u) \right]$$

$$\times \log\left[1 - J_0^2(u)\right]. \qquad (1.4\text{-}10)$$

The integration of Eq. (1.4-8) is then carried out by separating the range of integration into parts and making the appropriate approximations for the Bessel functions and the logarithm.

Region 1:

$$0 \leqslant \tau \leqslant \tau_A = \frac{1}{4\omega_m}, \qquad J_0(u) \doteq 1 - \left(\frac{u}{2}\right)^2, \qquad J_1(u) \doteq \frac{u}{2}.$$

$$S_1(f) \doteq -\omega_m \int_0^{\omega_m \tau_A} \log\left(\frac{u^2}{2}\right) \cos\left(\frac{\omega}{\omega_m} u\right) du$$

$$= -\frac{\omega_m^2}{\omega} \left[\log\left(\frac{\omega_m^2 \tau_A^2}{2}\right) \sin(\omega\tau_A) - 2 S_i(\omega\tau_A) \right], \qquad (1.4\text{-}11)$$

where $S_i(\omega\tau_A)$ is the sine integral function.

Region 3:

$$\tau_B \leqslant \tau < \infty: \qquad \log(1-J_0^2) \doteq -J_0^2(u), \qquad \tau_B \sim \frac{10}{\omega_m},$$

$$J_0(u) \sim \sqrt{\frac{2}{\pi u}} \, \cos\left(u - \frac{\pi}{4}\right),$$

$$J_1(u) \sim \sqrt{\frac{2}{\pi u}} \, \sin\left(u - \frac{\pi}{4}\right).$$

$$S_3(f) = \frac{8\omega_m}{\pi} \int_{\omega_m \tau_B}^{\infty} \left[\frac{1}{2u} + \frac{\cos(2u)}{(2u)^2} \right] \cos\left(\frac{\omega}{\omega_m} u\right) du$$

$$= \frac{2\omega_m}{\pi} \left[\frac{\cos(2\omega_m \tau_B)\cos(\omega\tau_B)}{\omega_m \tau_B} + \left(1 + \frac{\omega}{2\omega_m}\right) S_i(2\omega_m \tau_B + \omega\tau_B) \right.$$

$$+ \left(1 - \frac{\omega}{2\omega_m}\right) S_i(|2\omega_m \tau_B - \omega\tau_B|) - 2C_i(\omega\tau_B)$$

$$\left. - \frac{\pi}{2}\left(1 + \frac{\omega}{2\omega_m} + \left|1 - \frac{\omega}{2\omega_m}\right|\right) \right]. \tag{1.4-12}$$

Region 2: $\tau_A \leqslant \tau \leqslant \tau_B$. $S_2(f)$ must be evaluated by numerical integration. As ω approaches zero the spectrum has a logarithmic singularity due to the term $C_i(\omega\tau_B)$ in Eq. (1.4-12):

$$S_{\dot\theta}(f) \to -\frac{4\omega_m}{\pi} \log(k\omega), \qquad k = \text{a constant.}$$

The spectrum $S_{\dot\theta}(f) = S_1(f) + S_2(f) + S_3(f)$ is shown in Figure 1.4-2. Above $\omega = 2\omega_m$ the spectrum falls as $1/f$; thus $2\omega_m$ may be regarded as an approximate cutoff frequency. Note that this is the same as the cutoff frequency of the envelope spectra (Figure 1.3-1).

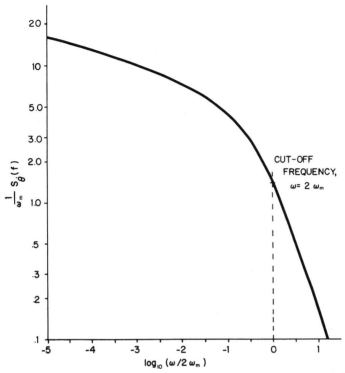

Figure 1.4-2   One-sided power spectrum of the instantaneous frequency, $\dot{\theta}$, of the E_z field.

Using Watson's lemma one can obtain the asymptotic form, as f approaches infinity, of the power spectrum of the random FM for an arbitrary Doppler spectrum[16]:

$$\lim_{f \to \infty} S_{\dot{\theta}}(f) \sim \left(\frac{b_2}{b_0} - \frac{b_1^2}{b_0^2} \right) f^{-1}, \qquad (1.4\text{-}13)$$

where b_0, b_1, b_2 are defined in Eq. (1.3-4). For low vehicle speeds and low carrier frequencies (e.g., 60 mph at UHF), the asymptotic form given here is accurate over the audio band from 300 to 3000 Hz.

1.5 COHERENCE BANDWIDTH

The characteristics of a single-frequency signal transmitted over the mobile radio propagation path have so far been explained on the basis of a fairly simple model. The only required knowledge was the angular distribution of the incident power, $p(\alpha)$, assumed to be carried by a large number of plane waves of random amplitudes, phases, and arrival angles. When we turn to questions concerning the statistics of several signals of different frequencies transmitted over the path, however, the model must be elaborated to include explicitly the fact that the path lengths of the constituent waves are different. The different path lengths give rise to different propagation time delays, of course. Typical spreads in time delays range from a fraction of a microsecond to many microseconds, depending on the type of environment. The longer delay spreads are usually found in metropolitan areas like New York City that contain many large buildings, whereas the shorter delay spreads are usually associated with suburban areas. In the latter the building structures are generally more uniform, consisting of one- or two-story houses.

The existence of the different time delays in the various waves that make up the total field causes the statistical properties of two signals of different frequencies to become essentially independent if the frequency separation is large enough. The maximum frequency difference for which the signals are still strongly correlated is called the coherence bandwidth of the mobile radio transmission path.

Besides providing a more complete physical model of the mobile transmission path, the study of the delay spreads and coherence bandwidth will be useful in assessing the performance and limitations of different modulation and diversity reception schemes, as we will see in later chapters.

1.5.1 Mathematical Model

We will now proceed to develop a mathematical model that includes the time delays explicitly. If a single signal of frequency ω is transmitted to the mobile unit, we assume the received field is the sum of a number of waves, as before:

$$E_z(\omega, t) = E_0 \sum_{n=1}^{N} \sum_{m=1}^{M} C_{nm} \cos(\omega t + \omega_n t - \omega T_{nm}). \qquad (1.5\text{-}1)$$

In this representation the nth wave at an arrival angle α_n is composed of M waves with propagation delay times T_{nm}. All of these M waves experience the same Doppler shift, $\omega_n = \beta v \cos \alpha_n$.* The amplitude coefficients C_{nm}

---

*The Frontispiece, supplied by Cox (*IEEE Proc. Lett.*, April, 1973), gives a vivid illustration of this model.

have been redefined to indicate the power associated with each individual wave:

$$C_{nm}^2 = G(\alpha_n)p(\alpha_n, T_{nm}) \, d\alpha \, dT. \qquad (1.5\text{-}2)$$

As before, $G(\alpha)$ is the horizontal directivity pattern of the receiving antenna. We interpret $p(\alpha, T) \, d\alpha \, dT$ to mean the fraction of the incoming power within $d\alpha$ of the angle α and within dT of the time T, in the limit with N and M very large.

We assume that the α_n and T_{nm} remain constant for motion of the mobile over distances of several tens of wavelengths and that they are frequency independent. It also seems reasonable that the difference between any two phase angles, represented by $\omega T_{ij} - \omega T_{nm}$, is much greater than 2π for $i \neq n, j \neq m$. At UHF frequencies and above, $\omega \geqslant 2\pi \times 10^9$; thus for $(T_{ij} - T_{nm})$ on the order of 0.1 to 10 μsec (the delay spread), the phase difference is on the order of several hundred times 2π, at least.

We consider the random process $E_z(\omega, t)$ to consist of sample functions corresponding to separate runs made by the mobile on the same section of street, and assume the phases $\omega_n t - \omega T_{nm}$ of the individual waves in the different sample functions are uniformly distributed independent random variables. The process $E_z(\omega, t)$ is then wide-sense stationary with respect to ensemble averages. It is not stationary with respect to time averages, however, and thus is nonergodic. But the difference between time and ensemble averages decreases as the number of waves becomes large; thus the statistical properties will be computed on the basis of ensemble averages. The results will eventually be compared with corresponding values derived from experiments, in which averages with time are usually used. The extent to which theory and experiment agree will govern the degree of confidence we place in the model.

To investigate the coherence bandwidth we will study the statistics and correlation properties of two signals received at frequencies ω_1 and ω_2, and, in particular, of their envelopes and phases. Only the E_z field will be explicitly treated; corresponding results can be easily derived for H_x and H_y by defining an appropriate antenna pattern as in Section 1.2. We start by writing the field at the two frequencies in terms of narrow-band in-phase and quadrature components:

At ω_1: $\quad E_z(\omega_1, t) = x_1(t) \cos \omega_1 t - x_2(t) \sin \omega_1 t,$

At ω_2: $\quad E_z(\omega_2, t) = x_3(t) \cos \omega_2 t - x_4(t) \sin \omega_2 t. \qquad (1.5\text{-}3)$

For large enough N and M the $x_i(t)$ are Gaussian random processes (by

the Central Limit Theorem), and are jointly Gaussian.
They are expressed as

$$\underset{2}{x^1}(t) = E_0 \sum_{n,m} C_{nm} \left\{ \begin{matrix} \cos \\ \sin \end{matrix} \right\} (\omega_n t - \omega_1 T_{nm}),$$

$$\underset{4}{x^3}(t) = E_0 \sum_{n,m} C_{nm} \left\{ \begin{matrix} \cos \\ \sin \end{matrix} \right\} (\omega_n t - \omega_2 T_{nm}). \qquad (1.5\text{-}4)$$

We will be interested in correlation properties as a function of both time
delay, τ, and frequency separation $s = \omega_2 - \omega_1$. Let us define the four
random variables x_1, x_2, x_3, x_4 for fixed time t as follows:

$$x_1 \overset{\Delta}{=} x_1(t), \qquad x_2 \overset{\Delta}{=} x_2(t),$$

$$x_3 \overset{\Delta}{=} x_3(t+\tau), \qquad x_4 \overset{\Delta}{=} x_4(t+\tau). \qquad (1.5\text{-}5)$$

The envelopes and phases are then defined by

$$\underset{2}{x^1} = \left\{ \begin{matrix} r_1 \cos\theta_1 \\ r_1 \sin\theta_1 \end{matrix} \right\}, \qquad \underset{4}{x^3} = \left\{ \begin{matrix} r_2 \cos\theta_2 \\ r_2 \sin\theta_2 \end{matrix} \right\}. \qquad (1.5\text{-}6)$$

The statistical properties will depend on moments of the type $\langle x_i x_j \rangle$, where
the brackets refer to ensemble averages. The moments are evaluated from
the series expansion of Eq. (1.5-4):

$$\langle x_1^2 \rangle = E_0^2 \sum_{n,m,p,q} \langle C_{nm} C_{pq} \cos(\omega_n t - \omega_1 T_{nm}) \cos(\omega_p t - \omega_1 T_{pq}) \rangle. \qquad (1.5\text{-}7)$$

The average will vanish unless $n = p$ and $m = q$, which gives

$$\langle x_1^2 \rangle = \frac{E_0^2}{2} \sum_{n,m} C_{nm}^2 = b_0 \sum_{n,m} G(\alpha_n) p(\alpha_n, T_{nm}) \, d\alpha \, dT. \qquad (1.5\text{-}8)$$

In the limit as N, $M \to \infty$:

$$\langle x_1^2 \rangle = b_0 \int_0^{2\pi} G(\alpha) \, d\alpha \int_0^\infty p(\alpha, T) \, dT. \qquad (1.5\text{-}9)$$

But, by definition,

$$\int_0^\infty p(\alpha, T) \, dT = p(\alpha); \qquad (1.5\text{-}10)$$

thus

$$\langle x_1^2 \rangle = b_0 \int_0^{2\pi} G(\alpha) p(\alpha) \, d\alpha. \qquad (1.5\text{-}11)$$

By similar arguments we can show

$$\langle x_1^2 \rangle = \langle x_2^2 \rangle = \langle x_3^2 \rangle = \langle x_4^2 \rangle, \qquad (1.5\text{-}12)$$

$$\langle x_1 x_2 \rangle = \langle x_3 x_4 \rangle = 0, \qquad (1.5\text{-}13)$$

$$\langle x_1 x_3 \rangle = \langle x_2 x_4 \rangle$$

$$= b_0 \int_0^{2\pi} G(\alpha) \, d\alpha \int_0^{\infty} p(\alpha, T) \cos(\beta v \tau \cos \alpha - sT) \, dT, \qquad (1.5\text{-}14)$$

$$\langle x_1 x_4 \rangle = -\langle x_2 x_3 \rangle$$

$$= b_0 \int_0^{2\pi} G(\alpha) \, d\alpha \int_0^{\infty} p(\alpha, T) \sin(\beta v \tau \cos \alpha - sT) \, dT. \qquad (1.5\text{-}15)$$

Using the general expression (Ref. 9, p. 255) for the joint density of the four Gaussian variables $x_1, \ldots, x_4$ and applying the transformation of variables of Eq. (1.5-6), we get the joint density of the envelopes and phases:

$$p(r_1, r_2, \theta_1, \theta_2) = \frac{r_1 r_2}{(2\pi\mu)^2 (1 - \lambda^2)}$$

$$\times \exp\left[-\frac{r_1^2 + r_2^2 - 2r_1 r_2 \lambda \cos(\theta_2 - \theta_1 - \phi)}{2\mu(1 - \lambda^2)} \right], \qquad (1.5\text{-}16)$$

where

$$\tan \phi = \frac{\mu_2}{\mu_1}, \qquad \lambda^2 = \frac{\mu_1^2 + \mu_2^2}{\mu^2}, \qquad (1.5\text{-}17)$$

$$\mu = \langle x_1^2 \rangle, \qquad \mu_1 = \langle x_1 x_3 \rangle, \qquad \mu_2 = \langle x_1 x_4 \rangle. \qquad (1.5\text{-}18)$$

All of the statistical properties of interest can now be derived from the four fold joint density function of Eq. (1.5-16) provided an explicit form for $G(\alpha)$ and $p(\alpha, T)$ is known. Interpretation of some measured data[7, 17, 18] indicates that an exponential distribution of the delay spreads is a good approximation. If we further assume a uniform distribution in angle of the incident power, the function $p(\alpha, T)$ can be expressed as

$$p(\alpha, T) = \frac{1}{2\pi\sigma} e^{-T/\sigma}, \qquad (1.5\text{-}19)$$

where σ is a measure of the time delay spread.

With this assumption, and the additional one of no antenna directivity so that $G(\alpha) = 1$, the quantities in Eqs. (1.5-17)–(1.5-18) may be worked out with the help of Eqs. (1.5-11)–(1.5-14):

$$\mu = b_0, \qquad \mu_1 = b_0 \frac{J_0(\omega_m \tau)}{1 + s^2 \sigma^2}, \qquad \mu_2 = -s\sigma\mu_1. \qquad (1.5\text{-}20)$$

$$\tan\phi = -s\sigma, \qquad \lambda^2 = \frac{J_0^2(\omega_m \tau)}{1 + s^2 \sigma^2}.$$

1.5.2 Envelope Correlation as a Function of Frequency Separation

The correlation of the envelopes of the signals at the two frequencies may now be calculated. We have

$$R_e(s, \tau) = \langle r_1 r_2 \rangle = \int_0^\infty \int_0^\infty r_1 r_2 p(r_1, r_2) \, dr_1 \, dr_2. \qquad (1.5\text{-}21)$$

We can get $p(r_1, r_2)$ by integrating Eq. (1.5-16):

$$p(r_1, r_2) = \int_0^{2\pi} \int_0^{2\pi} p(r_1, r_2, \theta_1, \theta_2) \, d\theta_1 \, d\theta_2$$

$$= \frac{r_1 r_2}{\mu^2(1 - \lambda^2)} \exp\left[-\frac{r_1^2 + r_2^2}{2\mu(1 - \lambda^2)} \right] I_0\left(\frac{r_1 r_2}{\mu} \frac{\lambda}{1 - \lambda^2} \right), \qquad (1.5\text{-}22)$$

where $I_0(x)$ is the modified Bessel function of zero order.

Substituting Eq. (1.5-22) into Eq. (1.5-21) the integration may be carried out exactly[14] to give

$$R_e(s, \tau) = \frac{\pi}{2} b_0 F(-\tfrac{1}{2}, -\tfrac{1}{2}; 1; \lambda^2), \qquad (1.5\text{-}23)$$

which may also be expressed as[14]

$$R_e(s,\tau) = b_0(1+\lambda)E\left(\frac{2\sqrt{\lambda}}{1+\lambda}\right), \qquad (1.5\text{-}24)$$

where $E(x)$ is the complete elliptic integral of the second kind. The expansion of the hypergeometric function yields a good approximation to Eq. (1.5-23):

$$R_e(s,\tau) \doteq \frac{\pi}{2}b_0\left(1 + \frac{\lambda^2}{4}\right). \qquad (1.5\text{-}25)$$

We know that $\langle r_1 \rangle = \langle r_2 \rangle = \sqrt{\tfrac{1}{2}\pi b_0}$ from Eq. (1.3-48) and also that $\langle r_1^2 \rangle = \langle r_2^2 \rangle = 2b_0$ from Eq. (1.3-14); thus the envelope correlation coefficient

$$\rho_e(s,\tau) = \frac{R_e(s,\tau) - \langle r_1 \rangle \langle r_2 \rangle}{\sqrt{[\langle r_1^2 \rangle - \langle r_1 \rangle^2][\langle r_2^2 \rangle - \langle r_2 \rangle^2]}}$$

becomes

$$\rho_e(s,\tau) = \frac{(1+\lambda)E\left(\dfrac{2\sqrt{\lambda}}{1+\lambda}\right) - \dfrac{\pi}{2}}{2 - \dfrac{\pi}{2}}$$

$$\doteq \lambda^2 = \frac{J_0^2(\omega_m\tau)}{1+s^2\sigma^2}. \qquad (1.5\text{-}26)$$

We see from this expression that the correlation between the envelopes decreases with increasing frequency separation s, as one would expect. One measure of coherence bandwidth corresponds to the frequency separation when the envelope correlation is 0.5. With $\tau = 0$ this occurs when $s\sigma = 1$; thus the coherence bandwidth is equal to $1/2\pi\sigma$. Equation (1.5-26) with $\tau = 0$ is plotted in Figure 1.5-1 for values of σ in the range $\tfrac{1}{8}$ to $\tfrac{3}{4}$ μsec. Also shown are a few measurements made in 1961 by Ossanna and Hoffman* in a suburban area at 836 MHz. A delay spread on the order of $\tfrac{1}{4}$ μsec appears to be appropriate in this case, corresponding to a coherence bandwidth of about 640 KHz. Measurements of the delay spread using a more direct pulse-type technique[19] tend to substantiate this value of time

---

*Bell Telephone Laboratories, unpublished work.

delay spread for suburban areas. Other delay distributions besides the exponential one of Eq. (1.5-18) could also be considered,[6] but they do not appreciably change the shape of the curves in Figure 1.5-1.

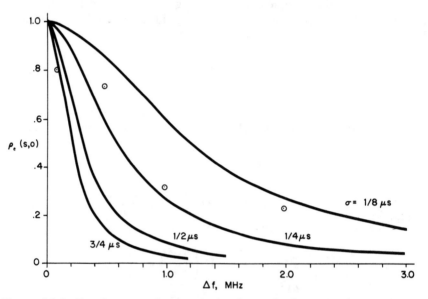

Figure 1.5-1 Envelope correlation of signals received at two frequencies for different time delay spreads, σ. Solid curves are theoretical for an exponential delay distribution. $\bigcirc$-measurements at 836 MHz in a suburban environment.

1.5.3 Phase Correlation as a Function of Frequency Separation

The statistics of the phases of the two signals are also of interest. (Note that we mean the phase angles of the signal phasors, θ_1 and θ_2, not the radio frequency phases.) One property is the correlation of the two phases:

$$R_\theta(s,\tau) = \langle \theta_1 \theta_2 \rangle$$

$$= \int_0^{2\pi} \int_0^{2\pi} \theta_1 \theta_2 p(\theta_1, \theta_2) \, d\theta_1 \, d\theta_2. \qquad (1.5\text{-}27)$$

Again we can obtain $p(\theta_1, \theta_2)$ by integrating Eq. (1.5-16):

$$p(\theta_1, \theta_2) = \int_0^\infty \int_0^\infty p(r_1, r_2, \theta_1, \theta_2) \, dr_1 \, dr_2.$$

The integration is straightforward, using known integrals of error functions[20] to obtain

$$p(\theta_1,\theta_2) = \frac{1-\lambda^2}{4\pi^2} \frac{\sqrt{1-B^2} + B\cos^{-1}(-B)}{(1-B^2)^{3/2}}, \qquad (1.5\text{-}28)$$

where

$$B = \lambda\cos(\theta_2 - \theta_1 - \phi), \qquad 0 < \cos^{-1}(-B) < \pi, \qquad (1.5\text{-}29)$$

and λ, ϕ are defined earlier in Eq. (1.5-20).

The integration of Eq. (1.5-27) with the expression for $p(\theta_1,\theta_2)$ substituted into the integrand cannot be carried out exactly, but integration by parts yields a fairly simple series expansion:

$$R_\theta(s,\tau) = \pi^2[1 + \Gamma(\lambda,\phi) + 2\Gamma^2(\lambda,\phi) - \tfrac{1}{24}\Omega(\lambda)], \qquad (1.5\text{-}30)$$

where

$$\Gamma(\lambda,\phi) = \frac{1}{2\pi}\sin^{-1}(\lambda\cos\phi),$$
$$(1.5\text{-}31)$$

$$\Omega(\lambda) = \frac{6}{\pi^2}\sum_{n=1}^{\infty}\frac{\lambda^{2n}}{n^2}, \qquad \Omega(1) = 1.$$

The phases θ_1 and θ_2 are random variables uniformly distributed from zero to 2π, that is, $p(\theta) = 1/2\pi$. Thus $\langle\theta_1\rangle = \langle\theta_2\rangle = \pi$, $\langle\theta_1^2\rangle = \langle\theta_2^2\rangle = 4\pi^2/3$. The correlation coefficient of the phases is then

$$\rho_\theta(s,\tau) = \frac{R_\theta(s,\tau) - \langle\theta_1\rangle\langle\theta_2\rangle}{\sqrt{[\langle\theta_1^2\rangle - \langle\theta_1\rangle^2][\langle\theta_2^2\rangle - \langle\theta_2\rangle^2]}}$$

$$= \frac{3}{\pi^2}[R_\theta(s,\tau) - \pi^2]. \qquad (1.5\text{-}32)$$

Substituting Eq. (1.5-30),

$$\rho_\theta(s,\tau) = 3\Gamma(\lambda,\phi)[1 + 2\Gamma(\lambda,\phi)] - \tfrac{1}{8}\Omega(\lambda). \qquad (1.5\text{-}33)$$

The dependence of this correlation coefficient on $s\sigma$ is shown in Figure 1.5-2, where again τ has been set equal to 0. If we choose as a measure of coherence bandwidth the frequency separation for which $\rho_\theta(s,0) = 0.5$ (analogous to the definition for the envelope correlation) we see that this occurs when $s\sigma = \frac{1}{2}$, corresponding to a coherence bandwidth equal to $1/4\pi\sigma$, or $\frac{1}{2}$ the value for the former case. Thus if $\sigma \sim \frac{1}{4}$ μsec, the coherence bandwidth for the phase is about 320 KHz.

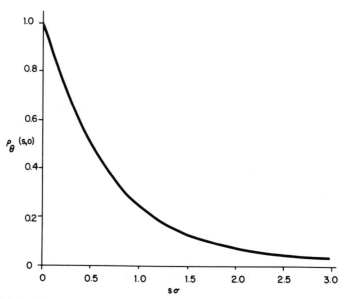

Figure 1.5-2 Dependence of the phase correlation coefficient on the frequency separation $s = \omega_2 - \omega_1$, and time delay spread σ.

1.5.4 Probability Distributions of the Phase Difference at Two Frequencies

Besides the correlation coefficient of the phases we will find the statistics of the phase difference, $\theta_2 - \theta_1$, to be of interest. Since both θ_1 and θ_2 can take any value from zero to 2π, the quantity $(\theta_2 - \theta_1)$ can have any value from -2π to $+2\pi$; thus we must be careful to avoid ambiguities of 2π in defining density functions and mean square values. First let us consider the probability density of the random variable

$$\zeta = \theta_2 - \theta_1. \tag{1.5-34}$$

From Ref. 9, p. 189, we find that

$$p(\zeta) = \int_0^{2\pi} p(\theta_1, \zeta + \theta_1)\, d\theta_1. \tag{1.5-35}$$

But θ_1 must satisfy two criteria:

$$0 \leqslant \theta_1 \leqslant 2\pi,$$

$$0 \leqslant \theta_1 + \zeta \leqslant 2\pi;$$

(1.5-36)

this defines two regions for $p(\zeta)$:

$$\zeta \geqslant 0: \quad p(\zeta) = \int_0^{2\pi - \zeta} p(\theta_1, \zeta + \theta_1) \, d\theta_1,$$

(1.5-37)

$$\zeta \leqslant 0: \quad p(\zeta) = \int_{-\zeta}^{2\pi} p(\theta_1, \zeta + \theta_1) \, d\theta_1.$$

From Eq. (1.5-28),

$$p(\theta_1, \zeta + \theta_1) = \frac{1 - \lambda^2}{4\pi^2}$$

$$\times \frac{\sqrt{1 - \lambda^2 \cos^2(\zeta - \phi)} + \lambda \cos(\zeta - \phi) \cos^{-1}[-\lambda \cos(\zeta - \phi)]}{[1 - \lambda^2 \cos^2(\zeta - \phi)]^{3/2}}. \quad (1.5\text{-}38)$$

For brevity it will be convenient in the following development to define a function F represented by Eq. (1.5-38):

$$p(\theta_1, \zeta + \theta_1) \overset{\Delta}{=} F[\cos(\zeta - \phi)]. \quad (1.5\text{-}39)$$

Thus $p(\theta_1, \zeta + \theta_1)$ is independent of θ_1, so that Eq. (1.5-37) may be immediately evaluated:

$$p(\zeta) = (2\pi + |\zeta|) F[\cos(\zeta - \theta)]. \quad (1.5\text{-}40)$$

As noted earlier, $-2\pi \leqslant \zeta \leqslant 2\pi$. It will be useful to define a new variable which is confined to the region from $-\pi$ to $+\pi$, and therefore corresponds to a physically measurable, unambiguous quantity. Let

$$\xi = \begin{cases} \zeta - 2\pi, & \pi \leqslant \zeta \leqslant 2\pi \\ \zeta, & -\pi \leqslant \zeta \leqslant \pi \\ \zeta + 2\pi, & -2\pi \leqslant \zeta \leqslant -\pi \end{cases}. \quad (1.5\text{-}41)$$

Then it can be shown[9] that under this transformation of variables,

$$p(\xi) = 2\pi F[\cos(\xi - \phi)], \quad -\pi \leqslant \xi \leqslant \pi. \quad (1.5\text{-}42)$$

We now consider an experiment designed to measure the statistics of the measurable phase difference. Assume two CW signals at frequencies ω_1 and ω_2 are transmitted from the base station to two separate antennas on the mobile unit; these antennas being spaced far enough apart so that all the statistics of the signals received on the two are independent. As shown in Figure 1.5-3, the signals at frequencies ω_1 and ω_2 from each antenna are multiplied together and the difference frequency components selected by the low-pass filters. These signals are at the same frequency, namely $\omega_2 - \omega_1$, and thus a phase detector can be used to measure the phase difference, $(\theta_{2b} - \theta_{1b} - \theta_{2a} + \theta_{1a})$. Let ξ_a correspond to the measurable phase difference $\theta_{2a} - \theta_{1a}$, ξ_b to $\theta_{2b} - \theta_{1b}$, and consider the statistics of $\xi_b - \xi_a$. First let

$$w = \xi_b - \xi_a, \qquad -2\pi \leqslant w \leqslant 2\pi. \tag{1.5-43}$$

The probability density $p(w)$ is then found in the same way as $p(\zeta)$, Eq. (1.5-35):

$$p(w) = \int_{-\pi}^{\pi} p(\xi_a, \xi_a + w)\, d\xi_a. \tag{1.5-44}$$

Under the assumption that ξ_a and ξ_b are independent (since antennas a and b are well separated), $p(\xi_a, \xi_b) = p(\xi_a)p(\xi_b)$; thus

$$p(w) = \begin{cases} \displaystyle\int_{-\pi}^{\pi - w} p(\xi_a)p(\xi_a + w)\, d\xi_a, & w \geqslant 0 \\[2mm] \displaystyle\int_{-\pi - w}^{\pi} p(\xi_a)p(\xi_a + w)\, d\xi_a, & w \leqslant 0 \end{cases}. \tag{1.5-45}$$

We now define a new variable ψ such that $-\pi \leqslant \psi \leqslant \pi$ to correspond to measurable phase angles:

$$\psi = \begin{cases} w - 2\pi & \pi \leqslant w \leqslant 2\pi \\ w, & -\pi \leqslant w \leqslant \pi \\ w + 2\pi, & -2\pi \leqslant w \leqslant -\pi \end{cases}. \tag{1.5-46}$$

Under this transformation of variables the probability density of ψ becomes

$$p(\psi) = \int_{-\pi}^{\pi} p(\xi_a)p(\xi_a + \psi)\, d\xi_a. \tag{1.5-47}$$

Substituting $p(\xi)$ from Eq. (1.5-42)

$$p(\psi) = 4\pi^2 \int_{-\pi}^{\pi} F[\cos(\xi_a - \phi)] F[\cos(\xi_a - \phi + \psi)] d\xi_a$$

$$= 4\pi^2 \int_{-\pi}^{\pi} F[\cos y] F[\cos(y + \psi)] dy$$

$$= 4\pi^2 \int_{0}^{\pi} F[\cos y] \{ F[\cos(y + \psi)] + F[\cos(y - \psi)] \} dy, \quad (1.5\text{-}48)$$

where we have made the variable change $y = \xi_a - \phi$. Inspection of Eq. (1.5-48) shows that $p(\psi)$ is independent of ϕ and symmetric in ψ, $p(-\psi) = p(\psi)$. The angle ψ thus corresponds to the magnitude of the phase difference $|(\theta_{2b} - \theta_{1b}) - (\theta_{2a} - \theta_{1a})|$ that would be measured by a phase detector in the range $-\pi$ to $+\pi$. The statistics of ψ, such as the density function $p(\psi)$ and the variance $\langle \psi^2 \rangle$ will agree with those one would measure by the experiment shown in Figure 1.5-3.

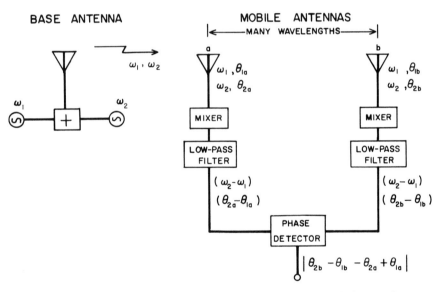

Figure 1.5-3 Experiment to measure the statistics of the phase difference between two signals at different frequencies.

The expression for $p(\psi)$ given by Eq. (1.5-48) cannot be integrated in closed form; numerical integration yields the curves of Figure 1.5-4 for various values of the parameter λ.* As λ increases, corresponding to a decrease in $s\sigma$, the phase difference tends to concentrate more about $\psi = 0$, as one would expect. Measurements[17] of the density show rough agreement with Figure 1.5-4, depending on the choice of σ.

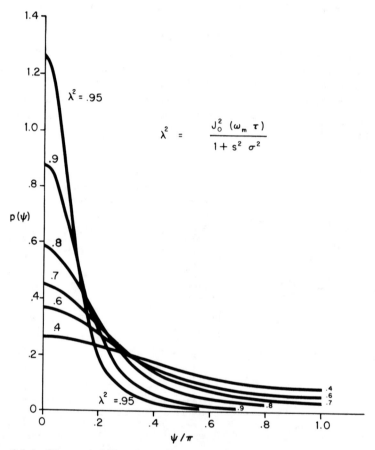

Figure 1.5-4 The probability density $p(\psi)$ of the "measurable" phase difference between signals of two different frequencies, ω_1 and ω_2. $s = \omega_2 - \omega_1$.

*An explicit expression can be obtained for the difference between $p(0)$ and $p(\pi)$, however:

$$p(0) - p(\pi) = \pi\lambda^2(4-\lambda^2)(1-\lambda^2)^{-1/2}/32.$$

This serves as a check on the numerical integration.

The variance of ψ may be easily found from $p(\psi)$:

$$\langle\psi^2\rangle = \int_{-\pi}^{\pi} \psi^2 p(\psi)\, d\psi. \tag{1.5-49}$$

Again numerical integration must be used, with the results shown in Figure 1.5-5. If $\lambda=0$ (signals at ω_1 and ω_2 uncorrelated), $P(\psi)=1/2\pi$; that is, ψ is uniformly distributed from $-\pi$ to $+\pi$. In this case $\langle\psi^2\rangle=\pi^2/3$. Some measured values of $\langle\psi^2\rangle$ are also shown in Figure 1.5-5.

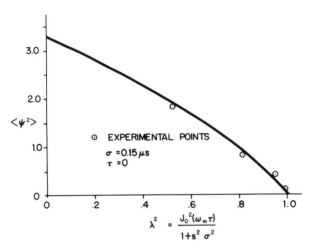

Figure 1.5-5 Mean square value of the "measurable" phase difference between signals of two different frequencies, ω_2 and ω_1. $s=\omega_2-\omega_1$.

1.5.5 Ratio of Signal Envelopes at Two Frequencies

As a final point of interest we will determine the probability that the signal envelope at frequency ω_2 exceeds that at ω_1 by a given amount. This would be of concern when considering interference between two signals at two frequencies transmitted from one base station. If the signals lie within the coherence bandwidth, the variation in their amplitudes due to multipath fading is appreciably correlated. A receiver that provides adequate discrimination between the desired signal (at ω_1) and the undesired signal under nonfading conditions would also do so under fading conditions in this case. However, as the frequency separation increases, the chance that the undesired signal occasionally exceeds the desired one by a given amount will increase, with a consequent increase in interference. The probability that $r_2 \geqslant a r_1$ may be obtained from the joint density of Eq. (1.5-22):

$$P[r_2 \geqslant a r_1] = \int_0^\infty dr_1 \int_{a r_1}^\infty p(r_1, r_2) \, dr_2$$

$$= \frac{1}{\mu^2(1-\lambda^2)} \int_0^\infty dr_1 \int_{a r_1}^\infty r_1 r_2 \exp\left[-\frac{r_1^2 + r_2^2}{2\mu(1-\lambda^2)}\right]$$

$$\times I_0\left[\frac{r_1 r_2}{\mu} \frac{\lambda}{1-\lambda^2}\right] dr_2. \tag{1.5-50}$$

By making a change of variables $r_2 = r \cos\theta$, $r_1 = r \sin\theta$ this integral may be easily evaluated[14]:

$$P[r_2 \geqslant a r_1] \stackrel{\Delta}{=} P(a, \lambda)$$

$$= \frac{1}{2} + \frac{1}{2} \frac{(1-a^2)}{\sqrt{(1+a^2)^2 - 4\lambda^2 a^2}}. \tag{1.5-51}$$

Equation (1.5-20) gives λ for the case of the exponential time delay distribution. Setting $\tau = 0$ we can express the probability in terms of $s\sigma$, obtaining the curves of Figure 1.5-6. For $\lambda \to 0$ the curves are asymptotic to $1/(1+a^2)$, and for small values of $s\sigma$ they approach

$$P(a, s\sigma) \doteq \left[\frac{a s\sigma}{1-a^2}\right]^2, \quad \text{if} \quad a > 1. \tag{1.5-52}$$

The two earlier definitions of coherence bandwidth correspond to $s\sigma = 1$ (50% amplitude correlation) or $s\sigma = 0.5$ (50% phase correlation). These values are shown on the figure, and we can see that if the frequency separation $s = \omega_2 - \omega_1$ is less than the coherence bandwidth the probability that r_2 exceeds r_1 by an appreciable amount is very small.

1.6 SPATIAL CORRELATIONS AT THE BASE STATION

The results of preceding sections have been obtained by considering the mobile unit as a receiver. At first thought it might seem that transmitting from mobile to base should not change matters; after all, radio transmission in a linear medium obeys the reciprocity theorem! However, the reciprocity theorem must be applied with care in a scattering medium.[21] The base station in a typical mobile radio system layout is usually located well above surrounding objects so that it has the best possible access to mobiles within its domain of coverage. The simplified model of this path places the important scattering objects (those that produce the multipath effects) within a small distance of the mobile, and more or less uniformly located around it. Up to now we have concentrated our attention on the resulting processes at the mobile; now we will examine the implications of this model with respect to the base station.

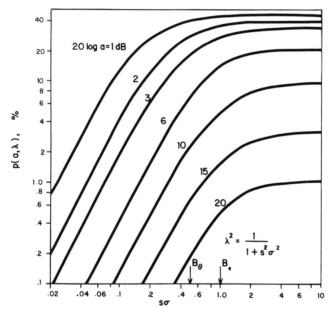

Figure 1.5-6 Probability that the signal envelope r_2 at frequency ω_2 exceeds a times that at ω_1; both signals transmitted from the same base station with equal power. B_e, envelope coherence bandwidth, B_θ, phase coherence bandwidth.

1.6.1 Mathematical Model

Referring to Figure 1.6-1 we assume that a ring of scattering objects whose bi-static scattering cross section is uniform are located in a circle of radius a around the transmitter at the mobile unit. The distance d from

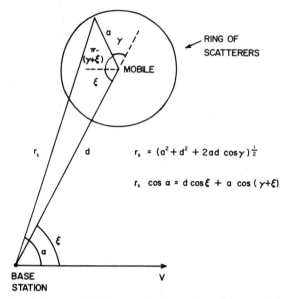

$$r_s = (a^2 + d^2 + 2ad \cos\gamma)^{\frac{1}{2}}$$

$$r_s \cos a = d \cos\xi + a \cos(\gamma+\xi)$$

Figure 1.6-1   Scattering model for examination of spatial correlations at the base station.

mobile to the base receiver will be assumed much greater than a, so that the base does not lie within this circle. It will be convenient to start with the power spectrum approach used in Section 1.2; to this end we artificially assume that the mobile is now fixed and the base moves along the x-axis with velocity v. We need then to introduce a further assumption that d is so large that the angle ξ between v and the direction to the mobile does not change significantly during observation times of interest (i.e., the movement of the base station along the x-axis is small compared to d).

Let the distribution of power radiated from the mobile with azimuth angle γ be denoted $p(\gamma)$. The power incident on the scatterers within a circumferential length dl is then $Kp(\gamma)d\gamma$. The proportionality constant K can be set equal to unity without loss of generality. This power is scattered uniformly in angle, so that the power, $p(\alpha)$, at the base station within $d\alpha$ of the angle α is, neglecting multiplicative constant factors independent of α and γ,

$$p(\alpha)\,d\alpha = p(\gamma)\,d\gamma. \tag{1.6-1}$$

This is also the power within df of the frequency f contained in the power spectrum $S(f)$ corresponding to the given α; thus

$$S(f)\,df = p(\alpha)\,d\alpha = p(\gamma)\,d\gamma. \tag{1.6-2}$$

A complex correlation function of time τ can be defined from Eqs. (1.3-2)–(1.3-3):

$$c(\tau) \overset{\Delta}{=} g(\tau) + ih(\tau)$$

$$= \int_{f_c-f_m}^{f_c+f_m} \exp\{i[2\pi\tau(f-f_c)]\}S(f)df. \quad (1.6\text{-}3)$$

In the present case $f=f_c+f_m\cos\alpha$, but α depends on γ through the relation

$$\alpha(\gamma)=\cos^{-1}\left[\frac{\cos\xi+k\cos(\gamma+\xi)}{\sqrt{1+2k\cos\gamma+k^2}}\right], \qquad k=\frac{a}{d} \quad (1.6\text{-}4)$$

which may easily be derived from the geometry of Figure 1.6-1. The integration on f in Eq. (1.3-21) is replaced by integration on γ, so that

$$c(\tau) = \int_{-\pi}^{\pi} e^{i\omega_m\tau\cos[\alpha(\gamma)]}p(\gamma)d\gamma. \quad (1.6\text{-}5)$$

Assuming now that the power transmitted from the mobile, b_0, is radiated uniformly in all directions, we get

$$c(\tau)=\frac{b_0}{2\pi}\int_{-\pi}^{\pi} e^{i\omega_m\tau\cos[\alpha(\gamma)]}d\gamma. \quad (1.6\text{-}6)$$

This integral, with Eq. (1.6-4) substituted for α, cannot be explicitly evaluated. But we have assumed $a\ll d$, or $k\ll1$, so an approximation may be obtained to various orders of k by expanding $\cos\alpha$ in powers of k:

$$\cos\alpha = \sum_{n=0}^{\infty} a_n k^n, \quad (1.6\text{-}7)$$

where the first few a_n are

$$a_0 = \cos\xi,$$

$$a_1 = -\sin\xi\sin\gamma,$$

$$a_2 = \tfrac{1}{2}\sqrt{1-\tfrac{3}{4}\cos^2\xi}\,\cos2(\gamma-\gamma_0)-\tfrac{1}{4}\cos\xi, \quad (1.6\text{-}8)$$

and

$$\tan2\gamma_0=2\tan\xi. \quad (1.6\text{-}9)$$

To second order in k the integral then becomes

$$c(\tau) \doteq \frac{b_0}{2\pi} e^{i\omega_m\tau(1-k/4)^2\cos\xi}$$

$$\times \int_{-\pi}^{\pi} \exp\left\{i\omega_m\tau\left[\tfrac{1}{2}k^2\sqrt{1-\tfrac{3}{4}\cos^2\xi}\ \cos2(\gamma-\gamma_0) - k\sin\xi\sin\gamma\right]\right\}d\gamma.$$

$$(1.6\text{-}10)$$

To carry out the integration we expand the exponential functions in terms of Bessel functions with the relation

$$e^{iz\cos x} = \sum_{n=-\infty}^{\infty} i^n J_n(z) e^{inx}.$$

$$(1.6\text{-}11)$$

The integral on γ can then be written

$$I = \int_{-\pi}^{\pi} \sum_{m,n=-\infty}^{\infty} (-1)^m i^n J_m(z_1) J_n(z_2) e^{i[(2n+m)\gamma - 2n\gamma_0]} d\gamma$$

$$= 2\pi \sum_{n=-\infty}^{\infty} J_{2n}(z_1) J_n(z_2) e^{in(\pi/2 - 2\gamma_0)},$$

$$(1.6\text{-}12)$$

where

$$z_1 = \omega_m \tau k \sin\xi,$$

$$(1.6\text{-}13)$$

$$z_2 = \tfrac{1}{2} k^2 \omega_m \tau \sqrt{1 - \tfrac{3}{4}\cos^2\xi}\ .$$

The envelope auto-covariance function may now be obtained from $c(\tau)$. From Eqs. (1.3-13) and (1.3-49) we see that

$$L_e(\tau) \doteq \frac{\pi}{8b_0} |c^2(\tau)| = \frac{b_0}{32\pi} II^*$$

$$= \frac{b_0\pi}{8} \sum_{m,n=-\infty}^{\infty} e^{i(\pi/2 - 2\gamma_0)(n-m)} J_{2n}(z_1) J_{2m}(z_1) J_n(z_2) J_m(z_2).\quad (1.6\text{-}14)$$

The term for $m=n=0$ is a good approximation to this expansion, for third-order accuracy in k (and for fourth-order accuracy in k when $\xi=0$); thus

$$L_e(\tau) \doteq \frac{\pi b_0}{8} J_0^2(z_1) J_0^2(z_2).$$

$$(1.6\text{-}15)$$

1.6.2   Envelope Correlation as a Function of Antenna Separation

Since $\omega_m \tau = 2\pi v \tau / \lambda$, Eq. (1.6-15) can be regarded as a function of spatial separation $\zeta = v\tau$. We can now abandon the artificial assumption of a moving base station, and instead consider that $L_e(\zeta)$ gives the correlation between the envelopes of signals received simultaneously on two antennas at the base, separated by a distance ζ. To third order in k, Eq. (1.6-15) is directly analogous to Eq. (1.3-50) giving the auto-covariance of the E_z field seen at the mobile. Comparing arguments, we see that the base antenna separation must be a factor $(k \sin \xi)^{-1}$ times greater than that at the mobile to obtain the same correlation. Also, for $\xi = 0$ the third-order approximation gives a constant value of correlation independent of separation and equal to the value for $\zeta = 0$. Thus the fourth-order approximation is needed in this case. Estimates of the scattering circle diameter vary, but it seems obvious that it must be at least equal to the distance between buildings on opposite sides of the street where the mobile is located. This is substantiated by some experimental measurements.[19] Thus $2a$ might typically be 100 ft; at a range of $d = 2$ miles, $k = 0.005$; thus the power series expansion in k appears justified.

Curves of the correlation coefficient $\rho_e \doteq J_0^2(z_1) J_0^2(z_2)$ for $k = 0.006$ are shown in Figure 1.6-2, along with some values measured at 836 MHz.[22] Comparison with Figure 1.3-6 illustrates how much more rapidly the signals at the mobile become decorrelated with antenna separation. It should be emphasized that the model used here assumes *no* scatterers in the immediate vincity of the base station; the presence of even a small number of local scatterers would have a strong effect on the correlation, particularly for $\xi = 0$.

The model also does not include the direction of motion of the mobile with respect to the line-of-sight to the base station. One would expect that motion along the line-of-sight would require greater base station antenna separation for the same correlation, compared to motion perpendicular to the line-of-sight. This effect could be included by assuming that the scatterers lie on an ellipse with major axis along the direction of motion. A refined model of this type would approach the actual disposition of the scatterers more closely.

1.7 LABORATORY SIMULATION OF MULTIPATH INTERFERENCE

The testing of mobile radio transmission techniques in the field is time-consuming and often inconclusive, due to uncertainty in the statistical signal variations actually encountered. Laboratory testing with signals that duplicate the assumed statistical properties of the signals encountered in

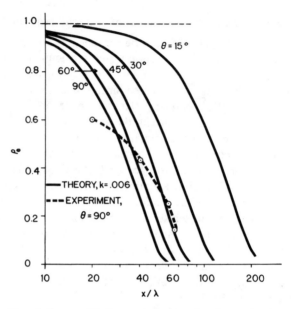

Figure 1.6-2 Correlation coefficient ρ_e between signals received on two antennas at a base station versus their separation and orientation angle θ.

the field is an attractive alternative, provided that all of the relevant properties can be simulated. Past approaches to the problem of simulating fading signals may be divided into three classes. First, tape recordings of the actual fading signals may be used.[23] In another method[24] a steady signal is split into several paths, each of which is then randomly phase modulated as shown in Figure 1.7-1(a). Uniformly distributed phase modulation is obtained by appropriately shaping the amplitude distribution of low-pass Gaussian noise. An approximation to Rayleigh fading is obtained by adding several such paths together. Frequency selective fading can also be produced by including path delay. However, the power spectrum of the output signal is very difficult to calculate or control. A third method[25] provides uniform phase modulation and Rayleigh envelope fading by amplitude modulation of the in-phase and quadrature components of a steady carrier with uncorrelated low-pass Gaussian noises, as shown in Figure 1.7-1(b). Frequency selective fading may be produced by combining several delayed fading signals. The different noise sources must have the same power spectrum to produce stationary fading, and the power spectrum of the fading signal will then be the same as the noise spectrum. The limitation with this approach is that only rational forms of the fading spectrum can be produced, whereas the spectra encountered in mobile radio are generally nonrational, as shown by Eqs. (1.2-11)–(1.2-13). A method[26] to simulate mobile radio fading that produces random phase modulation, a Rayleigh fading envelope, and a time-averaged, discrete approximation to the desired power spectrum will be discussed in the remainder of this section.

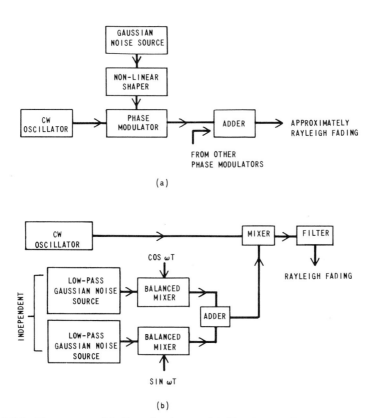

Figure 1.7-1 Two types of fading simulators. (*a*) Simulator using uniform phase modulation. (*b*) Simulator using quadrature amplitude modulation.

1.7.1 Mathematical Development

We start with an expression that represents the field as a superposition of plane waves:

$$E(t) = \mathrm{Re}[\, T(t) e^{i\omega_c t}\,], \qquad (1.7-1)$$

where

$$T(t) = E_0 \sum_{n=1}^{N} c_n e^{i(\omega_m t \cos \alpha_n + \phi_n)}, \tag{1.7-2}$$

and

$$c_n^2 = p(\alpha_n) \, d\alpha = \frac{1}{2\pi} \, d\alpha.$$

We assume that the arrival angles are uniformly distributed with $d\alpha = 2\pi/N$; thus $c_n^2 = 1/N$, and

$$\alpha_n = \frac{2\pi n}{N}, \qquad n = 1, 2, \ldots, N. \tag{1.7-3}$$

We further let $N/2$ be an odd integer; then the series can be rearranged to give

$$T(t) = \frac{E_0}{\sqrt{N}} \left\{ \sum_{n=1}^{N/2-1} \left[e^{i(\omega_m t \cos \alpha_n + \phi_n)} + e^{-i(\omega_m t \cos \alpha_n + \phi_{-n})} \right] \right.$$

$$\left. + e^{i(\omega_m t + \phi_N)} + e^{-i(\omega_m t + \phi_{-N})} \right\}. \tag{1.7-4}$$

The first term in the sum represents waves with Doppler shifts that progress from $+\omega_m \cos(2\pi/N)$ to $-\omega_m \cos(2\pi/N)$ as n runs from 1 to $N/2 - 1$, while the Doppler shifts in the second term go from $-\omega_m \cos(2\pi/N)$ to $+\omega_m \cos(2\pi/N)$. Thus the frequencies in these two terms overlap. The third and fourth terms represent waves with the maximum Doppler shift of $+\omega_m$ and $-\omega_m$, respectively. Without much loss of generality it will be convenient to represent the signal in terms of waves whose frequencies do not overlap:

$$T(t) = \frac{E_0}{\sqrt{N}} \left\{ \sqrt{2} \sum_{n=1}^{N_0} \left[e^{i(\omega_m t \cos \alpha_n + \phi_n)} + e^{-i(\omega_m t \cos \alpha_n + \phi_{-n})} \right] \right.$$

$$\left. + e^{i(\omega_m t + \phi_N)} + e^{-i(\omega_m t + \phi_{-N})} \right\}, \qquad N_0 = \frac{1}{2} \left(\frac{N}{2} - 1 \right) \tag{1.7-5}$$

where the factor $\sqrt{2}$ has been used so that the total power in $E(t)$ will be unchanged. The simulation should, among other things, provide a good approximation to Rayleigh fading. If N is large enough we may invoke the Central Limit Theorem to conclude that $T(t)$ is approximately a complex Gaussian process, so that $|T|$ is Rayleigh as desired. From the work of Bennett[27] and Slack[28] it follows that the Rayleigh approximation is quite good for $N \geqslant 6$, with deviations from Rayleigh confined mostly to the extreme peaks. Further information as to the value of N may be obtained by examining the autocorrelation of $E(t)$:

$$R(\tau) = \langle E(t)E(t+\tau) \rangle$$

$$= \tfrac{1}{2} \operatorname{Re}\left[\langle T(t)T(t+\tau)e^{i\omega_c(2t+\tau)} \rangle + \langle T^*(t)T(t+\tau)e^{i\omega_c\tau} \rangle \right]. \quad (1.7\text{-}6)$$

The expectations are taken over the random phases ϕ_n, ϕ_m, and they occur only as sums of differences. The only terms that contribute are those involving $\phi_n - \phi_m$ with $n = m$, so that

$$R(\tau) = \frac{b_0}{N} \cos \omega_c \tau \left[4 \sum_{n=1}^{N_0} \cos\left(\omega_m \tau \cos \frac{2\pi n}{N} \right) + 2 \cos(\omega_m \tau) \right]. \quad (1.7\text{-}7)$$

We note that Eq. (1.7-7) is of the form of a carrier factor multiplied by a low-frequency factor:

$$R(\tau) = g(\tau) \cos \omega_c \tau. \quad (1.7\text{-}8)$$

We also know, from Eq. (1.3-7), that for a uniformly scattered field $g(\tau) = b_0 J_0(\omega_m \tau)$. Although this expression was derived for a continuum of arrival angles, we may suspect that if N is large enough, the quantity in brackets in Eq. (1.7-7) will closely approximate $J_0(\omega_m \tau)$. Noting that $J_0(x)$ may be defined as

$$J_0(x) = \frac{2}{\pi} \int_0^{\pi/2} \cos(x \cos \alpha)\, d\alpha, \quad (1.7\text{-}9)$$

the bracketed factor of Eq. (1.7-7) may be put in the form of a discrete approximation (Riemann sum) to the integral (1.7-9). We thus expect that

$$2 \sum_{n=1}^{N_0} \cos\left(\omega_m \tau \cos \frac{2\pi n}{N} \right) + \cos(\omega_m \tau) = \frac{N}{2} J_0(\omega_m \tau). \quad (1.7\text{-}10)$$

Evaluation of Eq. (1.7-10) for various values of $\omega_m \tau$ and N shows that the series gives $J_0(\omega_m \tau)$ to eight significant digits for $\omega_m \tau \leqslant 15$ with $N = 34$. The number of frequency components needed is thus $\tfrac{1}{2}(\tfrac{34}{2} - 1) = 8$. The simulation will thus produce an RF spectrum which is a discrete approximation to the form

$$\left[1-\left(\frac{f-f_c}{f_m}\right)\right]^{-1/2}.$$

1.7.2   Realization of the Method

The simulation technique is now clear: N_0 low-frequency oscillators with frequencies equal to the Doppler shifts $\omega_m \cos(2\pi n/N)$, $n = 1, 2, \ldots N_0$, plus one with frequency ω_m are used to generate signals frequency-shifted from

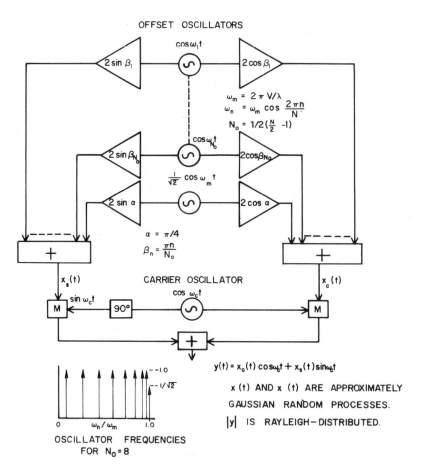

Figure 1.7-2   Simulator that duplicates mobile radio spectrum.

a carrier frequency ω_c using modulation methods. The amplitudes of all the components are made equal to unity except for the one with frequency ω_m, which is set equal to $1/\sqrt{2}$. The phases β_n are chosen appropriately so that the probability distribution of the resultant phase will be as close as possible to a uniform distribution, $1/2\pi$. A block diagram of such a simulator is shown in Figure 1.7-2 along with an illustration of the frequency spacings of the oscillators for $N_0 = 8$. By taking advantage of some trigonometric relationships, the proper oscillator phases are provided by amplifiers with gains set equal to $2\cos\beta_n$ or $2\sin\beta_n$. The outputs of the individual oscillators, with the appropriate gain factors, are first summed to produce in-phase (x_c) and quadrature (x_s) bands, which are then multiplied by in-phase and quadrature carrier components, respectively, and then summed to produce the final composite output signal $y(t)$. From the block diagram we get

$$x_c(t) = 2 \sum_{n=1}^{N_0} \cos\beta_n \cos\omega_n t + \sqrt{2} \cos\alpha \cos\omega_m t, \qquad (1.7\text{-}11)$$

$$x_s(t) = 2 \sum_{n=1}^{N_0} \sin\beta_n \cos\omega_n t + \sqrt{2} \sin\alpha \cos\omega_m t. \qquad (1.7\text{-}12)$$

The phase of $y(t)$ must be random and uniformly distributed from zero to 2π; this may be accomplished in several ways, provided $\langle x_c^2 \rangle \approx \langle x_s^2 \rangle$ and $\langle x_c x_s \rangle \approx 0$. We have

$$\langle x_c^2 \rangle = 2 \sum_{n=1}^{N_0} \cos^2\beta_n + \cos^2\alpha$$

$$= N_0 + \cos^2\alpha + \sum_{n=1}^{N_0} \cos 2\beta_n, \qquad (1.7\text{-}13)$$

$$\langle x_s^2 \rangle = 2 \sum_{n=1}^{N_0} \sin^2\beta_n + \sin^2\alpha$$

$$= N_0 + \sin^2\alpha - \sum_{n=1}^{N_0} \cos 2\beta_n, \qquad (1.7\text{-}14)$$

$$\langle x_c x_s \rangle = 2 \sum_{n=1}^{N_0} \sin\beta_n \cos\beta_n + \sin\alpha \cos\alpha. \qquad (1.7\text{-}15)$$

By choosing $\alpha = 0$, $\beta_n = \pi n/(N_0+1)$, we find $\langle x_c x_s \rangle \equiv 0$ and $\langle x_c^2 \rangle = N_0$, $\langle x_s^2 \rangle = N_0 + 1$. (Note that the brackets denote time averages now.) Thus $y(t)$ is a narrow-band signal centered on a carrier frequency ω_c, having Rayleigh fading characteristics, and with autocorrelation function approximately equal to $J_0(\omega_m \tau)$. Its spectrum is therefore the nonrational form given by Eq. (1.2-4), corresponding to a uniform antenna pattern, $G(\alpha) = 1$, and uniform distribution of the incident power, $p(\alpha) = 1/2\pi$. Random FM is also produced by this method. Since the carrier frequency is provided by one oscillator, it may be set to some convenient value, say 30 MHz, and voice-modulated either in amplitude or frequency for use with various reception techniques. The performance of a simulator built with nine offset oscillators ($N_0 = 8$) is illustrated in Figures 1.7-3 to 1.7-6, showing measured cumulative distribution of the envelope, autocorrelation function, RF spectrum, and random FM power spectrum. Comparison with the expected Rayleigh distribution, Bessel function autocorrelation, and theoretical RF and random FM spectra shows excellent agreement.

This technique may be extended to provide up to N_0 independently fading signals while still using the same offset oscillators. The nth oscillator is given an additional phase shift $\gamma_{nj} + \beta_{nj}$, with gains as before. By imposing the additional requirement that the output signals $y_j(t)$ be uncorrelated (or as nearly so as possible), the appropriate values for γ_{nj} and β_{nj} can be determined. The choices are not unique, but the following seems to be the simplest:

$$\beta_{nj} = \frac{\pi n}{N_0+1}, \tag{1.7-16}$$

$$\gamma_{nj} = \frac{2\pi(j-1)}{N_0+1}, \qquad n = 1, 2, \ldots, N_0. \tag{1.7-17}$$

By using two quadrature low-frequency oscillators per offset in place of the single oscillators shown in Figure 1.7-2, the use of phase shifters to perform the $\gamma + \beta$ shift can be eliminated. This leads to modified amplifier gains as sketched in Figure 1.7-7 for the nth offset amplifier of the jth simulator. The $N = 2$ curve in the $p(R)$ graph of Figure 1.7-3 shows the resulting combined envelope statistics of a simulated two-branch maximal ratio diversity combiner (cf. Section 5.2).

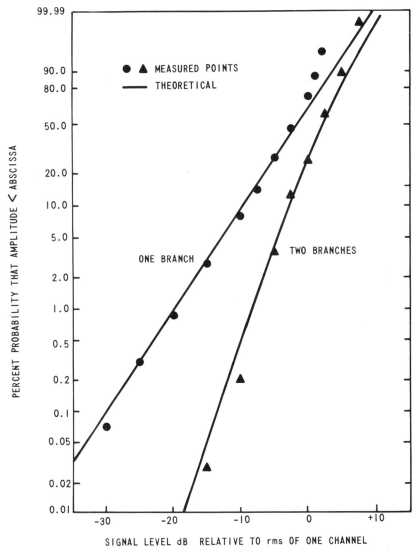

Figure 1.7-3 Probability distributions measured from the output of a fading simulator.

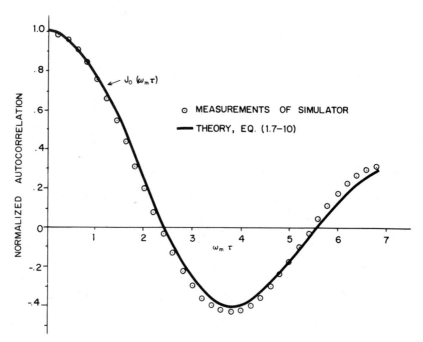

Figure 1.7-4 Comparison of theoretical autocorrelation function of the fading signal with data from a laboratory simulator.

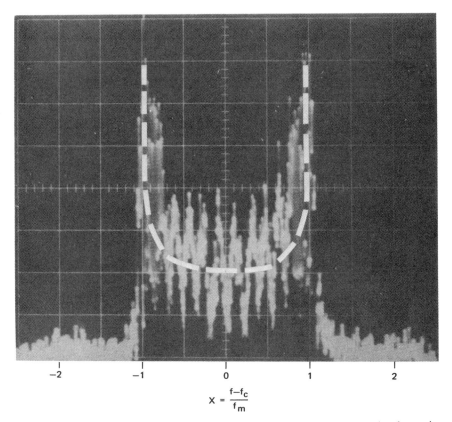

$$X = \frac{f - f_c}{f_m}$$

Figure 1.7-5 RF Spectrum of simulated fading carrier. Dashed line is the theoretical spectrum, $(1 - X^2)^{-1/4}$.

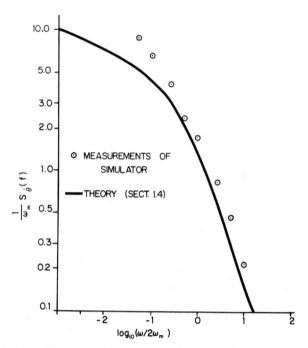

Figure 1.7-6 Comparison of theoretical spectrum of the instantaneous frequency with data from laboratory fading simulator.

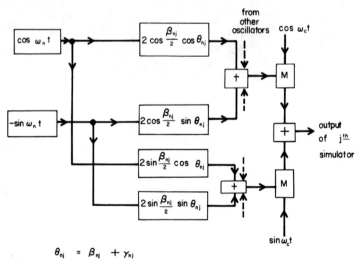

$$\theta_{nj} = \beta_{nj} + \gamma_{nj}$$

Figure 1.7-7 Use of quadrature low-frequency oscillators to provide uncorrelated fading carriers.

REFERENCES

1. W. C. Jakes, Jr. and D. O. Reudink, "Comparison of Mobile Radio Transmission at UHF and X-Bands," *IEEE Trans. Veh. Tech.* **VT-16**, October 1967, pp. 10–14.

2. W. R. Young, Jr., "Comparison of Mobile Radio Transmission at 150, 450, 900, and 3700 MHz," *Bell System Tech. J.* **31**, November 1952, pp. 1068–1085.

3. P. M. Trifonov, V. N. Budko, and V. S. Zotov, "Structure of USW Field Strength Spatial Fluctuations in a City," *Trans. Telecomm. Radio Eng.*, **9**, February 1964, pp. 26–30.

4. H. W. Nylund, "Characteristics of Small-Area Signal Fading on Mobile Circuits in the 150 MHz Band," *IEEE Trans. Veh. Tech.* **VT-17**, October 1968, pp. 24–30.

5. Y. Okumura, et al., "Field Strength and Its Variability in VHF and UHF Land-Mobile Radio Service," *Rev. Elect. Comm. Lab.*, **16**, September 1968, pp. 825–873.

6. M. J. Gans, "A Power-Spectral Theory of Propagation in the Mobile Radio Environment," *IEEE Trans. Veh. Tech.* **VT-21**, February 1972, pp. 27–38.

7. R. H. Clarke, "A Statistical Theory of Mobile Radio Reception," *Bell System Tech. J.* **47**, July 1968, pp. 957–1000.

8. S. O. Rice, "Mathematical Analysis of Random Noise," *Bell System Tech. J.* **23**, July, 1944, pp. 282–332; **24**, January 1945, pp. 46–156; "Statistical Properties of a Sine Wave Plus Random Noise," *Bell System Tech. J.* **27**, January 1948, pp. 109–157.

9. A Papoulis, *Probability, Random Variables, and Stochastic Processes*, McGraw-Hill, New York, 1965.

10. P. A. Bello and B. D. Nelin, "The Effect of Frequency Selective Fading on Intermodulation Distortion and Subcarrier Phase Stability in Frequency Modulation Systems," *IEEE Trans. Comm. Sys.*, **CS-12**, May 1964, pp. 87–101.

11. R. S. Kennedy, *Fading Dispersive Communications Channels*, Wiley-Interscience, New York, 1969, p. 18.

12. D. C. Cox, "Doppler Spectrum Measurements at 910 MHz Over a Suburban Mobile Radio Path," *Proc. IEEE*, **59** (Tech. Corres.), June 1971, pp. 1017–1018.

13. W. B. Davenport, Jr., and W. L. Root, *An Introduction to the Theory of Random Signals and Noise*, McGraw-Hill, New York, 1958.

14. I. S. Gradshteyn and I. W. Ryzhik, *Table of Integrals, Series, and Products*, Academic, New York, 1965.

15. J. F. Ossanna, Jr., "A Model for Mobile Radio Fading Due to Building Reflections: Theoretical and Experimental Fading Waveform Power Spectra," *Bell System Tech. J.* **43**, November 1964, pp. 2935–2971.

16. M. J. Gans, "A Study of the Upper Sideband Product from Mixing Gaussian Processes," to be published.

17. M. J. Gans and S. W. Halpern, "Some Measurements of Phase Coherence Versus Frequency Separation for Mobile Radio Propagation," to be published.

18. W. R. Young, Jr., and L. Y. Lacy, "Echoes in Transmission at 450 Mc from Land-to-Car Radio Units," *Proc. IRE*, **38**, March 1950, pp. 255–258.

19. D. C. Cox, "Delay-Doppler Characteristics of Multipath Propagation at 910 MHz in a Suburban Mobile Radio Environment," *IEE Trans. Ant. Prop.*, **AP-20**, September 1972, pp. 625–635.

20. E. W. Ng and M. Geller, "A Table of Integrals of the Error Functions," *J. Res. NBS-B*, **73B**, January–March 1969, pp. 1–20.

21. P. Beckmann and A. Spizzichino, *The Scattering of Electromagnetic Waves from Rough Surfaces*, The Macmillan Co., New York, 1963.

22. W. C. Y. Lee, "Antenna Spacing Requirement for a Mobile Radio Base-Station Diversity," *Bell System Tech. J.* **50**, July–August 1971, pp. 1859–1876.

23. B. Goldberg et al., "Stored Ionosphere," *IEEE First Annual Comm. Conf., Boulder, Colo.*, 1965, pp. 619–622.

24. R. C. Fitting, "Wideband Troposcatter Radio Channel Simulator," *IEEE Trans. Comm. Tech.*, **15**, August 1967, pp. 565–570.

25. R. Freudberg, "A Laboratory Simulator for Frequency Selective Fading," *IEEE First Annual Comm. Conf., Boulder, Colo., 1965*, pp. 609–614.

26. W. L. Aranguren and R. E. Langseth, "Baseband Performance of a Pilot Diversity System with Simulated Rayleigh Fading Signals and Co-Channel Interference," *Joint IEEE Comm. Soc.–Veh. Tech. Group Special Trans. on Mobile Radio Comm.*, November 1973, pp. 1248–1257.

27. W. R. Bennett, "Distribution of the Sum of Randomly Phased Components," Quart. Appl. Math., **5** January 1948, pp. 385–393.

28. M. Slack, "The Probability of Sinusoidal Oscillations Combined in Random Phase," *J. IEEE*, **93**, part III, 1946, pp. 76–86.

chapter 2

large-scale variations of the average signal

D. O. Reudink

SYNOPSIS OF CHAPTER

The principal methods by which energy is transmitted to a mobile, namely, reflection and diffraction, are often indistinguishable; thus it is convenient to lump the losses together and call them scatter (or shadow losses). In Chapter 1 it was shown that this scattering gives rise to fields whose amplitudes are Rayleigh distributed in space. The assumption of the Rayleigh model led to very powerful results and will be used again in the studies of modulation and diversity in Chapters 4 and 5. In this chapter it is shown that while the "local statistics" may be Rayleigh, the "local mean" varies because of the terrain and the effects of other obstacles. Indeed, observations of the local mean indicate that it can be characterized statistically.

Transmission paths from mobiles to base stations can be extremely varied, ranging from the occasional direct line-of-sight path to severely shadowed paths from large terrain obstructions. Under certain conditions, which are usually much simplified, the path loss can be calculated exactly between two antennas. In Section 2.1, the transmission loss for line-of-sight antennas is given. Then transmission over a plane earth is discussed, and finally the diffraction losses due to simple geometric objects in the path are calculated. There are other factors such as rain, water vapor, and oxygen in the atmosphere that attenuate microwave transmission. These factors are discussed in the latter part of Section 2.1.

In Section 2.2, signal losses are considered where terrain effects are not a factor. It is seen that there is a considerable difference in scattering losses between a suburban area and a highly built-up urban area. Antenna heights, separation distances, and frequency affect propagation in both cases. Within a city, buildings tend to channel energy parallel to the streets, strongly affecting the shadow losses.

Measurements over irregular terrain have also been made and are discussed in Section 2.3. The cases treated include rolling hills, large-scale slopes, land-sea paths, and transmission through tunnels and foilage.

A statistical representation of the local mean received signal is useful in estimating the coverage from a base station and for estimating cochannel interference. In Section 2.4 it is shown that the distribution of the local mean received signal is a log-normal distribution whose mean and variance depend on the environment. Section 2.5 concludes this chapter with predictions of path loss and estimates of the coverage surrounding a base station.

2.1 FACTORS AFFECTING TRANSMISSION

2.1.1 Free-Space Transmission Formula

The power received by an antenna separated from a radiating antenna is given by a simple formula, provided there are no objects in the region that absorb or reflect energy. This free-space transmission formula depends on the inverse square of the antenna separation distance, d, and is given by (see, for example, Ref. 1, Chapter 7)

$$P_0 = P_t \left(\frac{\lambda}{4\pi d} \right)^2 g_b g_m, \tag{2.1-1}$$

where P_0 is the received power, P_t is the transmitted power, λ is the wavelength, g_b is the power gain of the base station antenna, and g_m is the power gain of the mobile station antenna.

Thus the received radiated power decreases 6 dB for each doubling of the distance. On first inspection, one might conclude that higher frequencies might be unsuitable for mobile communications because the transmission loss increases with the square of the frequency. However, this usually can be compensated for by increased antenna gain. In mobile communications, it is often desirable to have antennas whose patterns are omnidirectional in the azimuthal plane; thus the increase in gain is required in the elevation plane. In the limit of the higher microwave frequencies, this additional gain may become impractical to realize for effective communications between an elevated base station and a mobile. This problem is considered in more detail in Chapter 3. Since the most ideal mobile radio

path involves line-of-sight propagation through the atmosphere with antennas located near the earth, we will consider the effects of both in the
following paragraphs.

2.1.2 Propagation Over a Plane Earth

Knowing the propagation characteristics over a smooth, conducting, flat
earth provides a starting point for estimating the effects of propagation
over actual paths. The complex analytical results for propagation over a
plane earth derived by Norton [2-4] have been simplified by Bullington [5,6] by
decomposing the solution of Norton into a set of waves consisting of
direct, reflected, and surface waves. The formula relating the power
transmitted to the power received following the approach of Bullington is

$$P_r = P_t \left[\frac{\lambda}{4\pi d} \right]^2 g_b g_m |1 + Re^{j\Delta} + (1 - R)Ae^{j\Delta} + \cdots|^2. \qquad (2.1\text{-}2)$$

Within the absolute value symbols, the first term (unity) represents the
direct wave, the second term represents the reflected wave, the third term
represents the surface wave, and the remaining terms represent the induction field and secondary effects of the ground.

The reflection coefficient, R, of the ground depends on the angle of
incidence, θ, the polarization of the wave, and the ground characteristics; it
is given by

$$R = \frac{\sin\theta - z}{\sin\theta + z} \qquad (2.1\text{-}3)$$

where

$$z = \frac{\sqrt{\epsilon_0 - \cos^2\theta}}{\epsilon_0} \qquad \text{for vertical polarization,}$$

$$z = \sqrt{\epsilon_0 - \cos^2\theta} \qquad \text{for horizontal polarization,}$$

$$\epsilon_0 = \epsilon - j60\sigma\lambda,$$

ϵ = the dielectric constant of the ground relative to unity in free space,

σ = the conductivity of the earth in mhos per meter.

The quantity Δ is the phase difference between the reflected and the direct paths between transmitting and receiving antennas, illustrated in Figure 2.1-1. Let h_b and h_m be the heights of the base and mobile antennas; then Δ is given by

$$\Delta = \frac{2\pi}{\lambda}\left[\left(\frac{h_b + h_m}{d}\right)^2 + 1\right]^{1/2}$$

$$- \frac{2\pi d}{\lambda}\left[\left(\frac{h_b - h_m}{d}\right)^2 + 1\right]^{1/2}. \tag{2.1-4}$$

For d greater than $5h_b h_m$,

$$\Delta \approx \frac{4\pi h_b h_m}{\lambda d}. \tag{2.1-5}$$

Since the earth is not a perfect conductor, some energy is transmitted into the ground, setting up ground currents that distort the field distribution relative to what it would have been over a perfectly reflecting surface. The surface wave attenuation factor, A, depends on frequency, polarization, and the ground constants. An approximate expression for A is given by

$$A \approx \frac{-1}{1 + j(2\pi d/\lambda)(\sin\theta + z)^2}, \tag{2.1-6}$$

which is valid for $|A| < 0.1$. More accurate values are given by Norton.[3] Since the effect of this surface wave is only significant in a region a few wavelengths above the ground, this effect can be neglected in most applications of microwave mobile communications.

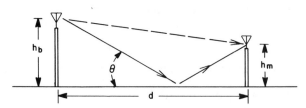

Figure 2.1-1   Propagation paths over a plane earth.

It is of interest to note that in the limit of grazing angle of incidence the value of the reflection coefficient, R, approaches -1 independent of the polarization. For frequencies above 100 MHz and for an "average" earth (see Table 1) and for vertical polarization, $|R|$ exceeds 0.9 for angles less that 10° above the horizon. For horizontal polarization above 100 MHz, $|R|$ exceeds 0.5 for angles less than 5°, but must be of the order of a degree or less for $|R|$ to exceed 0.9.[7]

Table 1 *Typical Ground Constants*

Type of Surface	$\sigma(\text{mho}/\text{m})$	ϵ
Poor ground	0.001	4
Average ground	0.005	15
Good ground	0.02	25
Sea water	5	81
Fresh water	0.01	81

Under the conditions where R equals -1 and A can be neglected, then (2.1-2) reduces to

$$P_r = 4P_0 \sin^2\left(\frac{2\pi h_b h_m}{\lambda d}\right), \qquad (2.1\text{-}7)$$

where P_0 is the expected power over a free space path. In most mobile radio applications, except very near the base station antenna, $\sin\frac{1}{2}\Delta \approx \frac{1}{2}\Delta$; thus the transmission loss over a plane earth is given by the approximation

$$P_r = P_t g_b g_m \left(\frac{h_b h_m}{d^2}\right)^2, \qquad (2.1\text{-}8)$$

yielding an inverse fourth-power relationship of received power with distance from the base station antenna.

The ground constants over the path of interest enter into both the calculations for line-of-sight and for diffraction attenuation. At microwave frequencies it is usually the dielectric constant, ϵ, which has the dominant effect on propagation. Table 1 gives values of typical ground constants. Applying these values to the formulas for the reflection coefficient over a plane earth just derived, we find that for frequencies above 100 MHz the effect of the ground constants are slight.

2.1.3 Rough Surface Criterion

At the higher microwave frequencies the assumption of a plane earth may no longer be valid, due to surface irregularities. A measure of the surface "roughness" that provides an indication of the range of validity of Eq. (2.1-2) is given by the Rayleigh criterion, which is

$$C = \frac{4\pi\sigma\theta}{\lambda},\qquad (2.1\text{-}9)$$

where σ is the standard deviation of the surface irregularities relative to the mean height of the surface, λ is the wavelength, and θ is the angle of incidence measured in radians from the horizontal. Experimental evidence shows that for $C < 0.1$ specular reflection results, and the surface may be considered smooth. Surfaces are considered "rough" for values of C exceeding 10, and under these conditions the reflected wave is very small in amplitude. Bullington[5] has found experimentally that most practical paths at microwave frequencies are relatively "rough" with reflection coefficients in the range of 0.2–0.4.

2.1.4 Refraction and Equivalent Earth's Radius

Because the index of refraction of the atmosphere is not constant, but decreases (except during unusual atmospheric conditions) with increasing height above the earth,[8] electromagnetic waves are bent as they propagate. The mean variation in refractive index can be considered linear with a constant gradient g of the form

$$n = n_0 + gh.\qquad (2.1\text{-}10)$$

In a medium where there are abrupt changes in index of refraction, Descarte's law applies:

$$n(a+h)\cos\alpha = n_0 a \cos\alpha_0,\qquad (2.1\text{-}11)$$

where α and α_0 are the angles at the discontinuity at height h, above the surface of the earth of radius a (see Figure 2.1-2). Note if the atmosphere is uniform the equation for rectilinear propagation is

$$\left(1 + \frac{h}{a}\right)\cos\alpha = \cos\alpha_0.\qquad (2.1\text{-}12)$$

When n has a constant gradient the propagation is given approximately by

$$\left[1 + h\left(\frac{1}{a} + g\right)\right]\cos\alpha \approx \cos\alpha_0.\qquad (2.1\text{-}13)$$

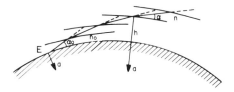

Figure 2.1-2   Ray bending from propagation through the atmosphere.

If we replace the earth's radius a by a fictitious value a', where

$$a' = \left(\frac{1}{a} + g\right)^{-1}, \qquad (2.1\text{-}14)$$

we now have an expression in the same form as that for rectilinear propagation.

Since the index of refraction in the troposphere is very nearly unity, the N-unit has been defined for convenience,

$$N_s = (n-1) \times 10^6, \qquad (2.1\text{-}15)$$

where n is the index of refraction in the atmosphere. Values of the minimum monthly mean value of N_s throughout the world have been published.[9] The most commonly used value for N_s is 301. This gives a value for the effective earth's radius a' which corresponds to four-thirds of the actual earth's radius. The empirical formula for a' is given by [10]

$$a' = 6370[1 - 0.04665 \exp(0.005577 N_s)]^{-1} \; km, \qquad (2.1\text{-}16)$$

where 6370 km is used for the earth's radius.

2.1.5 Transmission over a Smooth Spherical Earth

At microwave frequencies, diffraction due to the earth severely limits the amount of energy that propagates beyond the horizon. Considerable work has been done in an attempt to predict the signal attenuation over transhorizon paths.[10] Generally speaking, these predictions are semiempirical formulas which apply for frequencies below 1000 MHz. It is possible to obtain analytic expressions for the diffraction over a perfectly conducting sphere; however, the expressions are not simple relationships between the factors of frequency, conductivity of the earth, antenna height, and distance which govern the attenuation. A rigorous derivation of the diffraction over a spherical earth may be found in Chapter 8 of Jones.[11] Estimations of the attenuation due to diffraction over a smooth earth are particularly difficult in regions just beyond line-of-sight. Furthermore, surface roughness again seriously affects propagation. It is, of course, desirable to be able to estimate signal strengths beyond the horizon, particularly for cases where the same frequencies are being used at separated base stations. Bullington[6] has reduced the involved analytic relationships for the propagation over a smooth spherical earth to various asymptotic forms. Figure 2.1-3 is a nomograph, accurate to ± 2 dB, which was derived from his approximations. The distances d_1 and d_2 are the distances to the horizon, which can be written as

$$d_{1,2} = \sqrt{2a'h_{1,2}} \;, \tag{2.1-17}$$

where $h_{1,2}$ are the antenna heights and a' is the effective earth's radius.

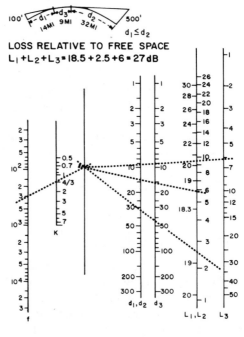

f = FREQUENCY IN MEGAHERTZ
K = RATIO OF EFFECTIVE EARTH'S RADIUS TO
TRUE EARTH'S RADIUS
d = DISTANCE IN MILES
L = ATTENUATION IN dB

Figure 2.1-3 Nomograph of signal attenuation from propagation over a smooth spherical earth.

2.1.6 Knife Edge Diffraction

Very often in the mobile radio environment a line-of-sight path to the base station is obscured by obstructions such as hills, trees, and buildings. When the shadowing is caused by a single object such as a hill, it is instructive to treat the object as a diffracting knife edge to estimate the amount of signal attenuation. The exact solution to the problem of diffraction over a knife edge is well known and is discussed in many textbooks (Ref. 11, for example).

Within the shadow region of the knife edge, the electric field strength E, can be represented as

$$\frac{E}{E_0} = A \exp(i\Delta), \qquad (2.1\text{-}18)$$

where E_0 is the value of the electric field at the knife edge, A is the amplitude, and Δ is the phase angle with respect to the direct path. The expressions for A and Δ are obtained in terms of the Fresnel integrals:

$$A = \frac{S + 1/2}{\sqrt{2}\,\sin(\Delta + \pi/4)}, \qquad (2.1\text{-}19)$$

$$\Delta = \tan^{-1}\left(\frac{S + 1/2}{C + 1/2}\right) - \frac{\pi}{4}, \qquad (2.1\text{-}20)$$

where

$$C = \int_0^{h_0} \cos\left(\frac{\pi}{2} v^2\right) dv, \qquad (2.1\text{-}21)$$

$$S = \int_0^{h_0} \sin\left(\frac{\pi}{2} v^2\right) dv, \qquad (2.1\text{-}22)$$

and

$$h_0 = h\sqrt{\frac{2}{\lambda}\left(\frac{1}{d_1} + \frac{1}{d_2}\right)}\,. \qquad (2.1\text{-}23)$$

For most microwave mobile radio applications several assumptions can be made to simplify the calculations. Consider an infinite completely absorbing (rough) half-plane that divides space into two parts as in Figure 2.1-4. When the distances d_1 and d_2 from the half-plane to the transmitting antenna and the receiving antenna are large compared to the height h, and

h itself is large compared with the wavelength, λ, that is,

$$d_1, d_2 \gg h \gg \lambda, \qquad (2.1\text{-}24)$$

then the diffracted power can be given by the expression

$$\frac{P}{P_0} = \frac{1}{2\pi^2 h_0^2} . \qquad (2.1\text{-}25)$$

This result can be considered independent of polarization as long as the conditions of Eq. (2.1-24) are met. In cases where the earth's curvature has an effect, there can be up to four paths. A simplified method of computing knife edge diffraction for such cases is treated by Anderson and Trolese.[12] Closer agreement with data over measured paths has been obtained by calculations that better describe the geometry of the diffracting obstacle.[13–15]

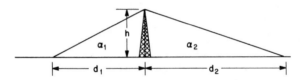

Figure 2.1-4   Geometry for propagation over a knife edge.

2.1.7 Effects of Rain and the Atmosphere

Microwave mobile radio signals are attenuated by the presence of rain, snow, and fog. Losses depend upon the frequency and upon the amounts of moisture in the path. At the higher microwave frequencies, frequency selective absorption results because of the presence of oxygen and water vapor in the atmosphere.[16] The first peak in the absorption due to water vapor occurs around 24 GHz, while for oxygen the first peak occurs at about 60 GHz, as shown in Figure 2.1-5.

The attenuation due to rain has been studied experimentally[17] by a number of workers. Figure 2.1-6 provides an estimate of the attenuating effect of rainfall as a function of frequency for several rainfall rates.[16] It should be noted that very heavy rain showers are usually isolated and not large in extent. Nevertheless, at frequencies above 10 GHz the effects of rain cannot be neglected.

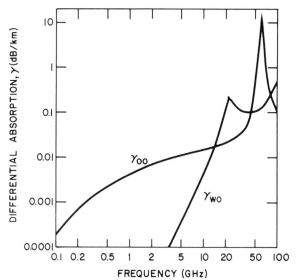

Figure 2.1-5 Signal attenuation from oxygen and water vapor in the atmosphere.

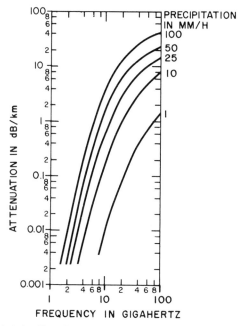

Figure 2.1-6 Signal attenuation for various rainfall rates.

89

2.1.8 Miscellaneous Effects

It is possible for microwave signals to propagate beyond the horizon by tropospheric scattering and by ducting due to diffraction. These phenomena could potentially cause interference problems; however, their effects usually can be neglected and the reader is referred to textbooks on the subject such as Ref. 8.

2.2 OBSERVED ATTENUATION ON MOBILE RADIO PATHS OVER SMOOTH TERRAIN

We will now depart from the realm of precise definitions and exact solutions to that of descriptive types of definitions and settle for results that are statistical at best. Signal attenuations over actual mobile radio paths result from a complicated dependence on the environment, and the formulas derived in the previous section apply only in special cases where the propagation paths can be clearly described. In this section we will concentrate our attention on propagation effects over relatively smooth terrain.

First, let us define the terrain to be "quasismooth" when the undulation of the ground about the average ground height is less than 20 meters. This definition would still allow the actual surface to be either "smooth" or "rough" according to the Rayleigh criterion. Furthermore, the undulations should be "gentle," that is, the distance between the peak and trough should be much larger than the height of the undulations. The average ground level should likewise remain constant to within 20 meters for distances of the order of kilometers.

The position of an antenna above the terrain affects transmission. A transmitting antenna, located 100 meters atop a mountain, certainly will propagate signals differently than an antenna located 100 meters above a flat plain. A definition of antenna height that accounts for this discrepancy is called for. Figure 2.2-1 illustrates a suitable definition for the base-station antenna height, b, which is the height of the antenna above the average ground level in the region of interest (usually considered to be greater than 3 km and less than 15 km).

Some sort of classification of the environment is also necessary since signal attenuation varies depending upon the type of objects that obstruct the path. Rather than attempt to precisely define many types of environments and then describe propagation characteristics for each we will restrict our definitions of environment to three types:

1. *Open areas*: areas where there are very few obstacles such as tall trees or buildings in the path, for instance, farm land or open fields.

2. *Suburban areas*: areas with houses, small buildings and trees, often near the mobile unit.

3. *Urban areas*: areas that are heavily built up with tall buildings and multistory residences.

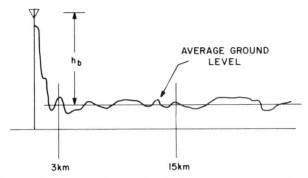

Figure 2.2-1 Definition of effective base station antenna height.

Transition regions will undoubtedly occur between the classification types and special cases can be imagined that do not fit any of the three classifications. However, the results presented for these classifications should apply in a great many instances and provide a basis to estimate results for other environments. The special problems of signal attenuation through foliage and tunnels can be isolated, and each is treated briefly at the end of this section.

Extensive measurements of radio transmission loss over various terrains in unpopulated areas have been made in the frequency ranges from 20 to 10,000 MHz by workers at the Institute for Telecommunications Sciences, and these data have been tabulated in several ESSA reports.[18-20] Mobile radio measurements in urban and suburban areas in the microwave region have been made by a number of workers.[21-33] To date, the most extensive work in the field has been reported by Okumura et al.[27] We will rely heavily upon their results to generate the prediction curves that are provided in the last portion of this chapter.

2.2.1 Field Strength Variation in Urban Areas

Since most vehicles are concentrated within large metropolitan centers, and likewise most mobile radio services will be initially provided in these areas, it is reasonable to use for a basis of comparison the median field strengths measured in quasismooth urban environments and then express suburban and open area measurements as departures from the baseline environment. Within the urban environment the received field strength is found to vary with the base station and mobile antenna heights, transmitting frequency, the distance from the transmitter, and the width and orientation of the streets. The median field strengths in a quasismooth urban area show a relatively continuous change with frequency, antenna height, and distance, while other effects appear less simply related.

With the imprecise definition of a quasismooth urban area and the variabilities of building heights and street widths that occur not only among cities but within them, one should be cautious about the range of applicability. However, as we proceed we shall see that there are many examples that indicate that there is consistency in characterizing propagation in complex environments. First, to appreciate the complexity of the situation let us examine in some detail the field strength measurements made in an urban area, and then we shall proceed to develop methods for statistically predicting signal coverage.

2.2.2 An Example of Signal Propagation

The following example is based upon measurements made by Black and Reudink[22] at 836 MHz in Philadelphia, Pennsylvania. Continuous field strength recordings were made of the signals received by a mobile unit as it moved about in the downtown area. The transmitter was located on the tallest building in the city at a height of 500 ft, and the areas covered on the test runs in the mobile unit were in the southwest quadrant of the city. Two sections studied in detail consisted of a smaller area extending about 1 mile west and 1 mile south of the transmitting location, and a larger area extending 1 mile west and 3 miles south, as indicated in Figure 2.2-2.

A one-quarter wavelength vertical whip antenna was mounted on the ground plane on the roof of the test vehicle and was used to detect the signals from the transmitter. The height above ground of the receiving antenna was approximately 3 meters. A multichannel FM magnetic tape recorder was used to record the signal strength, voice commentary, and a marker tone that was geared to one of the wheels of the vehicle so that locations could be determined accurately.

Figure 2.2-3 is an example of "local mean" signal level taken from recordings that were transferred from magnetic tape to pen recorder. The upper illustration shows a portion of Pine Street where the signal level was very low due to the shadowing effects of tall buildings. The gradual improvement going from VanPelt Street to 20th Street can be seen, and the calibration scale to the left gives an indication of the magnitude of this gradual change. The lower portion of the figure illustrates a peaking up of the signal level at street intersections such as Chestnut, Ranstead, and Ludlow Streets as the vehicle was traveling along 18th Street.

Figure 2.2-2 Aerial view of Philadelphia, Pennsylvania.

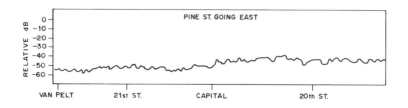

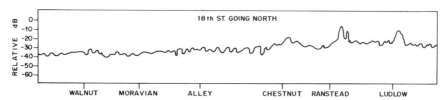

Figure 2.2-3 Typical average signal level recordings.

Figure 2.2-4 Approximate region covered in smaller area model.

94

The approximate region covered in the smaller test area as viewed from the base-station transmitter is seen in Figure 2.2-4. There are several buildings greater than ten stories in height within this area. Occasionally one can see directly to the street below, but most often the line-of-sight path was obscured. A cardboard model was constructed to correspond in size with the sections of the street map to give a visual picture of the variation in received signal strength within the area under study. The signal level for each street was plotted on a perpendicular strip of cardboard. The strips were then assembled with other sections to form a three-dimensional pattern. Figure 2.2-5 is a view of the field strength model as seen from behind the transmitter, which is indicated by the black vertical strip. This view shows the signal level variation along streets looking from the transmitted location in a southwest direction. Some items of interest are (1) the generally good signal level along the wide streets such as Market and Broad Street, (2) the variation of 20 dB or more along such streets near the transmitter caused by shadowing effects of the tall buildings near the transmitter, and (3) the general rise in signal level near the river as the signal path approached line-of-sight conditions. Figure 2.2-6 is a view from the south of the area looking in a northeast direction. There is a region of low signal extending out from the transmitter in a southwest direction, which can be seen clearly; the peaking up of the signal at street intersections is also evident at several points in this view.

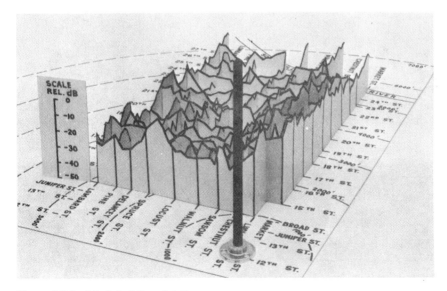

Figure 2.2-5 Model of fine detail test area.

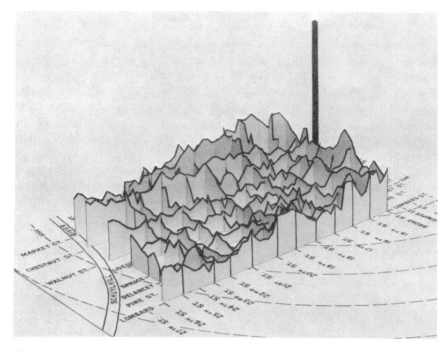

Figure 2.2-6 Model of fine detail test area.

Figure 2.2-7 Approximate region covered in larger area model.

96

The approximate area covered in the test runs from the larger area study is shown in the photograph in Figure 2.2-7, which is a view looking approximately southwest from the base-station transmitter. Figure 2.2-8 is a view of the field strength model from East Broad Street looking from the transmitter in a westerly direction. Figure 2.2-9 is a view of the model from the western side looking toward the transmitter in an easterly direction. The depressed signal area near 21st and Spruce Streets is seen to be small in comparison with the whole area.

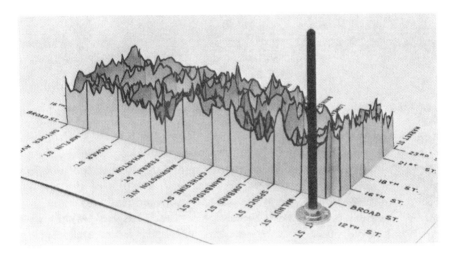

Figure 2.2-8 Model of larger test area.

There is considerably less severe variation of signal strength on the larger area model than on the smaller area model. Much more of this area is now out of the central portion of the city and the buildings are more uniform. As we shall see later, data such as these, when analyzed, have several consistent characteristics that allow statistical predictions of signal coverage.

2.2.3 Distance Dependence

One of the fundamental problems in the study of radio propagation is to describe the manner in which the signal strength attenuates as the receiving unit moves away from a transmitting base station. Obviously the signal level will fluctuate markedly (even when the Rayleigh fading is averaged out) since building heights, street widths, and terrain features are not constant. However, if we consider for a moment the behavior of the median values of the received signals, we find that there is a general trend

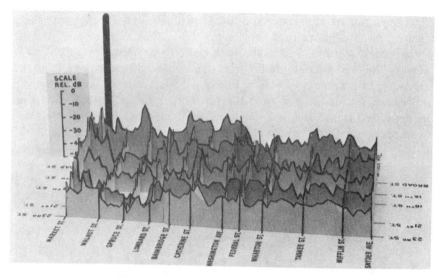

Figure 2.2-9   Model of larger test area.

for the signal levels to decrease more rapidly the further the mobile is separated from the base station. Figure 2.2-10 is a plot of the received signal power versus distance as measured by independent workers in three different cities—New York,[23] Philadelphia,[22] and Tokyo.[27] All the measurements were made at approximately 900 MHz from relatively high base-station locations. This remarkably consistent trend of both the falloff of the median signal value with distance and the excess attenuation relative to free space leads one to hope that signal strength parameters in different cities will likewise exhibit consistent traits.

The rate-of-signal decreases with distance does not appear to change significantly with increasing antenna height. However, raising the base-station antenna does tend to decrease the attenuation relative to free space. Figure 2.2-11 and 2.2-12 show measurements of these effects at 922 and at 1920 MHz by Okumura[27] in Tokyo. The distance dependence relative to free space, based on the measurements of Okumura, is shown in Figure 2.2-13 for an antenna height of 140 meters for frequencies of 453, 922, 1430, and 920 MHz. Up to distances of about 15 km the signal strength relative to free space falls off at a rate approximately proportional to the distance from the base-station antenna. At large separation distances, the signal level decreases at a much more rapid rate.

The relationship between distance dependence and antenna height (again based upon the measurements of Okumura) is shown in Figure 2.2-14. For antenna separation distances between 1 and 15 km the attenuation of median signal power with distance changes from nearly an inverse

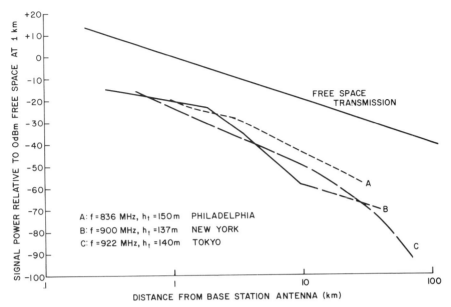

Figure 2.2-10 Examples of transmission loss with distance.

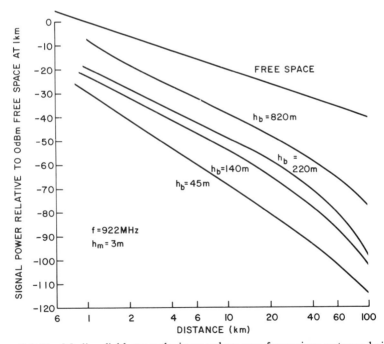

Figure 2.2-11 Median field strengths in an urban area for various antenna heights.

99

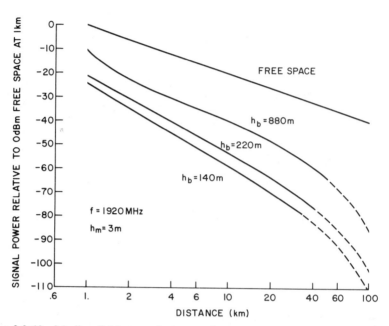

Figure 2.2-12 Median field strengths in an urban area for various antenna heights.

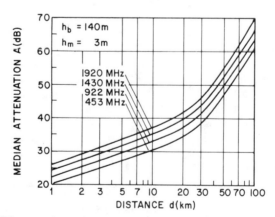

Figure 2.2-13 Distance dependence of median field strength attenuation (relative to free space) in urban area.

100

fourth power decrease for very low base-station antenna heights[25] to a rate only slightly faster than the free-space decrease for extremely high base-station antennas. For antenna separation distances greater than 40 km, the signal attenuation is very rapid.

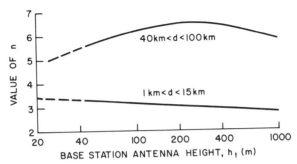

Figure 2.2-14 Distance dependence of median field strength in an urban area ($P_r \alpha d^{-n}$).

2.2.4 Frequency Dependence

Signal attenuation increases in urban areas as the frequency increases. For a fixed antenna height, the signal attenuation as a function of distance can be expressed in terms of n ($P_r \propto f^{-n}$, where P_r is the median received signal power) for varying frequencies. Figure 2.2-15 shows that n is roughly constant for distances under 10 km from the base station. As the separation increases, the decrease in signal strength with frequency becomes more rapid.

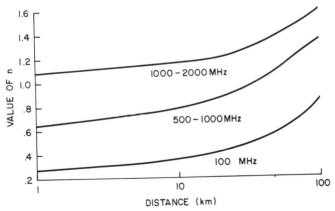

Figure 2.2-15 Frequency dependence of median field strength in an urban area ($P_r \alpha f^{-n}$).

Figure 2.2-16 is a prediction curve derived by Okumura[27] for the basic median signal attenuation relative to free space in a quasismooth urban area as it varies with both distance and frequency. These curves provide the starting point for predicting signal attenuation as discussed in Section 2.5. The curves assume a base-station antenna height of 200 meters and a mobile antenna height of 3 meters. Adjustments to these basic curves for different base station antenna heights and mobile antenna heights are considered in the paragraphs that follow.

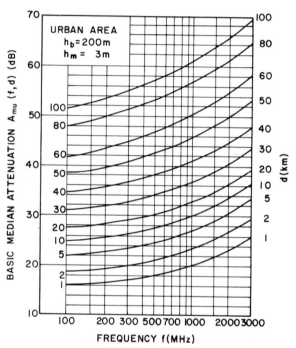

Figure 2.2-16 Prediction curve for basic median attenuation relative to free space in urban area over quasi-smooth terrain, referred to $h_b = 200$ meters, $h_m = 3$ meters.

2.2.5 Antenna Height Dependence

The formulas derived in Section 2.1 for transmission over a plane earth made no unique distinction between the effects of raising (or lowering) either the base-station or mobile-station antenna. Equation (2.1-8) predicts a 6 dB gain in received power for a doubling of the height of either antenna. In typical real-life situations the mobile antenna is likely to be buried within the confines of its surroundings while the base-station antenna will be elevated to some extent above local obstacles. The effect of a change of elevation is different in the two instances, and thus we shall treat the two cases separately.

Effect of Base-Station Antenna Height

Okumura[27] has found that the variation of received field strength with distance and antenna height remains essentially the same for all frequencies in the range from 200 to 2000 MHz. For antenna separation distances less than 10 km the received power varies very nearly proportional to the square of the base-station antenna height (6 dB per octave). For very high base-station antennas and for large separation distances (greater than 30 km), the received power tends to be proportional to the cube of the height of the base station antenna (9 dB per octave). Figure 2.2-17 is a set of prediction curves that give the change in received power (often called the height-gain factor) realized by varying the base-station antenna height. The curves are plotted for various antenna separation distances and predict the median received power relative to a 200-meter base-station antenna and a 3-meter mobile antenna. They may be used for frequencies in the range from 200 to 2000 MHz.

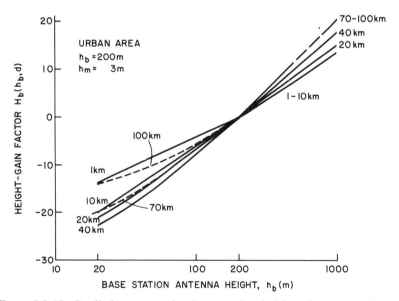

Figure 2.2-17 Prediction curves for base station height-gain factor referred to $h_b = 200$ meters.

Effect of Mobile Antenna Height

For obvious reasons mobile antenna heights are generally limited to no more than 4 meters. For a large range of frequencies and for several base station antenna heights Okumura observed a height-gain advantage of 3 dB for a 3-meter-high mobile antenna compared to a 1.5-meter-high mobile antenna. For special cases where antenna height can be above 5 meters, the height-gain factor depends upon the frequency and the environment. In a medium sized city where the transmitting frequency is 2000 MHz, the height-gain factor may be as much as 14 dB per octave, while for a very large city and a transmitting frequency under 1000 MHz the height-gain factor may be as little as 4 dB per octave for antennas above 5 meters. Prediction curves for the vehicle height-gain factor in urban areas are given in Figure 2.2-18.

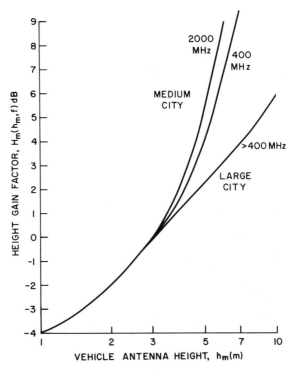

Figure 2.2-18 Prediction curve for vehicular height gain factor referred to $h_m = 3$ meters.

2.2.6 Correction Factor for Suburban and Open Areas

Suburban areas are generally characterized by lower buildings and generally less congestion of obstacles than in cities. Consequently, one should expect that radio signals would propagate better in such environments. Okumura has found that there is practically no change in the difference between urban and suburban median attenuation (suburban correction factor) with changes in base-station antenna height or with separation distances between the base and mobile antennas. The signal strength in suburban areas is weakly dependent on frequency and increases to some extent at the higher frequencies. A plot of the suburban correction factor is shown by the solid curve in Figure 2.2-19 for frequencies in the range of 100 to 3000 MHz. Recent data reported by Reudink[24] shows a 10 dB difference between urban and suburban values of the median received signal strength at a frequency of 11,200 MHz, slightly less than that predicted by Okumura in Figure 2.2-19.

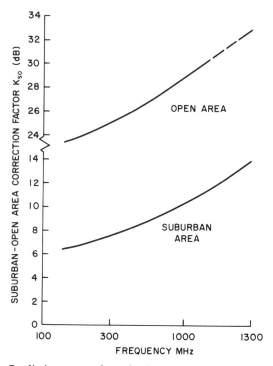

Figure 2.2-19 Prediction curves for suburban and open area correction factor, K_{s0}.

Open areas that occur rather infrequently tend to have significantly better propagation paths than urban and suburban areas, and typical received signal strengths run nearly 20 dB greater for the same antenna height and separation distances. The upper curve shown in Figure 2.2-19 provides a correction in dB that may be added directly to the prediction values for the urban case. Rural areas or areas with only slightly built-up sections have a median signal strength somewhere between the two curves.

2.2.7 Effects of Street Orientation

It has been observed that radio signals in urban areas tend to be channeled by the buildings so that the strongest paths are not necessarily the direct paths diffracted over the edge of nearby obstructing buildings, but are found to be from directions parallel to the streets. Streets that run radially or approximately radially from the base station are most strongly affected by this channeling phenomena. This causes the median received signal strength to vary by as much as 20 dB at locations near the transmitter, as shown in the signal strength model in Figure 2.2-16. Figure 2.2-20 is a sketch that indicates the way in which signal strengths may vary in an urban area because of street orientation. The density of arrows represents the relative signal strength along the various streets.

The distribution of the signal paths in the horizontal plane as seen from the mobile vehicle in an urban area is strongly affected by the street and

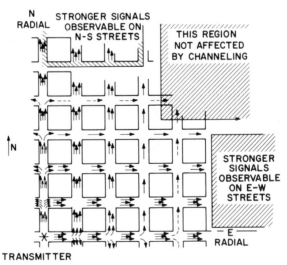

Figure 2.2-20   Idealized signal flow when channeling effects predominate.

building layout. Tests by Reudink[29] in New York City indicate that the signals arriving parallel to the direction of the street are typically 10–20 dB higher than waves arriving at other angles. These tests were carried out at 11.2 GHz by scanning with a highly directive antenna (beamwidth of 5°) at various locations in the city. Three examples of received signal strength versus angle are shown in Figure 2.2-21. In one case (on Madison Avenue) the direction in which the signal strength was strongest was nearly in the opposite direction to the base-station antenna.

2.2.8 Effects of Foliage

There are a great many factors that affect propagation behind obstacles such as a grove of trees. Precise estimates of attenuation are difficult because tree heights are not uniform; also, the type, shape, density, and distribution of the trees influence the propagation. In addition, the density of the foliage depends on the season of the year. However, some success has been obtained by treating trees as diffracting obstacles with an average effective height.

An experimental study of propagation behind a grove of live-oak and hackberry trees in Texas for several frequencies has been reported by Lagrone.[34] Height-gain measurements were made at several fixed distances behind the grove of trees for horizontally polarized waves. The measured results were compared to theoretical curves obtained assuming propagation over a smooth spherical earth and by assuming two-path diffraction over an ideal knife edge. At 82 MHz the trees were found to be fairly transparent, attenuating the signal approximately 1.6 dB per 100 ft. At a frequency of 210 MHz the absorption was found to be approximately 2.4 dB per 100 ft. At large distances of the order of 100 meters from the trees, whose heights were approximately 10 meters, the measured data fit the knife edge predictions fairly well, as seen in Figure 2.2-22. Here the measured signal strength at a frequency of 210 MHz and at a distance of 215 ft from the diffracting trees is compared to theoretical knife edge diffraction, assuming an effective height of the trees of 25 ft. At closer distances the agreement with theory is not as good. This is probably because the heights and distances from the trees cannot be clearly defined.

At frequencies from about 0.5 to 3 GHz and for distances greater than about five times the tree height, the measurements were in good agreement with the theoretical predictions of diffraction over an ideal knife edge, assuming distances and heights the same as those in the measurements. Figure 2.2-23 is a curve similar to Figure 2.2-22 but at a frequency of 2950 MHz. The measured data compared with the theoretical curves for a smooth spherical earth and with knife edge diffraction curves show better agreement with the knife edge diffraction theory. At shorter distances

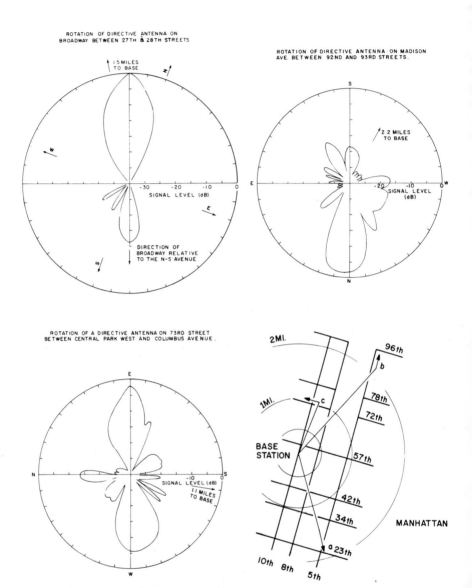

Figure 2.2-21 Examples of directions of signal arrivals.

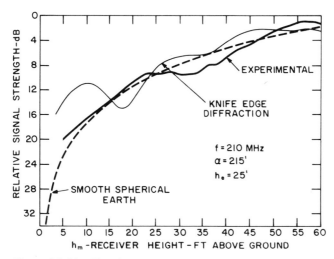

Figure 2.2-22   Signal propagation behind trees.

some propagation takes place through the trees, and this acts to reduce the effective height of the diffracting edge and at the same time increases the apparent distance from the diffracting edge to the antenna.

Recent measurements at 836 MHz and 11.2 GHz were made by Reudink and Wazowicz;[35] they compare the signal strength measured on the same streets in summer and in winter in suburban Holmdel, New Jersey.

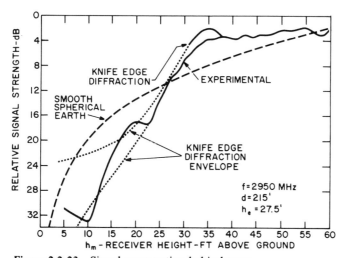

Figure 2.2-23   Signal propagation behind trees.

Figures 2.2-24 and 2.2-25 show signal strengths from data taken on a road that runs approximately perpendicular to a radial line 2 miles distant from a 400-ft base-station antenna site. The relative received signals are shown at a frequency of 836 MHz in both summer and winter in Figure 2.2-24. The corresponding data at 11.2 GHz are plotted in Figure 2.2-25. The curves for predicted values were derived from the knife edge diffraction formula and show reasonably good agreement. At the UHF frequency the average received signal strength in the summer when the trees were in full leaf was roughly 10 dB lower than for the corresponding locations in later winter. At X-band frequencies the losses during the summer appear to be greater in the areas where the signal levels were previously low.

2.2.9 Signal Attenuation in Tunnels

It is well known that frequencies in the VHF region commonly used for mobile communications are severely attenuated in tunnel structures.[36,37] Only by using special antennas are these frequencies usable in long (over 1000 ft) tunnels. However, at microwave frequencies tunnels are effective guiding or channeling mechanisms and can offer significant improvement over VHF for communications.

A test[38] was performed in the center tube of the Lincoln Tunnel, 8000 ft long, which connects midtown Manhattan to New Jersey under the Hudson River. The inside of the tunnel is roughly rectangular in cross section with a height of 13.5 ft and a width of 25 ft. Seven test frequencies roughly an octave apart were used to make signal attenuation measurements at the following frequencies: 153, 300, 600, 980, 2400, 6000, and 11,215 MHz. The transmitters were stationed 1000 ft inside the western portal in order to keep the test situation as simple as possible. This location cleared an initial curve at the entrance and allowed a line-of-sight path of nearly 2000 ft before an elevation change cut off the view. Beyond this point nearly another mile of tunnel remained before the eastern exit was reached.

The average loss of signal strength in dB against the antenna separation for the seven frequencies is plotted in Figure 2.2-26. For convenience in plotting the data, an arbitrary reference level of 0 dB at 1000 ft antenna separation was chosen. It is worth noting that the 153- and 300-MHz attenuation rates are nearly straight lines, implying that the signal attenuation has an exponential relationship to the separation. At 153 MHz the loss is extremely high (in excess of 40 dB per 1000 ft), where at 300 MHz the rate of attenuation is of the order of 20 dB per 1000 ft. At higher frequencies a simple exponential attenuation rate is not evident. In Figure 2.2-27 the data have been replotted on a logarithmic distance scale. Signal attenuations that depend upon distance raised to some power appear as

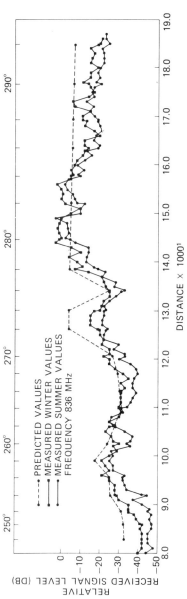

Figure 2.2-24 Signal propagation at nearly a constant radius from an elevated base station.

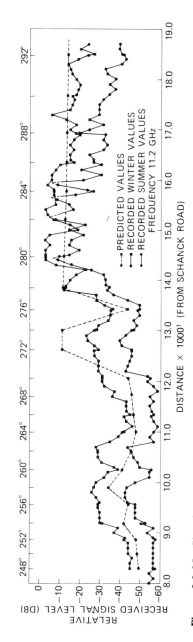

Figure 2.2-25 Signal propagation at nearly a constant radius from an elevated base station.

straight lines in this case. For the major portion of the length of the tunnel the received signal level at 900 MHz has an inverse fourth-power dependence upon the antenna separation, while at 2400 MHz the loss has an inverse square dependence. At frequencies above 2400 MHz, dependence of the signal strength with antenna separation is less than the free-space path loss (throughout most of the length of the tunnel). Roughly, the attenuation rates appear to be only 2–4 dB per 1000 ft for frequencies in the 2400–11,000 MHz range.

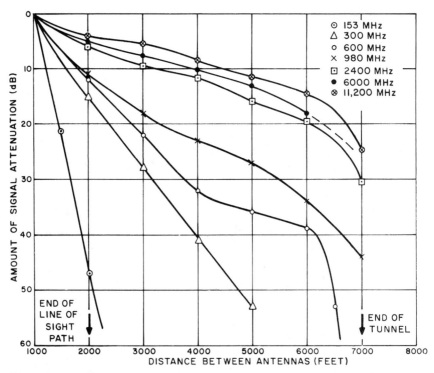

Figure 2.2-26 Signal loss versus antenna separation for seven frequencies.

2.3 EFFECTS OF IRREGULAR TERRAIN

The traditional approach of predicting attenuation from propagation over irregular terrain has been to approximate the problem to one that is solvable in closed form. This is usually done by solving problems dealing with smooth regular boundaries such as planes or cylinders.[39-48] Several workers have published approximations to the exact formulas for propagation over various obstacles. In Section 2.1 examples were given of the theoretical signal attenuation due to the earth's curvature and due to an ideal absorbing knife edge. Improved agreement with measurements for cases such as the propagation over an isolated mountain ridge are obtained by more accurately describing the obstacle in terms of more realistic geometries.

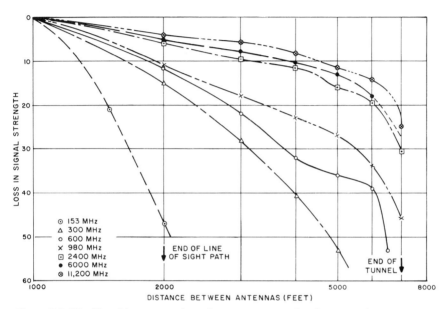

Figure 2.2-27   Signal loss versus log of antenna separation for seven frequencies.

2.3.1 Modeling Propagation by Frequency Scaling

To some extent propagation over more complicated obstacles can be determined by constructing models and performing laboratory experiments at optical frequencies or frequencies in the millimeter wavelength region.[49–51] Using a helium–neon gas laser operating at a wavelength of 632.8 nm with cylindrical diverging lens, Hacking[51] constructed a model transmitter and used a narrow-slit aperture to model a receiver. With this arrangement the wavelength scaling to UHF is approximately 10^6 to 1. Model hills could thus be constructed with workable dimensions. First obtaining agreement with theory by testing the system on smooth diffracting cylinders, Hacking has investigated more complicated obstacles such as double-hump or flat-top slabs. In addition to smooth objects, terrain roughness was simulated by wrapping abrasive papers around the cylinders. In this manner surface roughness whose rms deviation ranged from 2λ to 18λ were obtained. Figure 2.3-1 is a plot of the results obtained by this method. The solid curve is the theoretical calculation of the diffraction loss over a perfectly conducting smooth cylinder. Examples of the excess loss due to surface roughness are shown in the curves labeled A through D. In attempting to apply the rough surface to a real-life example, curve B would correspond to the field strength received at UHF in the shadows of hills on which there are irregular distributions of houses and trees. (Assume, for example, that there are two houses whose dimensions are $30 \times 20 \times 30$ ft and two trees whose dimensions are $15 \times 15 \times 30$ ft per acre. Then the rms surface height is calculated to be 6 ft.)

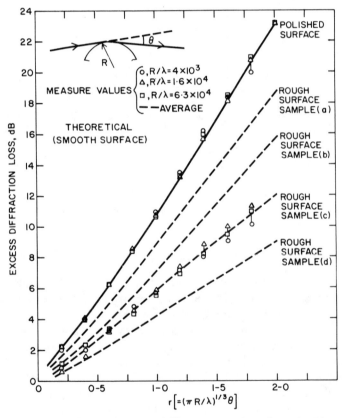

Figure 2.3-1   Examples of modelling propagation of rough objects.

Other ways of estimating signal attenuation over irregular terrain are empirical formulas derived from measurements. We shall briefly mention a computer method that predicts path loss in good agreement with experiment, and then present some prediction curves that provide quick but somewhat less accurate estimates of signal attenuation.

2.3.2 A Computer Method for Predicting Attenuation

A computer program has been published by Longley and Rice[9] that predicts the long-term median radio transmission loss over irregular terrain. The method predicts median values of attenuation relative to the transmission loss free space and requires the following: the transmission frequency, the antenna separation, the height of the transmitting and receiving antennas, the mean surface refractivity, the conductivity and dielectric constant of the earth, polarization, and a description of the terrain. This program was based on thousands of measurements and compares well with measured data[52] over the following ranges:

parameter	range
Frequency	20 MHz to 40 GHz
Antenna height	0.5 to 3000 meters
Separation distance	1 to 2000 kilometers
Surface refractivity	250 to 450 N units

The critical parameters necessary in any prediction of path loss are those which characterize the terrain. The "interdecile" (see illustration in Figure 2.3-2) range, $\Delta h(d)$, of terrain heights above and below a straight line is a parameter often used in prediction formulas and is calculated at fixed distances, d, along the path. Longley and Rice have found that the values of $\Delta h(d)$ increase with path length to an asymptotic value Δh according to the following formula:

$$\Delta h(d) = \Delta h[1 - 0.8 \exp(-0.02d)], \qquad (2.3-1)$$

where $\Delta h(d)$ and Δh are in meters and the distance d is in kilometers. For a particular path where profiles are available, $\Delta h(d)$ can be calculated precisely. In other cases or for area predictions, estimates of Δh are given in Table 2.

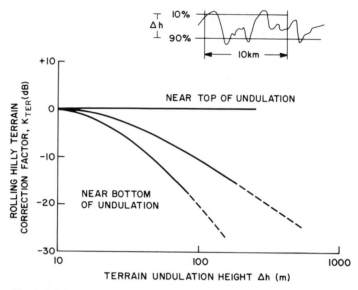

Figure 2.3-2   Rolling hilly terrain correction factor, K_{terr}.

The computer method of Longley and Rice provides both point-to-point and area predictions that agree well with the experiment, but a description of the calculation of the many parameters used in their method would be rather lengthy. The reader is referred to their work for precise calculations. We shall adopt a somewhat less accurate method in which we obtain correction factors to our basic median curves (Figure 2.2-16) for various terrain effects.

Table 2   *Estimates of Δh*

Type of Terrain	Δh (meters)
Water or very smooth plains	0–5
Smooth plains	5–20
Slightly rolling plains	20–40
Rolling plains	40–80
Hills	80–150
Mountains	150–300
Rugged mountains	300–700
Extremely rugged mountains	>700

2.3.3 Predictions of Propagation by Correction Factors: Undulating, Sloping, and Land-Sea Terrain

An approximate prediction curve for undulating terrain is given in Figure 2.3-2, which is based on work reported by the CCIR[16] and Okumura.[27] This estimates the correction factor to the basic median attenuation curves derived previously in Section 2.1 for quasismooth urban terrain. More exact predictions would probably have some dependence on frequency and on antenna separation distance. If the location of the vehicle is known to be near the top of the undulation, the correction factor in Figure 2.3-2 can be ignored. On the other hand, if the location is near the bottom of the undulation, the attenuation is higher and is indicated on the lower curve on Figure 2.3-2.

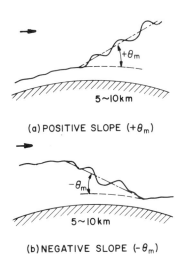

(a) POSITIVE SLOPE $(+\theta_m)$

(b) NEGATIVE SLOPE $(-\theta_m)$

Figure 2.3-3 Definition of average angle of general terrain slope.

In cases where the median height of the ground is gently sloping for distances of the order of 5 km, a correction factor may be applied. Let us define the average slope θ_m measured in milliradians as illustrated in Figure 2.3-3. Depending upon the antenna separation distance d, the terrain slope correction factor, K_{sp}, is given in Figure 2.3-4 in terms of the average slope θ_m. (It should be noted that these curves are based on rather scant data and should be considered as estimates applying in the frequency range of 450–900 MHz.)

Usually on propagation paths where there is an expanse of water between the transmitting and receiving stations the received signal strength

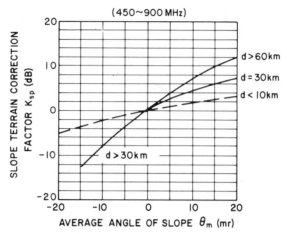

Figure 2.3-4 Measured value and prediction curves for "slope terrain correction factor."

tends to be higher than for cases where the path is only over land. The change in signal strength depends on the antenna separation distance and whether the water lies closer to the mobile receiver or the base-station transmitter, or somewhere in between. Let us define a ratio, β, which represents the fraction of the path that consists of propagation over water. Figure 2.3-5 illustrates two path geometries and the definition of β in each case. It has been observed that when the latter portion of the path from the

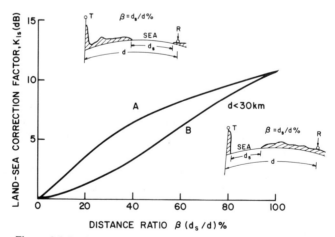

Figure 2.3-5 Prediction curves for land-sea correction factor.

base-station antenna to the mobile antenna is over water, the signal strength is typically 3 dB higher than for cases where the latter portion of the base mobile path is over land. Prediction curves for mixed land-sea paths have been obtained experimentally by Okumura,[27] as shown in Figure 2.3-5, which provides correction factors in terms of the percentage of the path over water.

2.4 STATISTICAL DISTRIBUTION OF THE LOCAL MEAN SIGNAL

Thus far we have obtained results based primarily on experimental evidence that has provided us with the behavior of median signal levels obtained by averaging received signals over a distance of 10–20 meters. Smooth curves were obtained relating the variation of the median received signal with distance, base-station antenna height, and frequency in urban, suburban, and rural areas. Consistent but less accurate predictions were found for dependence on street orientation, isolated ridges, rolling hills, and land-sea paths. Considerably fewer data are available to describe the fluctuations of the received signal about the median value. The dependence of the signal distribution upon the parameters mentioned above requires a good deal more investigation before definitive results are available.

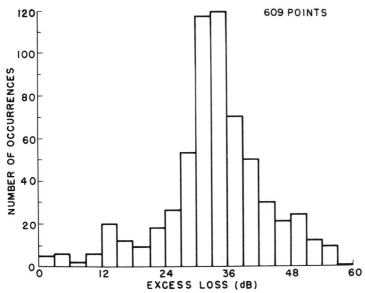

Figure 2.4-1 Histogram of excess path loss in New Providence, N.J.

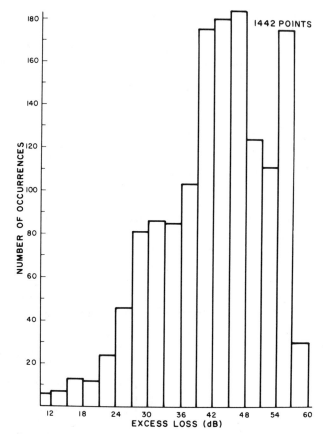

Figure 2.4-2 Histogram of excess path loss in New York City.

One consistent result that has been observed is that the distribution of the received signals at fixed base and mobile antenna heights, frequency, and separation distance from the base station within the same environment class (urban, for example) have very nearly a normal distribution when the distribution is plotted for the received signal measured in decibels. Such a probability distribution is often referred to as log-normal.[53] Also the excess path loss, that is, the difference (in decibels) between the computed value of the received signal strength in free space (Eq. 2.1-1) and the actual measured value of the local mean received signal has been observed to be log-normally distributed. Figures 2.4-1 and 2.4-2 are examples of histograms of excess path loss measured at 11.2 GHz in a suburban area of

New Jersey and in New York City.[24] The distributions of these two histograms are plotted in Figure 2.4-3. The median values of the received signal in the urban case is about 10 dB lower than for the suburban case. Both sets of data appear to fit straight lines rather well, corresponding to a standard deviation of 10 dB. The data were also sorted into various range slots and their corresponding distributions were calculated. For the suburban data no significant changes were observed. In the urban case, however, the standard deviation of the excess path loss was found to decrease to about 8 dB for locations less than 1 mile from the base-station transmitter and to increase to 12 dB for locations greater than 1 mile from the base station.[24]

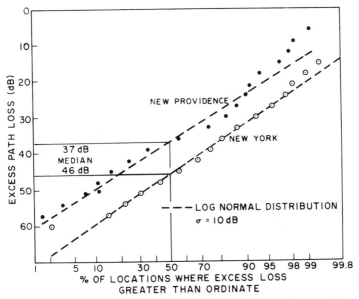

Figure 2.4-3 Distributions of excess path loss in New Providence, N.J. and New York City.

The distribution of excess path loss was also calculated for the signal strength models constructed from the measurements of the local mean signal received at 836 MHz in Philadelphia discussed in Section 2.2. Figure 2.4-4 is a plot of the distributions of the received signal levels for the models shown in Figures 2.2-5 and 2.2-8. In Philadelphia Black and Reudink[22] noted an increase in σ, the standard deviation, for distances close to the base station (see Figure 2.4-4). However, measured data in

New York[24] showed the opposite effect, namely, that the value of σ increased with distance. A partial explanation for these seemingly contradictory results is that in the Philadelphia case, for distances greater than 1 mile from the base station, the average building height was about 10–20 meters, tending to produce a low value of σ. On the other hand, in New York City at distances greater than 1 mile from the base station there are still many very tall buildings, some well over 100 meters, whose presence would tend to increase σ.

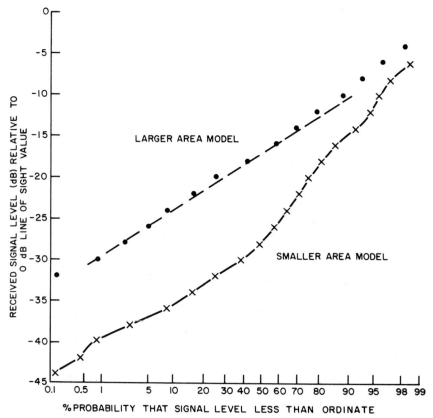

Figure 2.4-4 Distributions of excess path loss in Philadelphia.

Okumura[27] has measured the standard deviation of the median field strength variations in Tokyo and has found that the mean values of the standard deviation are not strongly dependent upon the base-station antenna height or antenna separation distance but do have a slight dependence with frequency. In these measurements base-station antenna heights ranged from 140 to 820 meters. For lower antenna heights where there are obstacles in the path comparable in height to the base-station antenna height, one would suspect that the standard deviation would increase somewhat. Measurements for base-station antenna heights ranging from 15 to 25 meters were made in the Philadelphia area by Ott.[25] His results show a standard deviation that decreased very slightly with distance from a nominal value of 8.2 dB at 1.5 km to 7 dB at 15 km.

Figure 2.4-5 is a prediction curve for the standard deviation, σ, of the log-normal distribution that describes the variation of the median signal strength values in suburban areas as given by Okumura. The data spreads at 850 MHz are from the data of Black[22] and Ott[25]; at 11.2 GHz the data are from Reudink.[24] The data points at 127 and 510 MHz are from the work of Egli.[54]

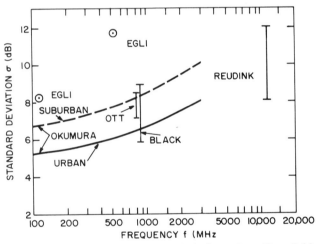

Figure 2.4-5 Prediction curves for standard deviation of median field strength variation in urban, suburban, and rolling hilly terrain.

2.5 PREDICTION OF FIELD STRENGTH

2.5.1 Prediction of Median Signal Strength

In order to predict the median power received by a mobile unit from a base-station antenna in a basic urban environment we may use the following equation (all quantities in decibels):

$$P_p = P_o - A_m(f,d) + H_b(h_b,d) + H_m(h_m,f), \qquad (2.5\text{-}1)$$

where

P_p represents the value of the predicted received power,

P_o is the power received for free-space transmission (Eq. 2.1-1)

$A_m(f,d)$ is the median attenuation relative to free space in an urban area where the effective base-station antenna height, h_b, is 200 meters and the vehicle antenna height, h_m, is 3 meters. These values are expressed as a function of distance and frequency and can be obtained from the curves of Figure 2.2-16.

$H_b(h_b,d)$ is the base-station height-gain factor expressed in decibels relative to a 200-meter-high base-station antenna in an urban area. This function is dependent upon distance and has been plotted in Figure 2.2-17.

$H_m(h_m,f)$ is the vehicle station height-gain factor expressed in decibels relative to a 3-meter-high vehicle-station antenna in an urban area. This factor is dependent upon frequency and has been plotted in Figure 2.2-18.

If the particular propagation path happens to be over a different environment type or to involve terrain that is not "quasismooth," we may amend our prediction formula for P_p by adding one or more of the correction factors that were described in earlier portions of the chapter. Thus, the "corrected" predicted received power P_c is

$$P_c = P_p + K_{so} + K_{sp} + K_{ls} + K_{ter}, \qquad (2.5\text{-}2)$$

where

K_{so} is the "correction factor for suburban and open terrain," which is plotted in Figure 2.2-19.

K_{sp} is the "correction factor for sloping terrain," which is obtained from the curve in Figure 2.3-4.

K_{ls} is the "land-sea correction factor," which provides a correction to the signal attenuation when there is an expanse of water in the propagation path. A description of this correction factor and a prediction curve are given in Figure 2.3-5.

K_{ter} is the "correction factor for rolling hilly terrain," which was discussed in Section 2.3 and may be obtained from Figure 2.3-2.

In addition to these correction factors, there are other factors such as isolated mountain ridges, street orientation relative to the base station, the presence or absence of foliage, effects of the atmosphere, and in the case of undulating terrain, the position of the vehicle relative to the median height. These additional effects together with the fact that the correction factors and indeed the basic transmission factors A_m, H_b, and H_m are average values based upon empirical data, should indicate that discrepancies between measured and predicted values are still possible. It is reassuring to point out, however, that these prediction curves, which are essentially the same as those of Okumura, have been compared to measured data with a great deal of success.[27] This was done over a variety of environments, antenna heights, and separation and for a number of frequencies.

2.5.2 Determination of Signal Coverage in a Small Area

Let us assume that the local mean signal strength in an area at a fixed radius from a particular base-station antenna is log-normally distributed. Let the local mean (that is, the signal strength averaged over the Rayleigh fading) in decibels be expressed by the normal random variable x with mean $\bar{x}$ (measured in dBm*, for example) and standard deviation σ (decibels). To avoid confusion, recall that $\bar{x}$ is the median value found previously from Eq. (2.5-2). As we have seen, $\bar{x}$ depends upon the distance (r) from the base station as well as several other parameters. Let x_0 be the receiver "threshold." We shall determine the fraction of the locations (at $r = R$) wherein a mobile would experience a received signal above "threshold." The "threshold" value chosen need not be the receiver noise threshold, but may be any value that provides an acceptable signal under Rayleigh fading conditions. The probability density of x is

$$p(x) = \frac{1}{\sigma\sqrt{2\pi}} \exp\left[\frac{-(x-\bar{x})^2}{2\sigma^2}\right]. \qquad (2.5\text{-}3)$$

The probability that x exceeds the threshold x_0 is

$$P_{x_0}(R) = P[x \geqslant x_0] = \int_{x_0}^{\infty} p(x)\,dx$$

$$= \frac{1}{2} - \frac{1}{2}\operatorname{erf}\left(\frac{x_0 - \bar{x}}{\sigma\sqrt{2}}\right). \qquad (2.5\text{-}4)$$

*The notation dBm used throughout the text stands for dB above one milliwatt.

If we have measured or theoretical values for $\bar{x}$ and for σ in the area of interest, we can determine the percent of the area for which the average signal strength exceeds x_0. For example, at a radius where the median (and hence the mean of the log-normal) signal strength is -100 dBm ($\bar{x} = -100$ dBm at some particular separation distance R and for some radiated power) and assume our system threshold happens to be -110 dBm, then, if we assume $\sigma = 10$ dB we have

$$P_{x_0}(R) = \frac{1}{2} + \frac{1}{2}\operatorname{erf}\left(\frac{1}{\sqrt{2}}\right) = 0.84.$$

2.5.3 Determination of the Coverage Area from a Base Station

It is also of interest to determine the percentage of locations *within* a circle of radius R in which the received signal strength from a radiating base-station antenna exceeds a particular threshold value. Let us define the fraction of useful service area F_u as that area, within a circle of radius R, for which the signal strength received by a mobile antenna exceeds a given threshold x_0. If P_{x_0} is the probability that the received signal, x, exceeds x_0 in an incremental area dA, then

$$F_u = \frac{1}{\pi R^2} \int P_{x_0} dA. \tag{2.5-5}$$

In a real-life situation one would probably be required to break the integration into small areas in which P_{x_0} can be estimated and then sum over all such areas. For purposes of illustration let us assume that the behavior of the mean value of the signal strength follows an r^{-n} law. Thus

$$\bar{x} = \alpha - 10n \log_{10} \frac{r}{R}, \tag{2.5-6}$$

where α, expressed in dB, is a constant determined from the transmitter power, antenna heights and gains, and so on. Then

$$P_{x_0} = \frac{1}{2} - \frac{1}{2}\operatorname{erf}\left[\frac{x_0 - \alpha + 10n \log_{10} r/R}{\sigma\sqrt{2}}\right]. \tag{2.5-7}$$

Then letting $a = (x_0 - \alpha)/\sigma\sqrt{2}$ and $b = 10n \log_{10} e/\sigma\sqrt{2}$, we get

$$F_u = \frac{1}{2} - \frac{1}{R^2} \int_0^R r \operatorname{erf}\left(a + b \log \frac{r}{R}\right) dr. \tag{2.5-8}$$

The integral above can be evaluated by substituting $t = a + b \log(r/R)$, so that

$$F_u = \frac{1}{2} + \frac{2e^{-2a/b}}{b} \int_{-a}^{\infty} e^{-2t/b} \mathrm{erf}(t)\,dt. \qquad (2.5\text{-}9)$$

From Ref. 55, page 6, No. 1,

$$F_u = \frac{1}{2}\left[1 - \mathrm{erf}(a) + \exp\left(\frac{1 - 2ab}{b^2}\right)\left(1 - \mathrm{erf}\,\frac{1 - ab}{b}\right)\right] \qquad (2.5\text{-}10)$$

For example, let us choose α such that $\bar{x} = x_0$ at $r = R$; then $a = 0$ and

$$F_u = \frac{1}{2} + \frac{1}{2}\exp\left(\frac{1}{b^2}\right)\left[1 - \mathrm{erf}\left(\frac{1}{b}\right)\right]. \qquad (2.5\text{-}11)$$

Let us further assume that $n = 3$ and $\sigma = 9$; then $F_u = 0.71$, or about 71% of the area within a circle of radius R has signal above threshold when half the locations on the circumference have a signal above threshold.

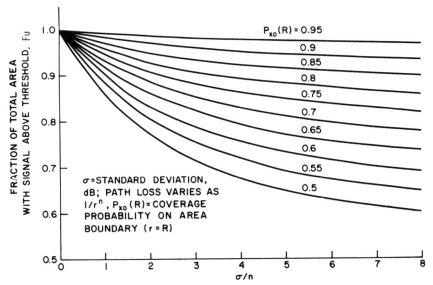

Figure 2.5-1   Fraction of total area with signal above threshold, F_u.

For the case where the propagation follows a power law, the important parameter is σ/n. Figure 2.5-1 is a plot of the fraction of the area within a circle of radius R that has a received signal above a threshold for various fractions of coverage on the circle.

A. P. Barsis[56] has outlined methods for determining coverage when the propagation law is not a simple r^{-n} relationship or when the service area is defined in terms of cochannel interference.

REFERENCES

1. H. H. Skilling, *Fundamentals of Electric Waves*, 2nd ed. Wiley, New York, 1948.

2. K. A. Norton, "The Propagation of Radio Waves Over the Surface of the Earth in the Upper Atmosphere," Part 1, *Proc. IRE*, **24**, October 1936, page 1367.

3. K. A. Norton, "The Propagation of Radio Waves Over the Surface of the Earth in the Upper Atmosphere," Part 2, *Proc. IRE*, **25**, September 1937, page 1203.

4. K. A. Norton, "The Calculation of Ground Wave Field Intensity Over a Fintely Conducting Spherical Earth," *Proc. IRE*, **29**, December 1941, page 63.

5. K. Bullington, "Radio Propagation at Frequencies Above 30 megacycles," *Proc. IRE*, **35**, October 1947, page 1122.

6. K. Bullington, "Radio Propagation Fundamentals," *Bell System Tech. J.* **36**, May 1957, page 593.

7. E. C. Jordan and Keith G. Balmain, *Electromagnetic Waves in Radiating Systems*, 2nd ed., Prentice-Hall, Englewood Cliffs, N. J., 1968, Chapter 16.

8. Francois Du Castel, *Tropospheric Radio Propagation Beyond the Horizon*, Pergamon, New York, 1966, Chapter 2.

9. B. R. Bean, J. D. Horn, and A. M. Ozanich, Jr., "Climatic Charts and Data of Radio Refractive Index for the United States and the World," NBS Monograph, No. 22, US Government Printing Office, Washington, D.C., 1960.

10. A. G. Longley and P. L. Rice, "Prediction of Trospospheric Radio Transmission Loss Over Irregular Terrain," A Computer Method—1968. ESSA Research Laboratories ERL79-ITS67, US Government Printing Office, Washington, D.C., 1968.

11. D. S. Jones, *The Theory of Electromagnetism*, MacMillan, New York, 1964, Chapter 9.

12. L. J. Anderson and L. G. Trolese, "Simplified Method for Computing Knife Edge Diffraction in the Shadow Region," *IRE Trans. Ant. Prop.*, **6**, July 1958, page 281.

13. N. P. Bachynski and M. G. Kingsmill, "Effect of Obstacle Profile on Knife Edge Diffraction," *IRE Trans. Ant. Prop.*, **10**, March 1962, page 201.

14. G. Millington, "A Note on Diffraction Round a Sphere or Cylinder," *Marconi Rec.*, **23**, 1960, page 170.

15. H. T. Dougherty and L. J. Maloney, "Application of Defraction by Convex Surfaces to Irregular Terrain Situations," *Radio Phone*, **68B**, No. 2, February 1964.

16. CCIR Report 370-1, Oslo, 1966 II.

17. D. C. Hogg, "Statistics on Attenuation of Microwaves by Intense Rain," *Bell System Tech. J.* **48**, November 1969, page 2949.

18. P. L. McQuate, J. M. Harman, and A. P. Barsis, "Tabulations of Propagation Data Over Irregular Terrain in the 230 to 9200 MHz Frequency Range, Part 1: Gun Barrel Hill Receiver Site," ESSA Technical Report ERL65-ITS52, March 1968.

19. P. L. McQuate, J. M. Harman, M. E. Johnson, and A. P. Barsis, "Tabulations of the Propagation Data Over Irregular Terrain in the 230 to 9200 MHz Frequency Range," Part 2: Fritz Peak Receiver Site, ESSA Technical Report ERL65-ITS58-2, December 1968.

20. P. L. McQuate, J. M. Harman, M. E. McClamaham, and A. P. Barsis, "Tabulations of Propagation Data Over Irregular Terrain in the 230 to 9200 MHz Frequency Range," Part 3: North Table Mountain–Golden, ESSA Technical Report ERL65-ITS58-3, July 1970.

21. W. C. Jakes, Jr., and D. O. Reudink, "Comparison of Mobile Radio Transmission at UHF and X-Band," *IEEE Trans. Veh. Tech.* **16**, October 1967, pp.10–14.

22. D. M. Black and D. O. Reudink, "Some Characteristics of Radio Propagation at 800 MHz in the Philadelphia Area," *IEEE Trans. Veh. Tech.* **21**, May 1972, pp. 45–51.

23. W. Rae Young, Jr., "Comparison of Mobile Radio Transmission at 150, 450, 900, and 3700 MC," *Bell System Tech. J.* **31**, November 1952, page 1068.

24. D. O. Reudink, "Comparison of Radio Transmission at X-Band Frequencies in Suburban and Urban Areas," *IEEE Trans. Ant. Prop.*, **AP-20**, July 1972, page 470.

25. G. D. Ott, "Data Processing Summary and Path Loss Statistics for Philadelphia HCMTS Measurements Program," unpublished work.

26. W. C. Y. Lee, "Preliminary Investigation of Mobile Radio Signal Fading Using Directional Antennas on the Mobile Unit," *IEEE Trans. Veh. Comm.*, **VC-15**, October 1966, pp. 8–15.

27. Y. Okumura, E. Ohmori, T. Kawano, and K. Fukuda, "Field Strength and its Variability in VHF and UHF Land Mobile Service," *Rev. Elec. Comm. Lab.*, **16**, September–October 1968, page 825.

28. D. C. Cox, "Time-and-Frequency Domain Characterizations of Multipath Propagation at 910 MHz in a Suburban Mobile-Radio Environment," *Radio Sci.*, **7**, December 1972, pp. 1069–1077.

29. D. O. Reudink, "Preliminary Investigation of Mobile Radio Transmission at X-Band in an Urban Area," 1967 Fall URSI Meeting at Ann Arbor, Mich.

30. W. C. Y. Lee and R. H. Brandt, "The Elevation Angle of Mobile Radio Signal Arrival," *IEEE Trans. Comm.*, **COM-21**, No. 11, November 1973, pp. 1194–1197.

31. H. L. Hanig, "The Characterization of Environmental Man-Made Noise at 821 MHz," Symposium on Microwave Mobile Communications, Boulder, Colo. 1972.

32. E. L. Caples, "Base Station Antenna Height Effects for the HCMTS Measurement Program," Symposium on Microwave Mobile Communications, Boulder, Colo., 1972.

33. S. Rhee and G. Zysman, "Space Diversity at Base Station," Symposium on Microwave Mobile Communications, Boulder, Colo., 1972.

34. A. H. LaGrone, P. E. Martin, and C. W. Chapman, "Height Gain Measurements at VHF and UHF Behind a Grove of Trees," *IRE Trans. Ant. Prop.*, **9**, 1961, pp. 487–491.

35. D. O. Reudink and M. F. Wazowicz, "Some Propagation Experiments Relating Foliage Loss and Diffraction Loss at X-Band at UHF Frequency," *Joint IEEE Comm. Soc.–Veh. Tech. Group Special Trans. on Mobile Radio Comm.*, November 1973, pp.1198–1206.

36. H. W. Nylund, and R. V. Crawford, "High-Speed Train Telephone System— Transmission Tests in Baltimore Railroad Tunnels," unpublished work.

37. R. A. Farmer, and N. H. Shepherd, "Guided Radiation... The Key to Tunnel Talking," *IEEE Trans. Veh. Comm.*, **14**, March 1965, pp. 93–98.

38. D. O. Reudink, "Mobile Radio Propagation in Tunnels," IEEE Veh. Tech. Group Conf., December 2–4, 1968, San Francisco, Calif.

39. S. O. Rice, "Diffraction of Plane Radio Waves by Parabolic Cylinder," *Bell System Tech. J.* **33**, March 1954, pp. 417–504.

40. J. R. Wait and A. M. Conda, "Diffraction of Electromagnetic Waves by Smooth Obstacles for Grazing Angles," *J. Res. NBS*, **63**, 1959, pp. 181–197.

41. J. C. Schelleng, C. R. Burrows, and E. B. Ferrell, "Ultra Shortwave Propagation," *Proc. IRE*, **21**, March 1933, pp. 427–463.

42. H. Selvidge, "Diffraction Measurements at Ultra-High Frequencies," *Proc. IRE*, **29**, Jan. 1941 pp. 10–16.

43. T. B. A. Senior, "The Diffraction of a Dipole Field by a Perfectly Conducting Half-Plane," *Quart. J. Mech. App. Math.*, **6**, 1953, pp. 101–114.

44. D. E. Kerr, *Propagation of Short Radio Waves*, McGraw-Hill, New York, 1951.

45. H. T. Dougherty, and L. J. Maloney, "Application of Diffraction by Convex Surfaces to Irregular Terrain Situations," *Radio Sci.*, **68**, 1964, pp. 284–305.

46. G. Millington, "A Note on Diffraction Round a Sphere or Cylinder," *Marconi Rev.*, **23**, 1960, pp. 170–182.

47. M. H. L. Pryce, "Diffraction of Radio Waves by the Curvature of the Earth," *Advan. Phys.*, **2**, 1953, pp. 67–95.

48. J. R. Wait and A. M. Conda, "Pattern of an Antenna on a Curved Lossy Surface," *IRE Trans. Ant. Prop.*, **AP-5**, 1958, pp. 187–188.

49. K. Hacking, "UHF Propagation Over Rounded Hills," *Proc. IEE*, **117**, March 1970, page 499.

50. M. P. Bachynski, "Scale Model Investigations of Electromagnetic Wave Propagation Over Natural Obstacles," *RCA Rev.*, **24**, March 1963, pp. 105–144.

51. K. Hacking, "Optical Diffraction Experiments Simulating Propagation Over Hills at UHF," Inst. Elect. Eng., London, Conf. Pub. #48, 1968.

52. A. G. Longley and R. K. Reasoner, "Comparison of Propagation Measurements with Predicted Values in the 20 to 10,000 MHz Range," ESSA Technical Report, ERL 148-ITS 97, January 1970.

53. J. Aitchison and J. A. C. Brown, *Lognormal Distribution with Special Reference to Its Uses in Economics*, Cambridge Univ. Pr. Cambridge, England, 1957.

54. J. Egli, "Radio Propagation Above 40 MC Over Irregular Terrain," *Proc. IRE*, October 1957, pp. 1383–1391.

55. Edward W. Ng., and Murray Geller, "A Table of Integrals of the Error Functions," *J. Res. of NBS-B*, **73B**, January–March 1969, page 1.

56. A. P. Barsis, "Determination of Service Area for VHF/UHF Land Mobile and Broadcast Operations Over Irregular Terrain," ITSOTTM56.

chapter 3
antennas and
polarization effects

Y. S. Yeh

SYNOPSIS OF CHAPTER

The mobile radio environment is typified by a multiple scattering process and the absence of direct line-of-sight paths between the base and the mobile. The conventional free-space antenna pattern is thus greatly modified, and antenna designs have to be tailored to the statistical nature of the environment. For this reason no specific antenna designs will be given in this chapter. Instead, the emphasis will be on the correlation between the antenna pattern and the statistics of the received signal strength.

Section 3.1 gives a general expression for the received voltage of a mobile antenna as a function of polarization, antenna pattern, and angle of arrival of the incident plane waves at the mobile. This expression is later simplified to the model described in Section 1.1. It is shown that azimuth directivity can reduce the fading rate but is rather ineffective in increasing the average signal strength. Vertical directivity, on the other hand, can be exploited to slightly increase the average signal strength but is ineffective in reducing the fading rate. The changes of antenna patterns due to the metallic body of the mobile are loosely examined and are shown to play a minor role in the determination of average signal strength. Antenna systems that respond to the magnetic field components are discussed at the end of this section.

Section 3.2 discusses base-station antennas. It is shown that with an elevated base antenna the azimuth directivity returns and the gain in average signal strength can be realized. A simple circular array arrangement that is suitable for base-station pattern shaping is reported.

In Section 3.3 the simultaneous transmission of vertically and horizontally polarized waves is examined. It is shown that the received instantaneous signals are uncorrelated and the average signal strengths of the two polarizations are within ± 3 dB about 90% of the time. This makes polarization diversity a very attractive means of achieving two diversity branches without excessive requirements on antenna spacing at the base station. (Diversity systems are discussed in Chapters 5 and 6.)

3.1 MOBILE ANTENNAS

In this section we shall consider the influence of the antenna pattern on the properties of the signal received by the mobile antenna. In the vicinity of the mobile, the electromagnetic fields can be represented by a large number of plane waves with random phases coming from all directions. The distribution of incident angles in the horizontal plane is usually considered to be uniformly distributed from 0 to 2π. The angular distribution in the vertical plane, however, is confined to small elevation angles depending on the building heights and distances from the base. Experiments in suburban and urban environments[1,2] indicate that distributions extending to 30° in elevation are quite common.

3.1.1 The Received Signal at the Mobile

The received signal of an antenna moving in a mobile radio environment has properties that depend on the joint contribution of the antenna pattern and the randomly scattered field. To demonstrate the relationship let the receiving antenna be situated at the origin of a spherical coordinate system and move with velocity $\mathbf{u}$ in the horizontal plane [x-y plane] as shown in Figure 3.1-1. The electric field pattern of the antenna is

$$\mathbf{E}_a(\Omega) = E_\theta(\Omega)\bar{a}_\theta(\Omega) + E\bar{a}_\phi(\Omega), \qquad (3.1\text{-}1)$$

where Ω is the coordinate point on a spherical surface given by (θ, ϕ), $\bar{a}_\theta, \bar{a}_\phi$ are unit vectors associated with Ω, and E_θ, E_ϕ are complex envelopes of the θ and ϕ components of the electric field pattern.

The electric field of the incident plane wave from a particular angle (θ, ϕ) is

$$\mathbf{A}(\Omega) = A_\theta(\Omega)\bar{a}_\theta(\Omega) + A_\phi(\Omega)\bar{a}_\phi(\Omega), \qquad (3.1\text{-}2)$$

where $A_\theta(\Omega)$ and $A_\phi(\Omega)$ are the random amplitude and phase of the incident E field in $\bar{a}_\theta$ or $\bar{a}_\phi$ direction, respectively.

The assumptions on the random variables are the following:

1. The phase angles of A_θ are independent for plane waves arriving from different angles Ω and Ω', hence we have

$$\langle A_\theta(\Omega)A_\theta^*(\Omega')\rangle = \langle A_\theta(\Omega)A_\theta^*(\Omega)\rangle\delta(\Omega - \Omega'). \qquad (3.1\text{-}3)$$

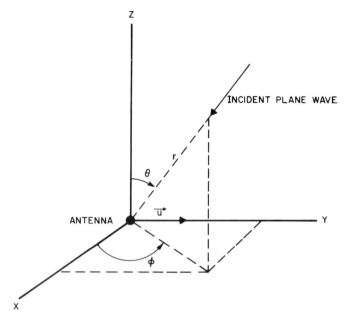

Figure 3.1-1   A spherical coordinate system showing incident wave and receiving antenna.

2.   The phase angles of A_θ and A_ϕ are independently distributed between 0 and 2π; therefore,

$$\langle A_\theta(\Omega)A_\phi^*(\Omega')\rangle = 0. \tag{3.1-4}$$

where the asterisk indicates complex conjugate.

The equivalent circuit of a matched receiving antenna is shown in Figure 3.1-2. The voltage across the unity load resistance is a random variable and is given by

$$v(t) = \mathrm{Re}\{V(t)e^{j\omega_c t}\}, \tag{3.1-5}$$

where ω_c is the carrier frequency.

The complex envelope is

$$V(t) = c\oint \mathbf{E}_a(\Omega)\cdot\mathbf{A}(\Omega)e^{-j\beta\mathbf{u}\cdot\bar{a}_r(\Omega)t}\,d\Omega, \tag{3.1-6}$$

where c is a proportionality constant, $e^{-j\beta\mathbf{u}\cdot\bar{a}_r(\Omega)t}$ is the Doppler shift caused by vehicle velocity **u**, $\oint d\Omega$ indicates integration over the spherical surface, that is, $\int_0^{2\pi}\int_0^{\pi}\sin\theta\,d\theta\,d\phi$, and $\bar{a}_r$ is a unit vector in the radial direction.

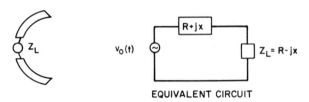

EQUIVALENT CIRCUIT

Figure 3.1-2   Equivalent circuit of the receiving antenna.

Equation (3.1-6) may be decomposed to summations over many small $\Delta\Omega$. Similar to arguments presented in Section 1.1, $V(t)$ may be approximated by a zero mean complex Gaussian process, that is,

$$V(t) = T_c(t) + jT_s(t). \tag{3.1-7}$$

To characterize $V(t)$ we need to know the correlation and cross-correlation functions of T_c and T_s.

Let us first obtain the autocorrelation function of $V(t)$, that is,

$$R_V(\tau) = \langle V(t)V^*(t+\tau)\rangle$$

$$= cc^*\oint\oint\langle\{\mathbf{E}_a(\Omega)\cdot\mathbf{A}(\Omega)\}\{\mathbf{E}_a^*(\Omega')\cdot\mathbf{A}^*(\Omega')\}\rangle$$

$$\times e^{-j\beta\mathbf{u}\cdot[\bar{a}_r(\Omega)-\bar{a}_r(\Omega')]t+j\beta\mathbf{u}\cdot\bar{a}_r(\Omega')\tau}\,d\Omega\,d\Omega'. \tag{3.1-8}$$

Substituting Eqs. (3.1-1) and (3.1-2) into (3.1-8) and making use of Eqs. (3.1-3) and (3.1-4) we have

$$R_V(\tau) = cc^*\oint\{E_\theta(\Omega)E_\theta^*(\Omega)\langle A_\theta(\Omega)A_\theta^*(\Omega)\rangle$$

$$+ E_\phi(\Omega)E_\phi^*(\Omega)\langle A_\phi(\Omega)A_\phi^*(\Omega)\rangle\}\cdot e^{j\beta\mathbf{u}\cdot\bar{a}_r(\Omega)\tau}\,d\Omega. \tag{3.1-9}$$

The power gain pattern of an antenna is related to its electric field pattern by

$$G(\Omega) = G_\theta(\Omega) + G_\phi(\Omega) \propto E_\theta(\Omega)E_\theta^*(\Omega) + E_\phi(\Omega)E_\phi^*(\Omega), \tag{3.1-10}$$

where $\propto$ stands for proportional.

The quantity $\langle A_\theta(\Omega)A_\theta^*(\Omega)\rangle$ represents the distribution of the incident

$\bar{a}_\theta$-polarized power as a function of Ω and may be normalized by

$$\langle A_\theta(\Omega) A_\theta^*(\Omega) \rangle = C_1 P_\theta(\Omega), \tag{3.1-11}$$

where $P_\theta(\Omega)$ is an angular density function with

$$\oint P_\theta(\Omega) \, d\Omega = 1. \tag{3.1-12}$$

Similarly we have

$$\langle A_\phi(\Omega) A_\phi^*(\Omega) \rangle = C_2 P_\phi(\Omega), \tag{3.1-13}$$

$$\oint P_\phi(\Omega) \, d\Omega = 1. \tag{3.1-14}$$

Substituting Eqs. (3.1-10), (3.1-11), and (3.1-13) into (3.1-9) we have

$$R_V(\tau) = 2 \oint \{ P_1 G_\theta(\Omega) P_\theta(\Omega) + P_2 G_\phi(\Omega) P_\phi(\Omega) \} \cdot e^{j\beta \mathbf{u} \cdot \bar{a}_r(\Omega)\tau} \, d\Omega. \tag{3.1-15}$$

With an isotropic receiving antenna polarized in the $\bar{a}_\theta$-direction only, $G_\theta(\Omega) = 1$ and we have at $\tau = 0$

$$R_V(0) = \langle V(t) V^*(t) \rangle = 2P_1. \tag{3.1-16}$$

Therefore P_1 is the average power that would be received by an isotropic antenna (polarized in $\bar{a}_\theta$) in the mobile radio environment. Similarly P_2 is the power that would be received by an $\bar{a}_\phi$-polarized isotropic antenna.

It is simple to show that [Eqs. (3.1-3) and (3.1-4)]

$$\langle V(t) V(t+\tau) \rangle = 0. \tag{3.1-17}$$

Therefore we have

$$\langle T_c(t) T_c(t+\tau) \rangle = \langle T_s(t) T_s(t+\tau) \rangle$$

$$= \oint \{ P_1 G_\theta(\Omega) P_\theta(\Omega) + P_2 G_\phi(\Omega) P_\phi(\Omega) \}$$

$$\times \cos(\beta \mathbf{u} \cdot \bar{a}_r(\Omega)\tau) \, d\Omega, \tag{3.1-18}$$

and

$$\langle T_c(t) T_s(t+\tau) \rangle = -\langle T_c(t+\tau) T_s(t) \rangle$$

$$= \oint \{ P_1 G_\theta(\Omega) P_\theta(\Omega) + P_2 G_\phi(\Omega) P_\phi(\Omega) \}$$

$$\times \sin(\beta \mathbf{u} \cdot \bar{a}_r(\Omega)\tau) \, d\Omega. \tag{3.1-19}$$

The correlation function of the instantaneous voltage $v(t)$ is

$$R_v(\tau) = \oint \{ P_1 G_\theta(\Omega) P_\theta(\Omega) + P_2 G_\phi(\Omega) P_\phi(\Omega) \} \cdot \cos(\omega_c \tau + \beta \mathbf{u} \cdot \bar{a}_r(\Omega) \tau) \, d\Omega.$$

$$(3.1-20)$$

The one-sided power spectrum of $v(t)$ is

$$S_v(f) = \int_0^\infty 4 R_v(\tau) \cos 2\pi f \tau \, d\tau$$

$$= 2 \oint \{ P_1 G_\theta(\Omega) P_\theta(\Omega) + P_2 G_\phi(\Omega) P_\phi(\Omega) \}$$

$$\times \delta \left[f - f_c - \frac{\beta}{2\pi} \mathbf{u} \cdot \bar{a}_r(\Omega) \right] d\Omega.$$

$$(3.1-21)$$

In the special case in which plane waves arrive only in the horizontal plane and are linearly polarized in the $\bar{z}$ [i.e., $-\bar{a}_\theta$] direction, Eq. (3.1-21) reduces to Eq. (1.2-3).

3.1.2 Antenna Directivity and Average Received Signal Power

The average received power of a mobile antenna is $\frac{1}{2} \langle V(t) V^*(t) \rangle_{av}$, and from Eq. (3.1-15) we have

$$P_{rec} = \oint \{ P_1 G_\theta(\Omega) P_\theta(\Omega) + P_2 G_\phi(\Omega) P_\phi(\Omega) \} \, d\Omega.$$

$$(3.1-22)$$

Since the gain patterns are subject to

$$\oint \{ G_\theta(\Omega) + G_\phi(\Omega) \} \, d\Omega = 4\pi,$$

$$(3.1-23)$$

Eq. (3.1-22) can be maximized for known distributions of $P_1 P_\theta(\Omega)$ and $P_2 P_\phi(\Omega)$ under the constraint of Eq. (3.1-23).

To illustrate the dependence of P_{rec} on the patterns, let us first consider the extreme case in which $P_1 = P_2$, $P_\theta(\Omega) = P_\phi(\Omega) = 1/4\pi$. This is the case that the density function of incoming plane waves is uniform in all incident angles. We then have

$$P_{rec} = \frac{P_1}{4\pi} \oint [G_\theta(\Omega) + G_\phi(\Omega)] \, d\Omega$$

$$= P_1.$$

$$(3.1-24)$$

Equation (3.1-24) indicates that the average received power is equal to that of an isotropic antenna and is independent of the actual antenna radiation pattern, in this case.

In mobile radio the component plane waves arriving at the mobile may come from all azimuth angles, but the distribution in elevation angles is restricted to small angular values around the horizontal plane ($\theta = \pi/2$). In the case of a vertically polarized base-station antenna, the incident plane waves may be considered as essentially vertically polarized. The average received power (Eq. 3.1-22) reduces to

$$P_{\text{rec}} = P_1 \int_0^{2\pi} \int_{\pi/2 - \delta}^{\pi/2 + \delta'} G_\theta(\theta,\phi) P_\theta(\theta,\phi) \sin\theta \, d\theta \, d\phi. \qquad (3.1\text{-}25)$$

In order to maximize P_{rec} we obviously require $G(\Omega)$ to consist of $G_\theta(\Omega)$ only; that is, the mobile antenna should also be vertically polarized. The constraint on G_θ is now

$$\int_0^{2\pi} \int_0^\pi G_\theta(\theta,\phi) \sin\theta \, d\theta \, d\phi = 4\pi. \qquad (3.1\text{-}26)$$

If $P_\theta(\theta,\phi)$ is not uniform we can adjust $G_\theta(\theta,\phi)$ such that the peaks of the gain pattern coincide with the peaks of $P_\theta(\theta,\phi)$ and hence increase the value of P_{rec}.

In the case in which P_θ is uniform in ϕ and θ within the region $[0 < \phi < 2\pi, \pi/2 - \delta < \theta < \pi/2 + \delta']$ we have

$$P_{\text{rec}} = \frac{P_1}{2\pi[\sin\delta' + \sin\delta]} \int_0^{2\pi} \int_{\pi/2 - \delta}^{\pi/2 + \delta'} G_\theta(\theta,\phi) \sin\theta \, d\theta \, d\phi. \qquad (3.1\text{-}27)$$

The maximum P_{rec} is $2P_1/(\sin\delta' + \sin\delta)$ and is obtained by confining G_θ within the same region of P_θ. One obvious conclusion is that as long as the radiation pattern is confined in the same region as P_θ, the actual distribution of G_θ is immaterial. Therefore pattern shaping in the azimuth plane would not increase P_{rec}. Experiments performed by Lee[3] at 800 MHz using linear monopole arrays with azimuth half-power beamwidth equal to 45°, 26°, and 13.5° indicated that the average received power is about the same compared to a monopole antenna, which is omnidirectional in the azimuth plane.

Since the elevation angular spread is small, radiation pattern shaping in the vertical plane may be expected to increase the average received power. Measurements made at 836 MHz in a suburban environment using a dipole (half-power beamwidth, 78° in the vertical plane) and a three-element colinear array (half-power beamwidth, 22° and 4 dB gain over dipole) indicated that P_{rec} of the array is about 2 or 3 dB higher than that of the dipole.[1] The failure to obtain the desired 4 dB improvement is an indication that the spread of the vertical incident angles from the horizontal is somewhat larger than 11° but is smaller than 39°.

In general, a practical design criterion as far as receiving maximum average signal strength is concerned is to try to confine the radiation pattern to those solid angles where scattered waves would be expected to arrive. This amounts to some mild shaping of the vertical gain pattern and no shaping in horizontal plane. Failing to do this would cause some power loss. However, overdoing this would not bring any further increase in the received average signal strength.

The mobile antennas are usually mounted either on the top or at the side of the mobile. The presence of the metallic vehicle influences the radiation pattern of the antenna used. For antennas mounted at the side of the mobile, the vehicle body blocks certain portions of the horizontal plane. The net effect is then a significant change of the horizontal gain pattern; nevertheless, the gain changes in the vertical plane might be expected to be small. According to the previous discussion we should therefore expect very little change in the average received signal strength.

The height difference of different vehicles may produce some small difference in the received average signal strength. Okumura's[4] data indicate a 3 dB increase obtained by raising the mobile antenna from 3 meters above the ground to 6 meters above the ground. Experiments performed by Ott of Bell Telephone Laboratories in Philadelphia using a monopole antenna mounted on top of a station wagon (5 ft above the ground) and a van (8 ft 8 in. above the ground) indicate that the latter had some slight increase in the average received signal strength.

3.1.3 Directivity and Level Crossing Rate

The instantaneous voltage received by a mobile antenna is given by

$$v(t) = \text{Re}\{(T_c + jT_s)e^{j\omega_c t}\}$$

$$= r\cos(\omega_c t + \phi). \tag{3.1-28}$$

It has been shown in Section 1.1 that T_c and T_s are zero mean Gaussian random variables; therefore r is Rayleigh and is time varying. The rapidity with which r fluctuates depends on the power spectrum of $v(t)$ and hence on the antenna power pattern.

The fading of r may also be considered to be produced by the motion of the receiving antenna through a standing-wave pattern formed by plane waves propagating in different directions. Since the vertical angular distribution of the incoming plane waves is confined to small elevation angles, to a good approximation we may simplify the problem by considering the distribution of incident waves in the azimuth plane only. This leads to the model presented in Section 1.1. The power spectrum of $v(t)$ (defined as E_z in Section 1.1) is given by Eq. (1.2-3) as

$$S_{E_z}(f) = \begin{cases} \dfrac{b}{\sqrt{f_m^2 - (f - f_c)^2}} [P(\alpha)G(\alpha) + P(\alpha)G(-\alpha)], & |f - f_c| \leqslant f_m \\ 0, & |f - f_c| > f_m, \end{cases}$$

(3.1-29)

where

$$\alpha = \cos^{-1}\left(\frac{f - f_c}{f_m}\right), \qquad 0 \leqslant \alpha \leqslant \pi,$$

$P(\alpha)$ is the angular density function of the incident plane waves, and $G(\alpha)$ is the antenna gain pattern in the $\theta = \pi/2$ plane.

For the case of uniform arrival angles, that is, $P(\alpha) = 1/2\pi$, the power spectra calculated from Eq. (3.1-29) for two different antenna patterns are presented in Figure 3.1-3. It is observed that different horizontal directivities serve to emphasize or deemphasize certain segments of the spectrum that would normally be received by an omnidirectional antenna. Comparison of the power spectra in Figure 3.1-3 reveals that the spread of the spectrum of a directive antenna is much less compared to that of an omnidirectional antenna. Therefore we would expect reductions in the rapidity of fading when a directional antenna is used. This will be observed in the level crossing rate calculations.

The rate at which r (the envelope of E_z) crosses an arbitrary level R in the negative direction is defined as level crossing rate (LCR). It is convenient to normalize R with respect to $\langle r^2 \rangle_{av}$, that is, define

$$\rho = \frac{R}{\sqrt{\langle r^2 \rangle}} = \frac{R}{\sqrt{2b_0}} \tag{3.1-30}$$

where b_0 is given by Eq. (1.3-4). The level crossing rate is given by (cf. footnote on page 34).

$$N_\rho = \frac{\rho}{\sqrt{\pi b_0}} \left(b_2 - \frac{b_1^2}{b_0}\right)^{1/2} \exp(-\rho^2), \tag{3.1-31}$$

where b_n is given by Eq. (1.3-4):

$$b_n = (2\pi)^n \int_{f_c - f_m}^{f_c + f_m} S_{E_z}(f)(f - f_c)^n \, df.$$

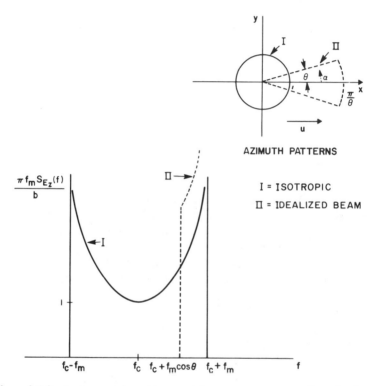

AZIMUTH PATTERNS

I = ISOTROPIC

Ⅱ = IDEALIZED BEAM

Figure 3.1-3   Antenna gain pattern and the power spectrum of received voltage.

Alternatively, by Eqs. (1.2-2) and (1.2-4) we have

$$b_n = (2\pi f_m)^n b \int_0^{2\pi} P(\alpha) g(\alpha) \cos^n \alpha \, d\alpha. \qquad (3.1\text{-}32)$$

Let us denote the LCR of an omnidirectional antenna by N_ρ and LCR of a directive antenna by N'_ρ; we have

$$\frac{N'_\rho}{N_\rho} = \left(\frac{\dfrac{b'_2}{b'_0} - \dfrac{b'^2_1}{b'^2_0}}{\dfrac{b_2}{b_0} - \dfrac{b^2_1}{b^2_0}} \right)^{1/2}. \tag{3.1-33}$$

Since b_n and b'_n are not functions of ρ, Eq. (3.1-33) tells us that the percentage reduction of LCR by a directive antenna from an omnidirectional antenna is independent of the particular level ρ. This means that the percentage reduction of LCR at $\rho = -10$ dB is the same as that at $\rho = -20$ dB or any other level.

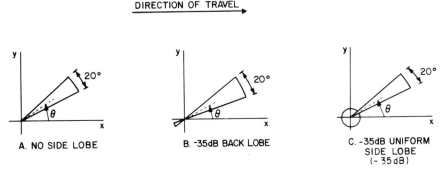

Figure 3.1-4   Gain patterns of three antennas.

To illustrate the reduction of N_ρ by directive antennas, N_ρ for the three different antenna patterns shown in Figure 3.1-4 are presented in Figure 3.1-5 as a function of θ, the beam pointing angle with respect to the direction of motion. The curves are normalized such that the N_ρ of an omnidirectional antenna is unity. It is observed that when the antenna is pointed 90° with respect to the direction of travel, N_ρ is reduced to 0.21 and is relatively insensitive to side-lobe structures. At $\theta = 0°$, however, N_ρ is extremely sensitive to side-lobe and back-lobe structures. Even with -35 dB side lobes, $N_\rho(0°)$ quickly increases to approach $N_\rho(90°)$.

The measured pattern of an eight-element array with back reflector is shown in Figure 3.1-6.[3] The calculated level crossing rates using either the measured pattern or the main beam only are presented in Figure 3.1-7. We notice that N_ρ based on the measured pattern is rather insensitive to beam pointing directions, whereas N_ρ based on the main beam alone is extremely sensitive to θ.

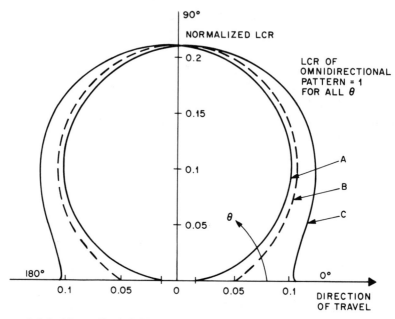

Figure 3.1-5 Normalized LCR of the received voltage of antennas shown in Figure 3.1-4.

Thus, for an environment where we can reasonably assume a uniform arrival angle for plane waves, directive antennas can be used to reduce the level crossing rate. The percentage reduction of N_ρ is rather insensitive to beam pointing angles due to the presence of side lobes. Calculating N_ρ with the main beam pointing at $\theta = 90°$ and free of all side lobes gives a lower bound on N_ρ. For example, by this assumption we would obtain

$$\frac{N_\rho}{N_{\rho \text{ omni antenna}}} = 0.056\delta \qquad (3.1\text{-}34)$$

where δ is the half-power bandwidth in degrees. For the pattern given by Figure 3.1-6, this estimate is smaller by a factor of 2 compared to exact calculation including side lobes.

So far $P(\alpha)$ has been considered to be uniform. In order to account for differences in LCRs, one needs to know something about $P(\alpha)$. In some instances, we might expect the presence of a direct beam from the base station, or we may expect some sort of channeling of waves down streets (as observed in New York City). In such cases, one might make some assumptions about $P(\alpha)$ and hopefully obtain better agreement with measurements.

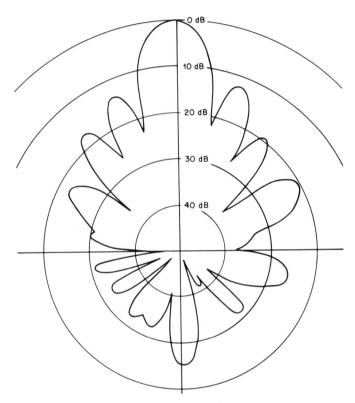

Figure 3.1-6 The measured gain pattern of an 8-element array with reflector (836 MHz).

Another approach* is to measure $P(\alpha)$ at enough points along a street to predict the LCR as accurately as desired. If $P(\alpha)$ does not change too rapidly with α, an antenna that has a narrow beam [narrow compared to

*Based on work done by D. O. Reudink of Bell Telephone Laboratories.

changes in $P(\alpha)$] and low side lobes [again lower side lobes than amplitude fluctuations in $P(\alpha)$] can act as a probe and measure $P(\alpha)$.

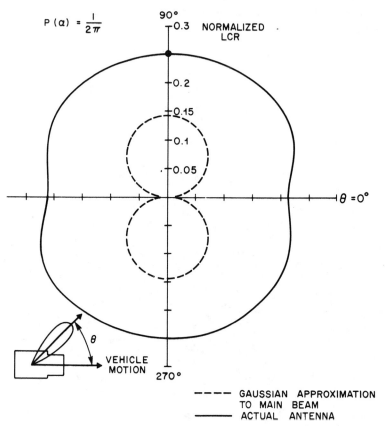

Figure 3.1-7 Comparison of theoretical LCRs for an eight element array antenna versus beam pointing angle.

In New York City, some experiments were made at 11.2 GHz where a 12° beamwidth antenna was rotated at fixed points on the city streets.[5] Assuming that these data, taken from a point at the center of a run, might give a reasonable approximation to the true $P(\alpha)$, it was sampled every 6° and used in the computation of the LCR. Also, the measured antenna pattern was used in the numerical integration. Figures 3.1-8 and 3.1-9 are the $P(\alpha)$ assumed for two streets in New York City. The predicted number of crossings is shown plotted against measured values of the level crossings in Figures 3.1-10 and 3.1-11. The agreement is surprisingly good, considering the assumptions about $P(\alpha)$.

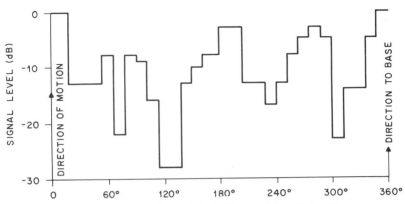

Figure 3.1-8   Assumed $P(\alpha)$ for Central Park West in New York City.

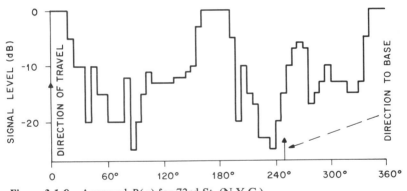

Figure 3.1-9   Assumed $P(\alpha)$ for 73rd St. (N.Y.C.).

We have thus shown that the theoretical model for calculation of LCRs is valid for directional antennas provided one describes in detail the antenna to be used in actual experiment. In addition, by making some crude assumptions about $P(\alpha)$, one can estimate the LCRs even better. Finally, it should be possible to attain a very low LCR as predicted by the dotted curve in Figure 3.1-7 if one uses a directional antenna with very low back lobes.

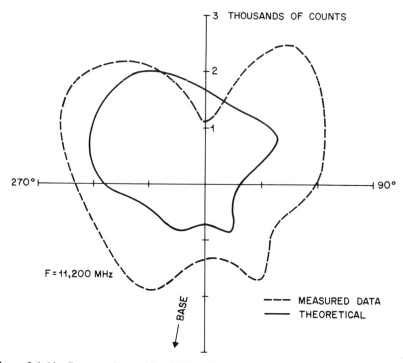

Figure 3.1-10   Run on Central Park West (N.Y.C.) level crossings 5 dB below rms.

3.1.4 Field Component Antennas

In Section 1.2 we have shown that E_z, H_x, and H_y are uncorrelated at the same point provided $P(\alpha)$ is uniformly distributed from 0 to 2π. Therefore, if we can design antennas that respond to E_z, H_x, and H_y separately, we would, in fact, have three diversity branches.[6] A vertical dipole responds to E_z. A horizontal magnetic dipole* responds to either H_x or H_y. The interesting question to ask is how the average received powers compare with each other. The radiation patterns of a vertical electric dipole and a magnetic dipole oriented along the $\bar{a}_x$ direction are shown in Figure 3.1-12. Note that both antennas have the same gain, but the patterns are different. For waves coming essentially from horizontal directions, interpretation of Eq. (3.1-22) would indicate a 3 dB weaker average power from the magnetic dipole. This can be explained after we look at the

*For example, a loop antenna.

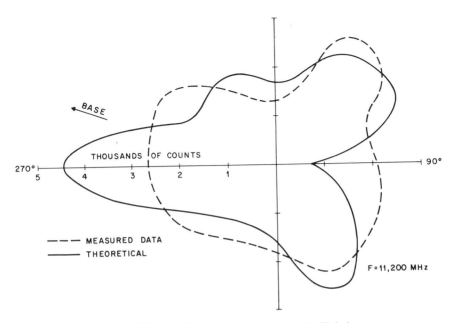

Figure 3.1-11 Run on 76th St. (N.Y.C.) level crossings, 5 dB below rms.

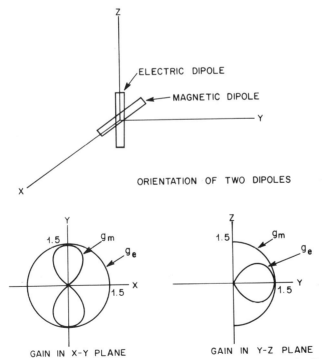

Figure 3.1-12 Gain patterns of electric dipole and magnetic dipole.

gain patterns in the vertical plane [x-z], noticing that the electric dipole has 3 dB vertical gain whereas the magnetic dipole does not have vertical gain. One way to bring the two dipole outputs to equal average power would be to use a vertical two-element array for the magnetic dipole.

3.2 BASE STATION ANTENNAS

It is quite obvious from the discussion of the previous section that the base station antenna pattern would play a minor role in the determination of the level crossing rate of the received signal if the plane-wave components at the mobile can still be reasonably assumed to be coming from all horizontal directions. When the base antenna height is of the same order as the mobile antenna, the horizontal directivity would again be lost. As the base antenna is elevated above the local scatterers, one would expect the angle of arrival of the component plane waves at the base to become more and more concentrated around the direct line path between the base and the mobile. Hence we would expect the gradual return of the base station antenna azimuth directivity.

Gusler* of Bell Telephone Laboratories carried out the following experiment designed to investigate the effect of directive base antennas. The experiment was performed at 960 MHz using a 60-ft dish antenna with a free-space beamwidth of 1.2° and located on top of a hill 300 feet above the surrounding terrain. Signal strength recordings were made by rotating the dish while the mobile remained stationary at 40 different locations ranging from 1.5 to 11 miles from the base, with heights ranging from 40 to 150 ft. A histogram of the smoothed 3-dB beamwidths is shown in Figure 3.2-1. This shows that a large number of patterns had beamwidths that were equal to or somewhat less than the free-space beamwidth of 1.2°; however, over half the recordings showed a rather significant beam spreading. The median effective 3 dB beamwidth was 1.7° and the average value was 2.6°. The spreading was generally in the form of lobes, which seem to be reflections of the main beam since their levels were above the side-lobe level for the antenna. In some business sections, beam spreadings up to 10° were observed.

In all but one case there were no apparent bearing shifts from the line-of-sight path. The received signal level, when compared with that from a dipole at the base, indicated most of the gain of the dish antenna had been realized at the base station.

The realization of horizontal directivity with elevated base antennas thus

*Unpublished work.

can be used to advantage in several aspects of mobile radio communication. For example, directivity could help in determining the direction of the mobile, or in tailoring base-station radiation patterns to accommodate severe shadowing along certain directions.

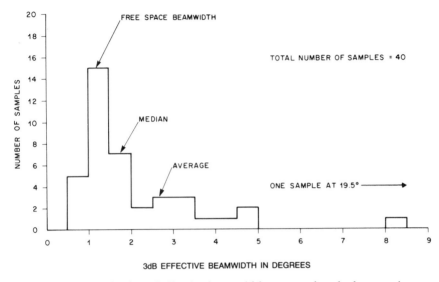

Figure 3.2-1   Distribution of effective beamwidth measured at the base station.

However, the height gain (discussed in Chapter 2) and the return of horizontal directivity are obtained at a price. We recall that since component plane waves arrive at the mobile from all azimuth angles, two mobile antennas spatially separated up to 1λ would have uncorrelated signals. The elevated base antenna would see very little beam spreading and thus require much wider separation to achieve spatial diversity. Experiments[7] performed at the same test site of the Gusler measurement indicated that spacings up to 30λ for broadside and 60λ for in-line incidence were required to reduce the correlation to 0.7. This amounts to separations up to 60 ft for 900-MHz mobile radio systems, which might cause hardships in the construction of base antennas.

These conclusions are based on measurements made with base-station antennas mounted at fairly high elevation in suburban environments. The required antenna spacings indicated should therefore be considered as an upper bound. Extensive measurements using lower antenna heights have recently been made by Rhee* of Bell Telephone Laboratories, both in urban and suburban areas. It was found that in suburban areas the spacing could be reduced to about 10λ for antennas at 100-ft elevation, and to 6λ for 50-ft elevation, still keeping interelement correlations below 0.7. Other measurements in an urban area indicated that horizontal directivity was preserved for a base-station antenna mounted at an elevation of 230 ft. In this case the front-to-back ratio was reduced from a free-space value of 17 dB to 10 dB in the urban environment.

Zachos[8] has reported a particularly simple circular array system that is suitable for azimuth beam shaping at base stations. The system consists of $N+1$ identical antennas, each omnidirectional in azimuth. N antennas are equally spaced on a circle of radius R and one antenna is located at the center. With the excitation current at the center element normalized to unity, the variables are the amplitudes and phases of the N antennas. For R sufficiently large that mutual couplings can be neglected, one is able to write down the antenna pattern. Given a particular pattern to be synthesized, the current excitations are continuously varied through an optimizing process, using a digital computer, until the root-mean-square error between the antenna pattern and the desired pattern reaches a local minimum. The success of this approach depends on the choice of the set of initial values that would lead to an acceptable local minima. Based on experience with pattern synthesis, an optimum set of initial values has been discovered, and successful synthesis of many patterns with reasonable accuracy has been reported. Figures 3.2-2 and 3.2-3 give typical pattern synthesis based on $N=6$ and $R=0.4\lambda$. The relative amplitudes and phases of the excitation currents are shown in the same figure.

3.3 POLARIZATION EFFECTS

It has been shown in Chapter 1 that in the transmission of vertically polarized waves, space diversity can be obtained at the mobile with very small antenna spacings, usually of the order of one wavelength or less. On the other hand, space diversity at an elevated base station may require antenna spacings up to 30λ for broadside incidence in a suburban environment. The key to the different spacing requirements is that the electromagnetic waves arrive at the mobile antennas from all angles; therefore a small spatial separation of the antennas would produce sufficient propagation delays to decorrelate the signals received at different antenna terminals. At the base station, however, the antenna is elevated above most local reflectors. The waves arrive through very narrow azi-

*Unpublished work.

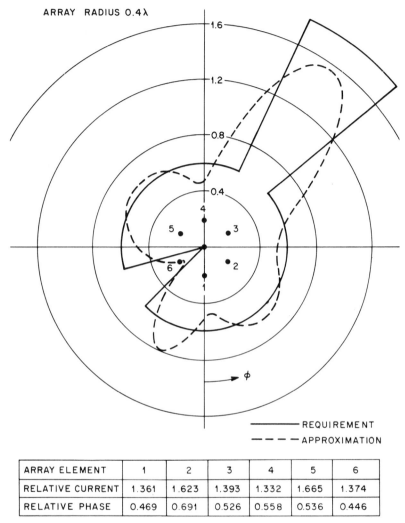

ARRAY ELEMENT	1	2	3	4	5	6
RELATIVE CURRENT	1.361	1.623	1.393	1.332	1.665	1.374
RELATIVE PHASE	0.469	0.691	0.526	0.558	0.536	0.446

Figure 3.2-2   Power pattern systhesis for a "wrench" shaped beam.

muthal angles and thus require much larger antenna spacing to decorrelate the signals.

In this section we shall examine the transmission of two orthogonally polarized waves and show their potential as two diversity branches.*

---

*Since the transmitted power is equally distributed in the two different polarizations, there is a 3-dB average power disadvantage for the polarization diversity system compared to space diversity.

ARRAY RADIUS 0.4λ

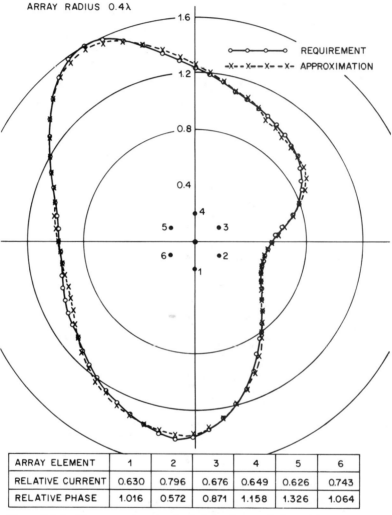

ARRAY ELEMENT	1	2	3	4	5	6
RELATIVE CURRENT	0.630	0.796	0.676	0.649	0.626	0.743
RELATIVE PHASE	1.016	0.572	0.871	1.158	1.326	1.064

Figure 3.2-3 Power pattern synthesis for a "potato" shaped beam.

Referring to Figure 3.3-1, let the base transmitting antennas be located at the same position. The antennas are ideal in the sense that one is a vertical electric dipole and the other one is a vertical magnetic dipole. The E_z and H_z field at the same point at the mobile can be represented by

$$E_z = \Gamma_{11} + \Gamma_{12}, \tag{3.3-1}$$

$$H_z = \Gamma_{22} + \Gamma_{21}, \tag{3.3-2}$$

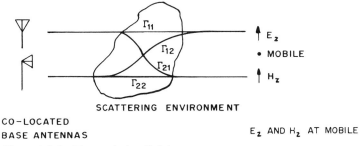

E_z AND H_z AT MOBILE

Figure 3.3-1   Transmission link between mobile and base.

where the Γ_{ij} are complex Gaussian random variables, representing the fading envelope and random phase. Γ_{11} represents the part of E_z produced by the electric dipole from the base. Γ_{22} is the part produced by the magnetic dipole. The quantities Γ_{12} and Γ_{21} represents the cross coupling produced by the scattering process. Since the scattering process, which produces cross coupling, is random, we would expect Γ_{12} and Γ_{21} to be independent. The fields at the mobile are again assumed to be decomposed into plane waves coming from all horizontal angles with random phases. Therefore, Γ_{11} can be represented by

$$\Gamma_{11} = \sum_{n=1}^{N} A_n e^{j\phi_n}, \qquad (3.3\text{-}3)$$

where A_n and ϕ_n represent the amplitude and phase of the component wave. Since the orthogonal antennas are colocated at the base (or may be located very close by), the transmitted waves would travel almost the same paths to reach the mobile. In each reflection process by vertical objects the reflection coefficients would be slightly different due to the nonperfect conducting nature of the building structures. In addition there will be a distinct π phase shift between E_z and H_z after each reflection. Thus the H_z field at the mobile may be expressed as

$$\Gamma_{22} = \sum_{n} A_n' e^{j\phi_n' + jk_n\pi}, \qquad (3.3\text{-}4)$$

where

$$k_n = \begin{cases} 1 & \text{odd number of reflections} \\ 0 & \text{even number of reflections} \end{cases}.$$

Even in the case where we may have $A'_n \cong A_n$, $\phi'_n \cong \phi_n$, the distinct π phase shift due to reflections would make Γ_{11} and Γ_{22} uncorrelated, as can be easily verified by examining the ensemble average of $\Gamma_{11}\Gamma_{22}$. Therefore, we conclude that E_z and H_z at the mobile are uncorrelated. By reciprocity, the E_z and H_z at the base are*

$$E_z = \Gamma_{11} + \Gamma_{21}, \tag{3.3-5}$$

$$H_z = \Gamma_{22} + \Gamma_{12}, \tag{3.3-6}$$

and are also uncorrelated.

From previous discussions we observed that independent E_z and H_z can be obtained at the same position, both at the base and the mobile. An antenna system making use of this property can be realized by a vertical dipole and a horizontal loop. In practical situations, the antennas would not be ideal and may produce cross coupling in their radiation pattern. Let g_p represent the gain pattern of the principal polarization, and g_c represent the gain pattern of cross polarization; it is reasonable to assume that g_c will not have the same shape as g_p. At the mobile, since waves are coming from all directions, the different patterns would operate to decorrelate the received cross-coupling components. At the base, the situation is quite different because waves are arriving through a very narrow angle. Under this condition, we expect the received signals to be of the following form:

$$V_v = R_1 e^{j\theta_1} + \alpha R_2 e^{j\theta_1}, \tag{3.3-7}$$

$$V_h = R_2 e^{j\theta_2} + \beta R_1 e^{j\theta_1}, \tag{3.3-8}$$

where the subscript v stands for vertical antenna and h for horizontal antenna. The α and β are complex quantities representing the cross coupling due to receiving antenna patterns. $R_1 e^{j\theta_1}$ and $R_2 e^{j\theta_2}$ are independent complex Gaussian variables representing the received signal from principal polarizations. The correlation between V_v and V_h can be calculated, and it can be shown that even if α and β amount to -5 dB, the maximum correlation between V_v and V_h is still less than 0.7.

Experiments were performed[9] at 836 MHz using vertical dipole and horizontal loop antennas at the mobile as transmitting antennas and two directive receiving horns (one polarized in each polarization sense) at the base. The results verified the Rayleigh nature of the horizontally polarized received signal and also verified that the vertically and horizontally polarized waves are uncorrelated.

---

*Notice that Eqs. (3.3-5) and (3.3-6) are different from Eqs. (3.3-1) and (3.3-2).

The average signal strengths received by the dipole and the loop were found to be almost equal.* They followed the general trend as a function of position, namely, maxima and minima almost simultaneously (Figure 3.3-2). The cumulative probability distribution of the ratio of the averaged signal strength of the two polarizations is shown in Figure 3.3-3. It is observed that both signal strengths are within 3 dB of each other about 90% of the time. It is further observed that this general behavior is independent of base-station antenna heights. This is in sharp contrast to the ordinary concept derived by consideration of vector addition of direct and ground-reflected waves. A possible explanation might be that since the multiple scattering nature of the mobile environment masks the phase associated with each component plane wave, the most important factor in determining the average signal strength is the total amount of power coming from all directions.

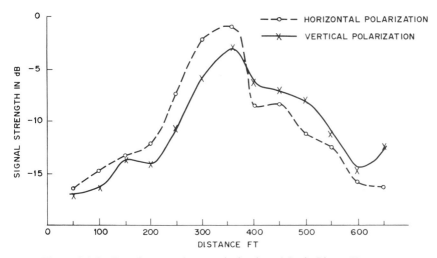

Figure 3.3-2 Local mean of two polarizations Maple Place, Keyport.

The location of the mobile antennas on the top of the vehicle causes different gain patterns in the vertical plane due to the image effect. It was observed, however, that as long as the antennas were about 1λ above the car roof, which was about 6λ by 10λ, the ground plane had very little effect on the distribution of the ratio of the two recorded average signal strengths.

*Note that the gain pattern of the two antennas are almost identical.

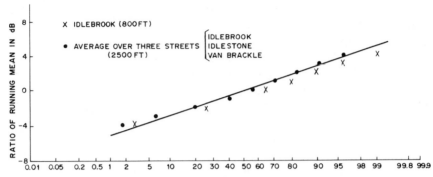

Figure 3.3-3   CPD of local mean ratio at Strathmore Development.

REFERENCES

1. W. C. Y. Lee and R. H. Brandt, "Elevation Angle of Mobile Radio Signal Arrival," *Joint IEEE Comm. Soc.-Veh. Tech. Group Special Trans. on Mobile Radio Comm.*, November 1973, pp. 1194–1197.

2. J. S. Bitler, "A Two-Channel Vertically Spaced UHF Diversity Antenna System," Microwave Mobile Radio Symp., Boulder Colorado, 1973.

3. W. C. Y. Lee, "Preliminary Investigation of Mobile Radio Signal Fading Using Directional Antennas on the Mobile Unit," *IEEE Trans. Veh. Comm.*, **VC-15**, October 1966, pp. 8–15.

4. Y. Okumura, E. Ohmori, T. Kawano, and K. Fukuda, "Field Strength and its Variability in VHF and UHF Land Mobile Service," *Rev. Elec. Comm. Lab.*, **16**, No. 9–10, September–October 1968.

5. D. O. Reudink, "Preliminary Investigation of Mobile Radio Transmission at X-Band in an Urban Area," URSI Commission 2, Ann Arbor, Michigan, October 16–18, 1967.

6. W. C. Y. Lee, "Energy-Density Antenna for Independent Measurement of the Electric and Magnetic Field," *Bell System Tech. J.* **46**, September 1967, pp. 1587–1600.

7. W. C. Y. Lee, "Antenna Spacing Requirement for a Mobile Radio Base-Station Diversity," *Bell System Tech. J.* **50**, July/August, 1971; pp. 1859–1876.

8. T. Zachos, "Synthesis of Base Station Antenna Patterns," Microwave Mobile Comm., Boulder, Colorado, March 2, 1972.

9. W. C. Y. Lee and Y. S. Yeh, "Polarization Diversity System for Mobile Radio," *IEEE Trans. Comm.*, **Com-20**, No. 5, October 1972.

part II
mobile radio systems

chapter 4

modulation, noise, and interference

M. J. Gans and Y. S. Yeh

SYNOPSIS OF CHAPTER

The first two chapters have firmly established the small-scale and large-scale properties of mobile transmission paths, and Chapter 3 discussed ways of coupling to this medium. We now turn our attention to matters of impressing useful information on the mobile path. Frequency modulation has long been the preferred modulation technique, and Section 4.1 contains an extensive treatment of this method. The effects of thermal noise and interference from other signals are first examined in the conventional, nonfading case for reference and illustration of the analytical techniques, then in the presence of the small-scale Rayleigh fading described in Chapter 1. This rapid fading alters the signal-to-noise (S/N) performance markedly, washing out the sharp threshold and capture properties of FM. Random FM caused by the fading imposes an upper limit on obtainable S/N ratios. Comparisons of FM with amplitude modulation in the fading case are then made, and show the generally disastrous effects of fading on AM or SSB. Finally, it is pointed out that the time-varying phase and amplitude transmission properties of the medium introduce distortion into the recovered baseband signal, and the magnitude of this effect is analytically related to the vehicle speed and coherence bandwidth of the medium.

Digital modulation has not found any appreciable use in commercial mobile telephone systems, but would give performance comparable to FM. In Section 4.2 various types of digital coding are described, followed by a thorough analysis of the expected error rate for two-level systems using several different modulation methods in the nonfading case. The effects of Rayleigh fading on these methods are then calculated from the standpoints of signal envelope variations, frequency selectivity of the transmission medium, and random FM.

161

Any high-capacity system necessarily involves a multiplicity of communication channels, and a variety of ways of multiplexing these channels for transmission over the mobile path are available to the system designer. The conventional approach is frequency-division multiplex, where the individual voice circuits occupy unique, identifiable frequency slots in the radio spectrum. Section 4.3 contains a thorough study of two alternative schemes with respect to the required transmitter power, the effects of fading, dispersion, interference, intermodulation, and preemphasis or deemphasis. Comparisons are made to the conventional method, which is shown to be preferred for most cases.

By the very nature of the use of mobile radio telephony, the system must operate in an environment contaminated to some degree by man-made electrical noise, that is, radiation from devices not intended to radiate. Section 4.4 briefly discusses the characterization of such noise, methods of measuring it, and includes a summary of urban noise measurements over the frequency range of interest. An extensive list of over 90 references on noise is also included for more specific information on actual measurements.

4.1 FREQUENCY MODULATION

Even when the mobile radio transmission is continuous wave (CW), the received signal is amplitude modulated due to the rapid fading experienced as the mobile unit moves through the interfering waves from the various scatterers (cf. Chapter 1). Early in the evolution of mobile radio systems, this amplitude fading was recognized as a serious interference present when communicating with amplitude modulated (AM) signals. The obvious solution was to employ angle modulated signals; for example, frequency modulated (FM) signals. The FM receiver suppresses the amplitude modulation, and thus the interference due to amplitude fading is considerably reduced.

Besides suppressing interference due to amplitude fading, FM systems have many other well-known characteristics; some are good, some bad. In this section we will review these characteristics of FM communication in the mobile radio environment.

4.1.1 Noise Performance in the Nonfading Case

One of the most outstanding characteristics of FM communication is that the audio output (baseband) signal-to-noise ratio (SNR) can be improved, with a fixed transmitter power, by increasing the frequency deviation of the modulation and consequently the intermediate frequency (IF) bandwidth. The penalty one pays for the improved baseband SNR is that the wider the IF bandwidth, the closer the received power is to the FM threshold. As the received power drops below the FM threshold, the baseband noise increases rapidly until, for received powers well below threshold, all of the FM improvement of baseband SNR is lost.

Rice[1] has computed the baseband noise performance of an FM receiver in terms of the IF signal power, S, IF noise spectral density, η, and IF filter characteristic $G(f)$. In sections 4.1.3 and 4.1.4 we will extrapolate the performance of FM receivers to different conditions, which will require a clear understanding of the properties of the various sources of noise in an FM receiver. Therefore, we will reconstruct Rice's FM noise derivations in considerable detail to provide the framework for later sections.

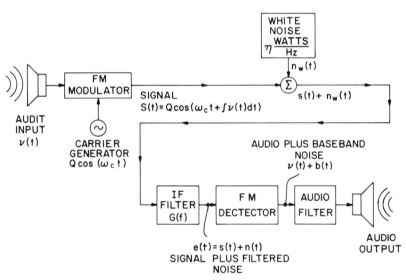

Figure 4.1-1   Simplified block diagram of an FM communication system (neglecting fading).

Consider an FM communication channel as depicted in Figure 4.1-1. For simplicity we compute output signal-to-noise ratios by approximating the output noise with a modulated carrier by that present when the carrier is unmodulated. This is the usual approximation used in FM calculations of SNR.[2] Rice[3] has shown that, for sine wave modulation, the baseband noise with modulation is closely approximated by that without modulation.

The voltage presented to the FM detector in Figure 4.1-1 is

$$e(t) = Q\cos\omega_c t + n(t)$$

$$= [Q + X_c(t)]\cos\omega_c t - X_s(t)\sin\omega_c t \qquad (4.1\text{-}1)$$

$$= R(t)\cos[\omega_c t + \theta(t)],$$

where Q is the peak amplitude of the carrier at frequency ω_c, $n(t)$ is the additive noise, and we have defined the in-phase, $X_c(t)$, and quadrature, $X_s(t)$, components of the noise, $n(t)$, by[4]

$$X_c(t) \overset{\Delta}{=} n(t)\cos\omega_c t + \hat{n}(t)\sin\omega_c t,$$

$$X_s(t) \overset{\Delta}{=} -n(t)\sin\omega_c t + \hat{n}(t)\cos\omega_c t, \qquad (4.1\text{-}2)$$

$$\hat{n}(t) \overset{\Delta}{=} \frac{1}{\pi}\int_{-\infty}^{\infty}\frac{n(\tau)}{t-\tau}d\tau \qquad \text{(Hilbert transform).} \qquad (4.1\text{-}3)$$

The amplitude, $R(t)$, and phase, $\theta(t)$, of the voltage $e(t)$, are given by

$$R(t) \overset{\Delta}{=} \left\{[Q + X_c(t)]^2 + X_s^2(t)\right\}^{1/2},$$

$$\theta(t) \overset{\Delta}{=} \arctan\left[\frac{X_s(t)}{Q + X_c(t)}\right]. \qquad (4.1\text{-}4)$$

The one-sided spectrum of $e(t)$ is shown in Figure 4.1-2. It consists of a Dirac delta function of area $Q^2/2$ at $f_c = \omega_c/2\pi$, representing the unmodulated carrier, and a noise spectrum which follows the shape of the IF filter, $\eta G(f)$. Without loss of generality we normalize $G(f_c) = 1$. The IF signal-to-noise ratio, ρ, of $e(t)$ is

$$\rho = Q^2\left[2\eta\int_0^{\infty}G(f)df\right]^{-1}. \qquad (4.1\text{-}5)$$

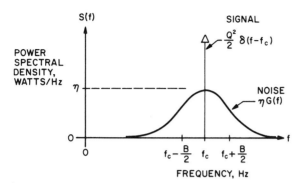

B = NOISE BANDWIDTH

Figure 4.1-2   Spectrum of unmodulated signal and noise after the IF filter.

As is well known, the filtered white noise, $n(t)$, is a stationary Gaussian stochastic process, as are its in-phase, $X_c(t)$, and quadrature, $X_s(t)$, components (Ref. 5, Section 3.7). Thus the probability densities of $n(t)$, $X_c(t)$, and $X_s(t)$ are given by

$$p_n(\alpha) = p_{X_c}(\alpha) = p_{X_s}(\alpha) = \frac{1}{\sqrt{2\pi \underline{b}_0}} \exp\left[-\frac{\alpha^2}{2\underline{b}_0} \right], \qquad (4.1\text{-}6)$$

where

$$\underline{b}_0 \stackrel{\Delta}{=} \langle n^2(t) \rangle, \qquad (4.1\text{-}7)$$

the average being taken over the ensemble of possible noise currents. From Eq. (4.1-2) and the properties of Hilbert transforms (Ref. 4, p. 48) we have

$$\langle X_c^2(t) \rangle = \langle X_s^2(t) \rangle = \underline{b}_0, \qquad (4.1\text{-}8)$$

as indicated in Eq. (4.1-6). Furthermore, we can derive from Eq. (4.1-2) the autocorrelation, $\underline{g}(\tau)$, cross correlation, $\underline{h}(\tau)$, and spectral moments, $\underline{b}_n$, identical to Eqs. (1.3-1)–(1.3-5) of Chapter 1, where $X_c(t)$ replaces $T_c(t)$,

$X_s(t)$ replaces $T_s(t)$, and $\eta G(f)$ replaces $S_i(f)$. Whence

$$\underline{g}(\tau) = \int_0^\infty \eta G(f) \cos 2\pi (f - f_c) \tau \, df, \qquad (4.1\text{-}9)$$

$$\underline{h}(\tau) = \int_0^\infty \eta G(f) \sin 2\pi (f - f_c) \tau \, df, \qquad (4.1\text{-}10)$$

$$\underline{b}_n = (2\pi)^n \int_0^\infty \eta G(f)(f - f_c)^n \, df, \qquad (4.1\text{-}11)$$

where the underlines are used to differentiate these noise correlations and moments from those related to the transmission coefficient as defined in Chapter 1.

The computation of the output noise is particularly simple when the SNR at IF is very large. When $\rho \gg 1$ we have

$$R(t) \doteq Q + X_s(t), \qquad (4.1\text{-}12)$$

$$\theta(t) \doteq \frac{X_s(t)}{Q}, \qquad (4.1\text{-}13)$$

and the detected frequency, $\theta'(t)$, is approximated by

$$\theta'(t) \doteq \frac{X_s'(t)}{Q}. \qquad (4.1\text{-}14)$$

The prime denotes differentiation with respect to time, and we assume 1(radian/second)^2 gives 1 watt output.

As mentioned above, the autocorrelation of $X_s(t)$ is $\underline{g}(\tau)$ given in Eq. (4.1-9) and the one-sided spectrum of $X_s(t)$ is the cosine transform of its autocorrelation (Ref. 5, p. 312):

$$W_{X_s}(f) = 4 \int_0^\infty \underline{g}(\tau) \cos 2\pi f\tau \, d\tau$$

$$= \eta [G(f_c - f) + G(f_c + f)]. \qquad (4.1\text{-}15)$$

Since the power spectrum of the derivative of a stationary stochastic process is $(2\pi f)^2$ times the spectrum of the process itself,[6] we have from

Eqs. (4.1-14) and (4.1-15) the one-sided spectrum of $\theta'(t)$,

$$\mathcal{W}_{\theta'}(f) = \left(\frac{2\pi f}{Q}\right)^2 \eta[G(f_c - f) + G(f + f_c)], \qquad (4.1\text{-}16)$$

which is shown in Figure 4.1-3. When the SNR at IF is large, the baseband noise spectrum is approximately parabolic in shape at low frequencies. Assume that the baseband filter cutoff frequency, W, is sufficiently less than the noise bandwidth of the IF filter, B,

$$W \ll B \triangleq \frac{1}{G(f_c)} \int_0^\infty G(f)\, df = \int_0^\infty G(f)\, df, \qquad (4.1\text{-}17)$$

so that the baseband noise spectrum is approximately parabolic within the baseband filter; that is, assume the system is designed for large index. Then in the baseband filter passband we may approximate

$$G(f_c - f) \doteq G(f_c + f) \doteq 1 \qquad \text{for} \quad f < W. \qquad (4.1\text{-}18)$$

The noise power out of the baseband filter then becomes (assuming a rectangular baseband filter)

$$N_{\theta'} = \int_0^W \mathcal{W}_{\theta'}(f)\, df \doteq \frac{2}{3}\left(\frac{2\pi}{Q}\right)^2 \eta W^3 \qquad \text{for} \quad \rho \gg 1. \qquad (4.1\text{-}19)$$

From Eq. (4.1-19), we see that the output noise is independent of the IF bandwidth B as long as the SNR at IF remains large. Thus the IF bandwidth may be increased to accomodate larger frequency deviations and consequently larger output signal powers while the output noise power remains constant. This is the property of FM communication systems that we mentioned above; namely, the baseband SNR can be improved, with a fixed transmitter power, by increasing the frequency deviation of the modulation and correspondingly the IF bandwidth. However, as B is increased, for constant transmitted power, $Q^2/2$, we see from Eqs. (4.1-5) and (4.1-17) that ρ decreases. As mentioned above, as the SNR at IF falls below a threshold, the baseband noise increases rapidly. Rice[3] has used the concept of "clicks" to give a simple description of the FM threshold effect. His approach gives accurate results for the baseband noise when ρ is near and above threshold. To calculate the performance in the presence of fading later on, however, we will need accurate predictions of baseband noise for all ranges of ρ, from well below threshold to well above. For this we will use Rice's calculations,[1] where he has computed the exact baseband noise spectrum for a Gaussian IF filter shape. Since the baseband filter width is usually narrow relative to the IF bandwidth ($W < B$), the exact shape of the IF filter has little effect on the output noise. Thus, we will lose little generality by including the assumption of a Gaussian shape for the IF filter. Rice divided the one-sided baseband noise spectrum, $\mathcal{W}(f)$, into three components:

$$\mathcal{W}(f) = \mathcal{W}_1(f) + \mathcal{W}_2(f) + \mathcal{W}_3(f). \tag{4.1-20}$$

$\mathcal{W}_1(f)$ has the same spectrum shape as the output noise spectrum when the carrier is absent. $\mathcal{W}_2(f)$ has the shape of the output noise spectrum when the carrier is very large,

$$\mathcal{W}_2(f) = (1 - e^{-\rho})^2 \mathcal{W}_{\theta'}(f), \tag{4.1-21}$$

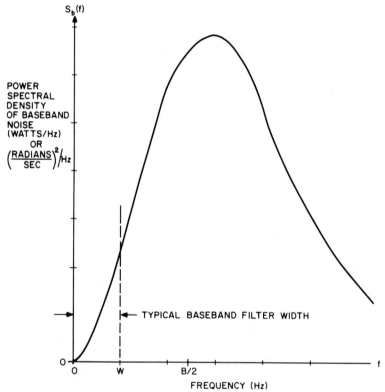

Figure 4.1-3   Spectrum of baseband noise for large SNR at IF ($\rho \gg 1$).

where $\mathcal{W}_{\theta'}(f)$ is the output noise spectrum when $\rho \gg 1$, given in Eq. (4.1-16). $\mathcal{W}_3(f)$ is a correction term that predominates in the threshold region of ρ.

Davis[7] has shown that in the frequency range from 0 to W, where $W < B$, the spectral components $\mathcal{W}_1(f) + \mathcal{W}_3(f)$ may be accurately approximated by

$$\mathcal{W}_1(f) + \mathcal{W}_3(f) \doteq 8\pi Be^{-\rho}[2(\rho + 2.35)]^{-1/2} \overset{\Delta}{=} \mathcal{W}_D(f). \quad (4.1-22)$$

The accuracy of the approximation (4.1-22) at $f = 0$ is illustrated in Figure 4.1-4 along with Rice's exact analysis[1] and his "clicks" approximation[3]:

$$\mathcal{W}_1(f) + \mathcal{W}_3(f) \doteq 4\pi B\sqrt{2\pi}\left(1 - \text{erf}\sqrt{\rho}\right) \overset{\Delta}{=} \mathcal{W}_c(f), \quad (4.1-23)$$

where the error function, $\text{erf}(y)$, is defined in Eq. (1.1-16). The above approximations may be combined to provide the approximation for the overall baseband noise spectrum for a Gaussian shaped IF filter:

$$\mathcal{W}(f) \doteq (1 - e^{-\rho})^2 \mathcal{W}_{\theta'}(f) + \mathcal{W}_D(f)$$

$$= \frac{[2\pi f(1 - e^{-\rho})]^2}{B\rho}e^{-(\pi f^2/B^2)} + \frac{8\pi Be^{-\rho}}{\sqrt{2(\rho + 2.35)}}, \quad (4.1-24)$$

where we have used

$$G(f) = e^{-\pi(f - f_c)^2/B^2}. \quad (4.1-25)$$

Approximation (4.1-24) is compared to the exact spectrum[1] in Figure 4.1-5. It is seen that the error in the approximation is zero at $f = 0$ and is still small at $f = B/2$. Since, for FM systems with modulation indices greater than unity, $W < B/2$, the approximation is extremely good.

The total noise out of the rectangular baseband filter is then

$$N(\rho) = \int_0^W \mathcal{W}_2(f)\,df + \int_0^W \mathcal{W}_D(f)\,df$$

$$= \frac{a(1 - e^{-\rho})^2}{\rho} + \frac{8\pi BWe^{-\rho}}{\sqrt{2(\rho + 2.35)}}, \quad (4.1-26)$$

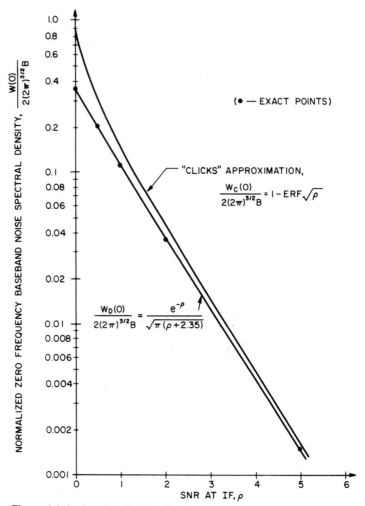

Figure 4.1-4   Baseband noise density versus IF SNR.

where, using the Maclaurin series expansion,

$$a \overset{\Delta}{=} \frac{(2\pi)^2}{B} \int_0^W f^2 e^{-\pi f^2/B^2} df$$

$$= \frac{4\pi^2 W^3}{3B} \left\{ 1 - \frac{6\pi}{10} \left(\frac{W}{B} \right)^2 + \frac{12\pi^2}{56} \left(\frac{W}{B} \right)^4 + \cdots \right\}. \tag{4.1-27}$$

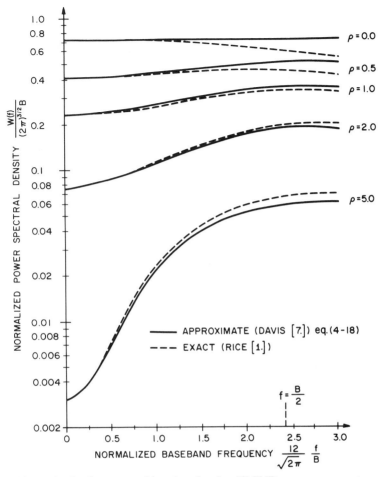

Figure 4.1-5 Spectrum of baseband noise. IF SNR, ρ, as a parameter.

As described in the appendix of Ref. 3, the baseband signal output, $v_0(t)$, is suppressed as the SNR at IF decreases. From the output signal in the absence of noise, $v(t)$ (see Figure 4.1-1),

$$v_0(t) = v(t)(1 - e^{-\rho}), \qquad (4.1\text{-}28)$$

so that the output baseband signal power, S_0, is related to the input modulation signal power, $S = \langle v^2(t) \rangle$, by

$$S_0 = (1 - e^{-\rho})^2 S. \qquad (4.1\text{-}29)$$

The output signal power (4.1-29) and noise power (4.1-26) are plotted versus IF SNR, ρ, for several IF to baseband bandwidth ratios, $B/2W$, in Figure 4.1-6. We have arbitrarily assumed that the rms frequency deviation is 10 dB less than half the noise bandwidth minus the top baseband frequency, so that signal deviation peaks do not often exceed the IF bandwidth (Carson's Rule):

$$10S = \pi^2(B-2W)^2 \quad \text{or} \quad S = \frac{\pi^2}{10}(B-2W)^2. \quad (4.1\text{-}30)$$

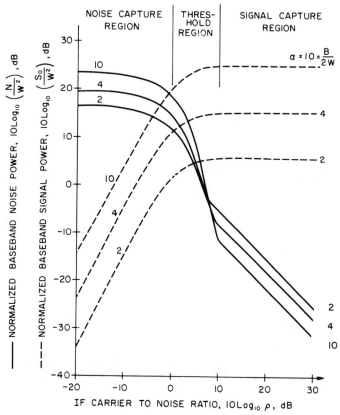

Figure 4.1-6 Noise quieting and signal suppression versus IF SNR and bandwidth ratios.

The threshold effect is very evident in Figure 4.1-6. Also evident in Figure 4.1-6 is the capture phenomenon of the FM receiver. When ρ is large the receiver captures on the signal modulation and suppresses the noise; when ρ is small the receiver captures on the noise and suppresses the signal modulation. As we will show later, the capture effect is even more pronounced when the interference is another signal instead of Gaussian noise.

If B increased, while the transmitter power is held constant, ρ will decrease. This effect is not clearly shown in Figure 4.1-6. Thus, to illustrate the system advantages of FM communication, we will compare the transmitter powers required to provide a 30-dB output SNR for an FM system versus an AM system with the IF to audio bandwidth ratio of the FM system as a parameter. For AM we make a similar assumption as used in FM, namely, the rms amplitude modulation is 10 dB less than 100%, so that modulation peaks do not often cause overmodulation. The AM wave has the form $A[1 + m(t)]\cos\omega_c t$, where we assume that

$$\langle m^2(t)\rangle = 0.1. \tag{4.1-31}$$

The output SNR of the AM detector is[2]

$$\left(\frac{S_0}{N}\right)_{\text{AM}} = \frac{A^2\langle m^2\rangle}{2\eta W}. \tag{4.1-32}$$

In Figure 4.1-7 we plot the carrier power of AM relative to that of FM, $20\log_{10}(A/Q)$, to provide a 30 dB output SNR in all cases. As seen in the figure the FM performance improves relative to AM as the FM bandwidth (and frequency deviation) is increased until the IF SNR of the FM drops below threshold and FM starts to lose its advantage over AM.

When the interference is due to another signal instead of Gaussian noise, both the amplitude of the signal and the interference are constant, since we are, as yet, neglecting fading. As a result, there is a much sharper capture effect evident in the FM receiver during the crossover when the interferer captures the output (because its amplitude becomes larger than that of the desired signal), or vice versa.

4.1.2   Cochannel Interference in the Nonfading Case

A primary object of high-capacity mobile radio systems is to conserve spectrum by re-using frequency channels in geographic areas located as close to each other as possible. The factor which limits the re-use of frequency channels is cochannel interference. Without fading, the capture effect of the FM receiver can suppress cochannel interference. We will follow the method of Prabhu and Enloe[8] for computing the cochannel interference at the FM receiver baseband output. The input to the FM detector in Figure 4.1-1 is assumed to consist of the desired signal, $s(t)$, and the interfering signal, $i(t)$. The Gaussian noise from the receiver front end is neglected in our cochannel interference calculations. The signals are angle modulated, so that

$$s(t) = \cos[\omega_c t + \phi(t) + \mu], \tag{4.1-33}$$

$$i(t) = R\cos[(\omega_c + \omega_d)t + \phi_i(t) + \mu_i], \tag{4.1-34}$$

where $f_c = \omega_c/2\pi$ is the carrier frequency of the desired signal and $f_d = \omega_d/2\pi$ is the difference between the carrier frequency of the interfering signal and that of the desired signal. For cochannel interference f_d is usually small, but if it is large enough to pass the bottom edge of the baseband filter, the offset, f_d, can have a strong effect. $\phi(t)$ is the phase modulation of the desired signal, while $\phi_i(t)$ is that of the interfering signal, and μ and μ_i are the phase angles of the desired and interfering signals, respectively. R is the amplitude of the interfering signal relative to that of the desired signal. The total voltage into the FM detector is

$$
\begin{aligned}
e(t) = {}& \mathrm{Re}\big\{\big[\exp[j(\phi(t)+\mu)] \\
& + R\exp[j(\omega_d t + \phi_i(t)+\mu_i)]\big]e^{j\omega_c t}\big\} \\
= {}& \mathrm{Re}\big\{\exp[j(\phi(t)+\mu)] \\
& \times [1 + R\exp[j(\omega_d t + \phi_i(t)-\phi(t)+\mu_i-\mu)]]e^{j\omega_c t}\big\}.
\end{aligned}
$$

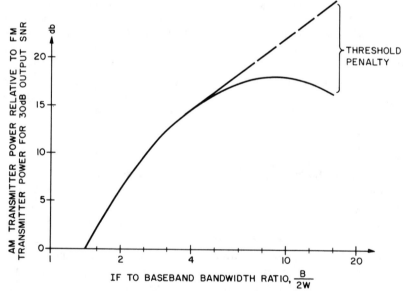

Figure 4.1-7 FM to AM transmitter power advantage versus bandwidth ratio.

The phase of this composite signal input to the FM detector, relative to $\omega_c t$, is $\phi(t) + \mu + \lambda(t)$, where

$$\lambda(t) = \text{Im}\{\log[1 + R \exp j(\omega_d t + \phi_i(t)$$
$$- \phi(t) + \mu_i - \mu)]\}$$

$$= \sum_{k=1}^{\infty} \frac{R^k}{k}(-1)^{k+1} \cos k[\omega_d t + \phi_i(t)$$

$$- \phi(t) + \mu_i - \mu], \qquad (R < 1)$$

is the phase noise due to cochannel interference. To calculate the spectrum of the baseband phase noise, we first compute the autocorrelation of $\lambda(t)$ by averaging over the random phase angles μ and μ_i, which are assumed to be uniformly and independently distributed from 0 to 2π:

$$R_\lambda(\tau) = \langle \lambda(t+\tau)\lambda(t) \rangle_{\mu, \mu_i}$$

$$= \frac{1}{2} \sum_{k=1}^{\infty} \frac{R^{2k}}{k^2} \cos k[\omega_d \tau + \phi_i(t+\tau) - \phi_i(t)$$

$$- \phi(t+\tau) + \phi(t)], \qquad (R < 1). \quad (4.1\text{-}35)$$

Thus, the autocorrelation and therefore the spectrum of the phase noise are equal to the autocorrelation and spectrum of a sum of angle modulated waves of frequency $k\omega_d$ with phase modulation $k[\phi_i(t) - \phi(t)]$ and with amplitude $R^K/k, k = 1,2,3, \ldots$. The spectrum of angle modulated waves has been the subject of much investigation (see, for example, Ref. 9). A simple method of spectrum calculation for angle modulated waves is presented in Appendix A; it is accurate in the spectral region near the carrier frequency. This is the main region of interest in determining the phase noise output. A comparison of the spectrum formula of Appendix A, Eq. (A-8), and the exact formula, Eq. (19) of Ref. 9, is shown in Figure 4.1-8. Note that the agreement is excellent near the carrier. Since Eq. (A-8) uses Poisson's sum formula,[11] it introduces an artificial periodicity not present in the actual spectrum. However, this periodicity in the spectrum does not significantly affect the spectral values near the carrier and therefore introduces negligible error in the computation of baseband output phase noise.

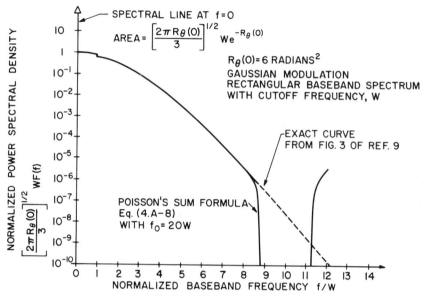

Figure 4.1-8   Spectrum of angle modulated wave.

For large index rectangular-spectrum Gaussian modulation, Eq. (21) of Ref. 9 provides an accurate estimate for the spectrum of angle modulated waves (less than 10% error if $R_\theta(0) \geqslant 10$ and $f < 2W\sqrt{R_\theta(0)}$):

$$S_v(f) = \frac{A^2}{4}[F(f-f_c) + F(f+f_c)], \tag{4.1-36}$$

where

$$F(f) \doteq \exp[-R_\theta(0)]\delta(f)$$

$$+ \sqrt{\frac{3}{2\pi R_\theta(0)}} \frac{\exp\{-[3f^2/2R_\theta(0)W^2]\}}{W}. \tag{4.1-37}$$

Equations (4.1-36) and (4.1-37) replace Eqs. (A-4) and (A-8) for the two-sided power spectrum of $v(t) = A\cos[\omega_c t + \theta(t) + \alpha]$ when the phase modulation is a stationary Gaussian process with a rectangular spectrum and has a mean square index of 10 or greater.

From Eq. (4.1-35) we have that the two-sided power spectrum of the output phase noise due to cochannel interference is

$$S_\lambda(f) = \sum_{k=1}^{\infty} S_{vk}(f),\qquad\qquad (4.1\text{-}38)$$

where $S_{vk}(f)$ is the two-sided power spectrum of the wave, $\cos\{k\omega_d t + k[\phi_i(t) - \phi(t)]\}$, as calculated by either Eqs. (A-4) and (A-8) or (4.1-36) and (4.1-37), depending on the modulation, $k[\phi_i(t) - \phi(t)]$. If $k^2 < 10/(\Phi^2 + \Phi_i^2)$, Eqs. (A-4) and (A-8) are used, while if $k^2 \geqslant 10/(\Phi^2 + \Phi_i^2)$, Eqs. (4.1-36) and (4.1-37) are used. Φ^2 is the mean square value of $\phi(t)$ and Φ_i^2 that of $\phi_i(t)$.

To illustrate the effects of cochannel interference we will compute the baseband output due to interference as a function of the IF interference to signal power ratio, R^2, and the mean square modulation indices, Φ^2 and Φ_i^2, of the signal and interferer, respectively. We model the modulations, $\phi(t)$ and $\phi_i(t)$, as independent stationary Gaussian stochastic processes with rectangular two-sided power spectra extending from $-W$ to W Hz. The two-sided spectra are expressed as

$$S_\phi(f) = \begin{cases} \dfrac{\Phi^2}{2W}(\text{radians})^2/\text{Hz}, & 0 \leqslant f \leqslant W \\[2mm] 0 & \text{otherwise} \end{cases}$$

and

$$S_{\phi_i}(f) = \begin{cases} \dfrac{\Phi_i^2}{2W}(\text{radians})^2/\text{Hz}, & 0 \leqslant f \leqslant W. \\[2mm] 0, & \text{otherwise} \end{cases} \qquad (4.1\text{-}39)$$

Thus,

$$\Phi^2 = \langle \phi^2(t) \rangle = \int_{-\infty}^{\infty} S_\phi(f)\,df,$$

and

$$\Phi_i^2 = \langle \phi_i^2(t) \rangle = \int_{-\infty}^{\infty} S_{\phi_i}(f)\,df. \qquad (4.1\text{-}40)$$

To relate this model for the modulation to conventional single channel voice FM, we note that the flat phase power spectrum corresponds to frequency modulation by a voice signal whose spectrum is flat from 0 to W Hz, the voice signal being preemphasized with a differentiator. The assumption that the modulation is a Gaussian stochastic process is only a rough approximation to the statistics of a single voice signal. This difference in statistics, however, affects mainly the tails of the IF spectrum of the FM wave, which are not of major importance in computing the cochannel baseband interference.

From Eqs. (4.1-38), (A-4), and (A-15) it is seen that the two-sided spectrum of the cochannel interference, with the assumed modulation, consists of delta functions at $f = \pm kf_d$, $k = 1, 2, 3, \ldots$ plus a continuous spectrum. If $f_d = 0$, the continuous spectrum is maximum at $f = 0$ and is discontinuous only at $|f| = W$ (see Ref. 8, p. 2345).

The spectrum of the baseband cochannel interference, with the assumed modulation, is plotted in Figure 4.1-9 assuming that the interfering signal carrier and the desired signal carrier are at the same frequency; that is, $f_d = 0$. The plot assumes $\Phi^2 + \Phi_i^2 = 13$ and four values of R^2 (IF ratio of interferer power to signal power). For $R^2 \ll 1$, the first term in (4.1-38) dominates and the baseband interference spectral density, $S_\lambda(f)$, is proportional to R^2. As seen from Figure 4.1-9, proportionality to R^2 maintains approximately even as R^2 approaches unity as long as f/W is small. Also shown in Figure 4.1-9 is the fact that the baseband interference spectrum is quite flat in the range $|f| < W$. Thus, the ratio of total signal to total interference in the baseband width is approximately equal to the ratio of interference power density to signal power density where the interference power density is greatest ($f/W = 0$).

When $R > 1$, we consider the entire baseband output to be interference. (Although the desired modulation is present in the baseband output, it is distorted.) The baseband output spectrum may again be computed as above except the roles of $\phi(t)$ and $\phi_i(t)$ are reversed and R is replaced with $1/R$. So for $R > 1$, the spectral density of the baseband interference is the sum of $\tilde{S}_\lambda(f) + S_{\phi_i}(f)$. Thus, without fading, the baseband interference power spectrum is

$$S_I(f) = \begin{cases} S_\lambda(f), & R < 1, \\ \tilde{S}_\lambda(f) + S_{\phi_i}(f), & R > 1 \end{cases} \qquad (4.1\text{-}41)$$

where $\tilde{S}_\lambda(f)$ is computed from (4.1-38) with $\phi(t)$ replaced by $\phi_i(t)$ and R replaced by $1/R$. Plots of baseband output interference, $2WS_I(0)$, are given in Figures 4.1-10 and 4.1-11 as a function of the IF signal to interference power ratio, R^2, and the mean square modulation indices, $\Phi^2 + \Phi_i^2$, assuming $\Phi^2 = \Phi_i^2$. For indices on the order of unity ($\Phi^2 + \Phi_i^2 = 2$) or larger, two properties are apparent from Figures 4.1-10 and 4.1-11. First, a strong capture effect is evident; that is, the baseband output interference increases abrubtly as the IF interferer strength exceeds that of the signal. Second, the functional dependence of the baseband output interference on the IF interferer to signal power ratio, R^2, is well approximated by two straight lines; that is,

$$2WS_I(0) \doteq \begin{cases} R^2\sqrt{\dfrac{3}{2\pi(\Phi^2 + \Phi_i^2)}}, & R<1 \\ \Phi_i^2, & R>1. \end{cases} \qquad (4.1\text{-}42)$$

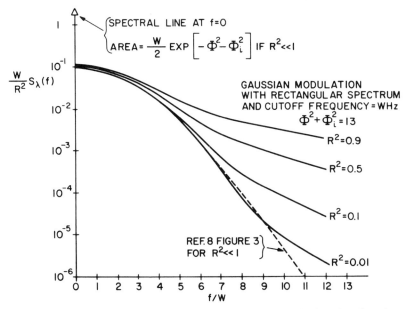

Figure 4.1-9 One half of the two-sided spectral density of the baseband cochannel interference.

The accuracy of Eq. (4.1-42) is better than 1 dB if $\Phi^2+\Phi_i^2 \geqslant 2$. Also, most of the inaccuracy occurs when $0.8 < R^2 < 1.2$. It is interesting to note that when the desired signal is stronger than the interference, the baseband output interference first increases then decreases as the modulation index is increased. This effect is shown more clearly in Figure 4.1-12, where it is seen that the output interference peaks at an index of $\Phi^2+\Phi_i^2 = 2.35$.

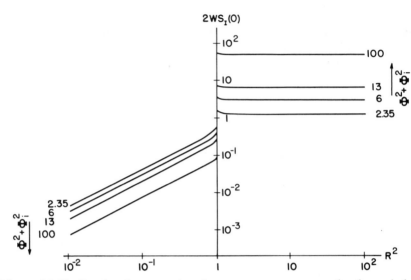

Figure 4.1-10 Baseband output interference versus power ratio (large index). Parameter, mean square index.

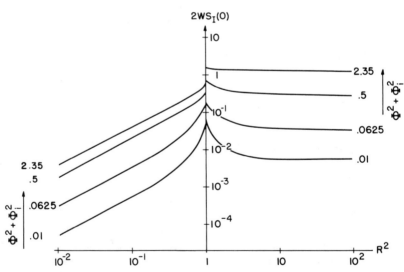

Figure 4.1-11 Baseband output interference versus power ratio (small index). Parameter, mean square index.

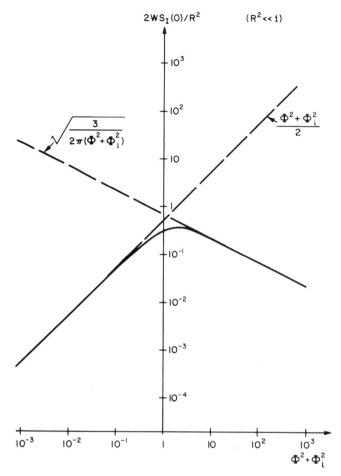

Figure 4.1-12 Baseband output interference versus modulation index.

If the desired signal carrier and the interfering signal carrier differ by an amount f_d, then the spectra $S_{vk}(f)$ in Eq. (4.1-38) are shifted by $\pm k f_d$ in the computation of baseband interference. If f_d is large compared to W, this offset can of course shift the major portion of the interference completely out of the baseband filter, resulting in the case of adjacent channel interference rather than that of cochannel interference. The continuous part of $S_{vk}(f)$ is relatively flat for $f < W$. Thus, for shifts $f_d < W$, the continuous portion of the spectrum within the baseband filter will remain approximately the same as in the case when $f_d = 0$. The major effect of the offset f_d is to move the interference spectral line at dc (see Figure 4.1-9) to spectral lines at kf_d ($k = 1, 2, ...$) introducing baseband interference; thus, from Eqs. (4.1-36), (4.1-37), and (4.1-38),

$$S_{I_{offset}}(f) \doteq S_I(f)$$

$$+ \sum_{k=1}^{\infty} \frac{R^{2k}}{4k^2} e^{-k^2(\Phi^2 + \Phi_i^2)} [\delta(f - kf_d) + \delta(f + kf_d)], \qquad (4.1\text{-}43)$$

for Gaussian modulation. If $\Phi^2 + \Phi_i^2 \geqslant 2$, less than 1 dB error in total baseband interference results from neglecting all spectral lines, and they are therefore neglected in the following calculations.

4.1.3   Adjacent Channel Interference in the Nonfading Case

In the case of adjacent channel interference, only the tails of the adjacent channel signal enter the FM demodulator from the IF filter. An example of the IF spectra of the desired signal and adjacent channel interferer after the IF filter is shown in Figure 4.1-13. The figure assumes rectangular-spectrum (0–W Hz) Gaussian phase modulation with mean square modulation index of 0.833 (radians)^2 and a channel center-to-center frequency spacing of $4.75W$. The IF filter is assumed to be a 14-pole Butterworth with a 3-dB bandwidth of $6W$. Because only the tail of the adjacent channel spectrum is passed by the IF filter, there is considerable phase-to-amplitude conversion, causing the adjacent channel signal to appear noiselike at the output of the IF filter. If one assumes that the adjacent channel signal strength is much less than that of the desired signal, so that the amplitude of the adjacent channel practically never exceeds that of the desired signal, then we can use the methods used in the above cochannel interference calculations for adjacent channel interference.

In Eq. (4.1-34), the amplitude ratio, R, becomes a function of time for the adjacent channel case because the time varying phase modulation, $\phi_i(t)$, produces amplitude variations, $R(t)$, via the IF filter. Assuming $R(t) \ll 1$, we keep only the first term of the expansion in (4.1-34):

$$\lambda(t) = R(t) \cos[\omega_d t + \phi_i(t) - \phi(t) + \mu_i - \mu]. \qquad (4.1\text{-}44)$$

Using the facts that μ_i, μ, and ϕ are independent of each other and of ϕ_i and R and that $\phi(-t)$ is as likely as $\phi(t)$, it is easy to show that the two-sided power spectrum of the output phase noise due to adjacent channel interference is given by the convolution

$$S_\lambda(f) = 2S_a(f) * S_v(f), \qquad (4.1\text{-}45)$$

where $S_v(f)$ is the two-sided spectrum of the signal, $\cos[\phi(t)+\mu]$, and $S_a(f)$ is the two-sided spectrum of the interferer, $R(t)\cos[\omega_d t + \phi_i(t) + \mu_i]$.

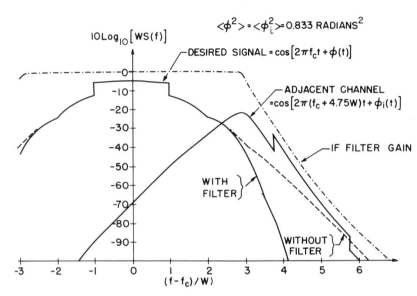

Figure 4.1-13 One-sided power spectral densities of desired and adjacent-channel signals after IF filtering.

The convolution of Eq. (4.1-45) was performed for the spectra given in Figure 4.1-13. The resulting baseband interference spectrum is shown in Figure 4.1-14 along with the corresponding baseband spectrum resulting if the interference had been cochannel rather than adjacent channel. As seen in Figure 4.1-14, the baseband interference due to an adjacent channel interferer is not only smaller than that due to a cochannel interferer of equal strength before the IF filter, but the interference is concentrated at the higher baseband frequencies with adjacent channel interference.

When the interference is not much smaller than the desired signal, we are no longer assured that $R(t)<1$, and the expansion given in Eq. (4.1-34) does not always converge. To compute the baseband interference when $R(t)$ is on the order of unity, we model the noiselike adjacent channel signal emerging from the IF filter as Gaussian noise and use the techniques given above for frequency demodulation in the presence of noise. However, this time the noise band is not centered on the carrier.

As suggested by the empirical approximation of Eq. (4.1-24), we approximate the one-sided baseband noise spectrum as the sum of three terms,

$$\mathcal{W}(f) = e^{-2\rho}\mathcal{W}_N(f) + (1 - e^{-\rho})^2 \mathcal{W}_c(f)$$

$$+ (1 - e^{-\rho})^2 \mathcal{W}_{\theta'}(f), \qquad (4.1\text{-}46)$$

where $\mathcal{W}_N(f)$ is the baseband output spectrum when no signal is present at IF ($\rho = 0$), $\mathcal{W}_c(f)$ is the baseband output spectrum of the "clicks," and $\mathcal{W}_{\theta'}(f)$ is the baseband noise spectrum when the IF signal is strong ($\rho \gg 1$). When the IF noise spectrum is Gaussian shaped with frequency and is centered on the signal carrier, we have, from Eq. (4.1-22),

$$\mathcal{W}_N(f) = \frac{8\pi B}{\sqrt{4.7}} \; ; \qquad (4.1\text{-}47a)$$

from Eq. (4.1-23)

$$\mathcal{W}_c(f) = 4\pi B\sqrt{2\pi}\left[1 - \mathrm{erf}\sqrt{\rho}\,\right]; \qquad (4.1\text{-}47b)$$

and from (4.1-16) and (4.1-25)

$$\mathcal{W}_{\theta'}(f) = \frac{4\pi^2 f^2}{B\rho} e^{-(\pi f^2 / B^2)}. \qquad (4.1\text{-}47c)$$

When compared with the exact results for a centered Gaussian-shaped noise band (as shown in Figure 4.1-4), the accuracy of Eqs. (4.1-46) and (4.1-47) is found to be better than 10% for all ρ, $0 < \rho < \infty$.

For a noncentered band of noise, as shown in Figure 4.1-15, new formulas for $\mathcal{W}_N(f)$, $\mathcal{W}_c(f)$, and $\mathcal{W}_{\theta'}(f)$ are required.

When no signal is present the instantaneous frequency detected from the noise band is equal to that given by a discriminator centered on the noise band plus the frequency offset, $2\pi(f_n - f_c)$. Thus, for a Gaussian-shaped noise band,

$$\mathcal{W}_N(f) = 4\pi^2(f_n - f_c)^2\delta(f) + \frac{8\pi B}{\sqrt{4.7}} . \qquad (4.1\text{-}48)$$

Rice's calculations[3] for click rate, N_c, are extended in Appendix B to nonsymmetrical noise bands giving the following approximation:

$$N_c \doteq (f_n - f_c)e^{-\rho}, \qquad (4.1\text{-}49)$$

and[3]

$$\mathcal{W}_c(f) = 8\pi^2 N_c \doteq 8\pi^2(f_n - f_c)e^{-\rho}. \qquad (4.1\text{-}50)$$

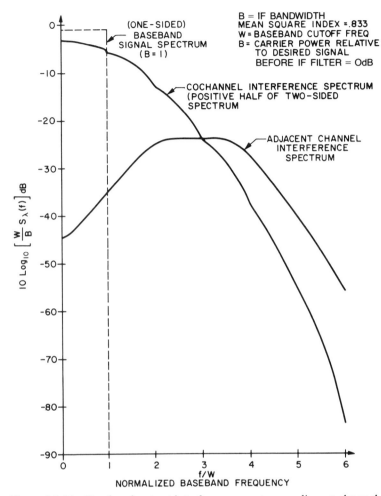

Figure 4.1-14   Baseband output interference spectrum; adjacent channel.

The large signal noise spectrum, $\mathfrak{W}_{\theta'}(f)$, is best computed from the adjacent channel interference formula,

$$\mathfrak{W}_{\theta'}(f) = 2(2\pi f)^2 S_\lambda(f), \tag{4.1-51}$$

because $S_\lambda(f)$ uses the actual IF spectra involved rather than an approximate shape that may not be accurate on the tail of the adjacent channel spectrum. The tail determines the output adjacent channel interference in the baseband filter. The factor $2(2\pi f)^2$ in Eq. (4.1-51) arises because

$\mathcal{W}_b(f)$ is the one-sided power spectrum of the instantaneous frequency, $\theta'(t)$, and $S_\lambda(f)$ is the two-sided power spectrum of the instantaneous phase, $\theta(t)$.

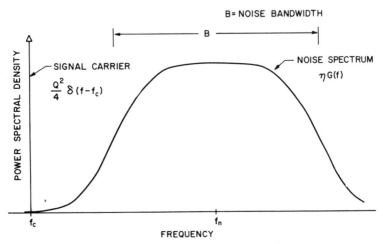

Figure 4.1-15   Signal carrier and off-center noise band.

A check on the accuracy of Eq. (4.1-46) is presented in Figure 4.1-16, for an off-center noise band as shown in Figure 4.1-15. The measurements shown in Figure 4.1-16 used a frequency offset of $f_n - f_c = 20$ kHz and a noise bandwidth of $B = 22$ kHz. The baseband output was passed through a filter with an approximately rectangular passband from 300 Hz to 2.9 kHz and integrated, so that the output corresponded to the portion of the spectrum of the instantaneous phase between 300 Hz and 2.9 kHz. Equation (4.1-48) could not be used for $\mathcal{W}_N(f)$ because the filter shape was not Gaussian. Deriving a new $\mathcal{W}_N(f)$ involves computing $\underline{g}(\tau)$ from Eq. (4.1-9) and the IF noise spectrum and substituting $\underline{g}(\tau)$ into Eq. (7.16) of Ref. 1 for the autocorrelation of the instantaneous frequency and Fourier transforming Eq. (7.16) to obtain $\mathcal{W}_N(f)$. In the comparison of theory and experiment shown in Figure 4.1-16, the value for $\mathcal{W}_N(f)$ was not computed but was set equal to the low signal asymptote of the measurements. Therefore, Figure 4.1-16 serves as a check on the parts of Eq. (4.1-46), excepting $\mathcal{W}_N(f)$. The agreement between theory and experiment is quite good, except for $\rho < 1$, where Eq. (4.1-46) predicts values about 1 dB less than measured.

Equations (4.1-46), (4.1-48), (4.1-50) and (4.1-51) were used to compute the adjacent channel interference for the case shown in Figure 4.1-13 with results shown in Figure 4.1-17. The adjacent channel IF spectrum after the IF filter is approximated by a Gaussian-shaped spectrum, Eq. (4.1-25), with $B = \sqrt{2\pi}\ W/4$ and $f_n - f_c = 3W$. Also shown in Figure 4.1-17 is the baseband interference that would result if the interferer were cochannel rather than adjacent channel.

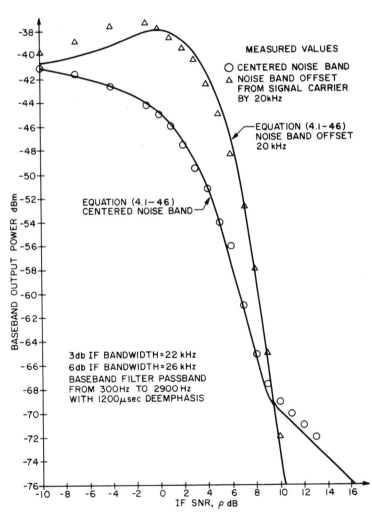

Figure 4.1-16 FM noise quieting with symmetrical and asymmetrical IF noise.

For the example shown in Figure 4.1-17, the adjacent channel interference is reduced first by a 21-dB reduction in its IF power by the IF filter. Also the spectrum of the adjacent channel near the signal carrier is low, so when the receiver is captured on the desired signal the baseband interference within the baseband filter is reduced relative to the cochannel case. However, the amplitude variations, introduced by the distortion of the adjacent channel by the IF filter, cause the receiver to start to lose capture ("clicks") on the desired signal when ρ (signal to interference ratio after the IF filter) is about 12 dB instead of the cochannel value of 0 dB.

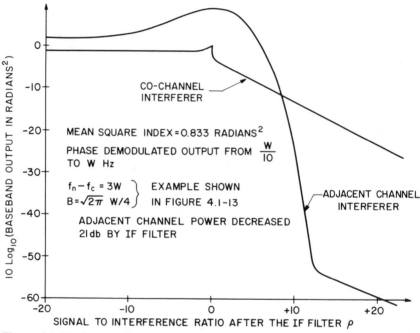

Figure 4.1-17 Example of baseband interference for adjacent and cochannel interferers.

4.1.4   Effects of Rayleigh Fading on Noise Performance

The rapid fading experienced as the mobile unit moves through the interfering waves from the various scatterers tends to wash out the capture properties of the FM receiver. Also, rapid phase changes accompany the deep fades, producing a random FM component in the output of the FM receiver (cf. Section 1.4). Furthermore, as the signal fades and the receiver loses capture on the signal the baseband output signal is suppressed (cf. Figure 4.1-6, for example). This rapid random suppression of the signal appears as an additional noise component in the baseband output.

Although the statistical properties of the various noise components appearing at the baseband output, in the presence of Rayleigh fading, are different, subjective tests[12] indicate that the average output signal to average noise ratio is an approximate measure of the quality of the mobile radio channel for voice communications with fading rates typical at microwave carrier frequencies (50–1000 fades per second).

To determine the average signal and noise in the presence of Rayleigh fading, Davis[7] used a quasistatic approximation that expresses the signal and noise as functions only of the instantaneous IF signal-to-noise ratio, ρ, and then averages over the statistics of ρ. As we shall see, the quasistatic approximation is accurate when the fading rate is small compared to the IF bandwidth.

We denote the average, over variations of ρ, of the output due to the information modulation as the signal output. The remainder of the output due to the information modulation is thus zero mean and uncorrelated with the signal output, and we classify it as signal suppression noise.

With Rayleigh fading the probability density of the IF SNR, ρ, is given by (cf. Chapter 1)

$$p_\rho(\alpha) = \frac{1}{\rho_0} \exp\left[-\frac{\alpha}{\rho_0}\right], \qquad (4.1\text{-}52)$$

where ρ_0 is the average IF SNR.

From Eq. (4.1-28) the signal output voltage (average over ρ) is

$$\bar{v}_0(t) = \langle v_0(t)\rangle_\rho = v(t) \int_0^\infty \frac{(1-e^{-\alpha})e^{-\alpha/\rho_0}}{\rho_0}\, d\alpha$$

$$= v(t)\frac{\rho_0}{\rho_0 + 1}. \qquad (4.1\text{-}53)$$

Thus the signal suppression noise, $n_s(t)$, is

$$n_s(t) = v_0(t) - \bar{v}_0(t) = v(t)\left[\frac{1}{\rho_0 + 1} - e^{-\rho}\right]. \qquad (4.1\text{-}54)$$

From Eq. (4.1-53), the signal output with Rayleigh fading is

$$\bar{S} = \langle \bar{v}_0^2(t)\rangle = \frac{\rho_0^2}{(\rho_0 + 1)^2}\langle v^2(t)\rangle = \frac{\rho_0^2 S}{(\rho_0 + 1)^2}. \qquad (4.1\text{-}55)$$

From Eqs. (4.1-26), (4.1-52), and (4.1-54), the average output noise is

$$\overline{N} = \int_0^\infty \frac{1}{\rho_0} e^{-\alpha/\rho_0} \left[S\left(\frac{1}{\rho_0 + 1} - e^{-\alpha} \right)^2 + \frac{a}{\alpha}(1 - e^{-\alpha})^2 \right.$$

$$\left. + \frac{8\pi BW}{\sqrt{2(\alpha + 2.35)}} e^{-\alpha} \right] d\alpha$$

$$= S\left(\frac{1}{2\rho_0 + 1} - \frac{1}{(\rho_0 + 1)^2} \right) + \frac{a}{\rho_0} \log \frac{(1 + \rho_0)^2}{1 + 2\rho_0} + 8\pi BW \sqrt{\frac{\pi}{2\rho_0(\rho_0 + 1)}}$$

$$\times \exp\left[2.35 \frac{\rho_0 + 1}{\rho_0} \right] \mathrm{erfc} \sqrt{2.35 \frac{\rho_0 + 1}{\rho_0}} \,, \qquad (4.1\text{-}56)$$

where a is defined in Eq. (4.1-27) and

$$\mathrm{erfc}(x) \overset{\Delta}{=} 1 - \mathrm{erf}(x). \qquad (4.1\text{-}57)$$

The first term in Eq. (4.1-56) is due to signal suppression noise, the second is due to "above-threshold" noise, and the last term is due to "threshold-and-below" noise.

In addition to the above noise components there is random FM noise in the baseband output. As described in Section 1.4 the one-sided random FM output spectrum may be approximated by

$$\mathcal{W}_\theta(f) \doteq \left(\frac{b_2}{b_0} - \frac{b_1^2}{b_0^2} \right) \Big/ f = 2\pi^2 \frac{f_m^2}{f} \,, \qquad (4.1\text{-}58)$$

and the corresponding baseband output noise is

$$N_\theta = \int_{0.1W}^W \mathcal{W}_\theta(f) \, df = 2\pi^2 f_m^2 \log 10. \qquad (4.1\text{-}59)$$

Equation (4.1-58) assumes a uniform angle of arrival Doppler spectrum, Eq. (1.2-11), and that baseband output frequencies, $0.1W$ to W, are greater than the maximum Doppler frequency shift, f_m. Although the output band used in computing Eq. (4.1-56) was 0 to W, it is changed negligibly by using $0.1W$ to W. The latter bandwidth allows the asymptotic form of the random FM spectrum to be used, Eq. (4.1-58), throughout the output band, if f_m is not too large. Furthermore, $0.1W$ to W corresponds to the usual voice bandwidth design: 300 Hz to 3 kHz.

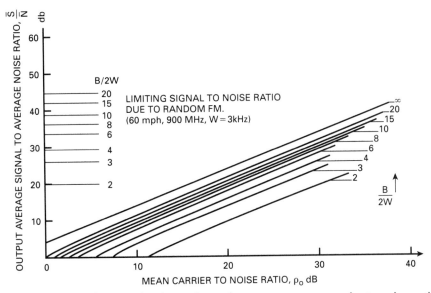

Figure 4.1-18 Output signal-to-noise ratio versus average carrier-to-noise ratio (Rayleigh fading).

In Figure 4.1-18 the output average signal to average noise, $\overline{S}/\overline{N}$, is plotted[7] versus the IF mean carrier to noise ratio, ρ_0, and the IF to baseband bandwidth ratio, $B/2W$. The sharp threshold due to capture (cf. Figure 4.1-6) is no longer apparent. As one increases the transmitter power to increase ρ_0, the output signal-to-noise ratio increases approximately linearly with ρ_0. Since threshold crossings are less frequent, there is less signal suppression and quadrature noise is suppressed. However, the random FM remains unchanged and eventually is the dominant noise component. Thus, depending on the maximum Doppler frequency and the signal frequency excursion, there is a limiting output signal-to-noise ratio. For example, with a carrier frequency of 900 MHz and a vehicle velocity of 60 miles/hr, the maximum Doppler frequency shift is $f_m = 80$ Hz. Note that $f_m < 0.1W$, with $W = 3$ kHz, so that (4.1-58) is valid. Using Eqs.

(4.1-30) and (4.1-59) we have the limiting output signal-to-noise ratio

$$\frac{S}{N_\theta} = \frac{(B-2W)^2}{20f_m^2 \log 10}.$$

(4.1-60)

The limiting output signal-to-noise ratios for $f_m = 80$ Hz and $W = 3$ kHz are marked on Figure 4.1-18 for various bandwidth ratios $B/2W$.

Curves similar to those of Figure 4.1-18 can be plotted for the case when one detects the instantaneous phase rather than instantaneous frequency, as assumed in the cochannel and adjacent-channel interference cases above. If instantaneous phase is detected, the noise spectra are equal to those for the instantaneous frequency divided by $4\pi^2 f^2$ and the signal is equal to the mean square index appropriate to the IF and audio bandwidths.

The above quasistatic approximation can be checked for accuracy and range of application by an exact analysis of the output noise of an FM discriminator responding to a Rayleigh fading signal in the presence of Gaussian noise. The exact method treats the fading signal and Gaussian noise as a fading signal whose Doppler spectrum is the composite of the actual Doppler spectrum and the noise spectrum as shown in Figure 4.1-19.

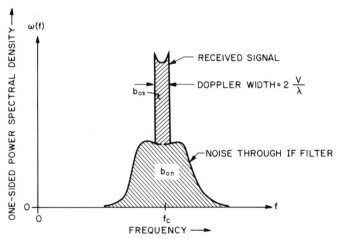

Figure 4.1-19 Total power spectrum of fading signal plus noise.

For a CW transmission the received signal phasor is a zero mean complex Gaussian process. The front-end noise phasor is also an independent zero mean complex Gaussian process. Thus the total input, signal plus noise, has a phasor that is a zero mean complex Gaussian process and its spectrum is the sum of the Doppler spectrum and the IF filtered noise, as shown in Figure 4.1-19. Assuming the spectrum of the total input is asymmetrical about f_c, the autocorrelation of the instantaneous frequency is given by Eq. (1.4-7):

$$R_{\dot{\theta}}(\tau) = -\frac{1}{2} \left\{ \left[\frac{g'(\tau)}{g(\tau)} \right]^2 - \frac{g''(\tau)}{g(\tau)} \right\} \log \left\{ 1 - \left[\frac{g(\tau)}{g(0)} \right]^2 \right\}, \quad (4.1\text{-}61)$$

where

$$g(\tau) = g_s(\tau) + \underline{g}_n(\tau). \quad (4.1\text{-}62)$$

The autocorrelation functions $g_s(\tau)$ and $\underline{g}_n(\tau)$ are defined in Eqs. (1.3-2) and (4.1-9) for the signal and noise, respectively. For an omnidirectional mobile radio antenna moving at velocity v through RF plane waves at frequency f_c, distributed uniformly in angle of arrival, we have from Eq. (1.3-7)

$$g_s(\tau) = b_{0_s} J_0(\omega_m \tau), \quad (4.1\text{-}63)$$

where b_{0_s} is the received signal power, ω_m is the maximum angular frequency Doppler shift $2\pi V f_c/c$, and c is the velocity of light. J_0 is the Bessel function of the first kind of order zero. If the IF filter has a Gaussian shape, given by Eq. (4.1-25), then from Eq. (4.1-9), we have

$$\underline{g}_n(\tau) = b_{0_n} \exp(-\pi B^2 \tau^2), \quad (4.1\text{-}64)$$

where b_{0_n} is the noise power and B is the noise bandwidth.

From Ref. 11, p. 28, the power out of the baseband filter is

$$N_{\dot{\theta}} = 2 \int_0^\infty f(\tau) R_{\dot{\theta}}(\tau) \, d\tau, \quad (4.1\text{-}65)$$

where $f(\tau)$ is the impulse response of the baseband circuit. For a unit gain rectangular baseband filter of cutoff frequency W, we have

$$f(\tau) = \frac{\sin 2\pi W \tau}{\pi \tau}. \quad (4.1\text{-}65a)$$

Equations (4.1-61), (4.1-63), (4.1-64), and (4.1-65a) were substituted into Eq. (4.1-65), which was evaluated by numerical integration. The results for output noise are compared to the noise (excepting signal suppression noise) obtained by the quasistatic approach, Eqs. (4.1-56) and (4.1-59), in Figures 4.1-20 and 4.1-21. In Figure 4.1-20 the IF bandwidth is 20 times the baseband cutoff frequency, while in Figure 4.1-21 it is 8 times the baseband cutoff frequency. The comparison between quasistatic and exact solutions is clarified by considering two regions of signal-to-noise ratio: $\rho_0 = b_{0_s} / \underline{b}_{0_n} > 10$ and $\rho_0 < 10$.

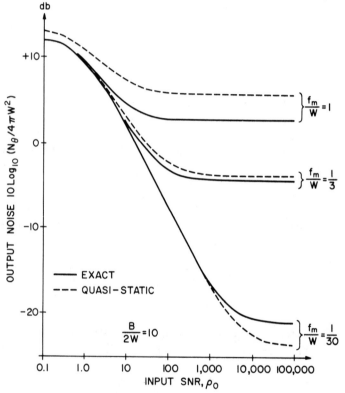

Figure 4.1-20   Output noise versus input SNR, $b/2w = 10$.

As ρ_0 increases above 10, the baseband noise eventually levels off at a value determined by the random FM of the Doppler spread signal. This level differs in the quasistatic approximation, for example, Eq. (4.1-59), from the exact for two reasons. First, Eq. (4.1-59) only includes the random FM spectrum from $W/10$ to W Hz. This effect is shown in Figures 4.1-20 and 4.1-21 by a lower quasistatic output when $f_m/W \ll 1$ and $\rho_0 \gg 1$. Second, when the maximum Doppler frequency increases to a significant fraction of the baseband cutoff frequency, the asymptotic form of the random FM spectrum used in Eq. (4.1-58), namely, $2\pi^2 f_m^2/f$, is no longer a good approximation to the true random FM power spectrum. This effect is shown in Figures 4.1-20 and 4.1-21 by a higher quasistatic output when $f_m/W \approx 1$ and $\rho_0 \gg 1$.

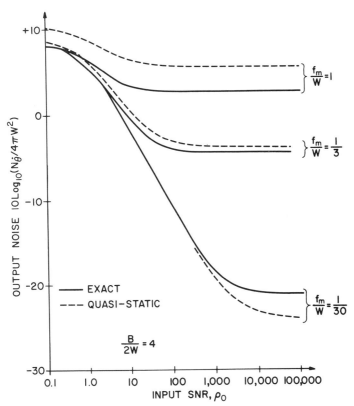

Figure 4.1-21   Output noise versus input SNR, $b/2w = 4$.

When $\rho_0 < 10$, the FM "click" noise[3] is present. The quasistatic approximation begins to fail when the fading duration becomes nearly as short as the "click" duration, which is on the order of the inverseband-width of the IF filter.[13] As the signal fades below FM threshold, a "click" may occur due to noise, and, at the same time, the random FM of the signal peaks to a value comparable to that of the "click." The FM receiver output is a nonlinear function of the noise "click" and the signal random FM peak, so that the output cannot be accurately approximated by addition of threshold noise and signal random FM as separate contributions.

As seen from Figures 4.1-20 and 4.1-21, the output noise of the quasistatic approximation approaches the sum of the random FM of the noise band and that of the fading signal as the input SNR approaches zero. The exact solution undergoes a transition from the random FM of the fading signal to that of the noise band as the input SNR goes from infinity to zero.

The quasistatic approximation is accurate, for $\rho_0 < 1$, whenever the maximum Doppler frequency is small compared to the IF bandwidth. For example, in Figure 4.1-20, $f_m / W = 1$ corresponds to a maximum Doppler frequency equal to $\frac{1}{20}$ of the IF noise bandwidth, and in Figure 4.1-21, $f_m / W = \frac{1}{3}$ corresponds to a maximum Doppler frequency equal to $\frac{1}{24}$ of the IF noise bandwidth.

Davis[7] has studied the advantages of squelching (shorting the baseband output) the FM receiver during the occurrence of a fade. He determined the optimum muting signal level, ρ_1 (below ρ_1 the baseband output is short circuited), as a function of the average level, ρ_0, which would minimize the quasistatic expression for baseband output noise, Eq. (4.1-56). Although the squelching reduces the noise due to threshold "clicks" as the receiver experiences a fade, it introduces more signal suppression noise. As a result Davis found insignificant improvement through the use of squelching. Rather than shorting the baseband output, some improvement may be gained by a sample-and-hold technique to minimize the signal suppression of the squelching operation.[14]

4.1.5 Effects of Rayleigh Fading on Cochannel Interference

Without fading, one could reduce the effect of cochannel interference by increasing the modulation index, Φ. In fact, the output signal to interference ratio improves as the cube of the index, since the signal output is proportional to Φ^2 and the interference output is inversely proportional to Φ [cf. Eq. (4.1-42)]. As shown below, this index cubed advantage *is lost* with Rayleigh fading for typical average IF signal-to-interference ratios. This is perhaps the most important effect of fading on cochannel interference; namely, although cochannel interference without fading could be markedly reduced by increasing the modulation index, it remains approximately constant with index, for indices greater than unity, when fading is present.*

Again, we apply the quasistatic approximation to determine the average baseband signal to average interference ratio. As in Eqs. (4.1-53) and (4.1-54), we denote the average, over variations in R^2 (ratio of IF interferer power to IF signal power), of the output due to the desired signal modulation as the signal output. The remainder of the output due to the desired signal modulation is thus zero mean and uncorrelated with the signal output, and we classify it as signal suppression noise.

With both the signal and interferer undergoing independent Rayleigh fading, the probability density of the IF interference to signal ratio, R^2, is given by the F distribution of two degrees of freedom[15]:

$$p_{R^2}(\alpha) = \frac{\Gamma}{(1+\Gamma\alpha)^2}, \qquad (4.1\text{-}66)$$

where Γ is the ratio of the average power of the desired signal to the average power of the interfering signal at IF. When captured on the signal, $R^2 < 1$, the full signal, $\phi(t)$, appears at the baseband output. When captured on the interferer, $R^2 > 1$, no undistorted signal appears at the baseband output. Therefore, the signal output voltage (average over R^2) is

$$\overline{\phi}_0(t) = \langle\phi_0(t)\rangle_{R^2} = \phi(t)\int_0^1 p_{R^2}(\alpha)\,d\alpha$$

$$= \phi(t)\int_0^1 \frac{\Gamma\,d\alpha}{(1+\Gamma\alpha)^2} = \frac{\Gamma\phi(t)}{1+\Gamma}. \qquad (4.1\text{-}67)$$

Thus the signal suppression noise, $n_\phi(t)$, is

$$n_\phi(t) = \phi_0(t) - \overline{\phi}_0(t)$$

$$= \begin{cases} \dfrac{1}{1+\Gamma}\phi(t) & \text{if} \quad R^2 < 1, \\[2mm] -\dfrac{\Gamma}{1+\Gamma}\phi(t) & \text{if} \quad R^2 > 1. \end{cases} \qquad (4.1\text{-}68)$$

*As shown in Section 5.4.4, the index-cubed advantage can be regained by using diversity.

From Eq. (4.1-67) the output signal power is

$$\overline{\Phi^2} = \langle \overline{\phi}_0^2(t) \rangle = \frac{\Gamma^2}{(1+\Gamma)^2} \Phi^2. \qquad (4.1\text{-}69)$$

From Eqs. (4.1-42) and (4.1-68) the baseband output interference averaged over the fading is

$$\overline{N_\phi} = \int_0^1 \frac{\Gamma}{(1+\Gamma\alpha)^2} \left\{ \alpha \sqrt{\frac{3}{2\pi(\Phi^2 + \Phi_i^2)}} + \frac{1}{(1+\Gamma)^2} \Phi^2 \right\} d\alpha$$

$$+ \int_1^\infty \frac{\Gamma}{(1+\Gamma\alpha)^2} \left\{ \Phi_i^2 + \frac{\Gamma^2}{(1+\Gamma)^2} \Phi^2 \right\} d\alpha$$

or

$$\overline{N_\phi} = \frac{1}{1+\Gamma} \left\{ \left[\left(1+\frac{1}{\Gamma}\right) \log(1+\Gamma) - 1 \right] \right.$$

$$\left. \times \sqrt{\frac{3}{2\pi(\Phi^2 + \Phi_i^2)}} + \Phi_i^2 + \frac{\Gamma\Phi^2}{1+\Gamma} \right\}. \qquad (4.1\text{-}70)$$

A plot of the average signal to average interference ratio at the baseband output is shown in Figure 4.1-22, assuming $\Phi_i^2 = \Phi^2$. As seen from the figure, practically no improvement in cochannel interference immunity is achieved by increasing the index beyond $\Phi = (\log \Gamma)^{1/3}$. Thus, contrary to the no-fading case, there is no cochannel interference advantage with large modulation indices for the range of average IF signal to interference power ratios of interest.

In the case where there are many cochannel interferers (six or more), each with different modulation signals, the central limit theorem suggests that the sum of interferers could be approximated by Gaussian noise with power equal to that of the sum of average interferer powers and with the spectral shape of the angle modulated waves. With this approximation the analysis used for Figure 4.1-18 applies to provide the average signal to average interference ratio in the baseband output.

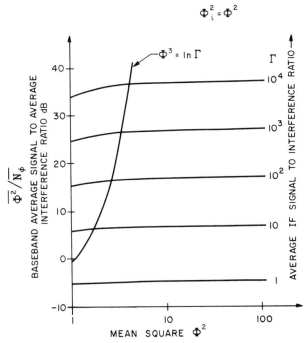

$$\Phi_i^2 = \Phi^2$$

Figure 4.1-22   Baseband average cochannel interference with Rayleigh fading.

4.1.6 Effects of Rayleigh Fading on Adjacent-Channel Interference

Adjacent-channel interference at the mobile receiver may result from a channel radiated from the same base station as the mobile's desired channel or from a different base station. Thus the Rayleigh fading of the desired signal and that of the adjacent-channel signal may be partially correlated. As described in Section 1.5, the time delay spread of the propagation medium determines the correlation of the received amplitudes of two signals transmitted from the same base station and received at the same mobile as a function of their frequency separation, $s/2\pi$ [cf. Eqs. (1.5-20) and (1.5-24)]. From Eq. (1.5-51), the probability density of the signal to adjacent-channel interference ratio, ρ (cf. Figure 4.1-17), after the IF filter is given in terms of the correlation amplitude, λ, by

$$p_\rho(\alpha) = -\frac{d}{d\rho} P[\alpha > \rho]$$

$$= -\frac{d}{da} P[r_2 \geqslant ar_1]\frac{da}{d\rho}, \tag{4.1-71}$$

where $a = \sqrt{\rho/G}$ and G is the power gain of the IF filter for the desired signal relative to the adjacent channel. Substituting (1.5-51) into (4.1-71) gives

$$P_\rho(\alpha) = \frac{(1-\lambda^2)(1+\alpha/G)}{G\left[(1+\alpha/G)^2 - 4\lambda^2\alpha/G\right]^{3/2}}, \qquad (4.1\text{-}72)$$

where, for an exponential time delay distribution, Eq. (1.5-19), the correlation amplitude λ is given in terms of the frequency separation, $s/2\pi$, by Eq. (1.5-20), with $\tau = 0$.

Again we divide the fluctuating output signal into average output signal and signal suppression noise:

$$\overline{\phi}(t) = \int_0^\infty P_\rho(\alpha)(1 - e^{-\alpha})\phi(t)\, d\alpha$$

$$= \phi(t)[1 - f(G,\lambda)], \qquad (4.1\text{-}73)$$

where

$$f(G,\lambda) \triangleq (1-\lambda^2)\int_0^\infty \frac{e^{-Gx}(1+x)}{\left[(1+x)^2 - 4\lambda^2 x\right]^{3/2}}\, dx \qquad (4.1\text{-}74)$$

and

$$\overline{n}_\phi(t) = (1 - e^{-\rho})\phi(t) - \overline{\phi}(t) = [f(G,\lambda) - e^{-\rho}]\phi(t). \qquad (4.1\text{-}75)$$

From Eq. (4.1-73), the signal output with Rayleigh fading is

$$\overline{S} = \langle \overline{\phi}(t)^2 \rangle = \Phi^2[1 - f(G,\lambda)]^2. \qquad (4.1\text{-}76)$$

From Eqs. (4.1-46), (4.1-48), (4.1-50), (4.1-51) and (4.1-75) the output noise, for a given ρ, is

$$N_\phi(\rho) = \int_{0.1W}^{W} \frac{\mathfrak{W}(f)}{4\pi^2 f^2}\, df + \Phi^2[f(G,\lambda) - e^{-\rho}]^2$$

$$= \frac{18}{W}\left[\frac{Be^{-2\rho}}{\sqrt{4.7}\,\pi} + (f_n - f_c)(1 - e^{-\rho})^2 e^{-\rho}\right]$$

$$+ (1 - e^{-\rho})^2 2\int_{W/10}^{W} S_\lambda(f)\, df + \Phi^2[f(G,\lambda) - e^{-\rho}]^2. \qquad (4.1\text{-}77)$$

The average output interference with Rayleigh fading is found by integrating $N_\phi(\rho)$ times the probability density of ρ:

$$\overline{N_\phi} = \int_0^\infty \frac{(1-\lambda^2)(1+\rho/G)N_\phi(\rho)}{G\left[(1+\rho/G)^2 - 4\lambda^2\rho/G\right]^{3/2}} \, d\rho. \qquad (4.1\text{-}78)$$

Equations (4.1-76) and (4.1-78) were evaluated for no correlation, $\lambda = 0$, and the case shown Figure 4.1-13, where

$$f_n - f_c = 3W, \qquad \Phi^2 = 0.833 \text{ radians}^2;$$

$$B = \frac{\sqrt{2\pi}\,W}{4}, \qquad G = 10^{2.1} = 126;$$

$$f(G,\lambda) = f(126,0) = \tfrac{1}{126} \qquad (4.1\text{-}79)$$

and from Figure 4.1-14

$$2\int_{W/10}^{W} S_\lambda(f)\,df = \frac{2.5 \times 10^{-4}}{\rho} \text{ rad}^2. \qquad (4.1\text{-}80)$$

The resulting average signal to average interference ratio for this case is 7.3 dB if the IF average signal to average interference ratio is unity before the IF filter. It is directly proportional to the average signal to interference ratio before the IF filter. It is interesting to note that, assuming that the interferer and signal have equal power before the IF filter, the adjacent-channel interference in the above example is reduced 38 dB further by the demodulation process, but is increased 51.7 dB due to the Rayleigh fading.

When there are many adjacent-channel interferers, the above analysis applies directly by using the composite interference spectrum after the IF filter.

4.1.7 Comparison with SSB and AM

Rapid Rayleigh fading generally has a disastrous effect on single-sideband (SSB) and AM communication systems. Unless corrective measures are taken, the distortion introduced by the fading is larger than the output signal, independent of how much the transmitter power is increased. Without fading, the signal-to-noise performance of SSB or AM systems may not be as good as FM (cf. Figure 4.1-7); however, neglecting cochannel interference, they enjoy a bandwidth advantage over FM. But with rapid fading, SSB and AM become unusable, unless it is feasible to correct for the fading.

As an example, consider the effect of fading on AM signals. If the modulation bandwidth is much smaller than the coherence bandwidth (cf. Chapter 1), the fading is "frequency flat"[16] over the signal band and the envelope of the received output signal is

$$a(t) = r(t)(1 + v(t)),   \tag{4.1-81}$$

where $v(t)$ is the information signal, $1 + v(t)$ is the envelope of the transmitted carrier, and $r(t)$ is the Rayleigh fading imposed by the medium. If the fading is extremely slow, the subjective effect on a listener may not be too bad; however, as the fading rate increases above even 1 Hz it becomes a definite detriment. We may estimate the distortion due to fading by calculating the signal suppression noise as above. The average, over the Rayleigh fading, of the output is taken as the true signal output, $\bar{v}(t)$: using Eq. (1.3-48), which shows that $\langle r \rangle = \sqrt{\pi b_0 / 2}$,

$$\bar{v}(t) \equiv \langle a(t) \rangle_r - \langle r \rangle = v(t)\langle r \rangle$$

$$= v(t)\sqrt{\frac{\pi b_0}{2}}   \tag{4.1-82}$$

The remainder of the output is zero mean and uncorrelated with $\bar{v}(t)$ and is taken to be the signal suppression noise,

$$n_v(t) = \left[r(t) - \sqrt{\frac{\pi b_0}{2}} \right](1 + v(t)).   \tag{4.1-83}$$

The signal to signal suppression noise ratio is then

$$\frac{\langle \bar{v}^2(t) \rangle}{\langle n_v^2(t) \rangle} = \frac{S}{1+S}\frac{\langle r \rangle^2}{\langle r^2 \rangle - \langle r \rangle^2} = \frac{S}{1+S}\frac{\frac{1}{2}\pi b_0}{(2 - \frac{1}{2}\pi)b_0},   \tag{4.1-84}$$

where $S = \langle v^2(t) \rangle$ is the signal modulation power which is usually on the order of $\frac{1}{10}$ to avoid overmodulation [cf. Eq. (4.1-31)] and we have used Eq. (1.3-14), $\langle r^2 \rangle = 2b_0$. Thus the signal suppression noise is on the order of 3 times that of the signal. Similar results are obtained for SSB. Here the transmitted signal is the one-sided spectrum of the information signal $v(t)$ translated to RF, and the received signal is the product of the complex Gaussian process of the multipath medium $r(t)e^{j\theta(t)}$, and the transmitted signal. If the spectrum of $r(t)$ lies well below the lowest baseband frequency, the signal-to-noise ratio can be improved by as much as 10.4 dB (cf. Section 6.6).

One method of correcting the real or complex envelope fading is to transmit a pilot or carrier that is used for gain and/or phase control at the receiver. The pilot must be separated sufficiently in time or frequency to avoid overlapping the signal due to multipath delay spread or Doppler frequency spread. However, as the separation is increased, error is introduced in estimating the envelope correction required. As an example, consider the attempt at amplitude correction in an AM system by transmitting a CW pilot, separated in frequency, along with the signal. The demodulated signal will be

$$a_c(t) = \frac{r(t)}{r_p(t)} [1 + v(t)], \tag{4.1-85}$$

where $r_p(t)$ is the envelope of the pilot. From Eq. (4.1-72) the probability density of the amplitude ratio, $A(t) \triangleq r(t)/r_p(t)$ is

$$p_A(\alpha) = \frac{2\alpha(1 - \lambda^2)(1 + \alpha^2/G)}{G\left[(1 + \alpha^2/G)^2 - 4\lambda^2\alpha^2/G\right]^{3/2}}, \qquad G = 1. \tag{4.1-86}$$

Following the same steps used in obtaining Eq. (4.1-84), we have for the signal to signal suppression noise ratio

$$\frac{\langle \bar{v}^2(t) \rangle}{\langle n_v^2(t) \rangle} = \frac{S}{1 + S} \frac{\langle A \rangle^2}{\langle A^2 \rangle - \langle A \rangle^2}. \tag{4.1-87}$$

From Eq. (4.1-86), it can be shown that $\langle A^2 \rangle$ is infinite unless $\lambda = 1$ (i.e., signal and pilot are perfectly correlated). Whereas, from Refs. 15 and 17, we have

$$A = 2 \int_0^\infty \frac{\alpha^2(1 - \lambda^2)(1 + \alpha^2)}{\left[(1 + \alpha^2)^2 - 4\lambda^2\alpha^2\right]^{3/2}} \, d\alpha$$

$$= \frac{2\sqrt{2\pi}\,(1 - \lambda^2)}{\lambda^3} \int_0^\infty \sqrt{t}\, e^{-(2/\lambda^2 - 1)t} I_0(t)\, dt$$

$$= E(\lambda), \tag{4.1-88}$$

where $I_0(t)$ is the modified Bessel function of the first kind of order zero, and $E(\lambda)$ is the complete elliptic integral of the second kind:

$$E(\lambda) = \frac{\pi}{2}\left\{ 1 - \sum_{n=1}^{\infty}\left[\frac{(2n-1)!!}{2^n n!}\right]^2 \frac{\lambda^{2n}}{2n-1}\right\},\tag{4.1-89}$$

$$(2n-1)!! \overset{\Delta}{=} 1\cdot 3\cdot 5\cdots (2n-1).$$

Thus $\langle A\rangle$, in contrast to $\langle A^2\rangle$, is finite:

$$\frac{\pi}{2} > [\langle A\rangle] > 1.\tag{4.1-90}$$

Therefore, gain control as specified by Eq. (4.1-85) results in swamping by signal suppression noise. To avoid this catastrophe, one can limit the amplifier gain, $1/r_p$, to a maximum value $1/r_0$. Then the limited gain times the signal amplitude is

$$A_L \overset{\Delta}{=} \begin{cases} r/r_p & \text{if } r_p > r_0 \\ r/r_0 & \text{if } r_p < r_0 \end{cases}.\tag{4.1-91}$$

From Eq. (1.5-22)

$$\langle A_L\rangle = \int_0^{\infty} dr \int_{r_0}^{\infty} dr_p \frac{r^2}{b_0^2(1-\lambda^2)} \exp\left[-\frac{r^2+r_p^2}{2b_0(1-\lambda^2)}\right] I_0\left(\frac{rr_p}{b_0}\frac{\lambda}{1-\lambda^2}\right)$$

$$+ \int_0^{\infty} dr \int_0^{r_0} dr_p \frac{r^2 r_p}{r_0 b_0^2(1-\lambda^2)} \exp\left[-\frac{r^2+r_p^2}{2b_0(1-\lambda^2)}\right] I_0\left(\frac{rr_p}{b_0}\frac{\lambda}{1-\lambda^2}\right)\tag{4.1-92}$$

and

$$\langle A_L^2\rangle = \int_0^{\infty} dr \int_{r_0}^{\infty} dr_p \frac{r^3}{r_p b_0^2(1-\lambda^2)} \exp\left[-\frac{r^2+r_p^2}{2b_0(1-\lambda^2)}\right] I_0\left(\frac{rr_p}{b_0}\frac{\lambda}{1-\lambda^2}\right)$$

$$+ \int_0^{\infty} dr \int_0^{r_0} dr_p \frac{r^3 r_p}{r_0^2 b_0^2(1-\lambda^2)} \exp\left[-\frac{r^2+r_p^2}{2b_0(1-\lambda^2)}\right] I_0\left(\frac{rr_p}{b_0}\frac{\lambda}{1-\lambda^2}\right).\tag{4.1-93}$$

From Refs. 15, 17, and 18, we have the evaluation of the integrations Eqs. (4.1-92) and (4.1-93):

$$\langle A_L \rangle = E(\lambda) - \sqrt{\pi} \; \frac{r_0}{\sqrt{2b_0(1-\lambda^2)}} \; \exp\left[\left(\frac{\lambda^2}{2}-1\right)\frac{r_0^2}{2b_0(1-\lambda^2)}\right]$$

$$\times \left[\frac{\lambda^2}{2} I_1\left(\frac{\lambda^2 r_0^2}{4b_0(1-\lambda^2)}\right) + \left(1-\frac{\lambda^2}{2}\right) I_0\left(\frac{\lambda^2 r_0^2}{4b_0(1-\lambda^2)}\right) \right]$$

$$+ \frac{\sqrt{\pi}}{\lambda^2} \sum_{k=0}^{\infty} (2^k k!)^{-2} \left\{ \frac{\sqrt{2b_0(1-\lambda^2)} \; \gamma\{2k+1,(\tfrac{1}{2}\lambda^2-1)[r_0^2/2b_0(1-\lambda^2)]\}}{r_0(2/\lambda^2-1)^{2k+1}} \right.$$

$$\left. - \frac{2^{3/2}(1-\lambda^2)\gamma\{2k+\tfrac{3}{2},(\tfrac{1}{2}\lambda^2-1)[r_0^2/2b_0(1-\lambda^2)]\}}{\lambda(2/\lambda^2-1)^{2k+3/2}} \right\} \qquad (4.1\text{-}94)$$

and

$$\langle A_L^2 \rangle = (1-\lambda^2)E_1\left(\frac{r_0^2}{2b_0}\right) + \frac{1-e^{-r_0^2/2b_0}}{r_0^2/2b_0}, \qquad (4.1\text{-}95)$$

where $\gamma(\alpha,x)$ is the incomplete gamma function,

$$\gamma(\alpha,x) = \sum_{n=0}^{\infty} \frac{(-1)^n x^{\alpha+n}}{n!(\alpha+n)}, \qquad (4.1\text{-}96)$$

and $E_1(x)$ is the exponential integral:

$$E_1(x) \overset{\Delta}{=} \int_x^{\infty} \frac{e^t}{t}\, dt, \qquad x>0.$$

The signal-to-signal suppression noise ratio,

$$\frac{\langle \bar{v}^2(t) \rangle}{\langle n_v^2(t) \rangle} = \frac{S}{1+S}\frac{\langle A_L \rangle^2}{\langle A_L^2 \rangle - \langle A_L \rangle^2}, \qquad (4.1\text{-}97)$$

is plotted in Figure 4.1-23 for $S=\tfrac{1}{10}$ and various correlation amplitudes, λ, versus the limiting gain of the automatic gain control, $(r_0^2/2b_0)^{-1}$. As seen in Figure 4.1-23 there is an optimum level at which to limit the automatic gain control (AGC) for each correlation amplitude, λ, as indicated by the dashed curve, although the optimum is rather broad. The optimum represents a compromise between too harsh a gain limit, which practically eliminates the gain control, and too little limiting, which allows the gain to be influenced by false pilot fades (where the pilot fades but the signal does not).

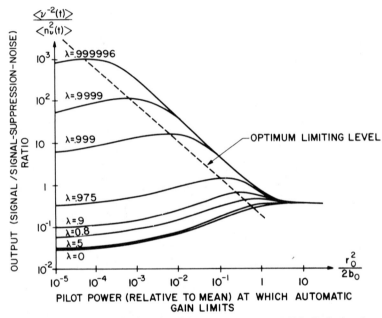

Figure 4.1-23 Output SNR of AM system versus AGC limit level and pilot correlation.

The most striking feature illustrated by Figure 4.1-23 is that very high pilot signal correlation is required to reduce the signal suppression noise to an acceptable level. For example, with optimum gain limiting and amplitude correlation of 0.9999, the output signal to signal suppression noise ratio is 20 dB. From Eq. (1.5-20), for a time delay spread of 1 μsec, a correlation amplitude of 0.9999 or greater requires that the frequency separation between pilot and signal be less than f_s, where

$$f_s = \frac{\sqrt{1-\lambda^2}}{2\pi\lambda\sigma} = \frac{\sqrt{1-(.9999)^2}}{2\pi(.9999)10^{-6}} = 2.25 \text{ kHz}. \qquad (4.1\text{-}98)$$

Even if the carrier of a SSB 3-kHz voice channel were used as a pilot for the AGC, it would not allow a 20 dB output signal to signal suppression noise ratio to be obtained in the above example of a 1-μsec spread in time delays.

If the fading rate is less than the lowest frequency of the voiceband, the signal suppression noise can be reduced somewhat by a high-pass audio filter in the output. However, the effect of fading is multiplicative and it produces intermodulation products with the signal modulation that cannot be eliminated by filtering.

Hopper[19] has reported analysis and experiments with a fast-acting AGC circuit operating on the total signal amplitude, rather than a pilot, for a two-path model of mobile radio propagation and with moderate fading (~3-dB fades). His results show that conventional AGC circuits (of the feedback type) acting fast enough to counteract rapid fading will, if stable, suppress neighboring voice frequencies. However, forward acting AGC is inherently stable, so it can be made very fast by inserting a sharp cutoff filter in the control path. This permits the suppression of 200 Hz fading with negligible effect on the 300–3000-Hz voiceband. His analysis and tests for the simple model (two-path, moderate fade) indicate that modulation products can be suppressed by about 20 dB with 200 Hz fading. He also points out that the control circuits and filters must have a minimum of delay distortion for successful operation of the AGC.

The above considerations indicate that for microwave mobile radio and deep fading, AM and SSB voice communication channels cannot provide telephone quality signals, even with fast-acting AGC of any type.

4.1.8   Distortion Due to Frequency-Selective Fading

In Section 1.5 it was shown that, due to the spread in time delays of the various paths of propagation between base and mobile stations, the received signal due to a CW transmission at one carrier frequency might be weak while at another carrier frequency it might be strong, if the separation of the carrier frequencies is on the order of the coherence bandwidth or greater. This phenomenon is often termed[16] frequency-selective fading.

For a CW transmission, one can picture the signal strength in the vicinity of the mobile receiver as a standing-wave pattern due to the various interfering reflected waves. Since the wavelength depends on the carrier frequency, the spatial position of the minima (fades) in the interference pattern will move as the carrier frequency is varied. Thus frequency modulation of the carrier by a voice signal causes the fades to move past the mobile receiving antenna. These fades introduce a random frequency fluctuation much the same as experienced for a CW transmission when the mobile moves through the fades; that is, the random FM described in Section 1.4, due to vehicle motion (Doppler fading).

Thus there is FM distortion due to frequency-selective fading just as there is FM distortion due to Doppler fading. Further, it has been shown[20] that the simultaneous presence of both Doppler fading and frequency-selective fading can increase the FM distortion an order of magnitude over that experienced when only one type of fading is present.

To analyze the FM distortion, we use the mathematical model for the mobile communication channel presented in Section 1.5. For an FM transmission, we represent the complex envelope of the transmitted signal, relative to a carrier frequency of f_c, as (see Figure 4.1-1)

$$z(t) = \exp\left\{ j \int v(t)\,dt \right\} \triangleq e^{j\psi(t)}, \qquad (4.1\text{-}99)$$

where $v(t)$ is the voice signal frequency modulation. From Eq. (1.5-1) the received signal is given by

$$E(t) = E_0 \sum_{n=1}^{N} \sum_{m=1}^{M} C_{nm} \exp j\left\{ -\omega_c T_{nm} + \omega_n t \right.$$
$$\left. + \psi\left(t - T_{nm} + \frac{\omega_n}{\omega_c} t \right) \right\}, \qquad (4.1\text{-}100)$$

where C_{nm} is given by Eq. (1.5-2), ω_c is the carrier frequency in radians per second, ω_n is the nth Doppler shift radian frequency, T_{nm} is the mth time delay associated with paths in the direction of the nth Doppler shift, and E_0 is a normalization constant. The Doppler shift to carrier frequency ratio, $\omega_n/\omega_c \leqslant v/c$, is on the order of 10^{-7} or less, where v is the vehicle speed and c is the speed of light. Since the voice modulation is restricted to relatively low frequencies (compared to the carrier frequency), the term $(\omega_n/\omega_c)t$ in the argument of ψ represents a negligible change in the phase of the terms in the summation of Eq. (4.1-100); therefore it will be neglected in the following calculations.

Vehicle Moving—Spread in Time Delays Small Compared to Inverse Signal Bandwidth

The case studied in Section 1.4 corresponds to the assumption that the spread in time delays, T_{nm}, is so small in comparison to the frequency excursions due to $v(t)$ that the time delays, T_{nm}, could be neglected in the argument of $\psi(t)$. In this case the expression for the received signal, (4.1-100), simplifies to

$$E(t) = E_0 e^{j\psi(t)} \sum_{n=1}^{N} \sum_{m=1}^{M} C_{nm}$$

$$\times \exp j\{-\omega_c T_{nm} + \omega_n t\}. \qquad (4.1\text{-}101)$$

From Eq. (4.1-101) it is seen that the received signal is just the transmitted signal times the time varying transmission coefficient of the medium evaluated at the carrier frequency, ω_c. The output of a frequency demodulator would be the sum of the voice signal plus the random FM of the time-varying transmission coefficient as computed in Section 1.4.

Vehicle Stationary—Spread in Time Delays on the Order of the Inverse Signal Bandwidth

In this case the frequency deviation of the wideband FM signal is no longer small in comparison to the coherence bandwidth of the channel. The term T_{nm} in the argument of ψ is no longer negligible, but all the ω_n are zero. Thus Eq. (4.1-100) becomes

$$E(t) = E_0 \sum_{n=1}^{N} \sum_{m=1}^{M} C_{nm}$$

$$\times \exp j\{-\omega_c T_{nm} + \psi(t - T_{nm})\}. \qquad (4.1\text{-}102)$$

A simple case of FM distortion due to frequency-selective fading is presented by Bennett, Curtis, and Rice[21] for a single echo small in amplitude relative to the main component. Their results show that the signal-to-distortion ratio at first degrades and then improves as the modulation index is increased.

Medhurst and Roberts[22] show a method of computing distortion in the case of a few large echoes by means of a Monte Carlo technique. However, the Monte Carlo method is similar to an experiment, in that its results are for the specific parameter tested and are difficult to generalize.

Bello and Nelin[23] have computed the distortion by using a power-series expansion for the transmission coefficient of the channel. This power-series model is particularly useful when the transmitted signal bandwidth is less than the coherence bandwidth of the channel, due to the resulting rapid convergence of the power series. In computing the distortion, Bello and Nelin confine their attention to intermodulation distortion [terms containing second-order or higher powers of the modulation, $v(t)$], and only consider terms linear in the modulation times a time-varying factor of the channel [e.g., $a(t)\cdot v(t)$, where $a(t)$ is a factor which varies at the Doppler rate and is independent of the signal, $v(t)$] when computing subcarrier phase stability. When the vehicle is stationary the linear distortion has negligible subjective effect on voice transmission, but, as will be shown later, Doppler shifts of a few hertz or more can make the linear distortion a major portion of the output interference.

Their results are, however, directly applicable to the case of a stationary vehicle. A useful formula presented in their Section V[23] expresses the median intermodulation to signal power spectral density ratio, ρ_{med}, at any baseband frequency f in terms of the voice spectra and the rms time delay spread, σ,

$$\rho_{\text{med}} = 1.2\sigma^4 \frac{S_{v\dot{v}}(f)}{S_v(f)} \qquad (4.1\text{-}103)$$

Here $S_{v\dot{v}}(f)$ is the power spectrum of $v(t)\,dv(t)/dt$, $S_v(f)$ is the power spectrum of $v(t)$, and if $p(T)$ is the power delay distribution (cf. Section 1.4), we have

$$\sigma^2 \overset{\Delta}{=} \int_{-\infty}^{\infty} p(T)(T-T_0)^2\,dT, \qquad (4.1\text{-}104)$$

where T_0 is the mean time delay:

$$T_0 \overset{\Delta}{=} \int_{-\infty}^{\infty} p(T)T\,dT. \qquad (4.1\text{-}105)$$

Equation (4.1-103) assumes a Gaussian-shaped power delay distribution, $p(T)$, to give the factor 1.2. However Bello and Nelin show that the factor is only slightly dependent on the shape of the power delay distribution. If $v(t)$ is Gaussian with a spectrum flat between aW and W Hz then the distortion-to-signal ratio is worst at the top baseband frequency and is given by[23]

$$\rho_{\text{med}} = 1.2(2\pi\sigma)^4 \frac{f_{\text{dev}}^2 W^2}{4}\left(\frac{1-2a}{1-a}\right), \qquad (4.1\text{-}106)$$

where f_{dev} is the rms value of $v(t)$.

Vehicle Moving—Spread in Time Delays on the Order of the Inverse Signal Bandwidth

This is the most general case. Both random FM and frequency-selective fading distortion contribute significantly to the receiver output. Furthermore, the combination of the two produces an additional distortion component that can be an order of magnitude greater than either produces individually. To analyze this case we employ Bello and Nelin's[23] power-series model for the channel. We introduce the Fourier transform pair

$$Z(f) = \int_{-\infty}^{\infty} z(t) e^{-j\omega t} \, dt$$

and

$$z(t - T_{nm}) = \frac{1}{2\pi} \int_{-\infty}^{\infty} Z(f) e^{j\omega(t - T_{nm})} \, d\omega. \qquad (4.1\text{-}107)$$

Substituting Eq. (4.1-107) into Eq. (4.1-100) we obtain

$$E(t) = \frac{1}{2\pi} \int_{-\infty}^{\infty} E_0 \sum_{n=1}^{N} \sum_{m=1}^{M} C_{nm}$$

$$\times \exp j \{ -\omega_c T_{nm} + \omega_n t - \omega T_{nm} \}$$

$$\times Z(f) e^{j\omega t} \, d\omega. \qquad (4.1\text{-}108)$$

Consider the channel to be a linear, time-variant, filter; the time-varying transfer function would be

$$T(f,t) = E_0 \sum_{n=1}^{N} \sum_{m=1}^{M} C_{nm}$$

$$\times \exp j \{ -2\pi f_c T_{nm} + 2\pi f_n t - 2\pi f T_{nm} \} \qquad (4.1\text{-}109)$$

and we have

$$E(t) = \int_{-\infty}^{\infty} T(f,t) Z(f) e^{j2\pi f t} \, df. \qquad (4.1\text{-}110)$$

We assume that the average time delay is zero. If it is not, we factor it out of the summation in Eq. (4.1-109) and include the factor, $e^{-j2\pi f_c T_0}$ in E_0, giving

$$T(f,t) = T_0(f,t) e^{-j2\pi f T_0}, \qquad (4.1\text{-}111)$$

where

$$T_0(f,t) = E_0 \sum_{k=1}^{K} C_k$$

$$\times \exp j \{ -2\pi (f_c + f)\tau_k + 2\pi f_k t \}. \tag{4.1-112}$$

In Eq. (4.1-112) we have renumbered the summations 1 to N and 1 to M for a single summation 1 to K, where $K = N \cdot M$, for simplicity. The purpose of the factorization (4.1-111) is to separate the part of the multipath transfer coefficient that represents an average time delay and introduces no distortion. The remaining portion of the time delays, τ_k, represent the variation of the time delay about the average time delay, T_0.

If the frequency deviation of the signal, $\dot{\psi}(t)$, is small relative to the coherence bandwidth, a power series expansion of the transmission coefficient, $T_0(f,t)$, around the transmitter carrier frequency, f_c (i.e., around $f = 0$) will converge rapidly with probability near one. We write the power-series expansion as follows:

$$T_0(f,t) = T_0(t) + j2\pi f T_1(t) + (j2\pi f)^2 T_2(t) + \cdots, \tag{4.1-113}$$

where

$$T_0(t) = T_0(0,t)$$

$$= \sum_{k=1}^{K} C_k \exp j \{ -2\pi f_c \tau_k + 2\pi f_k t \}, \tag{4.1-114}$$

$$T_1(t) = \frac{1}{2\pi j} \left. \frac{\partial T_0(f,t)}{\partial f} \right|_{f=0}$$

$$= - \sum_{k=1}^{K} \tau_k C_k \exp j \{ -2\pi f_c \tau_k + 2\pi f_k t \}, \tag{4.1-115}$$

$$T_2(t) = \frac{1}{2(2\pi j)^2} \left. \frac{\partial^2 T_0(f,t)}{\partial f^2} \right|_{f=0}$$

$$= \sum_{k=1}^{K} \frac{\tau_k^2}{2} C_k \exp j \{ -2\pi f_c \tau_k + 2\pi f_k t \}. \tag{4.1-116}$$

Substituting Eqs. (4.1-111) and (4.1-113) into Eq. (4.1-100) and carrying out the integration, we have

$$E(t) = T_0(t)z(t - T_0) + T_1(t)\dot{z}(t - T_0) + T_2(t)\ddot{z}(t - T_0) + \cdots . \quad (4.1\text{-}117)$$

As pointed out in Ref. 23, Eq. (4.1-117) shows that the physical implementation of the power series channel model is a time delay T_0, followed by a parallel combination of nth-order differentiators with time varying gains, $T_n(t)$, respectively, where $n = 0, 1, 2, 3, \ldots$. Upon substituting Eq. (4.1-99) into (4.1-117), the complex envelope of the received signal becomes

$$E(t) = e^{j\psi(t - T_0)} \{ T_0(t) + j\dot{\psi}(t - T_0)T_1(t)$$

$$+ [j\ddot{\psi}(t - T_0) - \dot{\psi}^2(t - T_0)]T_2(t) + \cdots \} . \quad (4.1\text{-}118)$$

The desired modulation is $\psi(t - T_0)$ and the distortions are contained in the terms in parenthesis of Eq. (4.1-118). If there is no modulation, the only remaining distortion term is $T_0(t)$, whose instantaneous frequency is the random FM, $\dot{\theta}(t)$, computed in Section 1.4. Thus, we may factor Eq. (4.1-118) to form

$$E(t) = |T_0(t)| e^{j[\psi(t - T_0) + \theta(t)]} \left\{ 1 + j\dot{\psi}(t - T_0)\frac{T_1(t)}{T_0(t)} \right.$$

$$\left. + [j\ddot{\psi}(t - T_0) - \dot{\psi}^2(t - T_0)]\frac{T_2(t)}{T_0(t)} + \cdots \right\} . \quad (4.1\text{-}119)$$

The phase of the received signal is

$$\phi(t) = \psi(t - T_0) + \theta(t) + \delta(t), \quad (4.1\text{-}120)$$

where $\psi(t - T_0)$ is the desired modulation, $\theta(t)$ is the random FM phase, and $\delta(t)$ is the remaining distortion:

$$\delta(t) = \arctan \left\{ \dfrac{ \begin{aligned} &\dot{\psi}(t-T_0)\operatorname{Re}\left\{\dfrac{T_1(t)}{T_0(t)}\right\} + \ddot{\psi}(t-T_0) \\ &\times \operatorname{Re}\left\{\dfrac{T_2(t)}{T_0(t)}\right\} - \dot{\psi}^2(t-T_0)\operatorname{Im}\left\{\dfrac{T_2(t)}{T_0(t)}\right\} + \cdots \end{aligned} }{ \begin{aligned} &1 - \dot{\psi}(t-T_0)\operatorname{Im}\left\{\dfrac{T_1(t)}{T_0(t)}\right\} - \ddot{\psi}(t-T_0) \\ &\times \operatorname{Im}\left\{\dfrac{T_2(t)}{T_0(t)}\right\} - \dot{\psi}^2(t-T_0)\operatorname{Re}\left\{\dfrac{T_2(t)}{T_0(t)}\right\} + \cdots \end{aligned} } \right\}$$

$$(4.1\text{-}121)$$

Following Ref. 23, we note that when the amount of selective fading is small, the arctangents in Eq. (4.1-121) may, with probability near unity, be accurately approximated by $\delta(t) \doteq \delta_1(t) + \delta_2(t)$, where

$$\delta_1(t) \overset{\Delta}{=} \dot{\psi}(t-T_0)\operatorname{Re}\left\{\frac{T_1(t)}{T_0(t)}\right\} + \ddot{\psi}(t-T_0)\operatorname{Re}\left\{\frac{T_2(t)}{T_0(t)}\right\}$$

and

$$\delta_2(t) \overset{\Delta}{=} \dot{\psi}^2(t-T_0)\left[\operatorname{Re}\left\{\frac{T_1(t)}{T_0(t)}\right\}\operatorname{Im}\left\{\frac{T_1(t)}{T_0(t)}\right\} - \operatorname{Im}\left\{\frac{T_2(t)}{T_0(t)}\right\}\right]. \quad (4.1\text{-}122)$$

When the vehicle is stationary, $T_0(t)$, $T_1(t)$, and $T_2(t)$ are constants. Thus $\theta(t)$ is a constant phase shift and

$$\dot{\psi}(t-T_0)\operatorname{Re}\left\{\frac{T_1(t)}{T_0(t)}\right\} + \ddot{\psi}(t-T_0)\operatorname{Re}\left\{\frac{T_2(t)}{T_0(t)}\right\}$$

represents linear distortion such as caused by a linear filter. It is usually not objectionable in normal voice communications and may be neglected. The only remaining distortion in Eq. (4.1-122) when the vehicle is stationary is the second-order distortion,

$$\dot{\psi}^2(t-T_0)\left[\operatorname{Re}\left\{\frac{T_1(t)}{T_0(t)}\right\}\operatorname{Im}\left\{\frac{T_1(t)}{T_0(t)}\right\} - \operatorname{Im}\left\{\frac{T_2(t)}{T_0(t)}\right\}\right].$$

This is the term that was used to compute the median intermodulation to signal power spectral density ratio at any baseband frequency, as given in Eq. (4.1-103).

As the vehicle increases its velocity, the linear distortion terms are modulated in time by the variations of $\mathrm{Re}[T_1(t)/T_0(t)]$ and $\mathrm{Re}\{T_2(t)/T_0(t)\}$. At Doppler frequencies of several Hz and larger, this modulation shifts signal frequency components in $\dot{\psi}(t-T_0)$ and $\ddot{\psi}(t-T_0)$ sufficiently to introduce distortion objectionable to the ear. The linear distortion can thus no longer be neglected. If the maximum Doppler frequency is small compared to the baseband cutoff frequency, then the variations in $T_n(t)$ are slow compared to those of $\psi(t)$, and we use a quasistatic approximation to compute the median spectrum of the distortion.

In the quasistatic approximation, the power spectrum of the distortion is assumed to be a slowly varying function of time due to the fluctuations in the $\{T_n(t)\}$. Assuming $\dot{\psi}(t)$ is a stationary, zero-mean, Gaussian process gives

$$S_\delta(f,t) = \left[\mathrm{Re}\left\{\frac{T_1(t)}{T_0(t)}\right\}\right]^2 S_{\dot\psi}(f) + \left[\mathrm{Re}\left\{\frac{T_2(t)}{T_0(t)}\right\}\right]^2 S_{\ddot\psi}(f)$$

$$+ \left[\mathrm{Re}\left\{\frac{T_1(t)}{T_0(t)}\right\}\mathrm{Im}\left\{\frac{T_1(t)}{T_0(t)}\right\} - \mathrm{Im}\left\{\frac{T_2(t)}{T_0(t)}\right\}\right]^2 S_{\dot\psi^2}(f), \quad (4.1\text{-}123)$$

where $S_{\dot\psi}(f)$, $S_{\ddot\psi}(f)$, and $S_{\dot\psi^2}(f)$ are the two-sided power spectra of the processes $\dot{\psi}(t)$, $\ddot{\psi}(t)$, and $\dot{\psi}^2(t)$, respectively. The mean square phase distortion is also a slowly varying function of time:

$$\langle\delta^2\rangle = \left[\mathrm{Re}\left\{\frac{T_1(t)}{T_0(t)}\right\}\right]^2 \langle\dot\psi^2(t)\rangle + \left[\mathrm{Re}\left\{\frac{T_2(t)}{T_0(t)}\right\}\right]^2 \langle\ddot\psi^2(t)\rangle$$

$$+ \left[\mathrm{Re}\left\{\frac{T_1(t)}{T_0(t)}\right\}\mathrm{Im}\left\{\frac{T_1(t)}{T_0(t)}\right\} - \mathrm{Im}\left\{\frac{T_2(t)}{T_0(t)}\right\}\right]^2 \langle\dot\psi_{(t)}^4\rangle. \quad (4.1\text{-}124)$$

The time average of the rms distortion as computed by Eq. (4.1-124) diverges, which indicates that the approximations used in obtaining (4.1-124) are not valid for obtaining average distortion. However, most of the time the approximations are valid. The conditions for validity can be seen by considering the median (with respect to channel variations) value of the rms distortion (averaging over modulation variations). Using the methods of Appendix C and assuming that the spectrum of $\psi(t)$ is flat from $W/10$ to W Hz, and that the modulation index is unity or greater, $f_{dev}W \geqslant 1$, then

$$\langle \delta^2 \rangle_{med} < 3(2\pi f_{dev}\sigma)^2, \tag{4.1-125}$$

for

$$2\pi f_{dev}\sigma < 0.3, \tag{4.1-126}$$

where f_{dev} is the rms frequency deviation and σ is the rms spread in time delays. This indicates that the small argument approximation used in obtaining (4.1-122) from (4.1-121) is accurate at least 50% of the time if the frequency deviation and spread in time delays satisfy the inequality (4.1-126). Under these conditions, the higher-order approximations used in obtaining (4.1-121) are also accurate with probability near unity. The reader is referred to Ref. 23 for further discussion of the convergence of the power-series model of fading dispersive channels.

The FM distortion for a particular case is plotted versus Doppler frequency in Figure 4.1-24. We have assumed frequency modulation with a baseband spectrum flat from 300 Hz to 3 KHz, an rms frequency deviation of 10 KHz, and a Gaussian power delay distribution with an rms time delay spread of five μsec. The distortion, as a function of Doppler frequency, exhibits three regions. At low Doppler frequencies, or vehicle speeds, the random FM [$\theta(t)$ in Eq. (4.1-120)] is negligible and the linear distortion $\delta_1(t)$ in Eq. (4.1-122) varies so slowly as to be unobjectionable. Thus only the intermodulation distortion [$\delta_2(t)$ in Eq. (4.1-122)] is included. It is computed by integrating over the median intermodulation distortion spectrum [Eqs. (132) and (77) of Ref. 23] from 300 Hz to 3 K Hz:

$$S_{\delta_2(t)}(f) = 1.2\sigma^4 S_{\ddot{\psi}\ddot{\psi}}(f), \tag{4.1-127}$$

where

$$S_{\ddot{\psi}\ddot{\psi}}(f) = 16\pi^6 f_{dev}^4 W \begin{cases} \dfrac{(f/W)^2}{1-a} - \dfrac{|f/W|^3}{(1-a)^2}, & 0 < |f| < 2aW \\[2ex] \dfrac{(f/W)^2(1-2a)-|f/W|^3/2}{(1-a)^2}, & 2aW < |f| < W(1-a) \\[2ex] \dfrac{|f/W|^3}{2(1-a)^2} - \dfrac{a(f/W)^2}{(1-a)^2}, & W(1-a) < |f| < W \end{cases}$$

$$\tag{4.1-128}$$

and $W = 3$ KHz, $a = \frac{1}{10}$.

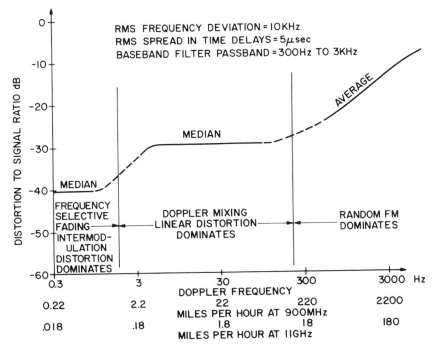

Figure 4.1-24   FM distortion versus Doppler frequency.

As the Doppler frequency increases above several Hz, the linear distortion becomes objectionable. The linear distortion is equivalent to passing the baseband signal through a filter whose gain characteristic is varying at a rate of the order of the Doppler frequency. As discussed in Appendix C, a tight lower bound on the median linear plus quadratic distortion is obtained from the median value of the first term of Eq. (4.1-123),

$$S_{\delta_{med}}(f) = \left[\text{Re}\left\{ \frac{T_1(t)}{T_0(t)} \right\}_{med} \right]^2 \omega^2 S_{\dot\psi}(f) = \frac{\sigma^2}{3}\omega^2 S_{\psi}(f), \quad (4.1\text{-}129)$$

where we have multiplied by ω^2 to obtain the spectrum of the instantaneous frequency from that of the instantaneous phase, assuming FM detection.

The spectrum of Eq. (4.1-129) is integrated from 300 Hz to 3 KHz to obtain the median output distortion in the range of intermediate Doppler frequencies shown in Figure 4.1-24.

As the Doppler frequencies increase further, the random FM, $\dot{\theta}(t)$, dominates. Since the random FM cannot be described as a slowly varying channel function times a rapidly varying modulation function, the quasi-static approximation cannot be used to obtain a median random FM spectrum. The average random FM power through the baseband filter can, however, be computed by integrating over the random FM spectrum of Section 1.4 from 300 Hz to 3 KHz. The result is shown in Figure 4.1-24 at the higher Doppler frequencies. The distortion-to-signal ratios, in the frequency band $W/10$ to W Hz, for frequency deviations and time delay spreads other than that shown in Figure 4.1-24 can be determined from the following formulas:

$$\left\{ \begin{array}{c} \text{median intermodulation} \\ \text{distortion to signal ratio} \end{array} \right\} = 0.108(2\pi W\sigma)^2(2\pi f_{\text{dev}}\sigma)^2, \quad (4.1\text{-}130)$$

$$\left\{ \begin{array}{c} \text{median linear} \\ \text{distortion to signal ratio} \end{array} \right\} = \frac{(2\pi W\sigma)^2}{8.1}, \quad (4.1\text{-}131)$$

$$\left\{ \begin{array}{c} \text{average random FM} \\ \text{distortion to signal ratio} \end{array} \right\} = 1.15\left(\frac{f_m}{f_{\text{dev}}}\right)^2, \quad (4.1\text{-}132)$$

where

$$2\pi f_{\text{dev}}\sigma < 0.3, \qquad \frac{W}{f_{\text{dev}}} < 1, \quad \text{and} \quad \frac{f_m}{W} < 0.1,$$

and f_m is the maximum Doppler frequency.

4.2   DIGITAL MODULATION

This section is not intended to be a treatise on all the aspects of digital modulation, for which there are many text books. For example, Lucky[25] and Bennet[26] have extensive discussions on baseband and modulation techniques. Sunde[27] and Schwartz[28] both have chapters on digital transmission through fading media. Instead, we will give a tutorial discussion on the various aspects of voice transmission by digital means over mobile radio channels.

4.2.1   Speech Coding

Speech signals first have to be coded into a format suitable for digital transmission. This can be done either with pulse code modulation (PCM) or delta modulation. It is customary to use 7-bit PCM sampled at 8 kHz rate for telephone channels.[29] This amounts to a bit rate of 56 kb/sec. With delta modulation, similar bit rates are required.[29] The bit rate can be reduced at the expense of additional baseband processing circuitry. Greefkes and Riemens reported a digitally controlled companding scheme[30] that appears to meet CCITT specification for a good telephone link with a bit rate of 32 kb/sec. For bit rates as low as 20 kb/sec, modulation schemes have been reported that were still able to provide intelligible speech signals. Curves showing the baseband SNR as a function of bit rate and average input signal level are presented in Figure 4.2-1 for delta modulation. It is seen that with a bit rate of 32 kb/sec we would have a baseband SNR of at least 30 dB for a dynamic range of 50 dB.

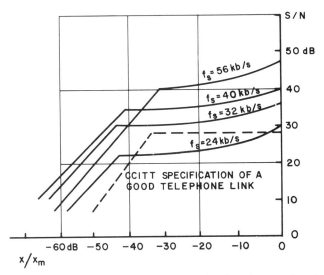

Figure 4.2-1 Signal-to-noise ratio at the output of a decoder as a function of input signal level x/x_m for companded delta modulation. f_s is the sampling rate (Ref. 6).

4.2.2 Modulation and Detection of Two-Level Systems Without Fading

The sampled speech data of bit rate f_s can be transmitted through the channel by AM, PM, or FM techniques. From the point of view of power economy and system simplicity we will only consider two-level systems. For equipment simplicity single-sideband AM is also not considered.

A double-sideband AM system is shown in Figure 4.2-2. Here filter shaping is done at the transmitter. The receiving filter is only for channel selection purposes and is assumed to be flat with a bandwidth of 2B, which is sufficient to pass the signal modulation without distortion. The transmitted waveform is

$$x(t) = \sum_{k=-\infty}^{\infty} \mathbf{a}_k h(t-kT)\cos 2\pi f_0 t, \qquad (4.2\text{-}1)$$

where f_0 is the carrier frequency, f_s is the sampling frequency, $T = 1/f_s$ is the sampling interval, $\mathbf{a}_k$ are information digits, and $h(t)$ is a baseband waveform band limited to bandwidth B.

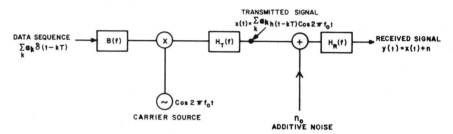

Figure 4.2-2   Double-sideband AM block diagram.

The received waveform is

$$y(t) = \sum_{k=-\infty}^{\infty} \mathbf{a}_k h(t-kT)\cos 2\pi f_0 t$$

$$+ n_c \cos 2\pi f_0 t - n_s \sin 2\pi f_0 t, \qquad (4.2\text{-}2)$$

where n_0 and n_s are the in-phase and quadrature noise components whose power is controlled by the front-end filter shape.

The requirement on $h(t)$ is that it be free of intersymbol interference, that is,

$$h(kT) = \begin{cases} 1 & k=0 \\ 0 & k\neq 0 \end{cases}. \qquad (4.2\text{-}3)$$

According to Nyquist's theorem,[25] any filter $H(f)$ that satisfies

$$\sum_{k=-\infty}^{\infty} H(f+kf_s) = T, \qquad |f| \leqslant \frac{f_s}{2} \qquad (4.2\text{-}4)$$

would satisfy Eq. (4.2-3). Here $H(f)$ is the Fourier transform of the filter impulse response $h(t)$. The filter that satisfies Eq. (4.2-4) and has a minimum bandwidth is

$$H(f) = \begin{cases} T, & |f| \leqslant \dfrac{f_s}{2} \\ 0, & \text{otherwise} \end{cases} . \tag{4.2-5}$$

The corresponding impulse response is

$$h(t) = \frac{\sin \pi t / T}{\pi t / T} . \tag{4.2-6}$$

Besides the practical difficulty of realizing the rectangular filter prescribed by Eq. (4.2-5), it is seen that since the tail of $h(t)$ dies off as $1/t$, the system would be very susceptible to intersymbol interferences caused by channel delay distortions or deviations of sampling timing.

It is therefore customary to widen the bandwidth of $H(f)$ to obtain a filter with smoother cutoff and hence faster roll-off of $h(t)$. The filter bandwidth B is usually chosen anywhere from $f_s/2$ to f_s. For example, a fully raised cosine filter is given by

$$H(f) = \begin{cases} \dfrac{T}{2}\left[1 + \cos \dfrac{2\pi f}{f_s}\right], & |f| < f_s \\ 0, & \text{otherwise} \end{cases} , \tag{4.2-7}$$

$$h(t) = \frac{\sin \pi t / T}{\pi t / T} \frac{\cos \pi t / T}{1 - 4t^2 / T^2} . \tag{4.2-8}$$

Here we note that $h(t)$ dies off as $1/t^3$. The responses of Eqs. (4.2-6) and (4.2-8) are shown in Figure 4.2-3. It is observed that the raised cosine shape has the additional feature that

$$h\left(\pm \frac{T}{2}\right) = \frac{1}{2},$$

$$h(\pm 1.5T) = h(\pm 2.5T) = \cdots + 0. \tag{4.2-9}$$

Therefore, the values of $\sum_k a_k h(t - kT)$ at $t = kT + T/2$ are also free of intersymbol interferences from all other pulses besides the one immediately following. This proves to be convenient in the clock recovery of on-off AM systems.

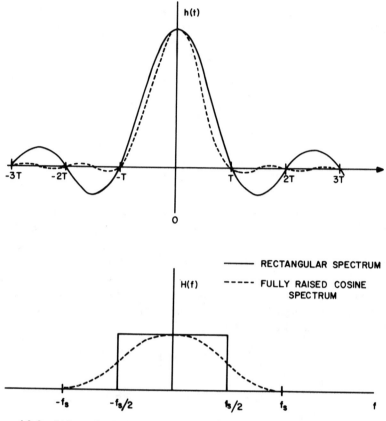

Figure 4.2-3 $h(t)$ and $H(f)$ of rectangular spectrum and fully raised cosine spectrum.

The required RF bandwidth for double-sideband AM systems is from f_s to $2f_s$. Due to the time delay spread in the mobile radio environment, to be discussed in Section 4.2.4, the choice of $2f_s$ is probably a necessity.

When the information digits $\mathbf{a}_k$ assume values of ± 1, we have binary AM or two-phase PM. The received waveforms can be detected coherently to determine the value of $\mathbf{a}_k$, that is, by multiplying Eq. (4.2-2) by $\cos 2\pi f_0 t$ and passing it through a low-pass filter to obtain the baseband output,

$$z(t) = \sum_{k=-\infty}^{\infty} \mathbf{a}_k h(t-kT) + n_c(t). \qquad (4.2\text{-}10)$$

At the sampling instant, that is, $t=kT$, we have

$$z(kT) = \mathbf{a}_k + n_c(kT). \qquad (4.2\text{-}11)$$

Since n_c is equally likely to be positive or negative, the decision threshold is set at zero; that is, if $z(kT)>0$ we decide that $\mathbf{a}_k = 1$. The error probability is

$$P_e = P\{1 + n_c < 0\}. \tag{4.2-12}$$

From Eq. (1.1-15), we obtain

$$\text{Binary AM:} \quad P_e = \frac{1}{2}\left[1 - \text{erf}\left(\frac{1}{\sqrt{2}\sigma}\right)\right] = \frac{1}{2}\,\text{erfc}\sqrt{\rho_p}, \tag{4.2-13}$$

where the noise power is

$$\sigma^2 = \langle n_c^2 \rangle = \langle n_s^2 \rangle \tag{4.2-14}$$

and the SNR at the sampling instant is

$$\rho_p = \frac{1}{2\sigma^2}. \tag{4.2-15}$$

The error rate P_e is presented in curve 1 of Figure 4.2-4 as a function of ρ_p. Coherent detection requires carrier phase recovery, which in the mobile radio environment with rapid phase fluctuations is not a simple task. An alternate detection scheme is differential phase shift keying (DPSK) detection, which uses the delayed signal as a phase reference. Detection of DPSK is accomplished by delaying $y(t)$ by one bit, multiplying it with the original waveform, and passing the resultant wave through a low-pass filter. This yields the baseband output

$$
\begin{aligned}
z(t) = &\left\{\left[\sum_k \mathbf{a}_k h(t-kT) + n_c(t)\right]\left[\sum_k \mathbf{a}_k h(t-kT-T) + n_c(t-T)\right]\right. \\
&\left. + n_s(t)n_s(t-T)\right\}\cos\omega_0 T \\
&-\left\{\left[\sum_k \mathbf{a}_k h(t-kT) + n_c(t)\right]n_s(t-T)\right. \\
&\left. -\left[\sum_k \mathbf{a}_k h(t-kT-T) + n_c(t-T)\right]n_s(t)\right\}\sin\omega_0 T. \tag{4.2-16}
\end{aligned}
$$

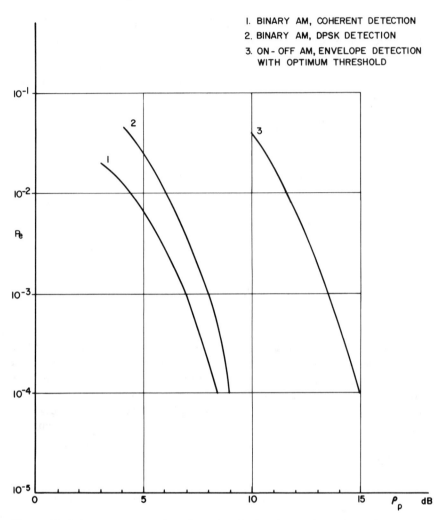

Figure 4.2-4 Error rate as a function of peak signal-to-noise ratio at the sampling instant.

At the sampling instant $t = kT$, and setting $\omega_0 T = 2n\pi$, we have

$$z(kT) = [\mathbf{a}_k + n_c(kT)][\mathbf{a}_{k-1} + n_c(kT - T)]$$

$$+ n_s(kT)n_s(kT - T). \tag{4.2-17}$$

The signal portion is given by $(\mathbf{a}_k)(\mathbf{a}_{k-1})$; it equals $+1$ if there is no sign change of two consecutive symbols and -1 if there is a sign change. For $\mathbf{a}_k$'s equally likely to be ± 1, the error probability is

$$P_e = P\{z(kT) < 0 | \mathbf{a}_k \mathbf{a}_{k-1} = 1\}, \tag{4.2-18}$$

where $P\{A|B\}$ is the probability of A under the condition B and is given by[26]

$$\text{DPSK:} \qquad P_e = \tfrac{1}{2} e^{-\rho_p} \tag{4.2-19}$$

where ρ_p is defined by Eq. (4.2-15).

The error rate is presented in curve 2 of Figure 4.2-4. It is seen that DPSK is inferior to coherent detection by a fraction of a dB at high ρ_p, but at low ρ_p the difference becomes bigger.

When the information digits $\mathbf{a}_k$ assume values unity or zero instead of ± 1, we have on-off AM. Referring to Eq. (4.2-2), the received wave can be detected by envelope detection. For example, assume we square $y(t)$ and pass it through a low-pass filter. The baseband output is

$$z(t) = \frac{1}{2} \left[\sum_k \mathbf{a}_k h(t - kT) + n_c \right]^2 + \frac{1}{2} n_s^2. \tag{4.2-20}$$

At the sampling instant $t = kT$, we have

$$z(kT) = \tfrac{1}{2} [\mathbf{a}_k + n_c]^2 + \tfrac{1}{2} n_s^2. \tag{4.2-21}$$

In order to determine $\mathbf{a}_k$ we may set a threshold q such that we decide

$$\mathbf{a}_k = 1 \quad \text{if} \quad z(kT) > q,$$

$$\mathbf{a}_k = 0 \quad \text{if} \quad z(kT) < q.$$

The optimum value q_{op} can be determined by varying q to make the error probability

$$P_e = \tfrac{1}{2} P\{z(kT) < q | \mathbf{a}_k = 1\} + \tfrac{1}{2} P\{z(kT) > q | a_k = 0\} \tag{4.2-22}$$

minimum. The reader is referred to Ref. 28 for a detailed analysis. It is sufficient to say here that q is a function of ρ_p and is approximately given by

$$q_{op} = \sigma \sqrt{2 + \frac{\rho_p}{2}}. \tag{4.2-23}$$

The error probability with q_{op} can be numerically calculated for different values of ρ_p and is shown in curve 3 of Figure 4.2-4.

We note that in all previous discussions the CNR ρ_p is defined as the peak carrier power to noise ratio after the receiver filter and the receiver filter is assumed to be flat. From the system design point of view we are more interested in comparing systems on the basis of required transmitter power. The use of a flat receiving filter is certainly not optimum in a noisy environment from the signal detection point of view. Referring to Figure 4.2-2, since the system is linear up to $y(t)$, it should be possible for us to divide the filters between the transmitter and receiver as depicted in Figure 4.2-5. For $y(t)$ to be free of intersymbol interference, we still require $h(t)$ to satisfy Eq. (4.2-3), which implies

$$B(f)H_{Tb}(f)H_{Rb}(f) = H(f), \qquad (4.2\text{-}24)$$

where

$$H_{Tb}(f) = H_T(f+f_0)$$

is the equivalent baseband filter of $H_T(f)$,

$$H_{Rb}(f) = H_R(f+f_0),$$

$H(f)$ is the spectrum of any $h(t)$ which satisfies Eq. (4.2-4).

The received noise power is

$$\sigma^2 = \int_0^\infty N_0(f)|H_R(f)|^2 df, \qquad (4.2\text{-}25)$$

where $N_0(f)$ is the one-sided power density of the front-end noise. The transmitted power is

$$P_T = \left\langle \lim_{M\to\infty} \frac{1}{2M} \int_{-M}^M \left[\sum_k a_k h'(t-kT)\cos\omega_0 t \right]^2 dt \right\rangle_{av}. \qquad (4.2\text{-}26)$$

The noise power σ^2 can be minimized while keeping P_T constant and Eq. (4.2-24) satisfied by varying $H_{Rb}(f)$ and $B(f)H_{Tb}(f)$. For white Gaussian noise, the result is extremely simple[25]:

$$H_{Rb}(f+f_b) = \alpha|H(f)|^{1/2}, \qquad (4.2\text{-}27)$$

$$B(f)H_{Tb}(f+f_b) = \frac{1}{\alpha}\frac{H(f)}{|H(f)|^{1/2}}, \qquad (4.2\text{-}28)$$

where α is an arbitrary constant.

Let us define

$$\rho = \frac{\text{average signal power before receiver filter}}{\text{front-end thermal noise power in bandwidth } f_s}$$

$$= \frac{P_T}{N_0 f_s} . \tag{4.2-29}$$

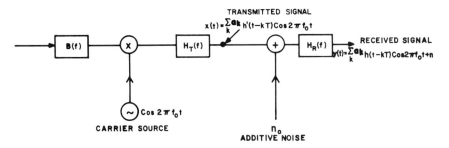

Figure 4.2-5   Double-sideband AM with optimum filter division.

With optimum filtering distribution prescribed by Eqs. (4.2-27) and (4.2-28), ρ is related to ρ_p by[26]

$$\rho = \begin{cases} \rho_p, & \text{binary AM} \\ \frac{1}{2}\rho_p, & \text{on-off AM} \end{cases} \tag{4.2-30}$$

Assuming optimum filter division, the error probabilities P_e of on-off AM with square law detection, binary AM with coherent, or DPSK detection are replotted in Figure 4.2-6 as functions of ρ. This then presents a comparison of the three system on the average transmitter power basis if there is no path loss. It is to be noted that in the on-off AM case, since the carrier is off half the time the transmitted peak power may far exceed the average power. The ratio of peak to average power depends on the specific shaping $H(f)$. For example, for a raised cosine spectrum given by Eq. (4.2-7),

$$H'(f) = \sqrt{T \cos^2 \frac{\pi f}{f_s}}$$

and the peak power is given by

$$P_{Tp} = \tfrac{1}{2} h'(0)^2$$

$$= \frac{1}{2} \left[\int_{-f_s}^{f_s} \sqrt{T} \cos \frac{\pi f}{f_s} \, df \right]^2 = \frac{8}{\pi^2} \frac{1}{T}. \tag{4.2-31}$$

The average power given by (4.2-26) can be obtained by noticing that

$$\int_{-\infty}^{\infty} h'(t - kT) h'(t - lT) \, dt = h((k - l)T)$$

$$= \begin{cases} 1, & k = l \\ 0, & k \neq l \end{cases}. \tag{4.2-32}$$

Therefore, we have

$$P_T = \frac{1}{4T}, \tag{4.2-33}$$

and the peak power given by Eq. (4.2-31) is 5.1 dB higher than the average transmitter power. In peak transmitter power limited systems, the binary on-off modulation thus suffers an additional loss of 5.1 dB. A similar calculation shows that binary AM suffers only 2.1 dB. Thus, there is an additional 3 dB difference between the two systems in a peak power limited situation.

Digital information can also be transmitted by FM, commonly known as FSK. The general waveforms are

$$x(t) = A_0 \cos [2\pi f_0 t + \theta(t)], \tag{4.2-34}$$

where

$$\theta(t) = 2\pi f_d \int_0^t s(t) \, dt \tag{4.2-35}$$

The baseband modulation $s(t)$ is given by

$$s(t) = \sum_k \mathbf{a}_k u(t - kT), \tag{4.2-36}$$

where

$$u(t) = \begin{cases} 1, & 0 \leqslant t \leqslant T \\ 0, & \text{otherwise} \end{cases}$$

and

$$\mathbf{a}_k = \pm 1.$$

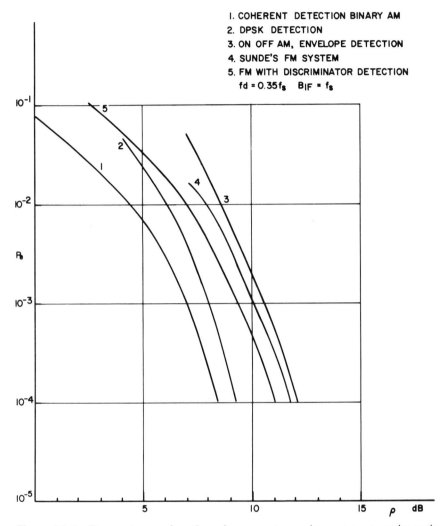

1. COHERENT DETECTION BINARY AM
2. DPSK DETECTION
3. ON OFF AM, ENVELOPE DETECTION
4. SUNDE'S FM SYSTEM
5. FM WITH DISCRIMINATOR DETECTION
   fd = 0.35fₛ   B_IF = fₛ

Figure 4.2-6 Error rate as a function of average transmitter power to noise ratio, $\rho = P_T / N_{0 f_s}$.

The peak frequency deviation is f_d. It is clear that $x(t)$ is no longer band limited. Calculations show[31, 32] that for small deviations, that is, $f_d \leqslant f_s/4$, the spectrum of $x(t)$ is essentially contained in a narrow RF bandwidth of $1.5 f_s$. A narrow-band system has been described by Salz[32, 33] with $f_d = f_s/4$ and a bandwidth of $1.5 f_s$. The transmitted wave was no longer strictly FM but had small amplitude fluctuations. With conventional limiter discriminator detection and optimum division of filters between transmitter and receiver it was reported that at a high carrier-to-noise ratio such a system theoretically suffers only 0.25-dB loss in comparison to ideal coherent detection of binary waveforms. Experimentally it was reported to be about 1 dB. Tjhung[34] has reported elaborate computer simulations considering the intersymbol interference due to the effect of limited bandwidth. He reported system performances about 3 dB poorer than coherent detection of binary AM. Tjhung's results are obtained by filtering at the receiver end; if optimum filter division between the transmitter and receiver is made, an improvement of 1–2 dB may be expected and would be close to the results reported by Salz. Tjhung's results with $f_d = 0.35 f_s$ and RF bandwidth equal to f_s are shown in curve 5 of Figure 4.2-6.

Sunde[27] has reported a minimum bandwidth FM, which is derived by linear superposition of double-sideband AM and carrier wave under the constraint that the discriminator output be free of intersymbol interference. The peak frequency deviation is $f_d = f_s/2$ and the required bandwidth is from f_s to $2f_s$ depending on how fast one wants the detected baseband signal to die off. The performance of such a system with optimum filter divisions is shown in curve 4 of Figure 4.2-6, where a fully raised cosine spectrum is used for the double-sideband AM.

4.2.3 Rayleigh Fading

If the time delay spread of the mobile environment is small in comparison to the signaling bandwidth, the received signal would only be corrupted by the multiplicative fading process. With the assumption that the fading is slow in comparison to the receiver front-end filter bandwidth so that the filter does not introduce distortions on the fading envelope,* we have, after the receiving filter, the received digital AM wave:

$$y(t) = r(t) \sum_k \mathbf{a}_k \big(h(t - kT) \cos[\omega_c t + \phi(t)] \big)$$

$$+ n_c \cos \omega_c t - n_s \sin \omega_c t,$$

(4.2-37)

where $r(t)$ is the Rayleigh fading envelope and $\phi(t)$ is the random phase fluctuation.

---

*Bello and Nelin[35] have considered the case when the front-end filter may also distort the fading for ideal square pulse DPSK and FSK. It is found that the representation (4.2-37) would provide an accurate estimate of error rate for the Gaussian type of correlation function, but not for an exponential correlation of fading.

The peak signal-to-noise ratio at sampling instant, instead of Eq. (4.2-15), is

$$\rho_p = \frac{r^2}{2\sigma^2}. \tag{4.2-38}$$

Defining the average signal-to-noise ratio before the receiving filter by

$$\rho = \frac{r^2 P_T}{N_0 f_s}, \tag{4.2-39}$$

ρ is still related to ρ_p by Eq. (4.2-30). Since r is Rayleigh, the probability density function of ρ is

$$f(\rho) = \begin{cases} \dfrac{1}{\rho_0} e^{-\rho/\rho_0}, & \rho \geqslant 0 \\ 0, & \text{otherwise} \end{cases} \tag{4.2-40}$$

and

$$\rho_0 = \frac{\langle r^2 \rangle_{av} P_T}{N_0 f_s} \tag{4.2-41}$$

is the average carrier-to-noise ratio averaged over the Rayleigh fading with noise bandwidth f_s.

For on-off AM, since the detection is performed by squaring Eq. (4.2-37), the random phase $\phi(t)$ disappears in the baseband output. The only effect of the fading is the fluctuation of $r(t)$, which produces varying ρ_p and ρ. The average error rate may be obtained by

$$P_1 = \int_0^\infty P_e(\rho) f(\rho) \, d\rho, \tag{4.2-42}$$

where $f(\rho)$ is given by Eq. (4.2-40), and $P_e(\rho)$ is shown in curve 3 of Figure 4.2-6,

Here we use notation P_1 to represent the error rate caused by the Rayleigh envelope fading. The resultant P_1 is presented in curve 3A of Figure 4.2-7 as a function of ρ_0. We note that in the on-off AM case, the optimum threshold q_{op} is a function ρ_p, the instantaneous peak carrier-to-noise ratio. This implies that q_{op} would have to be varied continuously according to the instantaneous fading envelope. In other words q_{op} is an optimum moving threshold. We may settle with a less optimal but much simpler strategy by selecting an optimum fixed threshold q'_{op}. The requirement now is that q'_{op} only respond to ρ_0 and does not change according to the instantaneous fading. With the above constraint, for each ρ_0 a q'_{op} can be found that minimizes P_1. The P_1 obtained[28] by using q'_{op} is presented in curve $3B$ of Figure 4.2-7 as a function of ρ_0. It is seen that under the Rayleigh fading condition the moving threshold system is considerably better than the fixed threshold system.

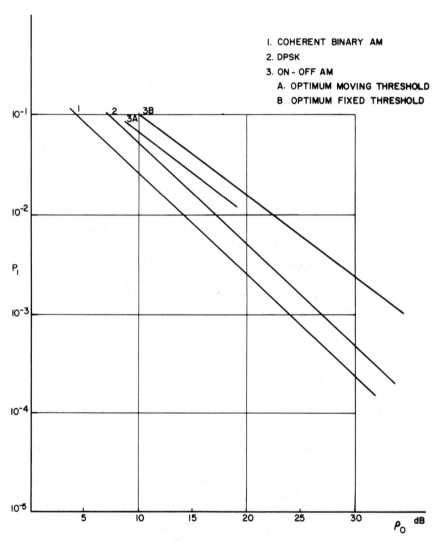

Figure 4.2-7 Error rate due to Rayleigh envelope fading.

For binary AM, if $\phi(t)$ of Eq. (4.2-37) is known exactly, coherent detection may be implemented. An integration over the Rayleigh density gives

$$P_1 = \int_0^\infty P_e(\rho) f(\rho) \, d\rho$$

$$= \int_0^\infty \frac{1}{2} (\text{erfc} \sqrt{\rho}) \frac{1}{\rho_0} e^{-\rho/\rho_0} d\rho$$

$$= \frac{1}{2} \left[1 - \frac{1}{\sqrt{1 + \frac{1}{\rho_0}}} \right] \qquad (4.2\text{-}43)$$

P_1 as a function of ρ_0 is shown in curve 1 of Figure 4.2-7, and this represents the theoretically best performance we can obtain in a Rayleigh fading environment. In practice, difficulties center around the carrier recovery techniques. For example, a common practice in carrier recovery is to square Eq. (4.2-37) and pass it through a narrow band filter centered around $2\omega_c$ to recover the carrier wave $\cos[2\omega_c t + 2\phi(t)]$. This wave is then frequency divided to obtain $\cos[\omega_c t + \phi(t) + n\pi]$. There is a phase ambiguity of π in the recovered carrier. To overcome this phase ambiguity, the transmitted digits would have to be differentially coded. Consequently each single error would become double errors, and the results of Eq. (4.2-43) would have to be multiplied by 2. The error performance then is about equal to that of DPSK system to be discussed next. Other carrier recovery techniques such as the use of a CW pilot, would require additional circuitry as well as additional pilot powers.

For DPSK detection of binary AM [Eq. (4.2-37)], the sampled value after the delay and multiplication process is

$$z(0) = \{ a_0 r(0) \cos[\phi(0)] + n_c(0) \} \{ a_{-1} r(-T) \cos[\phi(-T)] + n_c(-T) \}$$

$$+ \{ a_0 r(0) \sin[\phi(0)] - n_s(0) \} \{ a_{-1} r(-T) \sin[\phi(-T)] - n_s(-T) \}. \qquad (4.2\text{-}44)$$

The decision rules are

$$a_0 a_{-1} = 1 \quad \text{if} \quad z(0) > 0, \qquad (4.2\text{-}45)$$

$$a_0 a_{-1} = -1 \quad \text{if} \quad z(0) < 0. \qquad (4.2\text{-}46)$$

We note that since $r \cos \phi$ and $r \sin \phi$ are Gaussian with known correlation functions the conditional probability of $z(0)$ can be calculated. Assuming the noises are uncorrelated with their delayed version, Voelker[36] obtained the following expression for the error probability:

$$P_e = \frac{1 + \rho_0(1-g)}{2(\rho_0 + 1)} \qquad (4.2\text{-}47)$$

where ρ_0 is defined by Eq. (4.2-41) and $g = J_0(2\pi f_m T)$ is the autocorrelation of the in-phase or quadrature component of the complex Gaussian fading shifted by time T given by Eq. (1.3-7).

Note that Eq. (4.2-47) can be separated into two parts,

$$P_e = P_1 + P_2, \qquad (4.2\text{-}48)$$

where

$$P_1 = \frac{1}{2(\rho_0 + 1)}, \qquad (4.2\text{-}49)$$

$$P_2 = \frac{\rho_0(1-g)}{2(\rho_0 + 1)} \rightarrow \frac{1-g}{2} \quad \text{for} \quad \rho_0 \gg 1. \qquad (4.2\text{-}50)$$

P_1 is the error due to Rayleigh amplitude fading and can be obtained by integration of Eq. (4.2-19) over the Rayleigh density. In curve 2 of Figure 4.2-7, P_1 is shown as a function of ρ. For any reasonable error rate, P_2 quickly approaches its limit value given by $\rho \to \infty$, that is, $P_2 = (1-g)/2$. This is the irreducible error due to random FM. It is highly suggestive that for small error rates the errors produced by different mechanisms such as amplitude variation, phase variation, and time delay spread may be additive. The analysis of the system performance can then be greatly simplified by considering them separately.[27] Substituting $g = J_0(2\pi f_m T)$ into Eq. (4.2-50) the irreducible error rate due to random FM is approximately given by

$$P_2 \cong \frac{1}{2}\left(\frac{\pi f_m}{f_s}\right)^2, \qquad (4.2\text{-}51)$$

where

$$f_s = \frac{1}{T}$$

for small values of $2\pi f_m T$. For UHF mobile radio with $f_m = 100$ Hz, $f_s = 30$ kHz, P_2 is about 6×10^{-5} and is certainly negligible. Nevertheless, at higher frequencies, for example at 20 GHz, P_2 is about 2.4×10^{-2} for mobile speeds up to 60 mile/hr, and may no longer be negligible.

For FM, except for the case of an idealized square pulsed FSK System,[35] there are very few reports on the performance of discriminator detection schemes under fading conditions. The average error rate due to envelope fading alone may be obtained through an integration of the instantaneous error rates in curves 4 and 5 of Figure 4.2-6 over the Rayleigh density. Difficulties arise because of lack of knowledge of how P_e behaves at small P. By examining curve 5 of Figure 4.2-6, we notice that the error rate curve is horizontally displaced from the coherent detection curve by 2–3 dB even at low values of ρ. Thus we may expect as an approximation that in the fading case FM with discriminator detection would also require about 3 dB higher ρ_0. This would then put FM systems in the same category as DPSK in a fading environment.

Let the peak deviation of the FM system be f_d. The irreducible error due to random FM when $\rho \to \infty$ is given by

$$P_2 = P\left\{[2\pi f_d + \dot{\phi}] \leqslant 0\right\}$$

$$= P\left\{\dot{\phi} \leqslant -2\pi f_d\right\}. \tag{4.2-52}$$

The cumulative distribution of $\dot{\phi}$ is given by Eq. (1.4-5); thus we have

$$P_2 = \frac{1}{2}\left[1 - \sqrt{2}\frac{f_d}{f_m}\left(1 + 2\frac{f_d^2}{f_m^2}\right)^{-1/2}\right]. \tag{4.2-53}$$

For large f_d/f_m ratio we have

$$P_2 \cong \frac{1}{8}\left(\frac{f_m}{f_d}\right)^2. \tag{4.2-54}$$

Comparing Eq. (4.2-54) with (4.2-51) we can show that

$$P_{2_{\text{FM}}} = \frac{1}{4\pi^2}\left(\frac{f_s}{f_d}\right)^2 P_{2_{\text{DPSK}}}. \tag{4.2-55}$$

Thus FM is much more capable of coping with random FM than DPSK. For example with $f_d = \frac{1}{2}f_s$, there is almost a factor of 10 difference in the irreducible error rate. Such behavior has been reported experimentally[37] and is also found in the theoretical analysis by Bello and Nelin.[35]

4.2.4   Frequency-Selective Fading

When the time delay spread is no longer negligible in comparison to the modulation bandwidth, the received waveforms will be distorted due to different path length differences associated with the multipath incoming waves. This distortion causes intersymbol interference in the detection process.

Using complex notation, the transmitted wave can be written as

$$x(t) = R_e \{ \tilde{z}(t) e^{j\omega_0 t} \}, \qquad (4.2\text{-}56)$$

where the tilde indicates complex envelope. The received waveform similar to Eq. (4.1-108) is

$$y(t) = R_e \{ \tilde{E}(t) e^{j\omega_0 t} \} + n(t). \qquad (4.2\text{-}57)$$

Here $\tilde{E}(t)$ is related to the Fourier transform of $\tilde{z}(t)$ by

$$\tilde{E}(t) = \int_{-\infty}^{\infty} H_R(f + f_0) T(f, t) Z(f) e^{j2\pi f t} \, df, \qquad (4.2\text{-}58)$$

where $T(f, t)$ is the time-varying channel transfer function defined by (4.1-111), and $H_R(f)$ is the front-end filter at the receiver. Let us define a Fourier transform pair:

$$U(f) = Z(f) H_R(f + f_0), \qquad (4.2\text{-}59)$$

$$u(t) = \int_{-\infty}^{\infty} U(f) e^{j2\pi f t} \, df. \qquad (4.2\text{-}60)$$

As in Section 4.1.8, if the frequency selective fading is sufficiently small that $T(f, t)$ varies slowly in f over the range of $U(f)$; that is, the signal spectrum, a Taylor series expansion of $T(f, t)$ around $T(0, t)$ could be made and (4.2-58) becomes

$$\tilde{E}(t) = T_0 u(t - t_0) + T_1 \dot{u}(t - t_0) + T_2 \ddot{u}(t - t_0) + \cdots, \qquad (4.2\text{-}61)$$

where the $T_k(t)$ are defined in Eqs. (4.1-114–4.1-116) and are completely specified by the channel delay spread or frequency correlation functions, and t_0 is the average time delay.

In the case of digital AM, the receiving filter is matched to the transmitting filter, and we have

$$u(t) = \sum_k a_k h(t - kT), \qquad (4.2\text{-}62)$$

where $h(t)$ satisfies zero intersymbol interference criteria.

Therefore, Eq. (4.2-61) becomes, at $t = t_0$, that is, the sampling instant,

$$\tilde{E}(t_0) = T_0(t_0)\mathbf{a}_0 + T_1(t_0) \sum_{k=-\infty}^{\infty} \mathbf{a}_k \dot{h}(kT)$$

$$+ T_2(t_0) \sum_{k=-\infty}^{\infty} \mathbf{a}_k \ddot{h}(kT) + \cdots \tag{4.2-63}$$

where a dot over a character indicates the time derivative.

In Eq. (4.2-63) we see that the first term $T_0(t_0)\mathbf{a}_0$ is the desired envelope corrupted by the Rayleigh fading. The rest of the terms in Eq. (4.2-63) represent intersymbol interferences. Since T_1, T_2, are independent of T_0, these interferences make random contributions to the value of the first term. The shaping of $h(t)$ is quite important. It can be seen from Eq. (4.2-63) that if $h(t)$ dies off quickly, so will $\dot{h}$ and $\ddot{h}$, and hence the amount of intersymbol interference would be reduced.

Neglecting the variation of $T_k(t)$ as a function of t, that is, the case of a stationary vehicle, Bello and Nelin[38] have calculated the error probability due to frequency selective fading and Gaussian noise for square-pulse FSK and PSK systems with diversity combining. Due to their assumption of an ideal square pulse, only intersymbol interference from the immediately adjacent symbols is considered. Bailey and Lindenlaub[39] have extended Bello's results to DPSK detection of binary coding with fully-raised-cosine shaping for Gaussian delay spread density and rectangular delay spread density, considering only the interferences from immediately adjacent symbols. The two channel frequency correlations Bailey and Lindenlaub used are

$$R_T(f) = \langle T(\Omega)T^*(\Omega+f) \rangle_{av} = 2\sigma^2 \exp\left(-\frac{4f^2}{B_c^2}\right), \tag{4.2-64}$$

and

$$R_T(f) = 2R_0 T_m \frac{\sin 2\pi f T_m}{2\pi f T_m}. \tag{4.2-65}$$

The corresponding delay power density spectrums are the inverse Fourier transforms of $R_T(f)$; for the Gaussian delay spread we get

$$P(\tau) = \sigma^2 \sqrt{\pi}\, B_c \exp\left[-\left(\frac{\pi B_c \tau}{2}\right)^2\right] \tag{4.2-66}$$

and for the rectangular delay spread:

$$P(\tau) = \begin{cases} R_0 & -T_m \leqslant \tau \leqslant T_m \\ 0 & \text{otherwise.} \end{cases} \tag{4.2-67}$$

The mean square delay spreads defined by

$$\Delta^2 = \frac{\int_{-\infty}^{\infty} \tau^2 P(\tau)\, d\tau}{\int_{-\infty}^{\infty} P(\tau)\, d\tau}, \tag{4.2-68}$$

are, for the Gaussian,

$$\Delta^2 = \frac{2}{\pi^2 B_c^2}, \tag{4.2-69}$$

and for the rectangular,

$$\Delta^2 = \frac{T_m^2}{3}. \tag{4.2-70}$$

Results due to Bailey and Lindenlaub using square pulsing and DPSK detection through a Gaussian channel are presented in Figure 4.2-8 as a function of ρ_0 and the quantity d, defined as the product of the sampling frequency and rms delay spread

$$d = f_s \Delta. \tag{4.2-71}$$

It is observed that for small values of d, the error rate stays fairly constant and is dominated by the envelope fading. As d increases, the irreducible error due to frequency-selective fading becomes dominant, and can be approximated by

$$P_e = P_1 + P_3, \tag{4.2-72}$$

where P_3 is the irreducible error rate due to frequency-selective fading given by the curve $\rho_0 = \infty\,$dB, and P_1 is given by Eq. (4.2-49).

In Figure 4.2-9 the irreducible error rates P_3 of a square pulse through a Gaussian or a rectangular channel and fully raised cosine pulse through a rectangular channel are presented. Comparison of curves 1 and 2 immediately reveals that P_3 depends on the rms delay spread but is insensitive to the actual distribution of the delay spread. We may recall that similar behavior was also observed in analog FM systems subjected to frequency-selective fading reported in Section 4.1. However, pulse shaping does have some influence on P_3, as can be seen from the fact that fully cosine shaping can tolerate more delay spread in comparison to the square-pulse model. No reports on partial roll-off raised cosine spectra are available, although it could be expected to be worse than the fully raised cosine spectrum in coping with the frequency selective fading. As an example, the worst Δ reported in New York City around the downtown area is about 2.5 μsec.[40] For $f_s = 30$ kHz, d is 0.075 and from Figure 4.2-9 the irreducible error P_3 is 10^{-3} for fully raised cosine shaping and 2×10^{-3} for rectangular pulse. These values are too large to be acceptable. Nevertheless, diversity combining (reported in Section 6.7) can bring P_3 down to tolerable levels.

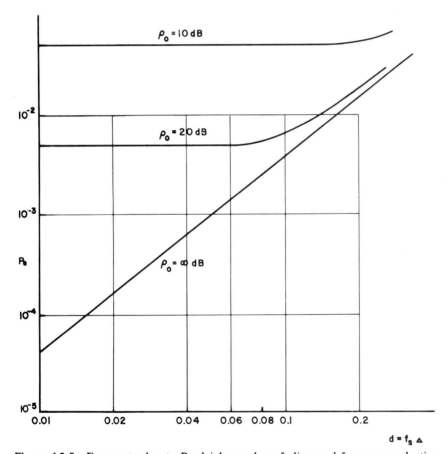

Figure 4.2-8 Error rate due to Rayleigh envelope fading and frequency-selective fading. (DPSK detection of square pulse through Gaussian channel).

Idealized FSK using rectangular pulse and matched filter detection has been reported by Bello and Nelin.[38] Their results with peak frequency deviation $f_d = f_s/2$ indicate that FSK is about equal to DPSK in a frequency-selective fading environment.

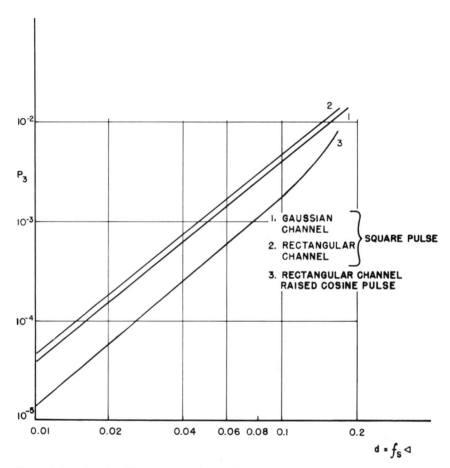

Figure 4.2-9 Irreducible error rate due to frequency-selective fading [DPSK detection].

4.3 CHANNEL MULTIPLEXING

Estimates of service requirements indicate that frequency allocations, even in the microwave bands, must be used extremely efficiently to provide service to a reasonable fraction of eligible users. Therefore, possible communications systems are carefully compared with particular emphasis on required spectrum space, even, perhaps, at the expense of required transmitter power. Spectrum conservation suggests the use of multiplex. Multiplex techniques have long been used to improve efficiency of spectrum usage between base stations and repeaters. However, multiplex is beset with several problems in the mobile radio environment, as will be described.

Frequency-division multiplex systems (as opposed to time-division multiplex) are most commonly used, and the most common frequency-division multiplex systems are FM/FM and SSB/FM. The notation A/B used to identify the frequency-division multiplex method denotes that the information is applied to the subcarriers by modulation type A to form subchannels. The subchannels are then added together and modulate the main carrier by modulation type B. For example, one would denote conventional FM or SSB channels with separate frequency slots as FM/SSB or SSB/SSB. FM/SSB and SSB/SSB were compared in Section 4.1.7. This section will compare the performance of the multiplex systems, FM/FM and SSB/FM, with the conventional FM channel arrangement, FM/SSB.

4.3.1 Description of FM/FM and SSB/FM

The RF spectrum of conventional base-station FM transmissions for mobile radio, FM/SSB, consists of separate subchannel FM spectra set side by side. As shown in Figure 4.3-1, these spectra may be separated to allow interleaving of FM/SSB channels of adjacent geographic cells. FM/SSB could be obtained by generating each of the FM subchannels independently at their specified RF center frequencies or by generating each FM subchannel at IF with the required frequency separations and translating the whole IF block of signals to the desired RF location by single-sideband modulation of an RF carrier of frequency f_c.

The RF spectrum of SSB/SSB consists of separate subchannel voice spectra, each with its reduced subcarrier on which the subchannel is single-sideband modulated, as shown in Figure 4.3-2. Again, the subchannels may be spaced to allow adjacent geographic cell signals to be interleaved, if desired.

Each subchannel of the SSB/FM system is a reduced amplitude subcarrier SSB-modulated by a voice signal. As shown in Figure 4.3-3, the power of each subchannel is weighted in proportion to the square of its baseband frequency, so that the signal-to-noise ratio after frequency modulation and demodulation will be the same for each subchannel, the same as the preemphasis used in conventional FM. This composite of SSB subchannels then frequency modulates the main carrier so that the signal transmitted from the base station is a single FM wave with a baseband bandwidth many times the normal voice bandwith.

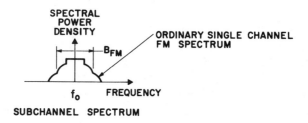

SUBCHANNEL SPECTRUM

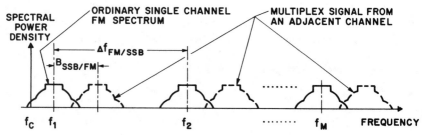

Figure 4.3-1   FM/SSB multiplex spectrum of M subchannels.

Each subchannel of the FM/FM system is a subcarrier, frequency modulated by its corresponding voice signal. As shown in Figure 4.3-4, the power of each subchannel is weighted in proportion to the square of its baseband frequency, so that the signal-to-noise ratio after frequency modulation and demodulation will be the same for each subchannel. This composite of FM subchannels then frequency modulates the main carrier so that the signal transmitted from the base station is, as in SSB/FM, a single FM wave with baseband bandwidth many times the normal voice bandwidth.

In many cases the lower-frequency portion of the noise spectrum after demodulation of the main carrier, when using SSB/FM or FM/FM, may be more flat than parabolic (due to noise in baseband oscillators, instabilities in carrier and local oscillators, threshold clicks, and so on). In such

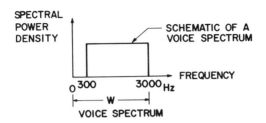

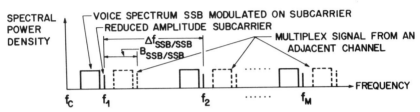

Figure 4.3-2   SSB/SSB multiplex spectrum of M subchannels.

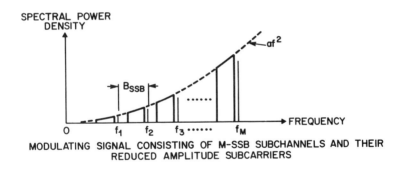

MODULATING SIGNAL CONSISTING OF M-SSB SUBCHANNELS AND THEIR
REDUCED AMPLITUDE SUBCARRIERS

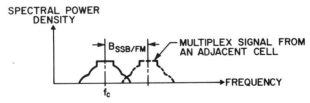

Figure 4.3-3   SSB/FM multiplex spectrum of M subchannels.

243

cases the subchannel weighting would be made flat over the spectrum for which the noise spectrum is flat and parabolic over the spectrum for which the noise spectrum is parabolic, in order to ensure an equal signal-to-noise ratio for each subchannel.

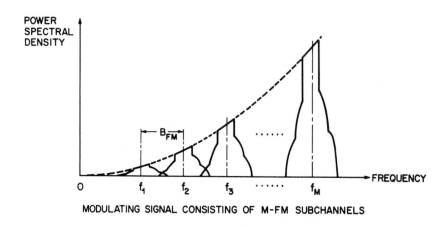

MODULATING SIGNAL CONSISTING OF M-FM SUBCHANNELS

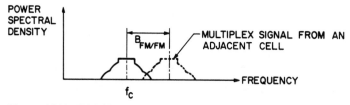

Figure 4.3-4   FM/FM multiplex spectrum of M subchannels.

We will restrict our discussion to SSB/FM and FM/FM, as compared with FM/SSB, since they are the common types of frequency-division multiplex that might be considered for use in mobile radio.

4.3.2   Required Transmitter Power

In this section we neglect the effects of fading and interference and compare the transmitter power requirements to provide an average output signal-to-noise ratio of 30 dB. Since the number of parameters affecting the performance of FM/FM, SSB/FM multiplex systems is large (modulation indices, number of channels, probability distribution of number of simultaneous users, allowable distortion, and so on), only representative systems[41-43] will be analyzed. This analysis will provide the basis for the reader to compute the performance of other system parameter choices of interest to him.

A basic factor affecting multiplex performance is the choice of subchannel spacing and the resulting subchannel filter bandwidths and overall bandwidth. The trade-off involved is noise bandwidth versus distortion. A good estimate of the distortion resulting from band-limiting an FM signal is provided by Anuff and Liou. Their formula[44] for the bandwidth required to provide a given signal-to-distortion ratio of spectral densities at the top baseband frequency is

$$B = 2\left\{ W\left[1 + 0.13\left(\frac{S/D}{20} \right) \right] + f_{\text{dev}}\left(\frac{S/D}{20} \right) \right\}, \qquad (4.3\text{-}1)$$

where B is the IF bandwidth, W is the top baseband frequency, f_{dev} is the rms frequency deviation, and S/D is the signal-to-distortion ratio in decibels. The above formula is an empirical fit to data obtained by Monte Carlo computer simulation of FM through a close approximation to a rectangular IF filter. Rice[45] has analytically verified Anuff and Liou's[44] results for rms frequency deviations up to about $f_{\text{dev}} \leqslant B/4$. Rice's analysis also indicates that spectral shaping of the baseband signal has only a small effect. Since the distortion tends to be worst at the top of the baseband frequencies, Eq. (4.3-1) can be used as a conservative estimate of the required bandwidth. As will be seen, Eq. (4.3-1) is less conservative than the usual criteria employed to estimate required bandwidth.

It is important to point out that Anuff and Liou's results and Rice's analysis in Ref. 45 compute cross-talk distortion (distortion produced at one baseband frequency due to signal at another baseband frequency), which occurs because the ideal filter eliminates portions of the FM spectrum that lie outside the filter bandwidth. This is in contrast to other treatments[46,47] that compute distortion mainly due to filter ripple inside the band. In other words we will assume the system filters have been well equalized so that the major portion of the distortion introduced by the filters is due to the elimination of FM spectral components outside the filter passband.

Perhaps the most common criterion used for determining FM bandwidth requirements is the Carson's rule formula:

$$B = 2(W + f_{\text{pk}}), \qquad (4.3\text{-}2)$$

where f_{pk} is the peak frequency deviation. The main objection to Carson's rule (4.3-2) is that it does not relate the bandwidth to the acceptable distortion, which should be the determining factor. Also, one must decide how the peak frequency deviation is determined from the modulation. The usual practice[41,43] is to assume the peak to average power ratio of a single voice signal is 10 dB, and the peak voltage of a combination of voice subchannels in a multiplex transmission is the sum of the peak voltages of the individual voice subchannels, plus the sum of the peak voltages of the sine wave subcarriers, if any. Thus from Figures 4.3-1, 4.3-3, and 4.3-4,

$$\text{FM/SSB:} \qquad f_{pk} = \sqrt{10}\, f_{dev}, \qquad (4.3\text{-}3)$$

where f_{dev} is the rms deviation of the voice modulation on the single FM carrier.

$$\text{SSB/FM:} \qquad f_{pk} \doteq \sqrt{10}\, m_v \sum_{i=1}^{M} \left(f_i - \frac{B_{SSB}}{2} + rf_i \right), \qquad (4.3\text{-}4)$$

where m_v is the rms modulation index of a voice subchannel (the same for all voice subchannels), r is the ratio of the peak modulation index of the subcarrier (the same for all subcarriers) to the peak modulation index of a voice channel ($\sqrt{10}\, m_v$), B_{SSB} is the frequency spacing of the subchannels, f_i is the frequency of each subcarrier, and M is the number of subchannels.

$$\text{FM/FM:} \qquad f_{pk} \doteq \sqrt{2}\, m_f \sum_{i=1}^{M} f_i, \qquad (4.3\text{-}5)$$

where m_f is the rms modulation index of an FM subchannel (the same for all subchannels) and f_i is the center frequency of each subchannel.

As the number of subchannels increases, the ratio of the peak frequency deviations defined in Eqs. (4.3-4) and (4.3-5) to their respective rms frequency deviations, f_{dev}, increases.

$$\text{SSB/FM:} \qquad f_{dev} = m_v \sqrt{ \sum_{i=1}^{M} \left[\left(f_i - \frac{B_{SSB}}{2} \right)^2 + 5r^2 f_i^2 \right] }, \qquad (4.3\text{-}6)$$

$$\text{FM/FM:} \qquad f_{dev} = m_f \sqrt{ \sum_{i=1}^{M} f_i^2 }. \qquad (4.3\text{-}7)$$

Although it is not necessarily true in all systems that the occupied subchannels in a multiplex transmission are those with the lowest subchannel frequencies, this is the usual[41-43] assumption for computing multiplex performance. Since click noise and random FM are much higher near $f = 0$, with phase demodulation, we will shift the band edge of the first subchannel by $B_{SSB} - W$ for SSB/FM and by $B_{FM}/2$ for FM/FM, respectively, up from $f = 0$. Thus,

$$\text{SSB/FM:} \qquad f_i \doteq i B_{SSB}, \qquad (4.3\text{-}8)$$

and

$$\text{FM/FM:} \qquad f_i \doteq i B_{FM}. \qquad (4.3\text{-}9)$$

Using (4.3-8) and (4.3-9), Eqs. (4.3-4) and (4.3-5) become

$$\text{SSB/FM:} \qquad f_{pk} = \sqrt{2.25} \, m_v B_{SSB} M [M + r(M+1)], \qquad (4.3\text{-}10)$$

$$\text{FM/FM:} \qquad f_{pk} = \sqrt{.5} \, m_f B_{FM} M (M+1), \qquad (4.3\text{-}11)$$

and Eqs. (4.3-6) and (4.3-7) become

$$\text{SSB/FM:} \qquad f_{dev} = \frac{m_v B_{SSB}}{\sqrt{12}} \sqrt{(1 + 5r^2)(4M^3 + 6M^2 + 2M) - 6M^2 - 3M}$$

$$(4.3\text{-}12)$$

$$\text{FM/FM:} \qquad f_{dev} = \frac{m_v B_{FM}}{\sqrt{12}} \sqrt{4M^3 + 6M^2 + 2M} \,. \qquad (4.3\text{-}13)$$

In Figure 4.3-5, the ratio f_{pk}/f_{dev} is plotted versus the number of subchannels, M, for SSB/FM and FM/FM.

It is interesting to compare the various bandwidth criteria versus the modulation index and the type of modulation (single voice channel or multiplex). Besides Anuff and Liou's formula, Eq. (4.3-1), and Carson's rule, Eq. (4.3-2) several other criteria that have been proposed will be compared. One proposal[2] recommends limiting the IF bandwidth to a value that would reject only those spectral components of an FM signal (with sinusoidal modulation at the top baseband frequency, W, and with the given rms deviation, f_{dev}) whose voltage amplitudes were less than 1% of the amplitude of the unmodulated carrier. (Note: This criterion becomes meaningless for indices greater than 89,496 since then no sidebands are greater than 1% of the unmodulated carrier.)

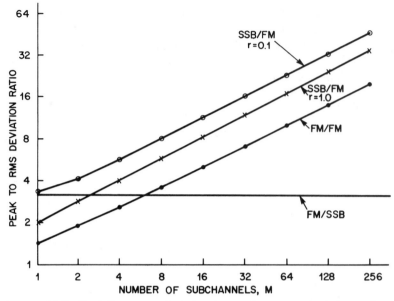

Figure 4.3-5   Peak frequency deviation to rms frequency deviation ratio.

Another proposal[43] limits the bandwidth to include 95% of the continuous (sideband) portion of the IF power spectrum of an FM signal with Gaussian phase modulation, the baseband spectrum of the phase being flat from zero to W Hz, and with the given rms frequency deviation, f_{dev}.

Wang[48] has calculated the distortion of sinusoidal modulation versus IF bandwidth and modulation index. He includes all harmonics of the modulation as distortion.

Figure 4.3-6 compares all of the above described bandwidth criteria: (a) Anuff and Liou's formula, (b) Carson's rule, (c) less than 1% sideband amplitude rule, (d) 95% of IF spectral power, and (e) Wang's sinusoidal modulation formula. In Anuff and Liou's formula and in Wang's sinusoidal modulation formula, a 40-dB signal-to-distortion ratio (SDR) is chosen so that the desired 30 dB overall signal-to-noise ratio for the mobile radio channel may be obtained without the IF filter distortion being a major factor. Carson's rule is plotted for single-channel and for 12-channel multiplex because it depends on the ratio of f_{pk} to f_{dev}, which in turn depends on the number of subchannels, as in Figure 4.3-5.

As seen in Figure 4.3-6, the 1% sideband amplitude criterion, Wang's 40-dB SDR for sinusoidal modulation, and Anuff and Liou's 40-dB SDR for Gaussian modulation agree quite closely. The Carson rule criterion, based on the peak deviation of single-channel Gaussian modulation and of 12-subchannel FM/FM and SSB/FM modulations, is seen to be far too conservative for large indices. The 95% sideband power criterion specifies too small a bandwidth and would thus result in a smaller SDR than 40 dB. This is to be expected, since Lynk[43] estimated the 95% sideband power criterion corresponded to about 5% baseband distortion, which is worse than the 40-dB SDR assumed for Anuff and Liou's and Wang's formulas.

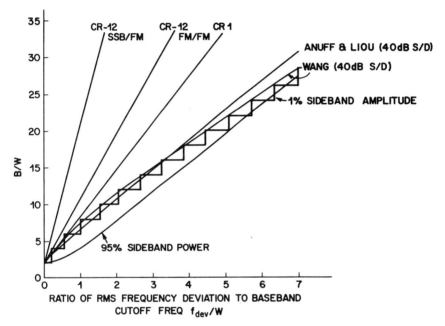

Figure 4.3-6   Comparison of bandwidth criteria.

Besides band-limiting distortion, the channel spacing and channel filters are dependent on distortion due to adjacent channels. In the multiplex case the adjacent channels will most likely not be located at the same base station. However, as a conservative estimate we will assume that the strength of the adjacent channels is equal to that of the desired channel at the receiver. Since the distortion is greatest at the top baseband frequency, we will require the distortion to be 40 dB below the signal at the top baseband frequency, so as to not be a major factor in the desired overall 30-dB SNR.

Lundquist[49] has computed adjacent-channel distortion based on Eq.

(4.1-45) for small modulation indices. Assuming two adjacent-channel interferers, one on either side of the desired channel, we show the channel spacing required for 40-dB SDR at the top baseband frequency in Figure 4.3-7. The calculation assumes no transmitter filter and assumes a receiver filter wide enough to provide 70 dB SDR due to band-limiting distortion (to allow comparison to Lundquist's results[49]). The adjacent-channel distortion is only slightly dependent on the presence of an equal bandwidth transmitter filter if the channel spacing is less than the filter bandwidth. Slightly smaller channel spacing results from assuming a filter bandwidth for 40-dB SDR due to band-limiting distortion. Note that the capture effect causes the adjacent-channel distortion to decrease with increasing index even though the spectra increasingly overlap. The main conclusion from Figure 4.3-7 is that no increase in channel spacing beyond the 40-dB SDR band-limiting-distortion bandwidth is required to provide 40-dB protection from adjacent-channel interference.

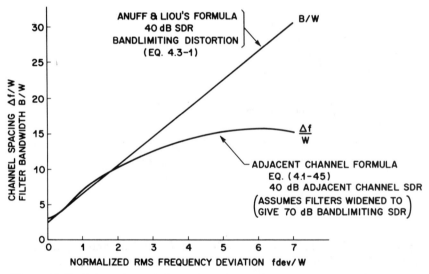

Figure 4.3-7 Comparison of channel spacing and bandwidth for 40-dB SDR.

This conclusion is in contrast to the results of Ref. 43, which recommends considerable increase in channel spacing to account for adjacent-channel interference. Lynk proposes[43] that satisfactory performance with respect to band-limiting distortion and adjacent-channel interference would be obtained if one spaced the adjacent channel at a frequency separation equal to (a) the frequency difference from the carrier at which the IF spectral power in a subchannel bandwidth is down 80 dB from the unmodulated carrier power plus (b) 1.25 times the half bandwidth that passes 95% of the continuous (sideband) portion of the IF spectrum. The bandwidth-widening factor of 1.25 was introduced to ensure that the filter would have negligible delay distortion over the 95% sideband spectrum width. Figure 4.3-8 compares the Anuff and Liou formula, at 40-dB SDR, and Lynk's results, both without the 1.25 bandwidth widening factor. (Lynk's curve assumes 12 subchannels.) Note that the 40-dB SDR criterion results in as much as a 35% narrower channel spacing than that proposed by Lynk.

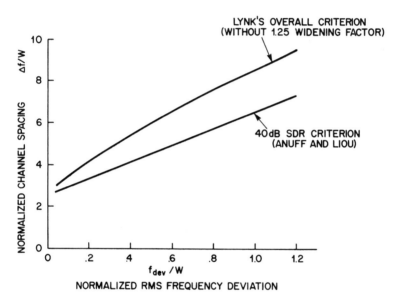

Figure 4.3-8   Comparison of channel spacing criteria.

In the following analyses, we will not add widening factors to allow for lack of delay equalization in filters or lack of frequency stability in channel oscillators, since these factors are highly dependent on the specific system design, and they can be introduced afterwards.

The above considerations outline the channel and subchannel frequency spacing requirements. We will use this information to compute the power required by each multiplex design to provide 30-dB SNR in the receiver output, assuming no fading.

In both SSB/FM and FM/FM the receiver first frequency demodulates the carrier and deemphasizes (i.e., phase demodulates the carrier). Assuming the demodulator is above threshold, the one-sided noise spectrum after the first demodulation process is obtained from Eqs. (4.1-13), (4.1-15), and (4.1-18):

$$\mathcal{W}_{\theta_n}(f) = \frac{2\eta}{Q^2} \frac{\text{rad}^2}{\text{Hz}}, \tag{4.3-14}$$

where Q is the amplitude of the received main carrier and η is the one-sided IF noise spectral density.

After the first demodulation the one-sided signal spectrum for SSB/FM is expressed as

$$\mathcal{W}_{\theta_s}(f) = \sum_{i=1}^{M} \mathcal{W}_v(f - f_i) m_v^2 \quad \text{(SSB/FM)}, \tag{4.3-15}$$

where m_v is the rms modulation index and $\mathcal{W}_v(f)$ is the one-sided normalized voice spectrum with top baseband frequency W_0,

$$\int_0^{W_0} \mathcal{W}_v(f) \, df = 1. \tag{4.3-16}$$

In SSB/FM, after the first demodulation, each subcarrier is filtered by a filter several orders of magnitude narrower than the voice signal bandwidth (to provide a local oscillator with negligible noise). It is then amplified for use as a local oscillator in a single-sideband mixer with the voice portion of its subchannel to shift the subchannel to its final output frequency (the natural voice frequency range). The resulting output SNR is the same as the SNR in the subchannel bandwidth after the first demodulation.

From Eqs. (4.3-14), (4.3-15), and (4.3-16), the output SNR is

$$\text{SSB/FM:} \quad \text{SNR}_{\text{out}} = \frac{m_v^2 Q^2}{2\eta W_0}. \tag{4.3-17}$$

Overall bandwidth is a more meaningful system parameter than the modulation index, m_v, and furthermore the condition (4.3-17) can be overridden by the requirement that the first angle demodulation remain above threshold (which in turn depends on overall bandwidth). We therefore compute the overall bandwidth for SSB/FM as follows. The subchannel bandwidth, B_{SSB}, is typically[41] chosen as $\frac{4}{3} W_0$ to facilitate SSB detection. Thus, from Eqs. (4.3-1) and (4.3-12), the overall IF bandwidth for a 40-dB signal to band-limiting distortion ratio is

$$\text{SSB/FM:} \qquad B_{40\text{ dB}} = 2\left\{ M\frac{4}{3}W_0(1.26) + m_v a W_0 \right\}, \qquad (4.3\text{-}18)$$

where

$$a \overset{\Delta}{=} \frac{4}{3\sqrt{12}}\sqrt{(1+5r^2)(4M^3+6M^2+2M)-6M^2-3M} \ . \qquad (4.3\text{-}19)$$

Thus

$$m_v = \frac{1}{2a}\left(\frac{B_{40\text{ dB}}}{W_0} - 3.36M \right). \qquad (4.3\text{-}20)$$

Equation (4.3-18) relates the overall bandwidth, B, to the modulation index, m_v, the number of subchannels, M, the subcarrier index ratio, r, and the voice bandwidth W_0. This relation is plotted in Figure 4.3-9. The minimum bandwidth obtainable for M subchannels is also shown in Figure 4.3-9.

The IF SNR is then given by

$$\text{SSB/FM:} \qquad \text{SNR}_{\text{IF}} = \frac{Q^2}{2\eta B} = \frac{Q^2}{2W_0[3.36M+2m_v a]} \ . \qquad (4.3\text{-}21)$$

We require that the SNR_{IF} remain above threshold as the IF bandwidth is varied. For the purpose of indicating the effect of various system parameters, it is sufficiently accurate to define threshold as $\text{SNR}_{\text{IF}} = 10$ dB. Then in order to remain above threshold the IF ratio of carrier power to noise in a voice bandwidth is

$$\text{SSB/FM:} \qquad \frac{Q^2}{2\eta W_0} > 10[3.36M+2m_v a] \quad \text{for threshold.} \quad (4.3\text{-}22)$$

As the IF bandwidth is increased, the required transmitter power is determined from Eq. (4.3-18) until threshold becomes the determining factor, at which point (4.3-22) takes over. As the bandwidth is increased beyond this point, the increasing transmitter power for threshold requirements results in an increasing baseband SNR_{out} over the 30 dB reference level.

The required IF carrier power is plotted versus bandwidth and number of subchannels in Figure 4.3-10. Figure 4.3-11 indicates the effect of the

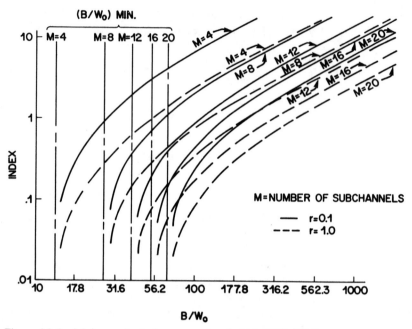

Figure 4.3-9 Main carrier index versus bandwidth (SSB/FM).

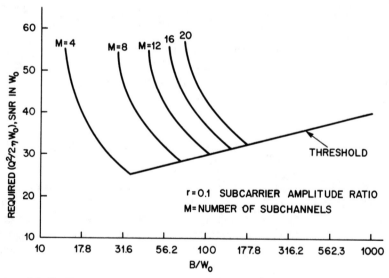

Figure 4.3-10 Required IF SNR in voice bandwidth to provide 30-dB output SNR(SSB/FM).

254

subcarrier index ratio on the required IF carrier power. Note that approximately an 8-dB power advantage may be obtained by reducing the subcarrier index ratio from unity to one-tenth when threshold is not the determining factor.

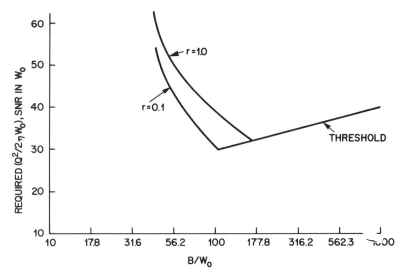

Figure 4.3-11 Comparison of required IF SNRs in voice bandwidth for two subcarrier index ratios (SSB/FM).

For FM/FM multiplex, the SNR after the first demodulation process, SNR_1, is still given by Eq. (4.3-17) except the rms modulation index on the main carrier is now denoted as m_f and the subchannel bandwidth is B_{FM}:

$$FM/FM: \qquad SNR_1 = \frac{m_f^2 Q^2}{2\eta B_{FM}} . \qquad (4.3\text{-}23)$$

Each subchannel is in turn an FM wave. The voice modulation on each subcarrier is assumed to be preemphasized to provide phase modulation. As shown in Section 4.3-7, among the types of preemphasis, phase modulation by the voice signal gives about the best overall performance. Thus we assume phase demodulation for the second detection, also.

If the rms index of the sine wave subcarrier on the main carrier is m_f, then the subcarrier amplitude is $\sqrt{2}\ m_f$. From Eq. (4.3-14) the noise density after the first demodulation is $2\eta/Q^2$. Thus, from Eqs. (4.1-13), (4.1-15), and (4.1-18), the one-sided noise spectral density after the second demodulation is

$$\text{FM/FM:} \qquad \mathcal{W}_{2\theta_n}(f) = \frac{2\eta}{Q^2 m_f^2}. \qquad (4.3\text{-}24)$$

Let m_{f0} denote the rms index of the voice on the subcarrier. Then the output signal spectrum is

$$\text{FM/FM:} \qquad \mathcal{W}_{f0}(f) = \mathcal{W}_v(f) m_{f0}^2, \qquad (4.3\text{-}25)$$

and integration over the spectra of Eqs. (4.3-24) and (4.3-25) yields the output SNR,

$$\text{FM/FM:} \qquad \text{SNR}_0 = \frac{m_{f0}^2 Q^2 m_f^2}{2\eta W_0}. \qquad (4.3\text{-}26)$$

To maintain an output SNR of 30 dB, the ratio at IF of the carrier power to the noise in a voice bandwidth is

$$\text{FM/FM:} \qquad \frac{Q^2}{2\eta W_0} = \frac{10^3}{m_{f0}^2 m_f^2}, \qquad (4.3\text{-}27)$$

for 30-dB SNR_0.

The above should be related to the overall bandwidth. If one assumes that the voice spectrum is flat over the voice bandwidth, W_0, then the rms frequency deviation of the subcarrier, f_{dev_0}, is related to the rms modulation index of the voice on the subcarrier, m_{f0}, by

$$f_{\text{dev}_0} = m_{f0} \frac{W_0}{\sqrt{3}}. \qquad (4.3\text{-}28)$$

From Eqs. (4.3-1) and (4.3-28), the subchannel bandwidth required to provide a 40-dB signal to bandlimiting distortion ratio is

$$B_{\text{FM}_{40\,\text{dB}}} = 2.52 W_0 + \frac{4 m_{f0}}{\sqrt{3}} W_0. \qquad (4.3\text{-}29)$$

Substituting Eq. (4.3-29) into Eq. (4.3-23) it follows that the requirement that the subcarrier be above threshold is

$$\text{FM/FM:} \qquad \frac{Q^2}{2\eta W_0} > \frac{10\left[2.52 + (4/\sqrt{3}) m_{f0}\right]}{m_f^2}, \qquad (4.3\text{-}30)$$

for subcarrier threshold. Again, we have defined threshold as a 10-dB SNR. From Eq. (4.3-9) the bandwidth of the first demodulation baseband is, using (4.3-30),

$$\text{FM/FM:} \quad W = (M + \tfrac{1}{2})B_{\text{FM}} = \left(2.52 + \frac{4m_{f0}}{\sqrt{3}}\right)(M + \tfrac{1}{2})W_0. \quad (4.3\text{-}31)$$

From Eqs. (4.3-1), (4.3-13), and (4.3-31), the overall bandwidth for a 40-dB signal to bandlimiting distortion ratio is

$$\text{FM/FM:} \quad B_{40 \text{ dB}} = \left[2.52(M + \tfrac{1}{2}) + 4m_f b\right]\left(2.52 + \frac{4m_{f0}}{\sqrt{3}}\right)W_0, \quad (4.3\text{-}32)$$

where

$$b \overset{\Delta}{=} \sqrt{\frac{4M^3 + 6M^2 + 2M}{12}} \qquad (4.3\text{-}33)$$

To keep the main carrier above threshold, the ratio at IF of the carrier power to noise in the voice bandwidth is

$$\text{FM/FM:} \quad \frac{Q^2}{2\eta W_0} > 10\left[2.52(M + \tfrac{1}{2}) + 4m_f b\right]\left(2.52 + \frac{4m_{f0}}{\sqrt{3}}\right), \quad (4.3\text{-}34)$$

for main carrier threshold.

Given an overall bandwidth, the system designer is faced with a choice of the relative strength of the sub- and main carrier modulation indices, m_{f0} and m_f, respectively. One choice optimizes the output signal-to-noise ratio, SNR_0, of Eq. (4.3-26). A plot of the sub- and main carrier modulation indices versus overall bandwidth and number of subchannels is plotted in Figure 4.3-12 for optimum SNR_0. Also shown in Figure 4.3-12 are the minimum overall bandwidths versus the number of subchannels. When the indices are chosen for optimum output SNR, as the transmitter power is decreased, the main carrier reaches threshold before the subcarrier.

Another choice of modulation indices that is sometimes employed[50] is that which causes the main carrier and subcarrier thresholds to occur at the same received carrier power. Equating (4.3-30) and (4.3-34), we obtain

$$4bm_f^3 + 2.52(M + 1)m_f^2 = 0.5. \qquad (4.3\text{-}35)$$

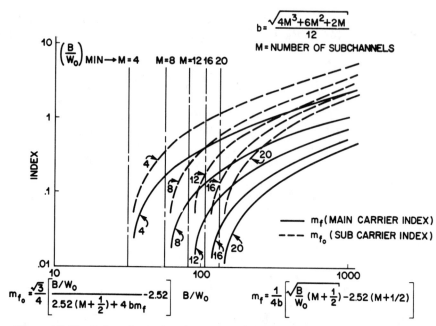

Figure 4.3-12 Subcarrier and main carrier indices for optimum SNR in FM/FM.

Thus the equal-threshold condition implies that the main carrier modulation index is independent of bandwidth and determined by the number of subchannels. From Eqs. (4.3-32) and (4.3-35) the subcarrier modulation index is determined as a function of bandwidth. Figure 4.3-13 shows the dependence of the modulation indices on bandwidth and number of subchannels for the equal threshold case. Note that since m_f is fixed versus bandwidth for the equal threshold case, and cannot go to zero as in the optimum SNR case, the minimum bandwidth for a given number of subchannels is larger in the equal threshold case than in the optimum SNR.

In FM/FM multiplex, the band-limiting distortion is complicated by the cumulative effect of the band-limiting distortion due to the main carrier filter and its resulting output distortion after the transformation of the second demodulation added to the band-limiting distortion of the subcarrier filter. If the subcarrier index, m_{f0}, is large enough, one may be able to relax the 40-dB band-limiting distortion requirement on the main filter, since FM advantage of the FM subcarrier will suppress the distortion by the capture effect described in Section 4.1.1. On the other hand, if m_{f0} is too small, the 40-dB signal-to-distortion ratio after the first demodulation may *not* be large enough to allow a 40-dB signal-to-distortion ratio after the second demodulation.

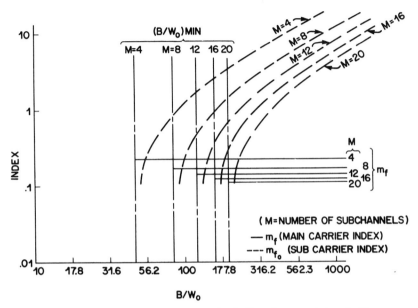

Figure 4.3-13 Subcarrier and main carrier indices for equal threshold.

Figure 4.3-14 shows the required ratio at IF of received carrier power to noise in the voice bandwidth versus overall bandwidth and number of subchannels for the case of optimum SNR modulation indices. Figure 4.3-15 compares required carrier power of the equal-threshold case to that of the optimum SNR case, when there are 12 subchannels. Except for small bandwidths, either criterion for choice of modulation indices (equal threshold versus optimum SNR) requires about the same carrier power.

An important question regarding multiplex is the following: "For equal overall bandwidths how much more, or less, transmitter power is required relative to that of a conventional single-channel system with the same number of channels?"

Assuming phase modulation of the voice on the single-channel carrier, the output SNR is computed just as for the second demodulation of the FM/FM multiplex, except that the input carrier amplitude is Q_{sc} instead of $\sqrt{2}\, m_f$ and the input noise spectral density is η instead of $2\eta/Q^2$. Thus

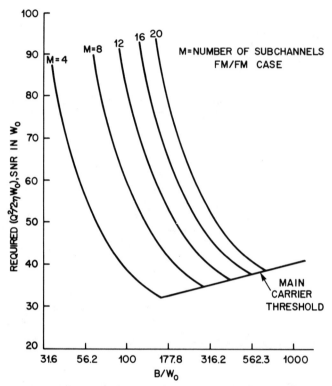

Figure 4.3-14 Required IF SNR in voice bandwidth for 30-dB output SNR(FM/FM) (optimum indices).

from Eqs. (4.3-1), (4.3-24), and (4.3-28) (40-dB band-limiting distortion),

$$\text{FM/SSB:} \qquad \text{SNR}_0 = \frac{3Q_{sc}^2}{2\eta W_0}\left(\frac{f_{dev}}{W_0}\right)^2$$

$$= \frac{3}{16}\left(\frac{Q_{sc}^2}{2\eta W_0}\right)\left(\frac{B_{sc}}{W_0}-2.52\right)^2. \quad (4.3\text{-}36)$$

For M channels, the overall bandwidth is MB_{sc} and the total peak square carrier amplitude is increased by M^2 at the transmitter and the total average square amplitude is increased by M at the transmitter. Thus the total ratios of peak carrier power and average carrier power to IF noise in the voice bandwidth required to maintain a 30-dB output SNR are

$$\text{FM/SSB:} \qquad \frac{Q_{pk}^2}{2\eta W_0} = \frac{16{,}000M^2}{3(B/MW_0-2.52)^2} \qquad \text{(peak)} \quad (4.3\text{-}37)$$

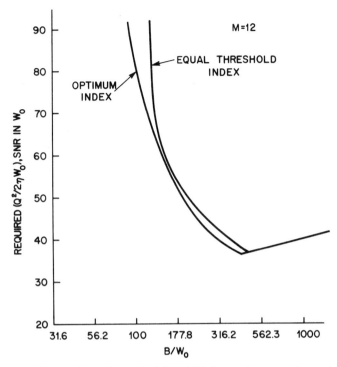

Figure 4.3-15 Comparison of required IF SNR for optimum and equal threshold indices (FM/FM).

and

$$\text{FM/SSB:} \qquad \frac{Q_{av}^2}{2\eta W_0} = \frac{16,000 M}{3(B/MW_0 - 2.52)^2} \qquad \text{(average)} \qquad (4.3\text{-}38)$$

respectively. Equations (4.3-37) and (4.3-38) are plotted versus bandwidth and number of channels in Figure 4.3-16. Usually the transmitter is peak power limited, but some system designs are more dependent on average transmitter power. The peak power above is computed from the sum of the peaks of the individual channels. If one allows some intermodulation distortion, the transmitter peak rating may be reduced somewhat from the above assumed value. This trade-off with intermodulation distortion is discussed in Section 4.3.6.

Figures 4.3-10, 4.3-14, and 4.3-16 may be compared to determine the relative power advantages of various multiplex systems versus conventional FM channel systems as a function of overall bandwidth and the number of

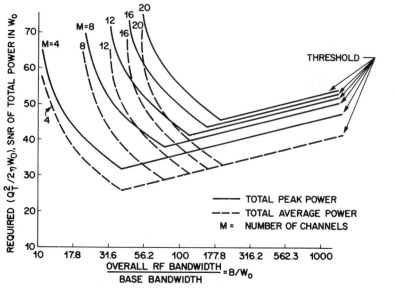

Figure 4.3-16 Required ratio of total RF power to noise in bandwidth w_0 for 30-dB output SNR, conventional FM channels.

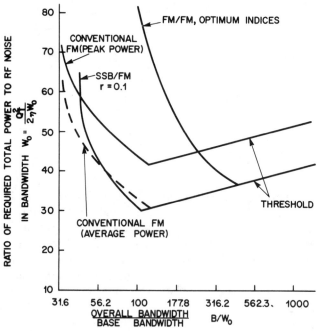

Figure 4.3-17 Comparison of required power for 30-dB output SNR, FM/FM, SSB/FM, and conventional FM ($m = 12$).

channels. Figure 4.3-17 shows this comparison for the case of 12 channels. In general, conventional FM channels require less average total transmitter power than SSB/FM and FM/FM, more total peak transmitter power than SSB/FM, and less total peak transmitter power than FM/FM, except when threshold limited. The increase in transmitter power required for FM/FM is usually prohibitive.

4.3.3 The Effect of Flat Fading on Multiplex Systems

As shown in Section 4.1.7, AM transmissions are distorted considerably more than FM transmissions by flat fading. It is interesting to compare the effect of flat fading on SSB/FM and FM/FM multiplex systems to that on conventional (FM/SSB) FM channels.

As shown in Section 4.1.4, the noise in a fading environment can be divided into four components: quadrature noise [$\mathcal{W}_2(f)$ from Eq. (4.1-21)], threshold noise [$\mathcal{W}_D(f)$ from Eq. (4.1-22)], signal suppression noise [$\eta_s(t)$ from Eq. (4.1-54)], and random FM noise [$\mathcal{W}_\theta(f)$ from Eq. (4.1-58)]. The threshold noise tends to dominate in most fading situations. Furthermore, the signal output is reduced during deep fades, as shown in Eq. (4.1-28).

First we will determine the required overall received signal power versus overall bandwidth requirements for the conventional (FM/SSB) channel system. In contrast to Section 4.1.4, phase modulation is assumed rather than frequency modulation, to allow a more direct comparison to the multiplex systems, and, as shown in Section 4.3.7, there are cases in which phase modulation is preferred. The quasistatic method will again be used to average over the fading to obtain average output signal and average output noise.

From Eq. (4.1-24) the spectrum of the quadrature plus threshold noise after phase demodulation is

$$\mathcal{W}_{\theta_n}(f) = \frac{\mathcal{W}(f)}{(2\pi f)^2}$$

$$\doteq \frac{(1 - e^{-\rho})^2}{B_\rho} + \frac{2Be^{-\rho}}{\pi f^2 \sqrt{2(\rho + 2.35)}}, \tag{4.3-39}$$

where the single-channel bandwidth, B, is related to the overall bandwidth, B_T, of M channels by

$$\text{FM/SSB:} \qquad B = \frac{B_T}{M}, \tag{4.3-40}$$

and the IF SNR is related to the ratio of the overall peak ($Q_{pk}^2/2$) and average $(Q_{av}^2/2)$ powers to noise in a voice bandwidth by

$$\text{FM/SSB:} \qquad \rho = \frac{Q^2}{2\eta B} = \frac{Q^2}{2\eta W_0}\left(\frac{W_0 M}{B_T}\right)$$

$$= \left(\frac{Q_{av}^2}{2\eta W_0}\right)\frac{W_0}{MB_T} = \left(\frac{Q_{av}^2}{2 W_0}\right)\frac{W_0}{B_T}. \qquad (4.3\text{-}41)$$

In Eq. (4.3-39), we have used $e^{-\pi(f/B)^2} \doteq 1$, since $f \ll B$ for frequencies in the output bandwidth and since the IF filter is closer to rectangular than Gaussian.

The input voice signal power is the mean square modulation index, which is related to the mean square frequency deviation by Eq. (4.3-28), which assumes a flat voice spectrum. The rms frequency deviation is in turn related to IF bandwidth by Eq. (4.3-1), assuming a 40-dB signal to band-limiting distortion ratio. Thus

$$\text{FM/SSB:} \qquad S_{in} = \frac{3}{16}\left(\frac{B_T}{M W_0} - 2.52\right)^2. \qquad (4.3\text{-}42)$$

From Eq. (4.1-54) the signal suppression noise is given by

$$N_{ss} = S_{in}\left[(1 - e^{-\rho}) - \langle 1 - e^{-\rho}\rangle\right]^2, \qquad (4.3\text{-}43)$$

and from (4.1-28) the output signal power is given by

$$\text{FM/SSB:} \qquad S_{out} = \frac{3}{16}\left(\frac{B_T}{M W_0} - 2.52\right)^2 (1 - e^{-\rho})^2, \qquad (4.3\text{-}44)$$

where the brackets $\langle\ \rangle$, indicate an average over Rayleigh fading. The random FM component of the output noise may be approximated by Eq. (4.1-58) when the maximum Doppler frequency shift is less than the lowest baseband output frequency, $W_0/10$:

$$\mathcal{W}_{RFM}^{(f)} = \frac{2\pi^2 f_m^2}{4\pi^2 f^2 \cdot f} = \frac{f_m^2}{2f^3}. \qquad (4.3\text{-}45)$$

The total output noise is obtained by integrating (4.3-39) and (4.3-44) over the output bandwidth, $W_0/10$ to W_0, and adding them to the signal suppression noise (4.3-43),

$$\text{FM/SSB:} \quad N_{\text{out}} = \frac{3}{16}\left(\frac{B_T}{MW_0} - 2.52\right)^2 \left[(1-e^{-\rho}) - \langle 1-e^{-\rho}\rangle\right]^2$$

$$+ \frac{9W_0 M(1-e^{-\rho})^2}{B_T \rho} + \frac{18 B_T e^{-\rho}}{\pi M W_0 \sqrt{2(\rho+2.35)}}$$

$$+ \frac{99}{4}\left(\frac{f_m}{W_0}\right)^2 \tag{4.3-46}$$

The average output signal to average output noise ratio is now calculated by averaging over the Rayleigh fading, where the probability density of ρ is given by Eq. (4.1-52). From Eq. (4.1-56) we have

$$\text{FM/SSB:} \quad \langle S_{\text{out}}\rangle = \frac{3}{16}\frac{\rho_0^2}{(1+\rho_0)^2}\left(\frac{B_T}{MW_0}-2.52\right)^2 \tag{4.3-47}$$

and

$$\text{FM/SSB:} \quad \langle N_{\text{out}}\rangle = \frac{3}{16}\left(\frac{B_T}{MW_0}-2.52\right)^2\left[\frac{1}{2\rho_0+1}-\frac{1}{(\rho_0+1)^2}\right]$$

$$+ \frac{0.9}{B_T/MW_0}\frac{1}{\rho_0}\log\left[\frac{(1+\rho_0)^2}{1+2\rho_0}\right] + \frac{18}{\pi}\frac{B_T}{MW_0}$$

$$\times \frac{e^{2.35(\rho_0+1)/\rho_0}}{\sqrt{2\pi\rho_0(\rho_0+1)}}\text{erfc}\sqrt{2.35\frac{\rho_0+1}{\rho_0}}$$

$$+ \frac{99}{4}\left(\frac{f_m}{W_0}\right)^2, \tag{4.3-48}$$

where ρ_0 is the average IF SNR.

The overall peak and average received powers averaged over the Rayleigh fading are given in terms of ρ_0 by Eq. (4.3-41),

$$\text{FM/SSB:} \qquad \left\langle \frac{Q_{pk}^2}{2} \right\rangle = \rho_0 M B_T, \qquad \left\langle \frac{Q_{av}^2}{2} \right\rangle = \rho_0 B_T. \qquad (4.3-49)$$

From (4.3-47) through (4.3-49) the ratio of average overall received power to noise in a voice bandwidth required to provide a 30-dB ratio of average signal out to average noise out is computed and is plotted in Figure 4.3-18.

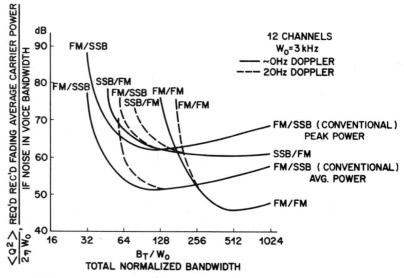

Figure 4.3-18 Required total received power versus bandwidth with fading (M = 12).

For SSB/FM multiplex, it is apparent from Eq. (4.3-39) that the threshold noise, which dominates in a fading environment, is largest in the lowest-frequency subchannel band, which lies in the range

$$B_{SSB} - W_0 \leqslant f \leqslant B_{SSB} - \frac{W_0}{10},$$

or

$$\frac{W_0}{3} \leqslant f \leqslant \frac{37}{30} W_0. \qquad (4.3-50)$$

Thus, the output SNR will be computed for this worst-case subchannel. The input signal is, from Eq. (4.3-19),

$$\text{SSB/FM:} \qquad S_{\text{in}} = m_v^2 = \frac{1}{4a^2}\left[\frac{B_T}{W_0} - 3.36M\right]^2, \qquad (4.3\text{-}51)$$

where a is defined in Eq. (4.3-19).

The signal suppression noise is related to the input signal by Eq. (4.3-43). Integrating (4.3-39) from $W_0/3$ to $37W_0/30$, we obtain the output quadrature and threshold noise:

$$\text{SSB/FM:} \qquad N_\theta = \frac{0.9(1-e^{-\rho})^2}{\rho(B_T/W_0)} + \frac{162(B_T/W_0)e^{-\rho}}{37\pi\sqrt{2(\rho+2.35)}}. \qquad (4.3\text{-}52)$$

From Eq. (4.3-45) it is seen that the random FM spectrum is highly concentrated at low frequencies, and the output due to random FM may, in some cases, require the lowest subchannel to remain unused. In the lowest subchannel the random FM noise is

$$\text{SSB/FM:} \qquad N_{\text{RFM}} = \frac{1}{4}\left[9 - \left(\frac{30}{37}\right)^2\right]\left(\frac{f_m}{W_0}\right)^2. \qquad (4.3\text{-}53)$$

The averaging over Rayleigh fading is performed as in Eqs. (4.3-47) and (4.3-48) to obtain the average output signal and noise:

$$\text{SSB/FM:} \qquad \langle S_{\text{out}}\rangle = \frac{1}{4a^2}\left[\frac{B_T}{W_0} - 3.36M\right]^2\frac{\rho_0^2}{(\rho_0+1)^2}, \qquad (4.3\text{-}54)$$

and

$$\text{SSB/FM:} \qquad \langle N_{\text{out}}\rangle = \frac{1}{4a^2}\left[\frac{B_T}{W_0} - 3.36M\right]^2\left[\frac{1}{2\rho_0+1} - \frac{1}{(\rho_0+1)^2}\right]$$

$$+ \frac{0.9}{B_T/W_0}\frac{1}{\rho_0}\log\left[\frac{(1+\rho_0)^2}{1+2\rho_0}\right] + \frac{162(B_T/W_0)}{37\pi}$$

$$\times \frac{e^{2.35(\rho_0+1)/\rho_0}}{\sqrt{2\pi\rho_0(\rho_0+1)}}\,\text{erfc}\sqrt{2.35\frac{\rho_0+1}{\rho_0}}$$

$$+ \frac{1}{4}\left[9 - \left(\frac{30}{37}\right)^2\right]\left(\frac{f_m}{W_0}\right)^2 \qquad (4.3\text{-}55)$$

where ρ_0 is the average IF SNR:

$$\rho_0 = \frac{\langle Q^2 \rangle}{2\eta B_T} = \frac{\langle Q^2 \rangle}{2\eta W_0}\left(\frac{W_0}{B_T}\right). \qquad (4.3\text{-}56)$$

Equations (4.3-54) through (4.3-56) are used to obtain the ratio of the received power to the IF noise in a voice bandwidth required to provide a 30-dB ratio of average signal out to average output noise. This required input ratio is plotted in Figure 4.3-18.

For FM/FM, the time-varying input SNR, ρ, due to fading, is transformed to a different time varying SNR, ρ_s, at the input to the subchannel demodulator. The output signal and noise are a function of ρ_s, so they are averaged over the variations of ρ_s. As in SSB/FM, the output signal and noise will be computed for the lowest-frequency subchannel, since it represents a worst case with the highest output noise. From Eq. (4.3-9) the lowest-frequency subchannel extends from $B_{FM}/2$ to $3B_{FM}/2$.

The signal power after the first demodulation is given by

$$\text{FM/FM:} \qquad S_b = (1 - e^{-\rho})^2 m_f^2, \qquad (4.3\text{-}57)$$

where, for optimum above threshold SNR, m_f is given by (see Figure 4.3-12)

$$\text{FM/FM:} \qquad m_f = \frac{1}{4b}\left[\sqrt{\frac{B_T}{W_0}\left(M + \frac{1}{2}\right)} - 2.52\left(M + \frac{1}{2}\right)\right], \qquad (4.3\text{-}58)$$

where b is given by Eq. (4.3-33). The noise presented to the second demodulator consists of the quadrature, threshold, and random FM noise. The signal suppression noise is a qualitative effect that does not appear until the output is measured. The random FM noise seen in a subchannel bandwidth is still somewhat correlated with the fading, but because of the effect of the subchannel filter (broad-band noise through a narrow-band filter), we will neglect this correlation and treat the filtered random FM as Gaussian noise, as an approximation. The same approximation is used for the filtered threshold noise. Thus, from Eqs. (4.3-39) and (4.3-45), the noise at the input to the second demodulator for the lowest-frequency subchannel is

$$\text{FM/FM:} \qquad N_b = \int_{B_{\text{FM}}/2}^{3B_{\text{FM}}/2} \left[\frac{(1-e^{-\rho})^2}{B_T\rho} + \frac{2B_T e^{-\rho}}{\pi f^2 \sqrt{2(\rho+2.35)}} + \frac{f_m^2}{2f^3} \right] df$$

$$= \frac{(1-e^{-\rho})^2}{(B_T/B_{\text{FM}})\rho} + \frac{8}{3\pi} \left(\frac{B_T}{B_{\text{FM}}} \right) \frac{e^{-\rho}}{\sqrt{2(\rho+2.35)}} + \frac{8}{9} \left(\frac{f_m}{W_0} \right)^2 \left(\frac{B_{\text{FM}}}{W_0} \right)^2. \quad (4.3\text{-}59)$$

The signal-to-noise ratio at the input to the subchannel demodulator is

$$\text{FM/FM:} \qquad \rho_s(\rho) = \frac{S_b}{N_b}. \qquad (4.3\text{-}60)$$

The output signal and output noise are computed as for FM/SSB, except that ρ is replaced by ρ_s, and there is no random FM introduced in the second demodulation. For the output signal,

$$\text{FM/FM:} \qquad \langle S_{\text{out}} \rangle = \langle 1 - e^{-\rho_s} \rangle^2 S_{\text{in}}, \qquad (4.3\text{-}61)$$

where from Figure 4.3-12,

$$S_{\text{in}} = m_{f0}^2 = \frac{3}{16} \left[\frac{B_T/W_0}{2.52(M+\frac{1}{2})+4bm_f} - 2.52 \right]^2. \qquad (4.3\text{-}62)$$

The signal suppression noise is

$$\text{FM/FM:} \qquad \langle N_{\text{SS}} \rangle = \langle (1-e^{-\rho_s})^2 - \langle 1-e^{-\rho_s} \rangle^2 \rangle S_{\text{in}} \qquad (4.3\text{-}63)$$

and the quadrature and threshold noise are

$$\text{FM/FM:} \qquad \langle N_\theta \rangle$$

$$= \frac{0.9}{B_{\text{FM}}/W_0} \left\langle \frac{(1-e^{-\rho_s})^2}{\rho_s} \right\rangle + \frac{18(B_{\text{FM}}/W_0)}{\pi} \left\langle \frac{e^{-\rho_s}}{\sqrt{2(\rho_s+2.35)}} \right\rangle, \qquad (4.3\text{-}64)$$

where

$$\text{FM/FM:} \qquad \frac{B_{\text{FM}}}{W_0} = 2.52 + \frac{4}{\sqrt{3}} m_{f0} \qquad (4.3\text{-}65)$$

and

$$\text{FM/FM:} \qquad \frac{B_T}{B_{FM}} = 2.52(M + \tfrac{1}{2}) + 4m_{fb}. \qquad (4.3\text{-}66)$$

The above averages are carried out by integrating over the probability density of ρ, Eq. (4.1-52), and using Eq. (4.3-60) to relate ρ_s to ρ. This integration was performed numerically to obtain

$$\frac{\langle S_{out} \rangle}{\langle N_\theta \rangle + \langle N_{SS} \rangle},$$

the ratio of average output signal to average output noise in terms of the average IF SNR, ρ_0, given by Eq. (4.3-56). The ratio of the received power to the IF noise in a voice bandwidth required to provide a 30-dB ratio of average output signal to average output noise is plotted in Figure 4.3-18.

In comparing the performance of various multiplex systems with fading (Figure 4.3-18) to that without fading (Figure 4.3-17), it is seen that fading causes the SSB/FM to perform worse relative to conventional FM, while fading causes FM/FM to perform better relative to conventional FM channels. This is because threshold noise dominates in a Rayleigh fading environment, and the threshold noise, with phase demodulation, is peaked at $f = 0$. Threshold noise is larger for FM/FM and SSB/FM than for conventional FM because of their higher click rate, due to their larger IF bandwidth. The FM/FM system, however, has a much higher low-frequency limit to its baseband,

$$\frac{B_{FM}}{2} \gg \frac{W_0}{10},$$

which causes the threshold noise in the subchannel to be less than that of the conventional FM channel due to the $1/f$ frequency dependence of the threshold noise spectrum. Thus the performance curves of Figure 4.3-18 can be changed considerably by leaving lower subchannels idle in the multiplex systems or by translating the baseband frequencies upward in the conventional FM channels.

4.3.4 Frequency-Selective Fading

As described in Section 4.1.8, wideband signals are subject to frequency-selective fading distortion. If the bandwidth is wide enough, the propagation characteristics of the multipath medium will have a deep fade at some frequency within the band most of the time. These notches in the transfer characteristic of the medium introduce distortion similar to filter distortion. Since FM/FM and SSB/FM multiplex systems have a much wider IF bandwidth than conventional FM channels, they experience greater frequency-fading distortion. In this section we will compare the effect of frequency-selective fading on the different FM multiplex systems.

First, we consider conventional FM channels as in Section 4.1.8, except that phase modulation will be assumed, as in the other comparisons with multiplex. From Eq. (4.1-128), the two-sided power spectrum of the frequency-selective fading distortion in the demodulated output phase for quadratic distortion (which dominates when the vehicle is stationary) is given by

$$S_{\delta_2}(f) = \frac{1}{(2\pi f)^2} S_{\dot{\delta}_2}(f)$$

$$= 4.8\sigma^4 \pi^4 \frac{f_{\text{dev}}^4}{W_0} \begin{cases} \dfrac{1}{1-a} - \dfrac{|f/W_0|}{(1-a)^2}, & 0 < |f| < 2aW_0 \\[3mm] \dfrac{(1-2a) - |f/2W_0|}{(1-a)^2}, & 2aW_0 < |f| < W_0(1-a), \\[3mm] \dfrac{|f/2W_0|}{(1-a)^2} - \dfrac{a}{(1-a)^2}, & W_0(1-a) < |f| < W_0 \end{cases}$$

$$(4.3\text{-}67)$$

where σ is the spread in time delays, f_{dev} is the rms frequency deviation, W_0 is the top baseband frequency, and aW_0 is the bottom baseband frequency.

As the vehicle moves, linear distortion becomes a factor, and the median power spectrum of the linear distortion in the demodulated phase outout is given, from Eq. (4.1-129), as

$$S_{\delta_{\text{med}}}(f) = \frac{\sigma^2}{3} S_{\dot{\psi}}(f) = \frac{2(\pi \sigma f_{\text{dev}})^2}{W_0} \left(\frac{f}{W_0}\right)^2. \qquad (4.3\text{-}68)$$

In both (4.3-67) and (4.3-68) we have assumed the signal phase modulation spectrum, $S_{\psi}(f)$, is flat. The baseband power spectra of Eqs. (4.3-67) and (4.3-68) are plotted in Figure 4.3-19.

The total distortion is obtained by integrating over the spectra of Eqs. (4.3-67) and (4.3-68). The output signal is equal to the mean square modulation index, which is related to the IF bandwidth, B_T/M, by Eqs. (4.3-28) and (4.3-29), assuming a flat baseband signal spectrum. The resulting output signal to frequency distortion ratios are

FM/SSB: $\quad \text{SDR}_2 = \left[0.2(\pi\sigma W_0)^4 \left(\frac{B_T}{W_0 M} - 2.52\right)^2 \right]^{-1} \quad$ (quadratic),

$$(4.3\text{-}69)$$

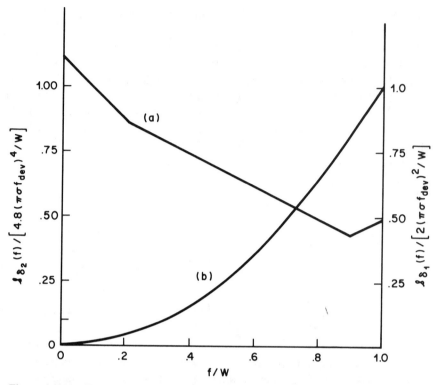

Figure 4.3-19 Spectra of linear and quadratic distortion due to frequency selective fading. (a) two-sided power spectrum of $S_{\delta_2}(f)$, quadratic distortion after first demodulation (dominates when vehicle is stationary). (b) two-sided power spectrum of $S_{\delta_{1med}}(f)$, linear distortion after first demodulation (dominates when vehicle is moving).

$$\text{FM/SSB:} \qquad \text{SDR}_{1_{med}} = \left[\frac{2\pi}{3} \sigma W_0 \right]^{-2} \qquad \text{(linear)}. \qquad (4.3\text{-}70)$$

These output signal-to-distortion ratios are plotted in Figure 4.3-20 for the 12-channel case and a 1-μsec spread in time delays. Note that the signal to quadratic distortion ratio is proportional to σ^{-4}, W_0^{-2}, and f_{dev}^{-2}, while the signal to linear distortion ratio is independent of IF bandwidth and is proportional to σ^{-2} and W_0^{-2}. In general, frequency-selective fading distortion is not much of a problem for conventional FM channels. However, as will be shown next, frequency-selective fading distortion can make multiplex systems unusable.

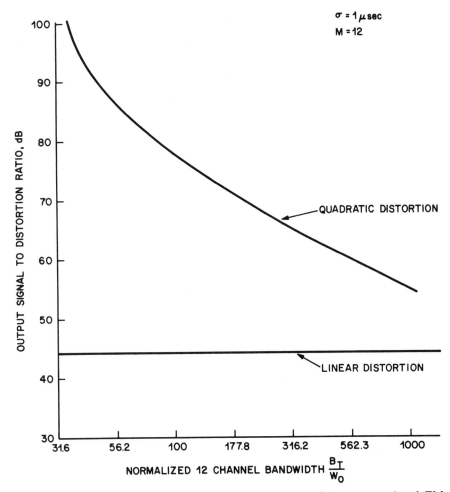

Figure 4.3-20 Frequency selective fading distortion FM/SSB (conventional FM channels).

With SSB/FM multiplex, only part of the distortion spectra of Figure 4.3-19 appears at the output of any one subchannel. As seen in Figure 4.3-19, the lowest-frequency subchannel has the most quadratic distortion, while the highest-frequency subchannel has the most linear distortion. Thus we will compute the signal-to-distortion ratios for these respective worst-case subchannels. The signal output is the mean square modulation index, m_v^2, which is related to the overall IF bandwidth by Eq. (4.3-20).

Thus from Eqs. (4.3-20), (4.3-67) and (4.3-68), the output signal-to-distortion ratios are

SSB/FM: $$SDR_2 = \frac{m_v^2 M}{7.2(\pi\sigma W_0)^4 (B_T/4W_0 - 10.08)^4}$$

(quadratic, bottom subchannel),   (4.3-71)

SSB/FM: $$SDR_{1_{med}} = \frac{m_v^2 M}{3(\pi\sigma W_0)^2 (B_T/4W_0 - 10.08)^2}$$

(linear, top subchannel),   (4.3-72)

where, from Eq. (4.3-20),

$$m_v = \frac{2}{a}\left[\frac{B_T}{4W_0} - 10.08\right],$$ (4.3-73)

and from Eq. (4.3-19)

$$a = \frac{4}{3\sqrt{12}}\sqrt{(1+5r)^2(4M^3+6M^2+2M) - 6M^2 - 3M}.$$ (4.3-74)

These signal-to-distortion ratios are plotted for $M = 12$ and $\sigma = 1$ μsec in Figure 4.3-21. For SSB/FM, as with conventional FM, the signal to quadratic distortion ratio is proportional to σ^{-4}, W_0^{-2}, and f_{dev}^{-2}, while the linear distortion is independent of IF bandwidth and proportional to σ^{-2} and W_0^{-2}. From Figure 4.3-21, we see that for the 12-channel case the quadratic distortion is acceptable for IF bandwidths less than about 150 W_0, but the linear distortion, experienced when the vehicle is moving, allows only a 22-dB output SDR.

The output distortion for FM/FM multiplex is somewhat more complicated to compute (than for FM/SSB and SSB/FM) because of the effect of the second frequency demodulation. The quadratic distortion presented to a given FM subchannel may be approximated as Gaussian noise, uncorrelated with the signal of that subchannel. The validity of this approximation follows from the fact that the distortion of the subchannel input is a narrow-band filtered output of the "spikey" frequency-selective

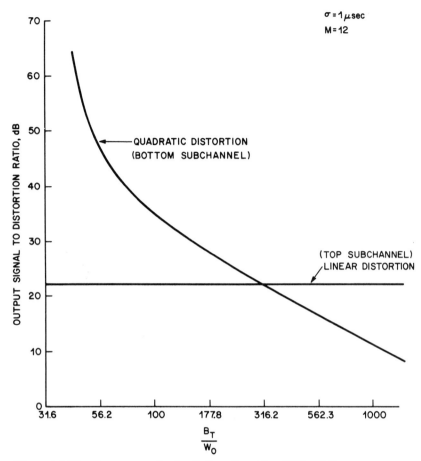

Figure 4.3-21 Frequency selective fading distortion (SSB/FM).

fading distortions experienced as the main carrier deviation (the sum of the deviations due to all subchannels) sweeps through the fades. Thus, the subchannel demodulator may be treated as in the conventional FM noise analysis, where its input SNR is determined by the quadratic distortion in the subchannel bandwidth. After the first demodulation, the signal power equals the mean square index of the subchannel on the main carrier, m_f^2, and the quadratic distortion (which is worst for the lowest subchannel) equals the spectrum of Eq. (4.3-67) integrated over the subchannel bandwidth, B_{FM}. Thus, from Eqs. (4.3-31) and (4.3-67), the SNR at the input to the bottom subchannel is

$$\text{FM/FM:} \quad \text{SNR}_2 \doteq \frac{m_f^2}{9.6(\pi\sigma W_0)^4(f_{\text{dev}}/W_0)^4 B_{\text{FM}}/W}$$

$$= \frac{m_f^2(M+1/2)}{9.6(\pi\sigma W_0)^4(f_{\text{dev}}/W_0)^4} . \qquad (4.3\text{-}75)$$

From Eqs. (4.3-13), (4.3-26), and (4.3-31), the FM/FM output SNR is related to its input SNR_2 as follows:

$$\text{FM/FM:} \quad \text{SNR}_0 = \frac{m_{f0}^2 B_{\text{FM}}}{W_0} \text{SNR}_2$$

$$= \frac{(M+1/2)m_{f0}^2}{9.6(\pi\sigma W_0 b)^4 m_f^2 \left(2.52 + 4m_{f0}/\sqrt{3}\,\right)^3} . \qquad (4.3\text{-}76)$$

Equation (4.3-76) assumes that the second demodulator is above threshold; that is, $\text{SNR}_2 > 10$ dB. For optimum SNR_0 above threshold we choose the indices m_f and m_{f0} as given by Eqs. (4.3-58) and (4.3-65),

$$m_f = \frac{1}{4b}\left[\sqrt{\frac{B_T}{W_0}(M+\tfrac{1}{2})} - 2.52(M+\tfrac{1}{2})\right],$$

$$m_{f0} = \frac{\sqrt{3}}{4}\left[\frac{B_T}{W_0}\left(\frac{1}{2.52(M+\tfrac{1}{2})+4m_f b}\right) - 2.52\right], \qquad (4.3\text{-}77)$$

$$b = \sqrt{\frac{4M^3 + 6M^2 + 2M}{12}} .$$

The output signal to quadratic distortion ratio is plotted versus B_T/W_0 for 12 subchannels and a time delay spread of 1 μsec in Figure 4.3-22. The output signal to quadratic distortion ratio for FM/FM is proportional to σ^{-4}, as with FM/SSB and SSB/FM, but has a more complicated dependence on W_0 and f_{dev}. It is seen that frequency-selective fading distortion is much more damaging to FM/FM than to conventional FM channels.

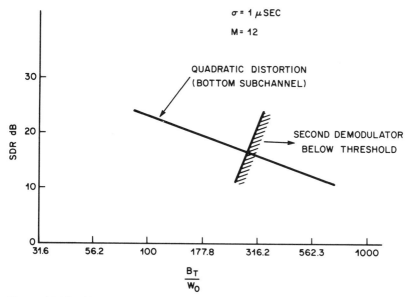

Figure 4.3-22   Frequency selective fading distortion (FM/FM).

The linear distortion appearing in a subchannel bandwidth is strongly correlated to the signal in that band and hence may not be modeled as Gaussian noise, as in the case of quadratic distortion. From Eq. (4.1-122), the second demodulator for the ith subchannel sees the voltage e_{in_2},

$$e_{\text{in}_2} = m_f \cos\left(\omega_i t + v(t) + \theta\right)$$

$$+ \operatorname{Re}\left\{\frac{T_1(t)}{T_0(t)}\right\}[\omega_i + \dot{v}(t)]m_f \sin\left(\omega_i t + v(t) + \theta\right), \qquad (4.3\text{-}78)$$

where ω_i is the radian frequency of the ith subcarrier, $v(t)$ is the voice modulation, and θ is an arbitrary constant phase shift; $T_0(t)$ and $T_1(t)$ are the zeroth- and first-order coefficients of the Taylor series expansion in frequency of the multipath transmission coefficient around the center RF carrier frequency. As seen in Eq. (4.3-78) the second term is in phase quadrature with the desired term. The second term is usually small relative to the first, so that the phase-demodulated output may be approximated by

$$v_0(t) \doteq v(t) + \operatorname{Re}\left\{\frac{T_1(t)}{T_0(t)}\right\}[\omega_i + \dot{v}(t)]. \qquad (4.3\text{-}79)$$

The spectrum of $\mathrm{Re}\{T_1(t)/T_0(t)\}$ is usually small at frequencies above the lowest output voice band frequency because the Doppler frequency is about 100 Hz or less in the usual application. Thus we will neglect the distortion, $\omega_i \mathrm{Re}\{T_1(t)/T_0(t)\}$, and compute only the median distortion arising from $\dot{v}(t)\mathrm{Re}\{T_1(t)/T_0(t)\}$. From Eq. (C-11)

$$\mathrm{Re}\left\{ \frac{T_1(t)}{T_0(t)} \right\}_{\mathrm{med}} = \sqrt{3}\,\sigma, \qquad (4.3\text{-}80)$$

where σ is the rms spread in time delays. The ratio of the spectral density of $\dot{v}(t)$ relative to that of $v(t)$ is ω^2. Thus the median output signal-to-distortion ratio at 3 kHz with $\sigma = 1$ μsec is

$$\mathrm{FM/FM:} \qquad \mathrm{SDR}_{\mathrm{med}} = \frac{1}{3\sigma^2(2\pi\cdot 10^3)^2 9} \doteq 10^3. \qquad (4.3\text{-}81)$$

Thus, the fact that the linear distortion allows us to suppress the $\omega_i \mathrm{Re}\{T_1(t)/T_0(t)\}$ portion of the output distortion below the lowest voice output frequency, 300 Hz, results in a lower contribution of linear distortion than quadratic distortion for the FM/FM case. However, the quadratic distortion output for FM/FM is much worse than for FM/SSB and SSB/FM.

4.3.5 Cochannel Interference

For efficient utilization of spectrum space, it is desirable to reuse frequencies at different locations. The limitation on this reuse is that two or more channels operating on the same frequency at different locations should not interfere (cochannel interference).

For SSB/SSB, the baseband output signal-to-interference ratio is the same as at RF. For the FM systems the output signal-to-interference ratio is a nonlinear function of the input signal-to-interference ratio. In Sections 4.1.2 and 4.1.5, methods of computing cochannel interference for a single interferer or for many interferers are given. In the usual high-capacity system a particular channel will most often be subject to many interferers. For example, in the hexagonal cell system (see Section 7.2), typically six closest interferers at approximately the same distance could be present. With six interferers of comparable power, the receiver responds as if the front-end noise had been raised to the level of the average total power of the interferers (cf. Section 4.1.5). Thus, Figure 4.3-18 may be used to estimate the required signal to average interference ratio (averaged over Rayleigh fading) to provide a 30-dB output signal-to-interference ratio. For example, in a 12-channel system with a total bandwidth of $256 \times W_0$

$= 768$ kHz (64 kHz per channel), the required ratios of average IF signal power to IF interference power are (approximating the interference IF spectrum by a flat spectrum)

FM/SSB: $\dfrac{S}{I} = 52.8 \text{ dB} - 10\log_{10}\dfrac{B_T}{W_0}$

$= 52.8\text{-}24.1 = 28.7\, dB$

SSB/FM: $\dfrac{S}{I} = 60.8 \text{ dB} - 24.1 \text{ dB} = 36.7 \text{ dB},$

FM/FM: $\dfrac{S}{I} = 52.3 \text{ dB} - 24.1 \text{ dB} = 28.2 \text{ dB}.$

To achieve the desired S/I, the designer can adjust the ratio of the distance from desired transmitter to the distance of the mobile unit from the interferer by increasing the number of cells with separate frequency assignments, and consequently the overall system bandwidth, as described in Section 7.4. In this way the overall system bandwidth requirements may be compared for different multiplex systems. In general, for systems employing less than 60 kHz per channel, conventional FM(FM/SSB) requires less overall system bandwidth than FM/FM and SSB/FM.

The computation of interference in the case of a single interferer is complicated for multiplex systems, relative to the single-channel case considered in Section 4.1.5, because the straight-line approximations to the curves of Figure 4.1-11 are not as applicable since low indices are usually employed in multiplex systems. However, the region of Figure 4.1-11 where R is near unity, that is, signal strength approximately equal to interferer strength, is the only region in which the straight-line approximation is inaccurate. If the probability density of R is not too peaked near unity, the straight-line approximation may still yield good estimates of average output signal-to-interference ratios even for the low-index cases. The interference spectrum after the first demodulation is approximately flat over the subchannel frequencies, 0 to W, so that the straight-line approximation of Eq. (4.1-42) gives

$$S_I(f) \doteq \begin{cases} \dfrac{R^2 g(\Phi^2 + \Phi_i^2)}{2W}; & R < 1, \\ \Phi_i^2/2W; & R > 1, \end{cases} \qquad 0 < |f| < W \qquad (4.3\text{-}82)$$

where $g(\Phi^2 + \Phi_i^2)$ is taken from Figure 4.1-12. For example,

$$g(\Phi^2 + \Phi_i^2) = \begin{cases} \sqrt{\dfrac{3}{2\pi(\Phi^2 + \Phi_i^2)}} & \text{large index} \\[2em] \dfrac{\Phi^2 + \Phi_i^2}{2} & \text{small index} \end{cases} \qquad (4.3\text{-}83)$$

Averaging over Rayleigh fading, as in Section 4.1.5, the average signal power spectrum and average interference power spectrum are given by Eqs. (4.1-69) and (4.1-70), respectively,

$$2WS_s(f) \doteq \frac{\Gamma^2}{(1+\Gamma)^2}\Phi^2 \qquad (4.3\text{-}84)$$

and

$$2WS_N(f) \doteq \frac{1}{1+\Gamma}\left\{\left[\left(1+\frac{1}{\Gamma}\right)\log(1+\Gamma)-1\right]g(2\Phi^2)\right.$$

$$\left. +\Phi^2\frac{1+2\Gamma}{1+\Gamma}\right\}, \qquad (4.3\text{-}85)$$

where Γ is the average IF signal to average IF interference power ratio, and we have assumed $\Phi_i^2 = \Phi^2$; that is, the index used on the interfering channel equals that on the desired signal.

The modulation index on the main carrier is related to the overall IF bandwidth, B, for M channels by the formulas of Section 4.3.2,

$$\Phi^2_{\text{FM/SSB}} = m_{f0}^2 = \frac{3}{16}\left(\frac{B}{MW_0}-2.52\right)^2, \qquad (4.3\text{-}86)$$

$$\Phi^2_{\text{SSB/FM}} = Mm_v^2 = \frac{M}{4a^2}\left(\frac{B}{W_0}-3.36M\right)^2, \qquad (4.3\text{-}87)$$

$$\Phi^2_{\text{FM/FM}} = Mm_f^2 = \frac{M}{16b^2}\left[\sqrt{\frac{B}{W_0}(M+\tfrac{1}{2})}-2.52(M+\tfrac{1}{2})\right]^2, \qquad (4.3\text{-}88)$$

where in the FM/FM case we have chosen the indices for optimum SNR, as described in Section 4.3.2.

For FM/SSB and SSB/FM the output signal-to-interference ratios are given directly by Eqs. (4.3-84) and (4.3-85). For FM/FM, there is a second demodulation process subject to the nonlinear effects of interference. When the desired signal dominates, the interference presented to the second demodulator is much like Gaussian noise. When the interferer dominates, it captures the receiver and the output is that of the corresponding subchannel in the interferer multiplex signal. The calculation of average output interference in the presence of Rayleigh fading can be simplified, however, by the fact that with Rayleigh fading the output interference is mainly due to the times when the interferer captures the receiver. This is equivalent to neglecting the portion of Eq. (4.3-85) due to interference when the desired signal has captured the receiver. Thus in this approximation, the average signal to average interference ratio at the output of the receiver is

$$\text{FM/FM:} \quad \frac{\langle S \rangle}{\langle I \rangle} = \Phi_2^2 \left[\frac{\Gamma_1^2}{(1+\Gamma_1)^2} \right] \Big/ \left(\frac{1+2\Gamma_1}{(1+\Gamma_1)^2} \right) \Phi_2^2$$

$$= \frac{\Gamma_1^2}{1+2\Gamma_1}, \qquad (4.3\text{-}89)$$

where Γ_1 is the average signal-to-interference ratio presented to the first demodulator and Φ_2 is the modulation index presented to the second demodulator.

Assuming that the output average signal to average interference is 30dB, the required IF average signal to interference power ratio is plotted in Figure 4.3-23 versus overall bandwidth for the various multiplex types for the 12-channel case. By comparison with Figure 4.3-18, it is seen that the requirements for a single interferer are much the same as those for multiple interferers.

4.3.6 Intermodulation

In this section we analyze only that distortion due to a nonlinear transmission coefficient in the transmitter and receiver. The nonlinearity is usually in the form of a saturation at high power levels. The carriers of this kind of intermodulation distortion will only superpose in the bandwidth of an FM wave if there are at least two other signals near the frequency of the FM wave. The larger the amplitudes of these FM waves, relative to the saturation level, the greater the intermodulation distortion.

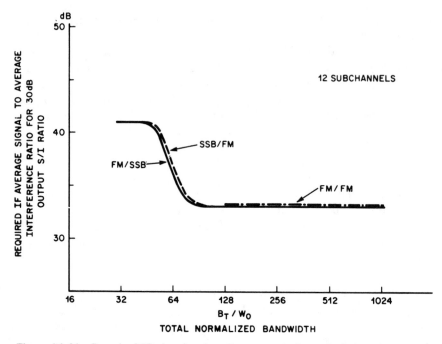

Figure 4.3-23   Required IF signal-to-interference ratio for a single interferer.

The only usual case in mobile radio, where three or more strong signals are likely to be present at the transmitter or receiver, is when the signals come from the same base station. One example is the conventional FM channel case (FM/SSB), where the typical cell might have 12 or more channels. One would normally not expect to use more than one multiplex signal per cell, since the purpose of the multiplex system is to combine the channels into a single FM wave. However, in one cell system design (see Section 7.3), the base stations are located at the cell boundary, so that one base station can be transmitting different multiplex signals to each of the adjacent hexagonal cells simultaneously. The possibility for intermodulation distortion also exists at the base-station receiver, which can be receiving many strong mobile signals simultaneously.

A typical saturation characteristic is shown in Figure 4.3-24. The most important properties of intermodulation distortion can be displayed by approximating the transmission curve by a cubic

$$v_{\text{out}} = a_1 v_{\text{in}} - a_3 v_{\text{in}}^3, \tag{4.3-90}$$

as shown in Figure 4.3-24.

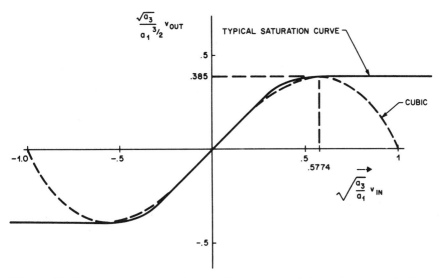

Figure 4.3-24 Approximation of a nonlinear transmission characteristic by a cubic.

If three multiplex signals are transmitted through the cubic nonlinearity of Eq. (4.3-90), Bennett[28] has shown the intermodulation products to be as shown in Eqs. (4.3-91) and (4.3-92) (assuming three equally spaced carrier frequencies):

$$v_{in} = A\cos\alpha(t) + B\cos\beta(t) + C\cos\gamma(t), \qquad (4.3\text{-}91)$$

$$
\begin{aligned}
v_{out} = \ & [a_1 A - \tfrac{3}{4}a_3 A(A^2 + 2B^2 + 2C^2)]\cos\alpha(t) && (\text{linear } \omega_c - \Delta) \\
& - \tfrac{3}{4}a_3 B^2 C \cos[2\beta(t) - \gamma(t)] && (\text{distortion } \omega_c - \Delta) \\
& + [a_1 B - \tfrac{3}{4}a_3 B(2A^2 + B^2 + 2C^2)]\cos\beta(t) && (\text{linear } \omega_c) \\
& - \tfrac{3}{2}a_3 ABC \cos[\alpha(t) + \gamma(t) - \beta(t)] && (\text{distortion } \omega_c) \\
& + [a_1 C - \tfrac{3}{4}a_3 C(2A^2 + 2B^2 + C^2)]\cos\gamma(t) && (\text{linear } \omega_c + \Delta) \\
& - \tfrac{3}{4}a_3 B^2 A \cos[2\beta(t) - \alpha(t)] && (\text{distortion } \omega_c + \Delta) \\
& - \{\text{terms at frequencies outside the range} \\
& \quad \omega_c - \Delta < \omega < \omega_c + \Delta\},
\end{aligned}
$$

$$(4.3\text{-}92)$$

where ω_c is the frequency of the center carrier and Δ is the separation between carrier frequencies. A quadratic nonlinearity could also be present without affecting the terms in the frequency range $\omega_c - \Delta < \omega < \omega_c + \Delta$. The largest distortion occurs for equal signal strengths and at the center frequency signal, $\beta(t)$. This distortion appears as a cochannel interferer with three times the mean square index as the desired signal, and whose modulation is approximately uncorrelated with that of the desired signal.

Let P_S be the power of a sine wave whose peak amplitude just equals the input that gives a peak output from the cubic, Eq. (4.3−90):

$$P_S = \frac{1}{2}v_{in_p}^2 = \frac{a_1}{6a_3}. \tag{4.3-93}$$

The IF signal-to-interference ratio can be computed from the linear and distortion terms in Eq. (4.3-92), respectively. The compression part of the linear term is generally negligible in computing large signal-to-interference ratios. Thus we have

$$\frac{S}{I} = \left[\frac{a_1 B}{\frac{3}{2}a_3 ABC} \right]^2 = \frac{4}{9}\left(\frac{a_1}{a_3} \right)^2 \frac{1}{A^4}. \tag{4.3-94a}$$

The total input power is

$$P_T = \frac{A^2}{2} + \frac{B^2}{2} + \frac{C^2}{2} = \frac{3}{2}A^2. \tag{4.3-94b}$$

Equations (4.3-93) and (4.3-94) combine to give

$$\left(\frac{S}{I} \right)_{IF} = 36\left(\frac{P_S}{P_T} \right)^2. \tag{4.3-95}$$

Since the signals are assumed to be coming from the same base-station transmitter at closely spaced frequencies, the three signals will fade in unison. This means that the signal-to-interference ratio will be constant if the distortion is introduced at the transmitter. If it is introduced at the mobile receiver, then P_T will be the sum of squares of three Rayleigh random variables and a time average over the fading would be appropriate. We will not carry out the averages over fading in this analysis.

The IF signal-to-interference ratio of Eq. (4.3-95) can be converted to a baseband signal-to-interference ratio by Eqs. (4.3-82) and (4.3-83), with $\Phi_i^2 = 3\Phi^2$ and $R < 1$:

$$\left(\frac{S}{I} \right)_{out} = \left(\frac{S}{I} \right)_{IF} \begin{cases} \Phi^3 \sqrt{\dfrac{8\pi}{3}}, & \text{large index,} \quad \Phi^2 > \dfrac{2.35}{4} \\[2ex] 1/2, & \text{small index,} \quad \Phi^2 < \dfrac{2.35}{4} \end{cases} \tag{4.3-96}$$

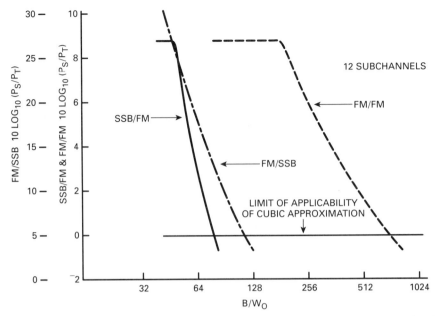

Figure 4.3-25 Maximum total power, p_T, allowed relative to saturation power level, p_S, which allows 30-dB output signal-to-distortion ratio.

The ratio of the maximum allowable total input power, P_T, to the saturation power level, P_S, is plotted for FM/FM and SSB/FM in Figure 4.3-25 versus overall multiplex bandwidth of one signal, with the requirement that the output signal-to-distortion ratio be equal to or better than 30 dB. The curves labeled FM/FM and SSB/FM in Figure 4.3-25 were obtained by setting $(S/I)_{out} = 1000$ in Eq. (4.3-96) and solving for $(S/I)_{IF}$, which was in turn used in Eq. (4.3-95) to obtain (P_S/P_T). For small modulation index the output signal-to-distortion ratio is half that at IF. At a large index the signal-to-distortion ratio improves with index. The second demodulation also improves the signal-to-distortion ratio in the FM/FM case when the index of the second demodulation is large. However, this effect does not show up in Figure 4.3-25 because the index of the second demodulation is small over the range of applicability of the cubic transfer function.

In the conventional FM case (FM/SSB), there are many channels producing intermodulation. We can model the many channel case (typically 12 or more channels at a base station) by Gaussian noise with a flat spectrum across the frequency band spanned by the channels. From Ref. 29, p. 272, the autocorrelation of the output of the cubic transfer function is

$$R_0(\tau) = \underbrace{a_1^2 R_i(\tau) + 9a_3^2 R_i^2(0) R_i(\tau)}_{\text{signal}} \quad \underbrace{+6a_3^2 R_i^3(\tau)}_{\text{intermodulation}} \quad (4.3\text{-}97)$$

where $R_i(\tau)$ is the autocorrelation of the input and $R_0(\tau)$ is that of the output.

When the signal-to-distortion ratio is large we can neglect the second term in the signal portion of Eq. (4.3-97). Let $S_i(f)$ denote the two-sided rectangular power spectrum of the input. Then the output signal and distortion spectra are, respectively,

$$S_{s_0}(f) = a_1^2 S_i(f)$$

$$S_{d_0}(f) = 6a_3^2 S_i(f)*S_i(f)*S_i(f), \qquad (4.3\text{-}98a)$$

where the * indicates convolution. These spectra are shown in Figure 4.3-26. As seen from the figure, the peak distortion is at the center of the band, where the signal to distortion spectral density ratio is

$$\left(\frac{S}{I}\right)_{IF} = \frac{16}{54}\frac{(a_1/a_3)^2}{P_T^2} = \frac{32}{3}\left(\frac{P_S}{P_T}\right)^2. \qquad (4.3\text{-}98b)$$

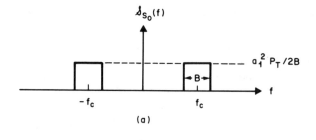

(a)

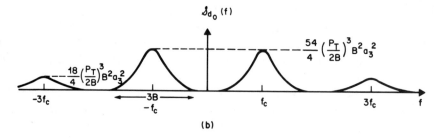

(b)

Figure 4.3-26 Output signal and distortion spectrum. (a) Output signal spectrum. (b) Output distortion spectrum.

By comparison with Eq. (4.3-95) we see that the signal-to-interference ratio is degraded by 5.28 dB in the multichannel case as compared to the single-channel case. Using Eq. (4.3-38) or Figure 4.3-16 one may determine the necessary $(S/I)_{IF}$ to maintain a 30-dB output signal-to-distortion ratio:

$$\left(\frac{S}{I}\right)_{IF} = \frac{16{,}000M}{3(B_T/W_0)(B_T/MW_0 - 2.52)^2}. \tag{4.3-99}$$

Combining (4.3-98b) and (4.3-99) we obtain a plot of the maximum allowable total power, P_T, versus the saturation level, P_S, in order that the output signal-to-distortion level is better than 30 dB for FM/SSB, as also shown in Figure 4.3-25.

4.3.7 Preemphasis and Deemphasis

As shown in Ref. 2 (pp. 387–388) the optimum signal-to-noise ratio in FM systems for a given signal spectrum is obtained by using a preemphasis gain, $H(f)$, given by

$$\text{preemphasis power gain} \overset{\Delta}{=} |H(f)|^2_{opt}$$

$$= \left[\frac{G_n(f)}{G_s(f)}\right]^{1/2}, \tag{4.3-100}$$

where $G_n(f)$ is the power spectrum of the baseband noise at the FM demodulator output and $G_s(f)$ is the power spectrum of the baseband signal before the preemphasis network at the FM modulator and after the deemphasis network at the FM demodulator. The result, (4.3-100), assumes that the mean square frequency deviation of the frequency modulation after the preemphasis network is held constant. Since the FM spectrum width is mainly determined by the mean square deviation [assuming the top baseband frequency is fixed; e.g., see Eq. (4.3-1)], fixing the mean square frequency deviation allows the RF bandwidth to remain fixed as various preemphasis networks are compared.

In the cases of SSB/FM and FM/FM where the block of subchannels frequency modulate the main carrier, the subchannels are of equal strength before preemphasis so that we may approximate the signal spectrum $G_s(f)$ as a constant, G_{s_0}, and the noise spectrum out of the demodulator when operating entirely above threshold is parabolic (neglecting effects of modulation on the noise spectrum, oscillator phase noise, and baseband amplifier noise):

$$G_n(f) = G_{n_0} f^2 \qquad \text{(Ref. 2, p. 483)} \tag{4.3-101}$$

Thus, for optimum output signal-to-noise ratio the preemphasis power gain would be linear in f

$$H(f)^2_{\text{opt}} = Kf. \tag{4.3-102}$$

However, equal signal-to-noise ratio on all subchannels is usually of primary importance, rather than optimum overall signal-to-noise ratio of the complete subchannel block, in such multiplex systems. Note that (4.3-101) and (4.3-102) result in an output noise spectrum that is linear in f while the signal spectrum, as mentioned above, is flat. Thus (4.3-102) results in a signal-to-noise ratio that varies from subchannel to subchannel. In order that each subchannel result in an equal signal-to-noise ratio, parabolic preemphasis is usually employed, as described in Section 4.3.1.

For an individual voice channel, however, the overall signal-to-noise ratio may be a suitable criterion of performance. The difficulty in applying (4.3-100) is that the signal spectrum shape varies from person to person and from time to time. We will examine five possible choices of preemphasis/deemphasis functions and compare their signal-to-noise ratio performance with some typical voice channel signals.

1. Preemphasis/deemphasis function No. 1 (straight FM, frequency modulation):

$$|H_1(f)|^2 = 1, \tag{4.3-103}$$

2. Preemphasis/deemphasis function No. 2 (straight PM phase modulation):

$$|H_2(f)|^2 = K_2 f^2, \tag{4.3-104}$$

3. Preemphasis/deemphasis function No. 3 (standard 75 μsec RC network):

$$|H_3(f)|^2 = K_3 \left[1 + \left(\frac{f}{2.1} \right)^2 10^{-6} \right], \tag{4.3-105}$$

4. Preemphasis/deemphasis function No. 4 (Standard Electronics Industries Association Land Mobile Receiver)[43]:

$$|H_4(f)|^2 = K_4 [(f^2 + 180^2)(f^2 + 190^2)(f^2 + 600^2)(f^2 + 800^2)] f^{-4}, \tag{4.3-106}$$

Equation (4.3-106) was obtained by polynomial curve fitting the typical response curve presented in Figure 3 of Ref. 43. As shown in Figure 4.3-27 the above polynomial fits the response curve to within 1 dB over the range from 200 Hz to 3 KHz.

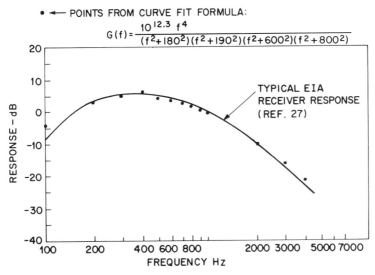

Figure 4.3-27 Curve fit of polynomial to typical electronics industries association land mobile radio system receiver response.

5. Preemphasis/deemphasis function No. 5 [optimum network for the average voice spectrum of men and women defined by Eq. (4.3-100)]:

$$H_5(f)^2 = K_5[(f^2 + 70^2)(f^2 + 180^2)(f^2 + 400^2)(f^2 + 700^2)]^{1/2}f^{-1},$$

$$(4.3\text{-}107)$$

Equation (4.3-107) was obtained by substituting the polynomial curve fit of the long-term average spectral density for continuous speech of six men and five women (as presented in Ref. 51, p. 78) for the signal spectrum G_s and Eq. (4.3-101) for the noise spectrum G_n into Eq. (4.3-100). The polynomial fit is shown in Figure 4.3-28.

The performance of the above preemphasis/deemphasis functions will be compared for five typical voice signal models. When comparing performance with different voice models we assume the powers in each voice model are equal, as a starting point. Thus, after preemphasis, unless automatic gain is used, all voice models will not have equal power.

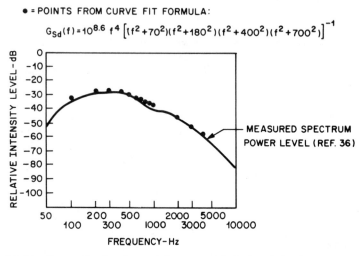

● = POINTS FROM CURVE FIT FORMULA:

$$G_{sd}(f) = 10^{8.6} \, f^4 \left[(f^2 + 70^2)(f^2 + 180^2)(f^2 + 400^2)(f^2 + 700^2)\right]^{-1}$$

Figure 4.3-28 Curve fit of polynomial to general shape of the long-term average spectral density for continuous speech (6 men, 5 women).

1. *Voice model* (a).   Sine wave at the low-frequency end of the voice band, 300 Hz:

$$G_{sa}(f) = C_a \delta(f - 300). \qquad (4.3\text{-}108)$$

where $\delta(f)$ is the Dirac delta function.

2. *Voice model* (b).   Sine wave at the high-frequency end of the voice band, 3 KHz:

$$G_{sb}(f) = C_b \delta(f - 3000). \qquad (4.3\text{-}109)$$

3. *Voice model* (c).   Flat voice spectrum from 300 Hz to 3 KHz:

$$G_{sc} = C_c [U(f - 300) - U(f\text{-}3000)], \qquad (4.3\text{-}110)$$

where $U(f)$ is the unit step function; zero for $f < 0$, unity for $f > 0$.

4. *Voice model* (d).   Long-term average spectral density for continuous

speech (Ref. 51, p. 78):

$$G_{sd} = C_d f^4 [(f^2 + 70^2)(f^2 + 180^2)(f^2 + 400^2)(f^2 + 700^2)]^{-1}. \quad (4.3\text{-}111)$$

5. *Voice model* (e). The vowel "a" as in "ate" at a pitch of 256 Hz (Ref. 51, p. 54):

$$G_{se} = C_e \{ 1.44\delta(f - 256) + 9\delta(f - 512) + 0.64\delta(f - 768)$$

$$+ 2.56\delta(f - 1024) + 0.16\delta(f - 2280) + 0.36\delta(f - 1536)$$

$$+ \delta(f - 1792) + 1.44\delta(f - 2048) + 1.69\delta(f - 2304)$$

$$+ \delta(f - 2560) + 0.04\delta(f - 2816) \}. \quad (4.3\text{-}112)$$

Equation (4.3-112) is taken from the line spectrum given in Ref. 51, p. 54. The output signal-to-noise ratio relative to that obtained for straight FM (preemphasis/deemphasis function No. 1) is computed for each preemphasis/deemphasis function, for each voice model. Given a set of preemphasis gains, the relative SNR performance of each respective function is independent of the voice model, since the signal out always equals the signal in. The criterion that the preemphasis gains be chosen to give the same mean square frequency deviation (and therefore approximately the same FM bandwidth) to compare performance will depend on which voice model is used to adjust gains. Therefore, we will repeat the above comparisons using each voice model in turn to determine the gain of each preemphasis function so that all functions give the same mean square frequency deviation for the chosen voice model. These comparisons are shown in Table 1.

Table 1 *Relative Output* SNR dB

Preemphasis/Deemphasis Function	Voice Signal Type				
	a $C_a = 1$	b $C_b = 1$	c $C_c = 0.37 \times 10^{-3}$	d $C_d = 5.22 \times 10^8$	e $C_e = .056$
1	0	0	0	0	0
2	15.7	−4.3	0	7.68	2.75
3	3.1	−1.65	0.75	3.2	1.8
4	13.1	−8.5	−2.32	7.96	1.26
5	16.3	−6.96	−1.4	8.1	1.75

Assume, in addition, that when the voice modulation is so large that the maximum frequency deviation allowed by the IF bandwidth will be exceeded, an automatic gain control (AGC) will adjust the preemphasis gain to limit the frequency deviation to the maximum level. When comparing two preemphasis/deemphasis systems, the system which limits first will degrade in output signal-to-noise ratio relative to the other by the number of decibels that the voice modulation is increased until the other system limits. Table 1 may also be interpreted as showing the relative performance for all the preemphasis/deemphasis functions under the condition that the voice modulation is assumed to have carried all systems into limiting. If one system is in limiting but another is not, their relative performance is between that indicated in the corresponding columns of Table 1.

In computing the table we make use of the following formulas:

$$\begin{bmatrix} \text{mean square} \\ \text{frequency deviation} \end{bmatrix} \overset{\Delta}{=} \sigma_f^2 = \int_{300}^{3000} |H(f)|^2 G_s(f)\, df, \quad (4.3\text{-}113)$$

$$\begin{bmatrix} \text{relative signal-to-} \\ \text{noise output without} \\ \text{limiting} \end{bmatrix} \overset{\Delta}{=} \gamma_0 = \frac{\displaystyle\int_{300}^{3000} f^2\, df}{\displaystyle\int_{300}^{3000} |H(f)|^{-2} f^2\, df}$$

$$= \frac{[\text{noise out of No. 1}]}{\begin{bmatrix} \text{noise out of} \\ \text{other system} \end{bmatrix}}, \quad (4.3\text{-}114)$$

$$\begin{bmatrix} \text{relative signal} \\ \text{at limiting} \end{bmatrix} \overset{\Delta}{=} \Gamma_r = \left[\int_{300}^{3000} |H(f)|^2 G_s(f)\, df \right]^{-1}, \quad (4.3\text{-}115)$$

$$\begin{bmatrix} \text{relative signal-to-} \\ \text{noise output with} \\ \text{limiting} \end{bmatrix} \overset{\Delta}{=} \gamma_L = \lambda_0 \Gamma_r. \quad (4.3\text{-}116)$$

The use of Table 1 may be described as follows: The large index assumption that the mean square frequency deviation determines the bandwidth of the FM wave implies that the bandwidth of the straight FM case (preemphasis/deemphasis function No. 1) is the same for all voice signal types (assuming equal signal powers; i.e., assuming proper choice of the C's). Thus the straight FM case gives the same mean square frequency deviation for all voice signal types and (neglecting the effect of modulation on output noise) the same output signal-to-noise ratio for all voice signal types. Table 1 uses straight FM as a reference for output signal-to-noise ratio. The other preemphasis/deemphasis functions would give different mean square frequency deviations for each voice signal type if the overall gain were not adjusted in each case. Thus, in order to compare their performance to the straight FM case, we assume in each column of Table 1 that the mean square frequency deviation resulting from the various preemphasis/deemphasis functions is equal to that of the straight FM case for the particular voice signal type heading the column (by adjusting the gain of the preemphasis/deemphasis networks). For example the decibel value given in row 3, column c, of Table 1 implies that the output signal-to-noise ratio of preemphasis/deemphasis system No. 3 is 0.75 dB higher than that for straight FM for all voice signal types when the gain of system No. 3 is adjusted to give the same mean square frequency deviation as does straight FM for voice signal type c.

If the modulation signal is strong enough to require AGC limiting in all cases to the maximum allowable mean square frequency deviation, the output signal-to-noise ratio depends on which voice signal type is transmitted. Another interpretation of Table 1 is the relative output SNR of each preemphasis/deemphasis system assuming that a given voice signal type is transmitted and the mean square frequency deviation is always set equal to the same limiting value (determined by the allowable bandwidth). Therefore, output SNR varies with voice model type. For example, the decibel value given in row 4, column d, of Table 1 implies that the output SNR of preemphasis/deemphasis system No. 4 is 7.96 dB higher than that of straight FM when their frequency deviations are equal to the same limiting value dictated by bandwidth, and voice signal, d, is being transmitted on both.

As seen from Table 1, in all cases preemphasis/deemphasis function No. 2 (straight phase modulation, PM) is better or within 2.65 dB of preemphasis/deemphasis function Nos. 3, 4, and 5. Only in one case (the 3-KHz sine wave voice signal, does it perform worse than straight FM, and then by only 4.3 dB, as compared to 15.7 dB better for a 300 Hz sine wave voice signal). For this reason and because of its simplicity, we choose straight PM as giving good overall performance of preemphasis/ deemphasis functions for angle modulation by voice-type signals.

Although the cases considered above are by no means a complete survey of possible preemphasis/deemphasis functions and voice signal types, they are representative enough to indicate the approximate performance that can be obtained and to justify the choice of straight PM for good overall performance with voice-type signals.

We note in passing that many authors (e.g., Refs. 2, 52, and 53) present preemphasis/deemphasis concepts in terms of the RC network (function No. 3) and do not place any restrictions on relative mean square frequency deviation when comparing to straight FM. In applying the formulas thus obtained (e.g., Refs. 41 and 54) to practical applications, one should be aware that the variations in output SNRs are partly due to preemphasis and partly due to IF bandwidth variations.

When receiving signals in a fading environment, the click noise experienced by FM receivers operating near threshold can reshape the baseband noise to be rather flat at low frequencies and cross over to the characteristic parabolic shaped noise at some frequency, f_{c_0}. Also, oscillator phase noise and baseband amplifier noise can flatten the noise spectrum at low frequencies.[29]

In contrast to the flat baseband noise spectrum, cochannel interference produces roughly a quadratic spectrum (Section 4.1.2) with frequency detection, as do linear and quadratic frequency-selective fading distortion (Section 4.1.8). Intermodulation distortion (Section 4.3.6) is similar to cochannel interference. However, adjacent-channel interference (Section 4.1.3) produces a higher than second power law baseband noise spectrum.

To see the effect of a flattened noise spectrum, Table 1 is recomputed assuming a flat noise spectrum (i.e., $f_{c_0} \gg 3$ kHz) and is shown in Table 2. When the cross-over frequency lies within the audio band (i.e., $300 < f_{c_0} < 3000$), the relative SNR performance will be between those figures given in Tables 1 and 2, respectively.

Table 2   *Relative Output SNR for flat Noise Spectrum* (dB)

Preemphasis/Deemphasis Function	Voice Signal Type				
	a	b	c	d	e
1	0	0	0	0	0
2	10	−10	−5.7	2	−2.95
3	1.9	−2.9	−0.5	2	0.5
4	6.7	−14.9	−8.75	1.5	−5.3
5	9.5	−13.6	−8.2	1.25	−5

In the case of a flat noise spectrum, as seen from Table 2, there is little difference between straight FM, straight PM, and the 75-μsec network, functions No. 1, 2, and 3, respectively. Since the noise spectrum will usually be closer to parabolic and since straight PM does not perform too badly (model b, a high pitched tone, is not too common in voice channels) with flat noise, we choose straight PM for good overall performance.

The quality of a voice channel is a subjective measure. Therefore, SNR may not be a valid criterion of comparison, and limiting the rms frequency deviation (syllabic companding) of the voice modulation may not be the most satisfactory amplitude control. However, in the absence of more complete information we have chosen these assumptions for our analysis.

4.4 MAN-MADE NOISE

By proper signal design and arrangement of frequencies in geographic cells, the mobile communications system designer may succeed in reducing the interference from undesired signals of other mobile units or base stations to an acceptable amount. The next major contributor to mobile communication interference is incidental man-made noise. By incidental man-made noise we mean noise that arises from the operation of equipment not designed or intended to emit radiation, such as automobile ignition systems, RF stabilized arc welders, plastic preheaters and welders, soldering machines, wood gluers, diathermy machines, microwave ovens, and the like.

In contrast, radio equipment that produces spurious signals, unintentionally in the band of a neighboring receiver, are classified as radio-frequency interference (RFI) sources. RFI problems are usually not widespread enough to warrant a statistical model, but are usually solved on an individual basis (e.g., locate the culprit and tell him to filter his transmitted signal). For this reason we will not consider RFI but will restrict our discussion to incidental man-made noise.

4.4.1 Characterization

In general, one would wish to have measurements or formulas that would allow him to compute the performance degradation due to man-made noise on any proposed communications system. What characteristics of the noise are needed for these computations? Typically, given the center frequency and the bandwidth of the proposed communications systems one might want to know the following properties of the noise: average total noise power, power spectrum, probability distribution of the noise voltage, probability distribution of pulse heights, rates, and widths, dependence on antenna polarization, direction, directivity, and height, and long-term variations with time and location. With thermal (Gaussian) noise the problem is much simpler; the power spectrum tells all.

Models for man-made noise are still embryonic. Typically[55] they indicate the average noise power per kilohertz of bandwidth as a function of frequency and distance from the center of a metropolitan area. The power is that received by a vertical short (compared to wavelength) dipole at a height of about 1 meter above street level. The magnitude and functional dependence of the expressions for the average power are not explicit enough to allow system calculations since they vary, for example, from city to city.

Hopefully, as more data are obtained, the models will progress to allow a more complete characterization of the man-made noise in a particular environment by measuring just a few parameters in that locale. Some models based on theory only[56] require comparison with measurement to determine parameter values and applicability.

4.4.2   Methods of Measurement

With so many variables describing the noise, the number of measurements required to characterize the noise becomes formidable. Also, the conditions under which each measurement is made must be completely specified in order reasonably to interpret the results.

The most common technique of measurement uses a receiver of fixed bandwidth (on the order of 10 kHz), with tunable center frequency, and the average received power is recorded. The average is over times on the order of 1 min at each location. This type of measurement is conveniently made with a spectrum analyzer.

Another method of measurement utilizes the quasipeak meter[57] as specified by C.I.S.P.R. (International Special Committee on Radio Interference) for the frequency range from 10 kHz to 1 GHz. Instead of averaging the video output of the receiver, as in the average power method, the video output is processed by a detector with different charging and discharging time constants; the ratio of these time constants is on the order of 1 : 600. As a result the response of the meter to periodic pulse trains is equal to the duty cycle times the time-constant ratio times the peak pulse voltage, if the duty cycle is small compared to the time-constant ratio. For large duty cycles the quasipeak meter reading approaches the pulse peak voltage. Originally the quasipeak meter was chosen because it gave a weighted response to pulse trains of different frequencies that correlated well with the subjective annoyance to the listener, based on the amplitude modulation broadcasting system.

More recently, measurement systems[58] attempt to record the instantaneous amplitude and phase of the noise in the RF receiver bandwidth. The receiver bandwidth is limited by the recorder bandwidth. Video tape recorders can allow the bandwidth to extend to several megahertz. The data can then be analyzed in the laboratory for most of the noise characteristics of interest: power spectra, probability distributions of amplitude, phase, pulse rate, pulse width, pulse height, and the like. In fact, if the communication system to be tested does not have a broader bandwidth than the recording, the recording may be used as a simulator to determine system performance (e.g., error rates) in the presence of man-made noise.

4.4.3 Some Compilations of Data

We will not attempt to present a complete description of the available data on measurements of man-made noise, but only summarize some of the urban data here. The list of references (55, 56, 59–146) will provide the reader with information on a large portion of previous measurements.

Figures 4.1-1 and 4.4-2 are presented to give some indication of the magnitude of man-made noise in an urban environment. Figure 4.4-1 represents a straight-line approximation to the average power measurements[147] (as described above) in New York City, Phoenix, Cleveland, Melbourne, and Tel Aviv. The spread in data is approximately 20 dB. The ordinate F_0 is the decibel difference between the received man-made noise spectral density and thermal noise spectral density, kT_0 where k is Boltzmann's constant and $T_0 = 290°$K. The ITT Handbook[148] provides data on man-made noise, reproduced in Figure 4.4-2. These curves are based on

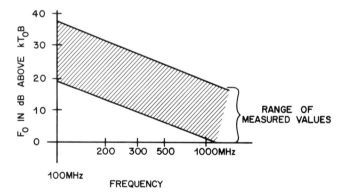

Figure 4.4-1 Composite urban average incidental man-made noise power.

measurements by Work Group 3 of the FCC Advisory Committee, Land Mobile Radio Service, and the Institute for Telecommunication Sciences and Aeronomy, Environmental Science Services Administration in Washington, D.C. Also shown in Figure 4.4-2 is the man-made noise for a suburban environment. Figures 4.4-1 and 4.4-2 are in rough agreement. In general, one may conclude from Figure 4.4-2 that man-made noise is negligible (relative to receiver front-end noise) for frequencies above 4 GHz.

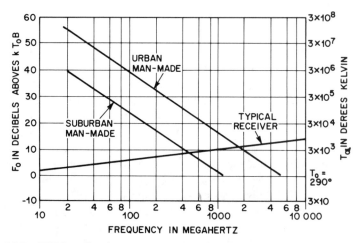

Figure 4.4-2 ITT handbook average man-made noise power curves.

REFERENCES

1. S. O. Rice, "Statistical Properties of a Sine Wave Plus Random Noise," *Bell System Tech. J.* **27**, pp. 109–157, January 1948.

2. M. Schwartz, *Information Transmission, Modulation and Noise*, 2nd ed., MGraw-Hill, New York, 1970, Sec. 6-11.

3. S. O. Rice, "Noise in FM Receivers," in *Proceedings, Symposium of Time Series Analysis*, M. Rosenblatt (ed.), Chapter 25, Figure 3, Wiley, New York, 1963.

4. H. E. Rowe, *Signals and Noise in Communications Systems*, Van Nostrand, Princeton, 1965, p. 17.

5. S. O. Rice, "Mathematical Analysis of Random Noise," *Bell System Tech. J.* **23**, July 1944, pp. 282–332; **24**, January 1945, pp. 46–156.

6. A Papoulis, *Probability, Random Variables, and Stochastic Processes*, McGraw-Hill, New York, 1965, p. 339.

7. B. R. Davis, "FM Noise with Fading Channels and Diversity," *IEEE Ttrans.*

Comm. Tech., **COM-19**, No. 6, Part II, December 1971, pp. 1189–1199.

8. V. K. Prabhu and L. H. Enloe, "Interchannel Interference Considerations in Angle Modulated Systems," *Bell System Tech. J.* **48**, No. 7, September 1969, pp. 2333–2358.

9. V. K. Prabhu and H. E. Rowe, "Spectral Density Bounds of a PM Wave," *Bell System Tech. J.* **48**, No. 3, March 1969, pp. 769–811.

10. D. Middleton, *An Introduction to Statistical Communication Theory*, McGraw-Hill, New York, 1960, p. 604.

11. A. Papoulis, *The Fourier Integral and Its Applications*, McGraw-Hill, New York, 1962, p. 47.

12. R. E. Langseth, private communication, October 1971.

13. W. H. Lob, "The Distribution of FM-Discriminator Click Widths," *Proc. IEEE*, **57**, No. 4, April 1969, pp. 732–733.

14. A. J. Rustako, Y. S. Yeh, and R. R. Murray, "Performance of Feedback and Switch Space Diversity 900 MHz FM Mobile Radio Systems with Rayleigh Fading," *Joint IEEE Comm. Soc.–Veh. Tech. Group Special Trans. on Mobile Radio Comm.*, November, 1973, pp. 1257–1268.

15. M. Abramowitz and I. A. Stegun, *Handbook of Mathematical Functions*, AMS-55, National Bureau of Standards, Washington, D.C., 1964, p. 946.

16. R. S. Kennedy, *Fading Dispersive Communication Channels*, Wiley Interscience, New York, 1969, p. 49.

17. I. S. Gradshteyn and I. M. Ryzhik, *Tables of Integrals, Series and Products*, Academic, New York, 1965.

18. Y. L. Luke, *Integrals of Bessel Functions*, McGraw-Hill, New York, 1962.

19. A. L. Hopper, "An Experimental Fast Acting AGC Circuit," *IRE International Convention Record (USA)*, **10**, Part 8, March 1962, pp. 13–20.

20. A. Y. S. Yeh, private communication, 1970.

21. W. R. Bennett, H. E. Curtis, and S. O. Rice, "Interchannel Interference in FM and PM Systems Under Noise Loading Conditions," *Bell System Tech. J.* **34**, No. 3, May 1955.

22. R. G. Medhurst and J. H. Roberts, "Evaluation of Distortion in FM Trunk Radio Systems by a Monte Carlo Method," *Proc. IEE*, **113**, No. 4, April 1966, pp. 570–580.

23. P. A. Bello and B. D. Nelin, "The Effect of Frequency Selective Fading on Intermodulation Distortion and Subcarrier Phase Stability in Frequency Modulation Systems," *IEEE Trans. Comm. Systems*, **COM-12**, March 1964, pp. 87–101.

24. V. K. Prabhu, "Some Considerations of Error Bounds in Digital Systems," *Bell System Tech. J.* **50**, No. 10, December 1971, pp. 3127–3151.

25. R. W. Lucky, J. Salz, and E. J. Weldon, "Principles of Data Communication," McGraw-Hill, New York, 1968.

26. W. R. Bennett and J. R. Davey, *Data Transmission*, McGraw-Hill, New York, 1965.

27. E. D. Sunde, *Communication Systems Engineering Theory*, Wiley, New York, 1969.

28. M. Schwartz, W. R. Bennett, and S. Stein, *Communication Systems and Techniques*, McGraw-Hill, New York, 1966.

29. *Transmission Systems for Communications*, 3rd ed., Technical Staff Bell Telephone Laboratories, 1964.

30. J. A. Greefkes and K. Riemens, "Code Modulation with Digitally Controlled Companding for Speech Transmission," *Phillips Tech. Rev.*, **31**, No. 11/12, 1970, pp. 335–353.

31. T. T. Tjhung, "Band Occupancy of Digital FM Signals," *IEEE Trans. Comm. Systems*, **Com-12**, December 1964, pp. 211–216.

32. J. E. Mazo and J. Salz, "Theory of Rrror Rates for Digital FM," *Bell System Tech. J.* **45**, November 1966, pp. 1511–1535.

33. J. Salz and V. G. Koll, "An Experimental Digital Multilevel FM Modem," *IEEE Trans. Comm. Tech.*, **Com-14**, No. 3, June 1966, pp. 259–265.

34. T. T. Tjhung and P. H. Wittke, "Carrier Transmission of Binary Data in a Restricted Band," *IEEE Trans. Comm. Tech.*, **Com-18**, No. 4, August 1970, pp. 295–304.

35. P. A. Bello and B. D. Nelin, "The Influence of Fading Spectrum on the Binary Error Probabilitites of Incoherent and Differentially Coherent Matched Filter Receivers," *IRE Trans. Comm. Systems*, **CS-10**, June 1962, pp. 160–168.

36. H. B. Voelker, "Phase-Shift Keying in Fading Channels," *Proc. IEE*, **107**, Part B, January 1960, p. 31.

37. A. T. Brennan, B. Goldberg, and A. Eckstein, "Comparison of Multichannel Radio Teletype Systems over a 5000-mile Ionosphere Path," *IRE Convention Record*, **8**, 1958, p. 254.

38. P. A. Bello and B. D. Nelin, "The Effect of Frequency Selective Fading on the Binary Error Probabilities of Incoherent and Differentially Coherent Matched Filter Receivers," *IEEE Trans. Comm. Systems*, **CS-11**, June 1963, pp. 170–186.

39. C. C. Bailey and J. C. Lindenlaub, "Further Results Concerning the Effect of Frequency Selective Fading on Differentially Coherent Matched Filter Receivers," *IEEE Trans. Comm. Tech.*, **Com-16**, October 1968, p. 749.

40. D. C. Cox, "910 MHz Urban Mobile Radio Propagation: Multipath Characteristics in New York City," *Joint IEEE Comm. Soc.–Veh. Tech. Group Special Trans. on Mobile Radio Comm.*, November 1973, pp. 1188–1194.

41. S. R. Lines, "Frequency-Division Multiplex Communications for the Land Mobile Radio Service," Report No. T-6801, Federal Communications Commission, Office of the Chief Engineer, June 20, 1968.

42. W. L. Behrend, "Multiplexing Land-Mobile Base Station Signals on the Carrier of an FM Broadcast Station," *IEEE Trans. Broadcasting*, **BC-13**, No. 2, April 1967, pp. 50–56.

43. C. N. Lynk, Jr., "Analysis of Multiplex Techniques for Land Mobile Radio Service," *IEEE Trans. Veh. Tech.* **VT-20**, No. 1, February 1971, pp. 1–12.

44. A. Anuff and M. L. Liou, "A Note on Necessary Bandwidth in FM Systems," *Proc. IEEE*, **59**, October 1971, pp. 1522–1523.

45. S. O. Rice, "Distortion Produced by Band Limitation of an FM Wave," *Bell System Tech. J.* **52**, May–June 1973, pp. 605–626.

46. R. K. Khatri and J. E. Wilkes, "Convolution Noise and Distortion in FDM/FM Systems," International Conference on Communications, Philadelphia Pennsylvania, June 19–21, 1972, Conference Record pp. 34–7 to 34–12.

47. E. Bedrosian and S. O. Rice, "Distortion and Crosstalk of Linearly Filtered, Angle-Modulated Signals," *Proc. IEEE*, **56**, No. 1, January 1968, pp. 2–13.

48. H. S. C. Wang, "Bandwidth Requirement for Frequency-Modulated Signals," *Proc. IEEE*, **53**, No. 8, August 1965, p. 1150.

49. L. Lundquist, "Channel Spacing and Necessary Bandwidth in FDM-FM Systems," *Bell System Tech. J.* **50**, No. 3, March 1971, pp. 869–880.

50. M. M. Nichols and L. L. Rauch, *Radio Telemetry*, 2nd ed., Wiley, New York, 1956, p. 90.

51. H. Fletcher, *Speech and Hearing in Communication*, Van Nostrand, New York, 1953.

52. P. F. Panter, *Modulation, Noise, and Spectral Analysis*, McGraw-Hill, New York, 1965, pp. 443–448.

53. H. S. Black, *Modulation Theory*, Van Nostrand, New York, 1953, pp. 228–231.

54. H. Magnuski and W. Firestone, "Comparison of SSB and FM for VHF Mobile Service," *Proc. IRE*, December 1956, p. 1837, Figure 5.

55. E. N. Skomal, "The Range and Frequency Dependence of VHF-UHF Man-Made Radio Noise in and above Metropolitan Areas," *IEEE Trans. Veh. Tech.* **VT-19**, No. 2, May 1970, pp. 213–221.

56. D. Middleton, "Man-Made Noise in Urban Environments and Transportation Systems: Models and Measurements," National Telecommunications Conference Record, Houston Texas, December 4-6, 1972, pp. 19C-1 to 19C-9.

57. F. L. H. M. Stumpers, "Interference to Communication Systems and the Work of C. I. S. P. R.," *Record of the 1971 IEEE International Conference on Communications*, **71C28-COM**, pp. 37–14 to 37–19.

58. A. D. Spaulding, "Amplitude and Time Statistics of Urban Man-Made Noise," *Record of the 1971 IEEE International Conference on Communications*, **71C28-COM**, pp. 37–8 to 37-13.

59. Y. Amamiya, "Research on Radio Noise in Medium and High Frequency Regions Generated by Electric Cars," *Bull. IRCA*, April 1962, pp. 192–203.

60. G. Anzic, "Radio Frequency Noise Measurements in Urban Areas at 480 and 950 MHz," NASA Tech. Memo TM-X-1972, March, 1970.

61. G. Anzic, "Measurement and Analysis of Radio Frequency Incidental Man-Made Noise," *1970 IEEE International EMC Symposium Record*, **70C28-EMC**, pp. 183–195.

62. G. Anzic and C. May, "Results and Analysis of a Combined Aerial and Ground Ultrahigh Frequency Noise Survey in an Urban Area," NASA Technical Memorandum TMX-2244, April 1971.

63. Automobile Manufacturers Association, "Ambient Electromagnetic Survey, Detroit, Michigan," Engineering Report 69-15, Radio Committee, AMA, Detroit, Michigan, October 13–24, 1969.

64. Automobile Manufacturers Association, "Microwave (1–2.5 GHz) Radiation from Moving Vehicles," Engineering Report, Radio Committee, AMA, Detroit, Michigan, September, 1970.

65. Automobile Manufacturers Association, "Data Package, Radio Interference Tests, South Lyon, Michigan, June 14–17, 197," AMA, Inc., Detroit, Michigan, June, 1971.

66. C. R. W. Barnard, "VHF Noise Levels Over Large Towns," Royal Aircraft Establishment, Leatherhead, Surrey, England, AD 827 640, Tech. Report 67213, August, 1967.

67. F. Baure, "Vehicular Radio Frequency Interference—Accomplishment and Challenge," *IEEE Trans. Veh. Tech.*, **VT-16**, October, 1967, pp. 58–68.

68. C. R. Bond, W. E. Pakala, R. E. Graham, and J. E. O'Neil, "Experimental Comparisons of Radio Influence Fields from Short and Long Transmission Lines," *IEEE Trans. Power Apparatus and Systems*, **82**, April 1963, pp. 175–185.

69. J. E. Bridges, L. W. Thomas, and W. E. Cory, "A Progress Report on G-EMC Standards Activities," *IEEE EMC Symposium Record*, **71C29-EMC**, July, 1971, pp. 23–29.

70. R. E. Buck, L. A. Frasco, H. D. Goldfein, S. Karp, L. Klein, E. T. Leonard, J. Liu, and P. Yoh, "Ground Vehicle Communications and Control," DOT Report No. DOT-TSC-UMTA-71-6, August 1971, pp. 76–83.

71. W. E. Buehler and C. D. Lunden, "Signature of Man-Made High-Frequency Radio Noise," *IEEE Trans. Electromagnetic Compatibility*, **EMC-8**, September 1966, pp. 143–152.

72. W. E. Buehler, C. H. King, and C. D. Lunden, "VHF City Noise," *1968 IEEE EMC Symposium Record*, **68012-EMC**, pp. 113–118.

73. C.C.I.R. Report 322, "World Distribution and Characteristics of Atomospheric Radio Noise," Geneva, 1964.

74. C.C.I.R. Report 413, "Operating Noise Threshold of a Radio Receiving System," Documents of the XIth Plenary Assembly, Oslo, 1966, Reports by the International Working Party III/1, Geneva, 1966.

75. D. B. Clark, "Evaluation of Interference Suppression of Fluorescent Lamps," U. S. Naval Civil Eng. Lab., Tech. Report 166, Oct., 1961.

76. R. T. Disney and A. D. Spaulding, "Interference Measurements and Analysis," Final Report to SAMSO, Vol. VI, prepared by ESSA under Contract

No. F04701-68-F-0072, Project 672A, ESSA, Item 19, CDRL, December 1968.

77. R. T. Disney and A. D. Spaulding, "Amplitude and Time Statistics of Atomospheric and Man-Made Radio Noise," ESSA ERL 150-ITS 98, U.S. Dept. of Commerce, Washington, D.C. 20402, February 1970.

78. R. T. Disney, "Estimates of Man-Made Radio Noise Levels Based on the Office of Telecommunications, ITS Data Base," *IEEE ICC Conf. Rec.*, **72CH0622-1-COM**, June 1972, pp. 20/13–20/19.

79. R. T. Disney, A. D. Spaulding, and D. H. Zacharisen, "Electromagnetic Interference from Man-Made Noise, Part I: Estimates of Business, Residential, and Rural Area Characteristics," to be Published as OT/ITS Telecommunications Research and Engineering Report (U.S. Government Printing Office, Washington, D.C. 20402).

80. R. T. Disney, A. D. Spaulding, and J. G. Osborn, "Electromagnetic Interference from Man-Made Radio Noise, Part II: Bibliography," to be published as an OT/ITS Telecommunications Research and Engineering Report (U.S. Government Printing Office, Washington, D.C. 20402).

81. J. L. Dolan, "EMC in Radio Astronomy," *1970 IEEE International EMC Symposium Record*, **70C28-EMC**, pp. 365–375.

82. A. C. Doty, Jr., "A Progress Report on the Detroit Electromagnetic Survey," *1971 IEEE International Electromagnetic Compatibility Symposium Record*, pp. 105–118.

83. D. D. Eaglesfield, "Motor-Car Ignition Interference," *Wireless Engineer*, October 1946, pp. 265–272.

84. C. Egidi, "Measurement and Suppression of VHF Radio Interference Caused by Motorcycles and Motor Cars," *IRE Trans. Radio Frequency Interference, New York*, May 1961, pp. 30–39.

85. A. G. Ellis, "Site Noise and its Correlation with Vehicular Traffic Density," *Proc. IRE (Australia)*, January 1963, pp. 45–52.

86. ESSA/ITS, "Final Report to Space and Missile Systems Organization, Communications Task—Project 672A," Contract AF D/O (04-694)-67-3, October 1967.

87. ESSA, "Interference Field Tests, Interim Evaluation Report to Space and Missile Systems," Contract F04701-68-F-0072, Communications Field Test FY68-69, Project 672A, Item 7, CRDL Supplemental, December 1968.

88. Fredrick Research Corp., "Factors for Predicting Radio Frequency Interference from Vehicles Ignition Systems," National Radio Astronomy Observatory Contract RAP-42, January 1964.

89. K. Furutsu and T. Ishida, "On the Theory of Amplitude Distribution of Impulsive Random Noise," *J. Appl. Phys. (Japan)*, **32**, No. 7, July 1961, pp. 1206–1221.

90. H. Garlan and G. L. Whipple, "Field Measurements of Electromagnetic Energy Radiated by RF Stabilized Arc Welders," FCC Tech. Div. Report T-6401, February 1964.

91. R. W. George, "Field Strength of Motor Car Ignition Between 40 and 450 Megacycles," *Proc. IRE*, **28**, September 1940, pp. 409–413.

92. J. P. German, "A Study and Forecast of the Electromagnetic Spectrum Technology. Part I and II," Texas A&M Research Foundation, College Station, Texas, 1971.

93. A. J. Gill and S. Whitehead, "Electrical Interferences with Radio Reception," *J. IEE (London)*, **83**, 1938, pp. 345–386.

94. A. A. Giordano, "Modeling of Atmospheric Noise," Doctoral Dissertation, Dept. of Electrical Engineering, Univ. of Pennsylvania, 1970.

95. V. N. Glinka and G. M. Myaskovskiy, "Some Time and Spectral Characteristics of the Over All Radio Interference in a Large City," *Telecommun. Radio Eng. (USSR)*, February 1967, pp. 40–44.

96. H. M. Hall, "A New Model for 'Impulsive' Phenomena: Application to Atomospheric Noice Communications Channels," Stanford Electronic Laboratories Technical Report Nos. 3412-8 and No. 7050-7, August, 1966.

97. E. G. Hamer, "Noise Performance of VHF Receivers," *Electronic Engineering (London)*, February 1953, pp. 68–70.

98. J. R. Herman, "Survey of Man-Made Radio Noise," *Progress in Radio Science 1966–1969*, Vol. I, G. M. Brown, N. D. Clarence, and M. J. Rycroft (eds.), URSI, 1970, pp. 315–348.

99. F. Horner, "Techniques Used for the Measurement of Atmospheric and Man-Made Noise," *Progress in Radio Science 1966–1969*, Vol. II, J. A. Lane, J. W. Findlay, and C. E. White (eds.), URSI, Brussels, 1970, pp. 177–181.

100. C. Huang, "Characteristics of Man-Made Noise in Taiwan," Dept. of Electrical Eng., National Taiwan University, July 1969.

101. A. G. Hubbard, "Man-Made Noise Measurements: An Unsolved Problem?", National Telecommunications Conference Record, Houston, Texas, December, 1972, pp. 19E-1 to 19E-3.

102. "Interference from Fluorescent Tubes," *Wireless World, (London)*, March 1960, pp. 93–84.

103. E. Jacobs, "Broadband Noise: Its Physical Nature and Measurement," *1970 IEEE International Symposium on EMC*, **70C28-EMC**, pp. 15–19.

104. A. de Jong: "Impulsive Interferences in Television Transmissions," *Het P.T.T. Bedrijf*, Part XVI, No. 2, March 1969, p. 112.

105. JTAC, "Spectrum Engineering—The Key to Progress, Man-Made Radio Noise," Report of Joint Technical Advisory Committee of the IEEE, Subcommittee 63.1.3 (Unintended Radiation), Supplement 9, Published by the IEEE, 1968, pp. 65–69.

106. F. D. Lewis and R. A. Soderman, "Radio-Frequency Standardization Activities," *Proc. IEEE*, **53**, June 1967, pp. 759–773.

107. R. J. Matheson, "Instrumentation Problems Encountered Making Man-Made Electromagnetic Noise Measurements for Predicting System Perfor-

mance," *IEEE Trans. on Electromagnetic Compatibility*, **EMC-12**, 4, November 1970, pp. 151–158.

108. D. Middleton, "Statistical-Physical Models of Urban Radio-Noise Environments, Part I: Foundations," *IEEE Trans. Electromagnetic Compatibility*, **EMC-14**, No. 2, May 1972, pp. 38–56.

109. D. Middleton, "Statistical-Physical Models of Man-Made Radio Noise. I. First-Order Probability Models," to be published as a Telecommunications Research and Engineering Report, ITS/OT, U.S. Dept. of Commerce, Boulder, Colo.

110. D. Middleton, "Statistical-Physical Models of Man-Made Radio Noise, Part II, First-Order Probability Models of Envelope and Phase.", to be published as a Telecommunications Research and Engineering Report, OT/ITS, U.S. Dept. of Commerce, Boulder, Colo.

111. A. H. Mills, "A Survey of Radio-Frequency Noise in Urban, Suburban, and Rural Areas," *1968 IEEE EMC Symposium Record*, **68C12-EMC**, pp. 108–112.

112. J. Meyer de Stadelhofen: "Measure des Perturbations Impulsives et Quasi Impulsives Affectant la Reception de la Télévision," *Bull. Tech. P.T.T.*, No. 5, 1967.

113. J. Meyer de Stadelhofen: "A New Device for Radio Interference Measurements at VHF: The Absorbing Clamp," *1969 IEEE Electromagnetic Compatibility Symp. Rec.*, pp. 189–193.

114. H. Myers, "Industrial Equipment Spectrum Signatures," *IEEE Transactions on Radio Frequency Interference*, **Vol. RFI-5**, March 1963, pp. 30–42.

115. T. Nakamura, M. Inoue, H. Suzuki, and K. Endo, "Characteristics of City Noise in the VHF Band," N.H.K. Laboratories Note, Number T-3, March, 1965.

116. W. Nethercot, "Car Ignition Interference," *Wireless Engineer*, August 1949, pp. 251–254.

117. M. Ottesen, "Electromagnetic Compatibility of Random Man-Made Noise Sources," Ph.D. Dissertation, University of Colorado, 1968.

118. G. P. Pacini, R. Gaudio, and F. Rossi Doria, "Experimental Investigation on Man-Made Noise in 850 MHz and 12 GHz Frequency Bands," *Alta Frequenza (English Issue)*, **XL**, No. 2, February 1971, pp. 132–139.

119. W. E. Pakala, E. R. Taylor Jr., and R. T. Harrold, "Radio Noise Measurements on High Voltage Lines From 2.4 to 345 KV," *1968 IEEE EMC Symposium Record*, **68C12-EMC**, pp. 96–107.

120. I. I. K. Pauling-Toth and J. R. Shakeshaft, "A Note on Radio Interference at a Frequency of 408 Mc/s," *Electronic Eng.*, July 1962, p. 488.

121. S. F. Pearce and J. H. Bull, "Interference From Industrial, Scientific and Medical Radio-Frequency Equipment," The Electrical Research Association, Leatherhead, Surrey, England, October 1964, Report 5033.

122. M. R. Pelissier, "Les Perturbations Radiophoniques Émises Par Les Lignes a Très Haute Tension," *Bull. Soc. Franc. Elec.*, July 1953, 409–418.

123. G. Ploussios, "City Noise and its Effect Upon Airborne Antenna Noise Temperatures at UHF," *IEEE Trans. Aerospace and Electronic Systems*, **AES-4**, January 1968, pp. 41–51.

124. B. G. Pressey and G. N. Ashwell, "Radiation from Car Ignition System," *Wireless Engineer*, January 1949, pp. 31–36.

125. L. C. Simpson, "Israel Intercity UHF Telephone Communication System," *RCA Rev.*, March 1953, pp. 100–124.

126. E. N. Skomal, "Distribution and Frequency Dependence of Unintentionally Generated Man-Made VHF/UHF Noise in Metropolitan Areas," Aerospace Corp. Report TDR-469 (S5805-45)-2, November 1964.

127. E. N. Skomal, "Distribution and Frequency Dependence of Unintentionally Generated Man-Made VHF/UHF Noise in Metropolitan Areas," *IEEE Trans. EMC*, **EMC-7**, September 1965, pp. 263–278.

128. E. N. Skomal, "Distribution and Frequency Dependence of Unintentionally Generated Man-Made VHF/UHF Noise in Metropolitan Areas, Part II, Theory," *IEEE Trans. EMC*, **EMC-7**, December 1965, pp. 420–427.

129. E. N. Skomal, "Analysis of the Frequency Dependence of Man-Made Radio Noise," *IEEE International Convention Record*, Part II, 1966, pp. 125–129.

130. E. N. Skomal, "Man-Made Noise, Pt. 1: Sources and Characteristics," *Frequency*, January–February 1967, pp. 14–19.

131. E. N. Skomal, "Comparative Radio Noise Levels of Transmission Lines, Automotive Traffic, and rf Stabilized Arc Welders," *IEEE Trans. EMC*, **EMC-9**, September 1967, pp. 73–77.

132. E. N. Skomal, "Analysis of Airborne VHF/UHF Incidental Noise Over Metropolitan Areas," *IEEE Trans. Electromagnetic Compatibility*, **EMC-11**, May 1969, pp. 76–84.

133. E. N. Skomal, "Distribution and Frequency Dependence of Incidental Man-Made HF/VHF Noise in Metropolitan Areas," *IEEE Trans. Electromagnetic Compatibility*, **EMC-11**, May 1969, pp. 66–75.

134. E. N. Skomal, "The Dimensions of Radio Noise," *1969 IEEE EMC Symposium Record*, pp. 18–28.

135. E. N. Skomal, "The Conversion of Area Distributed, Incidental Radio Noise Envelope Distribution Functions by Radio Propagation Processes," *IEEE Trans. Electromagnetic Compatibility*, **EMC-12**, August 1970, pp. 83–88.

136. E. N. Skomal, "Recent Extensions of Composite, Incidental Man-Made Noise Data and Their Relevance to the Hypothesis of the Noise Envelope Statistic Transformation," *1971 IEEE International EMC Symposium Record*, **71C29-EMC**, pp. 222–234.

137. E. N. Skomal, "Emerging Trends in UHF Incidental Noise," *1971 IEEE Mountain-West EMC Conference Record*, **71C60-MC**, pp. 155–166.

138. A. D. Spaulding, C. J. Roubique, and W. Q. Crichlow, "Conversion of the Amplitude Probability Distribution Function for Atmospheric Radio Noise From One Bandwidth to Another," *J. Res. Natl. Bur. Std.*, **66D**, November—December 1962, pp. 713–720.

139. A. D. Spaulding, R. T. Disney, and L. R. Espeland, "Noise Data and Analysis, Phase C Final Report to SAMSO," Prepared by ESSA Under Contract No. F04701-68-F-0072, Program 125B, ESSA Tech. Memo. ERLTM-ITS 184, June 1969.

140. A. D. Spaulding, "Measured Versus Theoretical Performance of Receivers of the Minuteman Launch Control Type," ESSA Tech. Memo. ERLTM-ITS 234, May 1970.

141. A. D. Spaulding, W. H. Ahlbeck, and L. R. Espeland, "Urban Residential Man-Made Radio Noise Analysis and Predictions," OT/ITS Telecommunications Research and Engineering Report 14, June 1971 (U.S. Government Printing Office, Washington, D.C. 20402).

142. F. L. H. M. Stumpers: "Progress in the Work of C.I.S.P.R.," *IEEE Trans. Electromagnetic Compatibility*, **EMC/12**, May 1970, pp. 29–32.

143. H. Suzuki, "Characteristics of City Noise in the UHF Band," *J. Inst. Elect. Commun. Eng. (Japan)*, **46**, February 1963, pp. 186–194.

144. W. I. Thompson III, "Bibliography on Ground Vehicle Communications and Control: A KWIC Index," Urban Mass Transportation Administration, U.S. Department of Transportation, Washington, D.C., Report No. DOT-UMTA-7-3, July, 1971.

145. W. R. Vincent, "Oberservations of Man-Made Noise and rfi in Urban and Suburban Areas," *IEEE 1971 Mountain-West EMC Conference Record*, **71C60-MC**, pp. 183–189.

146. W. R. Young, "Comparison of Mobile Radio Transmission at 150, 450, 900 and 3700 Mc/s," *Bell System Tech. J.* November 1952, pp. 1068–1085.

147. E. N. Skomal, "Characteristics of Urban Median Indicidental Man-Made Radio Noise," National Telecommunications Conference Record, Houston, Texas, December 4–6, 1972, pp. 19B-1 to 19B-5.

148. *ITT Reference Data for Radio Engineers*, 5th ed., Howard Sams, New York, 1968, pp. 27-1 to 27-2.

chapter 5

fundamentals of diversity systems

W. C. Jakes, Y. S. Yeh, M. J. Gans, and D. O. Reudink

SYNOPSIS OF CHAPTER

The first two chapters have shown in explicit detail the extreme and rapid signal variations associated with the mobile radio transmission path. We now turn our attention to methods of reducing these vicissitudes to a range that will permit acceptable transmission of voice. As the title indicates, the fundamental principles governing the operation of diversity systems are examined in this chapter. The first section contains a brief description of the basic diversity classifications: time, frequency, and space, with indications of restrictions appropriate to the mobile radio case. Space diversity appears to be favored for mobile radio use.

Section 5.2 contains a discussion of four different diversity combining methods, and develops theoretical results for the improvement in fading probability statistics that can be achieved by their use. The success of diversity techniques depends on the degree to which the signals on the different diversity branches are uncorrelated. Calculations are presented to show that correlation coefficients as high as 0.7 can still yield good improvement. Likewise, errors in the diversity combiner circuit will degrade performance, and this effect is analyzed.

Multi-element antenna arrays are generally the preferred technique for achieving the branch signals for space diversity. Parameters available to the designer include the array configuration (linear or two dimensional), interelement spacing, number of elements, and load impedance. These factors are analyzed in Section 5.3.

The user, of course, is only concerned with the signal he can hear, that is, the baseband signal. The ability of diversity to provide a clean, quiet signal in the presence of Rayleigh fading when FM is used is examined in

Section 5.4. The dependence of the baseband S/N ratio on the number of diversity branches, signal level, vehicle speed, and cochannel interference is analyzed. The use of squelching circuits to deaden the baseband during a fade is also considered.

The chapter is concluded by a study of the use of widely separated base stations to overcome the large-scale fading caused by terrain obstructions. Section 5.5 shows how significant improvements in average signal level and reductions in cochannel interference may be obtained using this technique.

5.1 BASIC DIVERSITY CLASSIFICATIONS

The principles of diversity combining have been known to the radio art for decades, with the first experiments being reported in 1927. The diversity method requires that a number of transmission paths be available, all carrying the same message but having independent fading statistics. The mean signal strengths of the paths should also be approximately the same. Proper combination of the signals from these transmission paths yields a resultant with greatly reduced severity of fading and correspondingly improved reliability of transmission. In the early experiments the independent paths were realized by transmitting from a single antenna to a number of receiving antennas. The distance between the receiving antennas was made large enough to ensure independent fading. This arrangement is called space-diversity reception; it will be shown in later sections that the roles of the transmitting and receiving antennas can be interchanged, resulting in space-diversity transmission. A greater degree of equipment sophistication is required in this case, as one might suspect.

With the principles of the diversity method in hand, it quickly became apparent that independent transmission paths suitable for the diversity method could be produced in other ways than using spaced antennas. Transmission paths at different frequencies or different times, for example, are acceptable under certain conditions.

The literature is replete with descriptions of the various kinds of diversity and their applications to conventional radio transmission systems from HF to SHF. Selected references are given for the reader who may wish to pursue the subject in greater depth.[1-6] In the following we will briefly describe a number of these alternatives and examine their applicability to mobile radio communications.

5.1.1 Space Diversity

This historical technique has found many applications over the years and is in wide use in a variety of present-day microwave systems. It is relatively simple to implement and does not require additional frequency spectrum. It is not surprising, therefore, that space diversity merits strong consideration for microwave mobile radio. The basic requirement is that the spacing of the antennas in the receiving or transmitting array be chosen so that the individual signals are uncorrelated. Design details will be elucidated more fully in the next section, but we can observe from Figure 1.3-6 that, ideally, spacings of $\lambda/2$ should be sufficient. (In the actual case the spacing depends on the disposition of the scatterers causing the multipath transmission.) Each of the M antennas in the diversity array provides an independent signal to an M-branch diversity combiner,* which then operates on the assemblage of signals to produce the most favorable result. A variety of techniques are available to perform the combining process, and these will be elaborated much more fully in subsequent sections.

The diversity array can be located either at the mobile unit, the base station, or both, depending on the particular combining technique used and degree of signal enhancement required. In principle, there is no limitation on M, the number of array elements, but the amount of improvement in the fading characteristics realized by adding one more element decreases as M grows larger.

5.1.2 Polarization Diversity

It has been shown in Section 3.3 that signals transmitted on two orthogonal polarizations in the mobile radio environment exhibit uncorrelated fading statistics. These signals thus become candidates for use in diversity systems. One might consider this a special case of space diversity, since separate antennas are used. The spacing requirements may be minimized by clever antenna configurations that capitalize on the fact that the field orthogonality suffices to decorrelate the signals. In this case, however, only two diversity branches are available, since there are only two orthogonal polarizations. There is also a 3-dB loss in signal, since the transmitter power is split between the two transmitting antennas needed.

5.1.3 Angle Diversity

A diversity method that has found some application in troposcatter ("beyond-the-horizon") systems involves the use of large reflector-type antennas equipped with a multiplicity of feeds that produce narrow secondary beams pointing in slightly different directions. It has been observed that the scattered signals associated with these directions are uncorrelated.

---

*In this book we have adopted the terminology of "branch" to designate each independent signal path in a diversity system, rather than the more ambiguous word "channel."

An analogous arrangement for mobile radio would be the use of directive antennas on the mobile pointing in widely different directions, since the scattered waves come from all directions. It was pointed out in Section 3.1 that the signal from a directive antenna already has less severe fading characteristics, and a number of such antennas could form an acceptable diversity array.

Alternatively, small antennas that respond specifically to the various field components (such as E_z, H_x, H_y) have certain directive properties as discussed in Section 3.1. These directive properties aid in producing independently fading signals for diversity processing, and the action can be interpreted as a form of angle diversity. The number of effective diversity branches is limited to three, however.

5.1.4 Frequency Diversity

Instead of transmitting the desired message over spatially separated paths as described in the preceding sections, one can employ different frequencies to achieve independent diversity branches. The frequencies must be separated enough so that the fading associated with the different frequencies is uncorrelated. The coherence bandwidth is a convenient quantity to use in describing the degree of correlation existing between transmission at different frequencies, as discussed in Section 1.5. Within the coherence bandwidth it is assumed that frequency-selective fading has no effect on transmission of the message itself; for frequency separations of more than several times the coherence bandwidth the signal fading would be essentially uncorrelated. In the mobile radio case, measurements indicate a coherence bandwidth on the order of 500 kHz; thus for frequency diversity the branch separations would have to be at least 1–2 MHz. The advantage of frequency over space diversity is the reduction of the number of antennas to one at each end of the path; on the other hand this method uses up much more frequency spectrum and requires a separate transmitter for each branch.

5.1.5 Time Diversity

It is clear that sequential amplitude samples of a randomly fading signal, if separated sufficiently in time, will be uncorrelated with each other, thus offering another alternative for realizing diversity branches. The required time separation is at least as great as the reciprocal of the fading bandwidth, or $1/(2f_m)$ in the mobile radio case (cf. Figure 1.2-1). For vehicle speeds of 60 mi/hr this time is on the order of 5–0.5 msec for frequencies in the 1–10 GHz range, respectively. Information storage for these times must be provided at both transmitter and receiver. Multiple diversity branches can be provided by successively transmitting the signal sample in each time slot; thus a delay of $M/2f_m$ seconds is incurred in transmitting M branches. The sampling rate must be at least $8M$ kHz for voice transmission. To keep the transmitted pulse width within the transmission bandwidth of the medium requires that M be limited to about 50 branches, at most.

The transmission delay and upper limit on M are not serious; a much more fundamental limitation probably rules out time diversity for mobile radio. This is the fact that the minimum time separation between samples is inversely proportional to the speed of the vehicle, since $f_m = v/\lambda$. In other words, for the vehicle stationary, time diversity is essentially useless. This is in sharp contrast to all of the other diversity types discussed in the preceding sections. In these cases the branch separations are not functions of the vehicle speed; thus the diversity advantages are realized equally well for the vehicle stationary or moving at high speed.

5.2 COMBINING METHODS

Over the years a number of methods have evolved to capitalize on the uncorrelated fading exhibited by separate antennas in a space-diversity array. In this section we will divide these methods into four generic categories, outline their operating principles, and derive some relationships that describe the improvement in the resulting signal statistics. Specific embodiments appropriate to microwave mobile radio systems will be detailed in Chapter 6.

5.2.1 Selection Diversity

This is perhaps the simplest method of all. Referring to Figure 5.2-1, that one of the M receivers having the highest baseband signal-to-noise ratio (SNR) is connected to the output.* As far as the statistics of the output signal are concerned, it is immaterial where the selection is done. The antenna signals themselves could be sampled, for example, and the best one sent to one receiver. To derive the probability density and distribution of the output signal we follow the approach given by Brennan.[7]

We assume that the signals in each diversity branch are uncorrelated and Rayleigh distributed with mean power b_0. The density function of the signal envelope is given in Eq. (1.1-14):

$$p(r_i) = \frac{r_i}{b_0} e^{-r_i^2/2b_0}.$$

---

*In practice, the branch with the largest $(S + N)$ is usually used, since it is difficult to measure SNR.

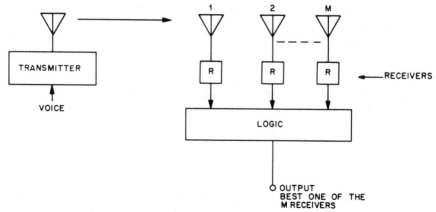

Figure 5.2-1   Principles of selection diversity.

where r_i is the signal envelope in the ith branch. We will be interested in SNR; thus it is convenient to introduce new variables. The local (averaged over one RF cycle) mean signal power per branch is $r_i^2/2$. Let the mean noise power per branch $\overline{n_i^2}$ be the same for all branches, $\overline{n_i^2} = N$, and let

$$\gamma_i \overset{\Delta}{=} \frac{\text{local mean signal power per branch}}{\text{mean noise power per branch}}$$

$$= \frac{r_i^2}{2N}, \tag{5.2-1}$$

$$\Gamma \overset{\Delta}{=} \frac{\text{mean signal power per branch}}{\text{mean noise power per branch}}$$

$$= \langle \gamma_i \rangle = \frac{b_0}{N}. \tag{5.2-2}$$

Then

$$p(\gamma_i) = \frac{1}{\Gamma} e^{\gamma_i/\Gamma}. \tag{5.2-3}$$

The probability that the SNR in one branch is less than or equal to some specified value γ_s is

$$P[\gamma_i \leqslant \gamma_s] = \int_0^{\gamma_s} p(\gamma_i)\, d\gamma_i$$

$$= 1 - e^{-\gamma_s/\Gamma}. \tag{5.2-4}$$

The probability that the γ_i in all M branches are simultaneously less than or equal to γ_s is then

$$P[\gamma_i \cdots \gamma_M \leqslant \gamma_s] = (1 - e^{-\gamma_s/\Gamma})^M = P_M(\gamma_s). \qquad (5.2\text{-}5)$$

This is the distribution of the best signal, that is, largest SNR, selected from the M branches. $P_M(\gamma_s)$ is plotted in Figure 5.2-2 for diversity systems with 1, 2, 3, 4, and 6 branches. The potential savings in power offered by diversity are immediately obvious: 10 dB for two-branch diversity at the 99% reliability level, for example, and 16 dB for four branches.

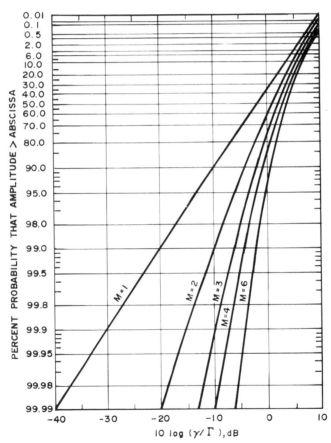

Figure 5.2-2 Probability distribution of SNR γ_s for M-branch selection diversity system. $\Gamma = $ SNR on one branch.

The mean SNR of the selected signal is also of interest. This may be conveniently obtained from the probability density function of γ_s from the integration:

$$\langle \gamma_s \rangle = \int_0^\infty \gamma_s P_M(\gamma_s) d\gamma_s, \tag{5.2-6}$$

where $P_M(\gamma_s)$ is obtained from

$$P_M(\gamma_s) = \frac{d}{d\gamma_s} P_M(\gamma_s)$$

$$= \frac{M}{\Gamma}(1 - e^{-\gamma_s/\Gamma})^{M-1} e^{-\gamma_s/\Gamma}. \tag{5.2-7}$$

Substituting $P_M(\gamma_s)$ into Eq. (5.2-6),

$$\langle \gamma_s \rangle = \Gamma \sum_{k=1}^M \frac{1}{k}. \tag{5.2-8}$$

The dependence of $\langle \gamma_s \rangle$ on M is shown in Figure 5.2-3.

The selection diversity system shown in Figure 5.2-1 is a "receiver" diversity type that can be used at either the base station or the mobile, the only difference being the somewhat larger antenna separation required at the base station (cf. Section 1.6). It is possible to conceive of a selection diversity scheme where the diversity antenna array is at the transmitting site, as shown in Figure 5.2-4. The transmitters operate on adjacent frequency bands centered at $f_1, f_2, ..., f_M$. These bands are separated in a branching filter at the receiving site; each signal is then separately detected and the best one chosen as before. Although more frequency bandwidth is required, the transmitter antenna array spacing may be slightly reduced by taking advantage of a certain amount of frequency diversity. Of course, if the transmitted bands were separated widely enough (cf. Section 1.5), one could completely exchange frequency diversity with space diversity and use only one antenna.

5.2.2 Maximal Ratio Combining

In this method, first proposed by Kahn,[8] the M signals are weighted proportionately to their signal voltage to noise power ratios and then summed. Figure 5.2-5 shows the essentials of the method. The individual signals must be cophased before combining, in contrast to selection diversity; a technique described in Chapter 6 does this very simply. Assuming

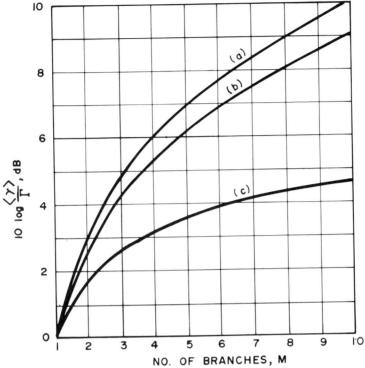

Figure 5.2-3 Improvement of average SNR from a diversity combiner compared to one branch. (*a*) maximal ratio combining, (*b*) equal gain combining, (*c*) selection diversity.

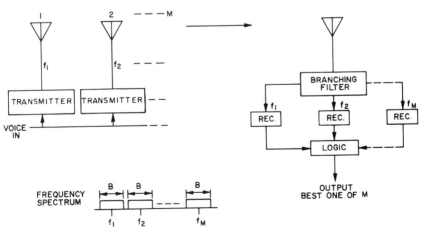

Figure 5.2-4 Selection diversity scheme with antenna array at the transmitting site.

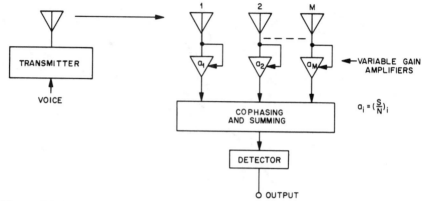

Figure 5.2-5 Principles of maximal ratio combining (note that if the $a_i = 1$, equal-gain combining results).

this cophasing has been accomplished, the envelope of the combined signal is

$$r = \sum_{i=1}^{M} a_i r_i, \tag{5.2-9}$$

where the a_i are the appropriate branch gains. Likewise the total noise power is the sum of the noise powers in each branch, weighted by the branch gain factors:

$$N_T = N \sum_{i=1}^{M} a_i^2, \tag{5.2-10}$$

where, again, it has been assumed that $\overline{n_i^2} = N$ for all i. The resulting SNR is

$$\gamma_R = \frac{r^2}{2N_T}. \tag{5.2-11}$$

It can be shown that if the a_i are chosen as stated above, that is, $a_i = \gamma_i / \overline{n_i^2} = r_i / N$, then γ_R will be maximized with a value

$$\gamma_R = \frac{\left(\sum r_i^2 / N \right)^2}{2N \sum (r_i/N)^2} = \sum_{i=1}^{M} \frac{r_i^2}{2N} = \sum_{i=1}^{M} \gamma_i. \tag{5.2-12}$$

Thus the SNR out of the combiner equals the sum of the branch SNRs. Now we know that

$$\gamma_i^2 = \frac{1}{2N} r_i^2 = \frac{1}{2N}(x_i^2 + y_i^2), \qquad (5.2\text{-}13)$$

where x_i and y_i are independent Gaussian random variables of equal variance b_0 and zero mean (cf. Section 1.1.2). Thus γ_R is a chi-square distribution of $2M$ Gaussian random variables with variance $b_0/2N = \frac{1}{2}\Gamma$. The probability density function of γ_R can then be immediately written down:

$$p(\gamma_R) = \frac{\gamma_R^{M-1} e^{-\gamma_R/\Gamma}}{\Gamma^M(M-1)!}, \qquad \gamma_R \geqslant 0. \qquad (5.2\text{-}14)$$

The probability distribution function of γ_R is given by integrating the density function,

$$P_M(\gamma_R) = \frac{1}{\Gamma^M(M-1)!} \int_0^{\gamma_R} x^{M-1} e^{-x/\Gamma} dx$$

$$= 1 - e^{-\gamma_R/\Gamma} \sum_{k=1}^{M} \frac{(\gamma_R/\Gamma)^{k-1}}{(k-1)!}. \qquad (5.2\text{-}15)$$

The distribution $P_M(\gamma_R)$ is plotted in Figure 5.2-6. This kind of combining gives the best statistical reduction of fading of any known linear diversity combiner. In comparison with selection diversity, for example, two branches give 11.5 dB gain at the 99% reliability level and four branches give 19 dB gain, improvements of 1.5 and 3 dB, respectively, over selection diversity.

The mean SNR of the combined signal may be very simply obtained from Eq. (5.2-12):

$$\langle \gamma_R \rangle = \sum_{i=1}^{M} \langle \gamma_i \rangle = \sum_{i=1}^{M} \Gamma = M\Gamma. \qquad (5.2\text{-}16)$$

Thus $\langle \gamma_R \rangle$ varies linearly with M, whereas for selection diversity it increases much more slowly, as shown in Figure 5.2-3.

5.2.3 Equal Gain Combining

It may not always be convenient or desirable to provide the variable weighting capability required for true maximal ratio combining. Instead,

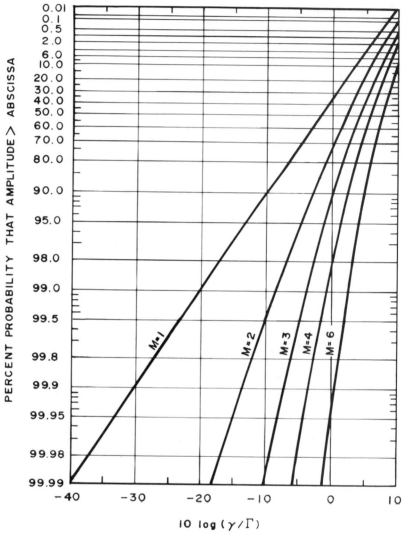

Figure 5.2-6 Probability distribution of SNR γ_r for M-branch maximal ratio diversity combiner. Γ, SNR on one branch.

the gains may all be set equal to a constant value of unity, and equal-gain combining results. The envelope of the combined signal is then given by Eq. (5.2-9) with $a_i = 1$:

$$r = \sum_{i=1}^{M} r_i. \tag{5.2-17}$$

The SNR of the output is

$$\gamma_E = \frac{r^2}{2NM},\tag{5.2-18}$$

again assuming equal noise in the branches.

The combined output r is a sum of M Rayleigh variables. The problem of finding the distribution of the square of this sum (γ_E) is an old one, going back even to Lord Rayleigh, but has never been solved in terms of tabulated functions for $M \geqslant 3$. However, Brennan[7] has obtained values by computer techniques, and his results for $P(\gamma_E)$ are shown in Figure 5.2-7. The distribution curves fall in between the corresponding ones for maximal ratio and selection diversity, and generally only a fraction of a decibel poorer than maximal ratio.

In contrast to the distribution function, the mean value of γ_E can be simply obtained:

$$\langle \gamma_E \rangle = \frac{1}{2NM} \left\langle \left(\sum_{i=1}^{M} r_i \right)^2 \right\rangle$$

$$= \frac{1}{2NM} \sum_{i,j=1}^{M} \langle r_i r_j \rangle.\tag{5.2-19}$$

Now $\langle r_i^2 \rangle = 2b_0$, $\langle r_i \rangle = \sqrt{\pi b_0/2}$ from Chapter 1. Furthermore, since we have assumed that the signals from the various antennas are uncorrelated, $\langle r_i r_j \rangle = \langle r_i \rangle \langle r_j \rangle$, $i \neq j$. Thus Eq. (5.2-19) can be evaluated:

$$\langle \gamma_E \rangle = \frac{1}{2NM} \left[2Mb_0 + M(M-1)\frac{\pi b_0}{2} \right]$$

$$= \Gamma \left[1 + (M-1)\frac{\pi}{4} \right].\tag{5.2-20}$$

The dependency of γ_E on M is also shown in Figure 5.2-3, and it is seen to be only a little poorer than maximal ratio combining. In fact, the difference is only 1.05 dB in the limit of an infinite number of branches.

5.2.4 Feedback Diversity

A very elementary type of diversity reception, called "scanning" diversity,[7] is similar to selection diversity except that instead of always using the best one of M signals, the M signals are scanned in a fixed sequence until

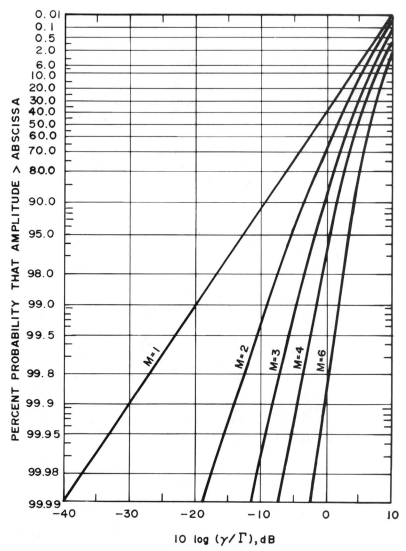

Figure 5.2-7 Probability distribution of SNR γ for M-branch equal-gain diversity combiner. Γ, SNR on one branch.

one is found above threshold. This signal is used until it falls below threshold, when the scanning process starts again. The resulting fading statistics are somewhat inferior to those from other diversity systems; however, a modification of this technique appears promising for mobile radio applications. The principles of operation are shown in Figure 5.2-8

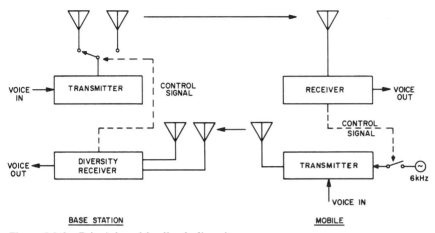

Figure 5.2-8   Principles of feedback diversity.

for two-branch base transmitter diversity. In this system the fact that every base-mobile contact is a two-way affair is exploited by using the mobile-to-base path as a signaling channel in addition to carrying the voice modulation. It is assumed that the mobile-to-base path is reliable, using some sort of base-receiver diversity. The base transmitter is connected to one of its two antennas by a switch, and remains there until the received signal at the mobile falls below a preset threshold level. It signals this fact over the mobile-to-base path, and the transmitter then switches to the other antenna and remains there until the new signal again falls below the threshold. Since the chance of having both transmission paths poor simultaneously is smaller than either one being weak, there should be an average improvement in the signal received by the mobile. The performance is affected by the total time delay in actuating the switch, which is the sum of the round-trip propagation time and the time delay corresponding to the bandwidth of the control channel. If this delay is too great, the signal at the mobile could continue into a fade below the threshold before the transmitter switches and the new signal arrives. At UHF and vehicle speeds of 60 mi/hr, however, the expected signal drop is only 1 or 2 dB; thus the technique appears promising for use in this frequency range. A variation on this scheme can provide very simple mobile receiver diversity, with the receiver switching between two antennas using the same logic principles previously described. In this case the time delay in the switching process can be made very small, with a consequent improvement in the signal fading statistics. Specific embodiments of these schemes and signal distribution curves are given in Section 6.2.

5.2.5 Impairments due to Branch Correlation

In the preceding sections the analysis has been based on the assumption that the fading signals in the various branches are uncorrelated. It may happen in some cases that this is difficult to achieve, for example, if the antennas in the diversity array are improperly positioned, or if the frequency separation between diversity signals is too small. It is thus important to examine the possible deterioration of performance of a diversity system when the branch signals are correlated to a certain extent. Intuitively one might expect that a moderate amount of correlation would not be too damaging. Since deep fades are relatively rare in any event, it would require a very high degree of correlation between two fading signals to bring about a higher correspondence between the deep fades.

The effect of correlation in diversity branches has been studied by many workers. For maximal ratio combining it has been shown possible to derive the probability distribution of the combined signal for any number of branches. In the case of selection diversity, however, it does not appear that one can handle more than two branches. In either case the analysis is beyond the scope of this book; results described by Stein[1] will simply be presented for $M = 2$.

In the following the quantity ρ stands for the magnitude of the complex cross-covariance* of the two fading Gaussian signals (assumed also to be jointly Gaussian); ρ^2 is very nearly equal to the normalized envelope covariance of the two signals.

Selection Diversity

$$P_2(\gamma_s) = 1 - e^{-\gamma_s/\Gamma}[1 - Q(a,b) + Q(b,a)], \qquad (5.2\text{-}21)$$

where

$$Q(a,b) = \int_b^\infty e^{-\frac{1}{2}(a^2 + x^2)} I_0(ax) x \, dx, \qquad (5.2\text{-}22)$$

$$b = \sqrt{\frac{2\gamma_s}{\Gamma(1-\rho^2)}}, \qquad a = b\rho. \qquad (5.2\text{-}23)$$

For $\rho = 0$ the distribution reduces to $(1 - e^{-\gamma_s/\Gamma})^2$, using the fact that $Q(b, 0) = 1$ and $Q(0,b) = e^{-b^2/2}$. For $\gamma_s \ll \Gamma$,

$$P_2(\gamma_s) \doteq \frac{\gamma_s^2}{\Gamma^2(1-\rho^2)}. \qquad (5.2\text{-}24)$$

*Defined in Section 1.3. See Eqs. (1.3-1), (1.3-12), (1.3-13), and (1.3-16).

Maximal Ratio Combining

$$P_2(\gamma_R) = 1 - \frac{1}{2\rho}\left[(1+\rho)e^{-\gamma_R/\Gamma(1+\rho)} - (1-\rho)e^{-\gamma_R/\Gamma(1-\rho)}\right]. \quad (5.2\text{-}25)$$

Again for $\rho = 0$ the distribution reduces to that for uncorrelated fading, and for $\gamma_R \ll \Gamma$,

$$P_2(\gamma_R) \doteq \frac{\gamma_R^2}{2\Gamma^2(1-\rho^2)}. \quad (5.2\text{-}26)$$

The above distributions are shown in Figures 5.2-9 and 5.2-10. We can easily see that the intuitive feeling expressed earlier is borne out in the numerical results. For example, with selection diversity one finds the combined signal to be 9.3 dB better than Rayleigh fading for 98% of the time with uncorrelated signals. Even for a correlation as high as 80% the combined signal is 6.3 dB better than no diversity for 98% of the time.

5.2.6 Impairments due to Combining Errors

The description of the various combining methods so far has implied that the combining mechanism, whatever it may be, operates perfectly. Since the information needed to operate a combiner is extracted in some way from the signals themselves, there is the possibility of making an error and thus not completely achieving the expected performance. This effect has been studied in detail by Gans[9] for the particular case of the maximal ratio combiner, and his results will be summarized here. Similar results were also obtained earlier by Bello.[10]

We recall (Section 5.2.2) that in the maximal ratio combiner the signals are cophased and then summed, with the amplitude of each branch signal being weighted by its own SNR. A particular embodiment of this method is described in Section 6.4, and involves use of a CW "pilot" signal transmitted adjacent to the message band. The phase and amplitude of the pilot signal are sensed and used to adjust the complex weighting factors of the individual branches so that true maximal ratio combining results. Now the fading of the pilot may not be completely correlated with that of the message, possibly because the pilot frequency is too far removed from the message. In this case the complex weighting factors would be somewhat in error, and degraded performance results. The effects can be completely described in terms of the quantity ρ, defined as the magnitude of the complex cross correlation between the transmission coefficients of the medium associated with the pilot and message frequencies. Approximately,

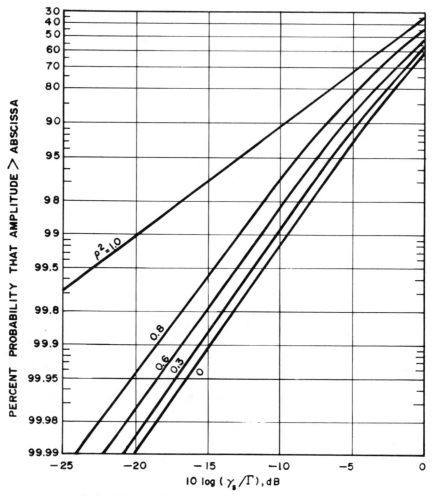

Figure 5.2-9 Probability distribution for a two-branch selection diversity combiner with correlated branch signals. Γ, SNR on one branch; $\rho^2 \doteq$ envelope correlation.

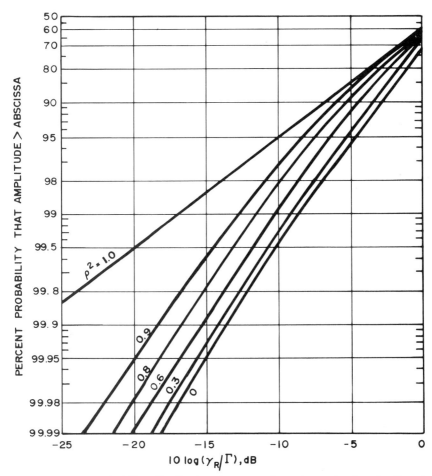

Figure 5.2-10 Probability distribution for a two-branch maximal ratio diversity combiner with correlated branch signals. Γ, SNR on one branch; $\rho^2 \doteq$ envelope correlation.

we can also identify ρ^2 as the correlation between the envelopes of two single sinusoids sent over the transmission path at the frequencies of the pilot and message. Gans[9] has then shown that the probability density of the combiner output signal can be written as

$$P_M(\gamma_R) = \frac{1}{\Gamma}(1-\rho^2)^{M-1}e^{-\gamma_R/\Gamma}\sum_{n=0}^{M-1}\binom{M-1}{n}$$

$$\times\left[\frac{\gamma_R\rho^2}{\Gamma(1-\rho^2)}\right]^n\frac{1}{n!}, \qquad (5.2\text{-}27)$$

where the binomial coefficient is

$$\binom{M-1}{n} \triangleq \frac{(M-1)!}{(M-n-1)!n!}. \qquad (5.2\text{-}28)$$

The mean SNR may now be simply calculated by integrating:

$$\langle\gamma_R\rangle = \int_0^\infty \gamma_R p(\gamma_R)\,d\gamma_R$$

$$= \Gamma[1+(M-1)\rho^2]. \qquad (5.2\text{-}29)$$

To get the probability distribution of the output signal, Eq. (5.2-27) is integrated:

$$P_M(\gamma_R) = \int_0^{\gamma_R} p_M(x)\,dx.$$

$$= 1 - e^{-\gamma_R/\Gamma}\sum_{n=0}^{M-1}\binom{M-1}{n}\rho^{2n}(1-\rho^2)^{M-n-1}\sum_{k=0}^{n}\frac{(\gamma_R/\Gamma)^k}{k!}. \qquad (5.2\text{-}30)$$

In the worst case (for diversity action), the pilot and message signals are uncorrelated, $\rho = 0$, and $P_M(\gamma_R) = 1 - e^{-\gamma_R/\Gamma}$, the same as for no diversity at all. In this case, also, $\langle\gamma_R\rangle = \Gamma$, the SNR for one branch. On the other hand, if $\rho = 1$ (pilot and message signals perfectly correlated) the distribution function and mean SNR reduce to the expressions previously derived:

$$P_M(\gamma_R) = 1 - e^{-\gamma_R/\Gamma}\sum_{k=0}^{M-1}\frac{(\gamma_R/\Gamma)^k}{k!}, \qquad \langle\gamma_R\rangle = M\Gamma.$$

As an example of the effect of combiner error, the distribution and mean value of the SNR are plotted in Figures 5.2-11 and 5.2-12, respectively, for a four-branch maximal ratio combiner. Examining these figures we see that combiner errors have significant impact for the deeper fades for relatively small decorrelation. For example, if $\rho^2 = 1$ the combined signal is above -7 dB for 99.99% of the time, whereas if $\rho^2 = 0.75$ it is above this value only 99.6% of the time. On the other hand, the mean value only drops 0.9 dB if ρ^2 changes from 1.0 to 0.75. Thus combiner errors affect the mean SNR negligibly in comparison to their effect on deep fades.

The specific results presented have been for maximal ratio combining. It appears reasonable, however, that for equal gain or selection diversity, if the combiner is controlled by sampling a pilot, decorrelation of the pilot and message channel would have similar effects on the probability distribution and mean SNR.

5.3 ANTENNA ARRAYS FOR SPACE DIVERSITY

In Section 5.2 the diversity combining of Rayleigh fading branches was discussed. It has been shown that diversity combining can improve the probability distribution of the resultant carrier-to-noise ratio (CNR) over that of a single branch system. Since the E_z field at the mobile has a spatial correlation coefficient $\rho = J_0(\beta d)$, by separating antennas a sufficiently large distance d such that $\rho \to 0$, we should expect to have the required independent branches for space-diversity application.

The problem becomes complicated if we want to place dipoles closer together because of space limitations on the mobile. By placing dipoles close to each other we must take into account the mutual coupling among antennas and also the finite correlations between the received fading signals. These effects can reduce the diversity advantage as mentioned in Section 5.2.5, and can also present complicated antenna matching problems.

In this section we shall examine the antenna spacing requirements of the linear and planar monopole arrays shown in Figure 5.3-1, assuming maximal ratio diversity combining. The results presented are based on the work of Lee.[11,12] The emphasis will be on the average and the cumulative probability distribution (CPD) of the combiner output CNR. The CNR of an equal-gain system[13,14] is very close to the maximal ratio case; therefore it is not reported here. The level crossing rates and power spectrum of linear and planar diversity arrays can also be found in Refs. 13 and 14.

It should be mentioned that in the mobile radio environment, the diversity array gains are obtained by coherent combining of the random signals received from each array element. Since the random phase

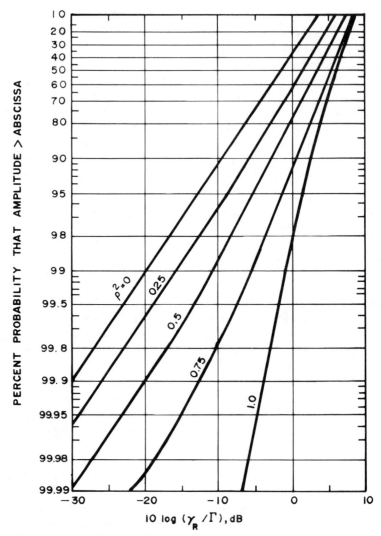

Figure 5.2-11 Probability distribution for a four-branch maximal ratio combiner with combiner errors. Γ, SNR on one branch; $\rho^2 \doteq$ envelope correlation between pilot and signal.

associated with each branch is varying as the mobile moves, the phase adjustments on the carriers from each branch must be performed continuously to achieve coherent combining. The combining is therefore signal dependent. This contrasts sharply with that of a conventional directive array for which the carriers are combined at a preset phase difference. Directive arrays are treated in Chapter 3 and have been shown to have

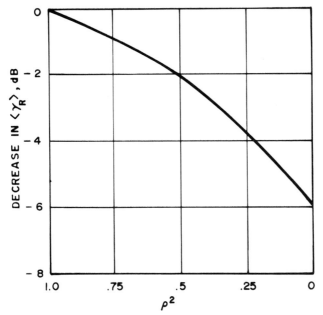

Figure 5.2-12 Decrease in mean SNR in a 4-branch maximal ratio diversity combiner due to combiner errors. ρ^2 is the correlation of the pilot and signal.

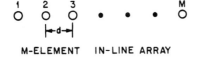

M-ELEMENT   IN-LINE ARRAY

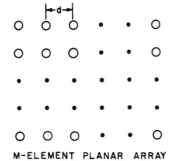

M-ELEMENT PLANAR ARRAY

Figure 5.3-1 Two diversity array configurations.

little advantage in increasing the average CNR or improving the CPD over that of a dipole antenna, although the fading rate can be markedly reduced.

5.3.1   Equivalent Circuit of a Diversity Array

The equivalent circuit of an M-element diversity array is shown in Figure 5.3-2(a). Here $[Z]$ is the impedance matrix of the array. The element Z_{ii} is the self-impedance of the ith element and Z_{ij} is the mutual impedance between the ith and jth elements (with all other elements open circuited).

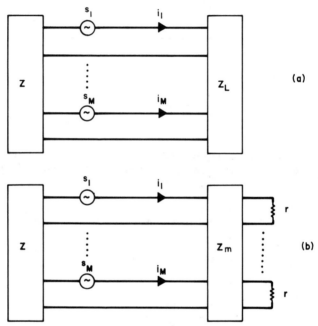

Figure 5.3-2   Equivalent circuit of a receiving diversity array. (a) Load impedance matrix z_l. (b) Load resistance r fed through a matching impedance matrix z_m.

The impedance matrix is symmetric, that is,

$$[Z]=[Z]^t, \tag{5.3-1}$$

where t indicates transpose. Let s_i be the open-circuit voltage of the ith antenna (with all antennas open circuited.) Since a monopole responds only to E_z, s_i is a complex Gaussian random variable and is represented by

$$s_i = T_{ic} + jT_{is}. \tag{5.3-2}$$

Assume uniform distribution of arrival angles for plane waves at the mobile; the correlation coefficients are

$$R_{ij} = \langle s_i s_j^* \rangle = 2b_0 J_0(\beta d_{ij}),$$ (5.3-3)

where d_{ij} is the distance between the ith and jth element.

Referring to Figure 5.3-2, let us introduce the following vectors:

$$S = \begin{pmatrix} s_1 \\ s_2 \\ \vdots \\ s_M \end{pmatrix},$$ (5.3-4)

$$I = \begin{pmatrix} i_1 \\ i_2 \\ \vdots \\ i_M \end{pmatrix}$$ (5.3-5)

The load impedance matrix is Z_L. Haus and Adler[15] have shown that if $Z_L = Z^\dagger$ (the dagger stands for the complex conjugate transposed), the total power delivered to the load matrix by the antenna array is maximized, and is given by

$$P_{rec} = \tfrac{1}{4} S^\dagger ([Z] + [Z]^\dagger)^{-1} S.$$ (5.3-6)

They further showed that there exists a matching network Z_m, indicated in Figure 5.3-2(b), such that the total power delivered to the resistances r is equal to the available power in the $Z^\dagger$ loading condition.

In practice, the matching network is hard to realize and we shall settle for the less optimum case of resistive loading shown in Figure 5.3-3. The front end device* has a power gain G, and a real input impedance r. The choice of real impedance is made because the self-impedance of a $\lambda/4$ monopole is considered to be almost real.

Let the CNR at the output of each front end be γ_i and assume that the output noise is predominantly produced by the internal noise. This would be the case if the front end has a noise figure several times greater than

---

*The front end can either be a pre-amplifier or a mixer-IF amplifier combination which serves to establish the output carrier-to-noise ratio.

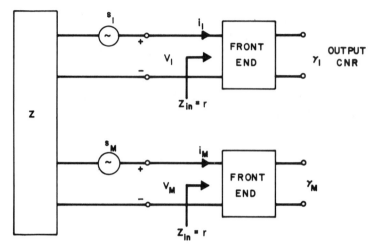

Figure 5.3-3   Equivalent circuit of a diversity array with resistive loading.

unity. The CNR γ_i may be expressed in terms of the input power by

$$\gamma_i = \frac{G}{N_0} \frac{1}{2} Re\{V_i I_i^*\}, \qquad (5.3\text{-}7)$$

where N_0 is the noise power in each branch. For maximal ratio combining, the combined carrier-to-noise ratio is

$$\gamma = \sum_{i=1}^{M} \gamma_i = \frac{G}{N_0} \sum_{i=1}^{M} \frac{1}{2} Re\{V_i I_i^*\}, \qquad (5.3\text{-}8)$$

and is directly proportional to the total power received by the diversity array.

The circuit matrix equations are

$$S = ([Z] + r[I])I, \qquad (5.3\text{-}9)$$

$$V = rI,$$

where $[I]$ is the identity matrix.

The received power P_R is then given by

$$P_R = \frac{1}{2} \sum_{i=1}^{M} Re\{V_i I_i^*\} = \tfrac{1}{2} S[C]S^*, \qquad (5.3\text{-}10)$$

where

$$[C] = ([Z] + r[I])^{-1} r ([Z] + r[I])^{-1\dagger},$$

and -1 indicates the inverse of the matrix.

Given an array configuration, we should be able to compute [Z] and also the correlation coefficients $\langle s_i s_j^* \rangle$. Since P_R is a quadratic form of Gaussian variables, the distribution and the average values of P_R can be calculated.[1,11] In the following numerical examples, the monopoles are assumed to have a length-to-diameter ratio of 36.5. The mutual impedances are taken from results computed by Tai.[16] The self-impedance is $Z_{ii} = 34 + j15$. The real part of the self-impedance will be denoted by r_0 and the imaginary part by x_0.

5.3.2 Average Carrier-to-Noise Ratio

The average CNR of a diversity array is given by

$$\langle \gamma \rangle = \frac{G}{2N_0} \sum_{i=1}^{M} \sum_{j=1}^{M} C_{ij} \langle s_i s_j^* \rangle. \qquad (5.3\text{-}11)$$

We define a normalization factor $\langle \gamma \rangle_{\text{ind}}$, the average CNR of an m-branch diversity array with large spacings. This is the case where mutual couplings and correlations among different antennas are zero. The load impedance is taken to be Z_{ii}^*. This then represents the best $\langle \gamma \rangle$ we expect to obtain. Numerical results[12] of $\langle \gamma \rangle / \langle \gamma \rangle_{\text{ind}}$ for two-branch and four-branch in-line arrays are shown in Figure 5.3-4. It is observed that for each particular spacing there exists a different r/r_0 ratio that would yield the maximum CNR. Nevertheless, r/r_0 can vary over a broad range without causing any significant degradation of $\langle \gamma \rangle / \langle \gamma \rangle_{\text{ind}}$. As the antenna separation d approaches $\lambda/2$, the gain of a diversity array is almost equal to the output of an array consisting of M independent branches.

To show the dependence of $\langle \gamma \rangle$ on array spacing, $\langle \gamma \rangle / \langle \gamma \rangle_{\text{ind}}$ for an in-line array is shown in Figure 5.3-5 as a function of antenna spacing. In each case, the load resistance r has been chosen to maximize $\langle \gamma \rangle$; hence we replace $\langle \gamma \rangle$ by $\langle \gamma \rangle_{\text{or}}$. We observed that arrays with a higher order of diversity require slightly larger spacing to obtain the same value of $\langle \gamma \rangle_{\text{or}} / \langle \gamma \rangle_{\text{ind}}$. For spacings larger than 0.3λ the loss of CNR from that of an ideal independent array is only 2 dB.

The cauculated $\langle \gamma \rangle_{\text{or}} / \langle \gamma \rangle_{\text{ind}}$ of planar arrays is shown in Figure 5.3-6. We note that in comparison to a linear array, the planar array requires slightly larger spacings to achieve the same $\langle \gamma \rangle_{\text{or}} / \langle \gamma \rangle_{\text{ind}}$. As the separation increases we observe a peak in gain around 0.4λ and a dip in gain around 0.55λ. Since the correlation between the antennas is governed by $J_0(\beta d_{ij})$, the peak in gain corresponds to the first zero of the Bessel function and the dip in gain corresponds to the first minimum of the Bessel function.

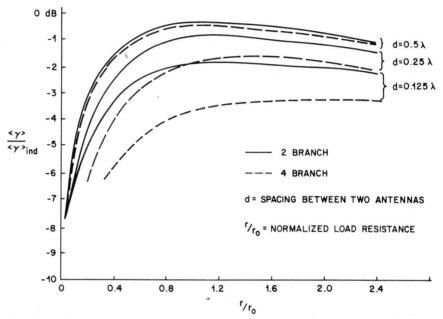

Figure 5.3-4 $\langle\gamma\rangle/\langle\gamma\rangle$ind of an in-line array with maximal ratio combining and resistive loading.

Nevertheless, for $d > 0.3\lambda$ the reduction of gain of a nine-element array is still less than 2 dB.

5.3.3   Probability Distribution of the Carrier-to-Noise Ratio

From Eqs. (5.3-8) and (5.3-10) we obtain

$$\gamma = \frac{G}{2N_0} S[C]S^*. \tag{5.3-12}$$

Define Γ to be the average CNR of a single-branch receiver system with load impedance Z_{11}^*, we have

$$\Gamma = \frac{Gb_0 r_0}{N_0(4r_0^2 + x_0^2)}. \tag{5.3-13}$$

The CPD of γ_{or}/Γ, that is, the load resistance optimized to yield the maximum γ, is[11]

$$P\left\{\frac{\gamma_{or}}{\Gamma} \leqslant x\right\} = 1 - \sum_{j=1}^{M} \frac{\lambda_j^{M-1} e^{-x/\lambda_j}}{\displaystyle\prod_{j \neq k} (\lambda_j - \lambda_k)}, \tag{5.3-14}$$

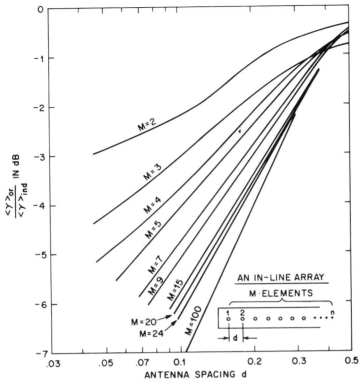

Figure 5.3-5 The optimum SNR, $\langle\gamma\rangle_{or}$, of a resistive load network for an in-line array with M elements.

where λ_j are the eigenvalues of $[R][C]$, and $[R]$ is the correlation matrix of s and is given by

$$R_{ij} = 2b_0 J_0(\beta d_{ij}).$$

The CPD of γ_{or}/Γ for a four-branch linear array is presented in Figure 5.3-7. The results are calculated for cases with and without mutual impedance. It is observed that mutual impedance plays a minor role in the determination of the diversity improvements. For example, when $d = 0.05\lambda$ the inclusion of mutual impedance only slightly perturbs the CPD. The loss of diversity advantage, in comparison to the CPD of four independent branches, is primarily due to the correlation between the received signals. As d approaches 0.5λ, the effects of mutual impedance and correlation both diminish and the CPD of γ_{or}/Γ is as good as that of four independent branches combined. Similar results are obtained for planar arrays and are shown in Figure 5.3-8.

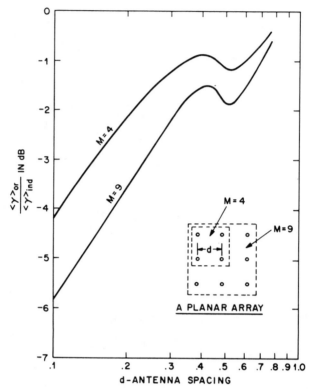

Figure 5.3-6 The optimum SNR, $\langle\gamma\rangle_{or}$, for a resistive load network for a planar array with M elements.

5.3.4 Second-Order Statistics

In previous sections the diversity array has been shown to provide significant improvement in the receiver output CNR. The amount of improvement depends on antenna spacing but is independent of the direction of travel. In other words, the different orientations of a linear array with respect to the direction of travel will not change the results of previous sections.

The power spectrum of γ, however, does depend on the relative orientation of the array and the direction of motion. For example, the power spectra of γ when the vehicle is traveling in a direction perpendicular to the linear array or along the linear array are markedly different.[13] For planar arrays the differences become smaller.

The level crossing rates of γ for several linear and planar arrays have been calculated by Lee.[14] It is observed that although the power spectrum

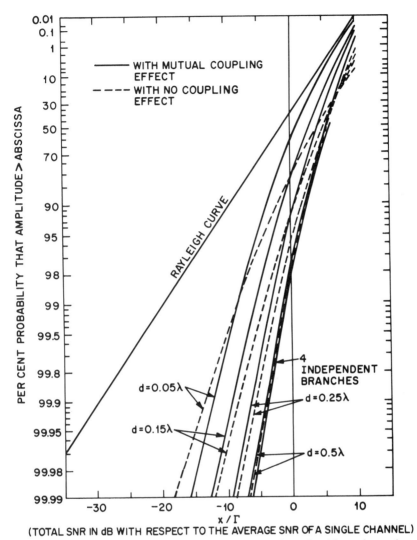

Figure 5.3-7 Comparison of the cumulative distribution of a 4-branch maximum-ratio combiner with no coupling effect and with coupling effect for an in-line antenna array.

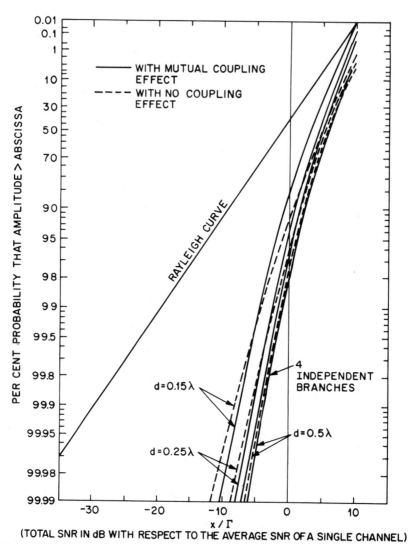

Figure 5.3-8 Comparison of the cumulative distribution of a 4-branch maximum-ratio combiner with no coupling effect and with coupling effect for a planar antenna array.

is different for different directions of travel, the level crossing rates of diversity arrays are insensitive to the direction of travel. Therefore, for all practical purposes the orientation of the diversity array can be chosen arbitrarily.

5.4 EFFECT OF DIVERSITY ON FM NOISE AND INTERFERENCE

The main effect of diversity is to change the probability distribution of the IF signal-to-noise ratio. These probability distributions have been computed for various types of diversity combining methods in Section 5.2. By using these distributions and the quasistatic approximation formulas of Chapter 4 one is able to determine the resulting output noise and interference.

5.4.1 Thermal Noise

In Section 4.1.4 the effects of Rayleigh fading on the output noise resulting from thermal input noise were derived by averaging over the IF SNR Rayleigh distribution for the three noise components: (a) quadratic noise, which was described in Section 4.1.1 as representing the output noise spectrum when the IF SNR is large, $\mathcal{W}_2(f)$, and whose frequency dependence at low baseband frequencies is quadratic ($\sim f^2$); (b) threshold noise, which appears sharply as the input SNR drops below threshold level and which saturates at a finite level as the receiver "captures" on the noise rather than the signal [the spectrum of this noise was denoted $\mathcal{W}_D(f)$ in Section 4.1.1]; and (c) signal suppression noise, which is only present when the IF carrier is modulated. It results from suppression of the signal as the receiver captures on the noise rather than the signal when the input carrier drops below threshold. The signal suppression noise, n_s, is a function of the probability distribution of the fading IF SNR as described in Eq. (4.1-54).

In this section we will assume conventional FM detection (i.e., no preemphasis or deemphasis) and will follow the derivations in Ref. 17. From Eq. (4.1-26), the quadratic output noise expressed as a function of the IF SNR, ρ, is

$$N_2(\rho) = \frac{a(1 - e^{-\rho})^2}{\rho}, \tag{5.4-1}$$

where the constant a is defined in terms of the IF bandwidth, B, and the baseband cutoff frequency, W, by Eq. (4.1-27):

$$a = \frac{4\pi^2 W^3}{3B} \left\{ 1 - \frac{6\pi}{10}\left(\frac{W}{B}\right)^2 + \frac{12\pi^2}{56}\left(\frac{W}{B}\right)^4 + \cdots \right\}. \tag{5.4-2}$$

Also from Eq. (4.1-26), the threshold noise is given as

$$N_D(\rho) = \frac{8\pi BW e^{-\rho}}{\sqrt{2(\rho+2.35)}} \, . \tag{5.4-3}$$

If $p(\rho)$ denotes the probability density of the IF SNR resulting from a given diversity combining method, then the average output quadratic and threshold noises are given by

$$\langle N_2 \rangle = \int_0^\infty p(\rho) N_2(\rho) \, d\rho \tag{5.4-4}$$

and

$$\langle N_D \rangle = \int_0^\infty p(\rho) N_D(\rho) \, d\rho, \tag{5.4-5}$$

respectively. From Eqs. (4.1-28) and (4.1-54) the average output signal sup-pression noise is

$$\langle N_s \rangle = \langle (V_0(t) - \overline{V}_0(t))^2 \rangle, \tag{5.4-6}$$

where the instantaneous output signal, $V_0(t)$ is given by

$$V_0(t) = V(t)(1 - e^{-\rho}), \tag{5.4-7}$$

and the effective instantaneous output signal is obtained by averaging over the fading,

$$\overline{V}_0(t) = V(t) \int_0^\infty p(\rho)(1 - e^{-\rho}) \, d\rho. \tag{5.4-8}$$

Thus

$$\langle N_s \rangle = S \left\{ \int_0^\infty e^{-2\rho} p(\rho) \, d\rho - \left[\int_0^\infty e^{-\rho} p(\rho) \, d\rho \right]^2 \right\}, \tag{5.4-9}$$

where $V(t)$ is the input signal modulation in radians per second and S is the mean square input signal power, $\langle V^2(t) \rangle$.

The average output signal power is given by

$$\langle S \rangle = \langle \overline{V}_0^2(t) \rangle = S \left[1 - \int_0^\infty p(\rho) e^{-\rho} \, d\rho \right]^2. \tag{5.4-10}$$

Maximal Ratio Combining

For an M-branch maximal ratio combiner, the probability density of the IF SNR is given by Eq. (5.2-14),

$$p_{MR}(\rho) = \frac{\rho^{M-1} e^{-\rho/\rho_0}}{\rho_0^M (M-1)!}, \qquad \rho \geq 0, \qquad (5.4-11)$$

where ρ_0 is the mean carrier-to-noise ratio on each of the branches.

Substituting Eq. (5.4-11) into Eq. (5.4-4) and integrating gives the average output quadratic noise with a maximal ratio combiner ($M \geq 2$):

$$\langle N_2 \rangle_{MR} = \frac{a}{(M-1)\rho_0} \left[1 - \frac{2}{(1+\rho_0)^{M-1}} + \frac{1}{(1+2\rho_0)^{M-1}} \right]. \qquad (5.4-12)$$

The average output threshold noise is obtained by using Eq. (5.4-11) in Eq. (5.4-5),

$$\langle N_D \rangle_{MR} = \frac{8\pi BW}{(M-1)! \sqrt{2} \, \rho_0^M} \int_0^\infty \frac{\rho^{M-1}}{\sqrt{\rho + 2.35}} \exp\left[-\rho\left(1 + \frac{1}{\rho_0} \right) \right] d\rho.$$

$$(5.4-13)$$

Davis[17] has shown that the integral

$$I_M = \int_0^\infty \frac{\rho^{M-1} e^{-\beta\rho}}{\sqrt{\rho + \gamma}} d\rho \qquad (5.4-14)$$

may be conveniently evaluated by means of the recurrence relation:

$$I_{M+1} = \left(\frac{2M-1}{2\beta} - \gamma \right) I_M + (M-1)\frac{\gamma}{\beta} I_{M-1}, \qquad M \geq 2, \quad (5.4-15)$$

where

$$I_1 = \sqrt{\frac{\pi}{\beta}} \, e^{\gamma\beta} \operatorname{erfc} \sqrt{\gamma\beta} \qquad (5.4-16)$$

and

$$I_2 = \left(\frac{1}{2\beta} - \gamma \right) I_1 + \frac{\sqrt{\gamma}}{\beta}. \qquad (5.4-17)$$

Thus,

$$\langle N_D \rangle_{MR} = \frac{8\pi BW}{(M-1)!\sqrt{2}\,\rho_0^M} I_M, \tag{5.4-18}$$

with $\beta = (1+1/\rho_0)$ and $\gamma = 2.35$.

Using Eq. (5.4-11), we have

$$\int_0^\infty p_{mr}(\rho)e^{-\rho}\,d\rho = \frac{1}{\rho_0^M(M-1)!}\int_0^\infty \rho^{M-1}e^{-\rho(1+1/\rho_0)}\,d\rho$$

$$= \frac{1}{(\rho_0+1)^M} \tag{5.4-19}$$

and

$$\int_0^\infty p_{mr}(\rho)e^{-2\rho}\,d\rho = \frac{1}{(2\rho_0+1)^M}. \tag{5.4-20}$$

Equations (5.4-9), (5.4-10), (5.4-19), and (5.4-20) give the following results for the average output signal and signal suppression noise:

$$\langle S \rangle_{MR} = S\left[1 - \frac{2}{(\rho_0+1)^M} + \frac{1}{(\rho_0+1)^{2M}}\right] \tag{5.4-21}$$

and

$$\langle N_s \rangle_{MR} = S\left[\frac{1}{(2\rho_0+1)^M} - \frac{1}{(\rho_0+1)^{2M}}\right]. \tag{5.4-22}$$

The output average signal to average noise ratio, SNR_{MR}, is obtained by inserting Eqs. (5.4-12), (5.4-18), (5.4-21), and (5.4-22) into

$$SNR_{MR} = \frac{\langle S \rangle_{MR}}{\langle N_2 \rangle_{MR} + \langle N_D \rangle_{MR} + \langle N_s \rangle_{MR}}. \tag{5.4-23}$$

Using the criterion of Eq. (4.1-30),

$$S = \frac{\pi^2}{10}(B-2W)^2, \tag{5.4-24}$$

to relate the signal modulation to the IF and baseband bandwidths, Figures 5.4-1 through 5.4-3 show the output SNR_{MR} performance for two-, three-, and four-branch maximal ratio combiners, versus the mean IF carrier-to-noise ratio on each branch input, ρ_0, and with the IF to baseband bandwidth ratio $B/2W$, as a parameter.

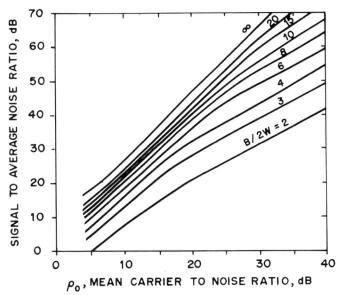

Figure 5.4-1 Average output signal to average output noise ratio $\langle\text{SNR}\rangle$ versus mean if carrier to noise ratio, ρ_0, for two-branch maximal ratio combining $(M=2)$.

It is clear from the figures that large improvements of output SNR are possible through the use of maximal ratio combining, just as the IF SNR is improved. Detailed system implications of the effects of diversity combining on output SNR are described in Section 6.8.

Selection Diversity

The probability density of the IF carrier-to-noise ratio, ρ, with an M-branch selection diversity combiner is, from Eq. (5.2-7),

$$p_S(\rho) = \frac{M}{\rho_0}\left(1 - e^{-\rho/\rho_0}\right)^{M-1} e^{-\rho/\rho_0}, \tag{5.4-25}$$

where ρ_0 is again the mean IF carrier to noise ratio on each branch. As

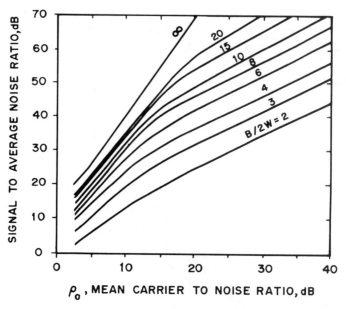

Figure 5.4-2 ⟨SNR⟩ versus ρ_0 for three-branch maximal ratio combining (M=3).

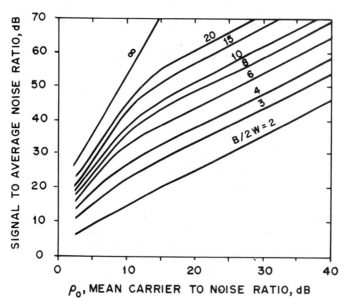

Figure 5.4-3 ⟨SNR⟩ versus ρ_0 for four-branch maximal ratio combining (M=4).

shown in Section 5.2, the distribution of Eq. (5.4-25) is nearly identical to that of the maximal ratio combiner, Eq. (5.4-11), except the value of ρ_0 must be adjusted the proper amount, depending on the number of branches. Substituting Eq. (5.4-25) into Eqs. (5.4-4), (5.4-5), (5.4-9), and (5.4-10) yields the average output noise components and average output signal for the selection diversity case. Davis[17] has evaluated the resulting integrals to obtain

$$\langle N_2 \rangle_S = \frac{\alpha M}{\rho_0} \sum_{k=0}^{M-1} \binom{M-1}{k} (-1)^k \log \left[\frac{(k+1+\rho_0)^2}{(k+1)(k+1+2\rho_0)} \right], \quad (5.4\text{-}26)$$

$$\langle N_D \rangle_S = \frac{8\pi B W M \sqrt{\pi}}{\sqrt{2}\,\rho_0} \sum_{k=0}^{M-1} \frac{\binom{M-1}{k}(-1)^k e^{\gamma(k+1+\rho_0/\rho_0)}}{\sqrt{(k+1+\rho_0)/\rho_0}}$$

$$\times \operatorname{erfc}\sqrt{\gamma(k+1+\rho_0)/\rho_0}, \quad (5.4\text{-}27)$$

$$\langle N_s \rangle_S = S \left\{ \left[M \sum_{k=0}^{M-1} \binom{M-1}{k} \frac{(-1)^k}{k+1+2\rho_0} \right] \right.$$

$$\left. - \left[M \sum_{k=0}^{M-1} \binom{M-1}{k} \frac{(-1)^k}{k+1+\rho_0} \right]^2 \right\}, \quad (5.4\text{-}28)$$

and

$$\langle S \rangle_S = S \left[1 - M \sum_{k=0}^{M-1} \binom{M-1}{k} \frac{(-1)^k}{k+1+\rho_0} \right]^2, \quad (5.4\text{-}29)$$

where S is given in Eq. (5.4-24), $\binom{M-1}{k}$ is the binomial coefficient and $\gamma = 2.35$.

Equation (5.4-27) is not satisfactory for computing $\langle N_D \rangle_s$ for large values of ρ_0, as the terms in the sum are several orders of magnitude greater than the sum. To compute $\langle N_D \rangle_s$ with large values of ρ_0, Davis[17] used a power-series expansion of Eq. (5.4-27) in powers of $1/\rho_0$.

Table 1   *Shift in IF CNR, ρ_0, from Maximal Ratio to Selection*

M	1	2	3	4
Shift in ρ_0 (dB)	0.0	1.4	2.5	3.3
$(M!)^{1/M}$ (dB)	0.0	1.5	2.6	3.5

Plotting the output average signal to average noise ratio,

$$\text{SNR}_S = \frac{\langle S \rangle_S}{\langle N_2 \rangle_S + \langle N_D \rangle_S + \langle N_s \rangle_S}. \qquad (5.4\text{-}30)$$

Davis[17] found that to within the accuracy of the plots (better than 0.5 dB) the curves of SNR_S versus ρ_0 for selection diversity were identical to those for maximal ratio combining, Figures 5.4-1, 5.4-2, and 5.4-3, except for a shift in the abscissa, ρ_0. This congruence is a result of the aforementioned congruence of the probability distributions of the IF carrier-to-noise ratio, ρ, for selection diversity and maximal ratio combining with the same shift in ρ_0, the mean IF carrier-to-noise ratio on each branch. The offset in ρ_0 when converting the maximal ratio combining curves of Figures 5.4-1, 5.4-2, and 5.4-3 into the corresponding curves for selection diversity is shown in Table 1.

Selection diversity requires a ρ_0 greater than that for maximal ratio combining for the same output SNR. This is to be expected, since maximal ratio combining provides an IF carrier-to-noise ratio that is the sum of that on each branch and is, therefore, larger than the largest CNR of any individual branch, which is provided by selection diversity. Since major noise contributions occur for small ρ, the offsets in ρ_0 should not be greatly different from those computed by aligning the selection and maximal ratio IF CNR distributions for small ρ. This gives offset factors $(M!)^{1/M}$ also shown in Table 1.

The curves corresponding to Figures 5.4-1, 5.4-2, and 5.4-3 are thus obtained for selection diversity by adding the corresponding decibel offset factor from Table 1 to the abscissa in each of Figures 5.4-1, 5.4-2, and 5.4-3.

Equal-Gain Combining

As described in Section 5.2.3, equal-gain combining provides an IF carrier-to-noise ratio, ρ, which is the square of the sum of Rayleigh variables,

$$\rho = \frac{1}{M}\left(\sum_{k=1}^{M}\sqrt{\rho_k}\right)^2, \tag{5.4-31}$$

where ρ_k is the IF CNR on the kth branch of an M-branch combiner. The distribution of ρ in Eq. (5.4-31) is not known in terms of elementary functions for $M > 2$ but tables of the cumulative distribution of $\sqrt{\rho}$ exist[18] and curves are presented in Figure 5.3-7. Since the noise during deep fades predominates, the approximate distribution for small ρ may be used,[1]

$$p_{EG}(\rho) = \frac{\rho^{M-1}}{\rho_x^M (M-1)!} e^{-\rho/\rho_x}, \tag{5.4-32}$$

where

$$\rho_x = \frac{2\rho_0}{M}\left[\frac{\Gamma(M+\frac{1}{2})}{\Gamma(\frac{1}{2})}\right]^{1/M}, \tag{5.4-33}$$

and $\Gamma(x)$ is the gamma function.

As Davis[17] points out, the probability density of Eq. (5.4-32) is identical to the maximal ratio probability density of Eq. (5.4-11), except that ρ_x is used instead of ρ_0. The relation between ρ_0 and ρ_x is that obtained if the distributions for equal gain and maximal ratio are aligned for small ρ. Equal gain requires more mean IF CNR, ρ_0, on each branch than maximal ratio for the same output SNR, the offset factors appearing in Table 2.

By comparing Tables 1 and 2, we see that equal-gain combining is superior to selection diversity.

5.4.2 Random FM

The random frequency modulation, $\dot{\theta}$, experienced by a mobile receiver

Table 2   *Offset in IF CNR, ρ_0, from Maximal Ratio to Equal Gain*

M	1	2	3	4
Offset in ρ_0 (dB)	0.0	0.6	0.9	1.0

when moving through the pattern of randomly scattered waves from the base station was described in Section 1.4. The probability density of the random FM of a single branch is given by Eq. (1.4-1):

$$p(\dot{\theta}) = \frac{1}{2}\sqrt{\frac{b_0}{b_2}}\left[1 + \frac{b_0}{b_2}\dot{\theta}^2\right]^{-3/2}, \qquad (5.4\text{-}34)$$

for the case of symmetric Doppler spectra. As pointed out in Eq. (1.4-2) the mean square value of the random FM is infinite. From Eq. (1.4-7), for symmetric Doppler spectra, the autocorrelation of the random FM of a single branch is

$$R_{\dot{\theta}}(\tau) = -\frac{1}{2}\left\{\left[\frac{g'(\tau)}{g(\tau)}\right] - \left[\frac{g''(\tau)}{g(\tau)}\right]\right\}\log\left\{1 - \left[\frac{g(\tau)}{g(0)}\right]^2\right\}. \qquad (5.4\text{-}35)$$

The correlation function, $g(\tau)$, and the spectral moments b_0 and b_2 of the Doppler spectrum are defined in Section 1.3.

The power spectrum of the random FM for a uniform angle of arrival Doppler spectrum is shown in Figure 1.4-2. The asymptotic form of the two-sided power spectrum of the random FM for high frequencies is given in Section 1.4:

$$\lim_{f\to\infty} S_{\dot{\theta}_1}(f) = \frac{1}{2}\left(\frac{b_2}{b_0} - \frac{b_1^2}{b_2^2}\right)\bigg/f, \qquad (5.4\text{-}36)$$

where, for symmetric Doppler spectra, $b_1 = 0$.

Equations (5.4-34), (5.4-35), and (5.4-36) review the characteristics of random FM without diversity. The effects of diversity on these characteristics will now be described.

Maximal Ratio and Equal-Gain Combining

As described in Chapter 6, conventional means of providing maximal-ratio and equal-gain combining employ a pilot or carrier that has experienced the same phase distortion due to the randomly scattering medium that has been experienced by the signal. This pilot or carrier is then mixed with the signal, and the resulting difference frequency component is cophased with all other branches because the propagation phase fluctuations are canceled by the mixing process. On the other hand, the signal phase fluctuations (common to all branches) remain. We assume that the pilot or carrier is close enough in frequency and time to the signal so that its phase fluctuations due to the propagation paths are the same as those of the signal (cf. coherence bandwidth, section 1.5 and coherence time, Section 1.2).

Since the random FM is produced by the propagation phase fluctuations, even a single branch of a maximal-ratio or equal-gain combiner has no random FM. Thus maximal-ratio or equal-gain combining using such a well-correlated pilot would eliminate random FM.

Selection Diversity

Davis[19] has shown that although selection diversity does not eliminate random FM, it does substantially reduce it. We will briefly summarize his analysis here.

When the base station transmits a CW carrier, the kth branch of a mobile receiver receives a narrow-band Gaussian process with amplitude $r_k(t)$ and phase $\theta_k(f)$. The random FM at the output of a selection diversity system is thus

$$\dot{\theta}(t) = \dot{\theta}_m(t), \quad \text{where } r_m(t) \geqslant r_k(t) \text{ for all } k; \tag{5.4-37}$$

that is, the output random FM is the random FM of the branch with largest envelope at time t. As t changes, m will change.

For M branches, Davis integrates over the $2M$-dimensional joint Gaussian distribution of the branch signals to obtain the following expression for the probability density function of the output random FM[19]:

$$p_S(\dot{\theta}) = \frac{M}{2\sqrt{\dfrac{b_2}{b_0}}} \sum_{k=0}^{m-1} \binom{M-1}{k} (-1)^k \left[k+1+\frac{\dot{\theta}^2 b_0}{b_2} \right]^{-3/2}. \tag{5.4-38}$$

Figure 5.4-4 shows the probability density function of $\dot{\theta}$ computed from Eq. (5.4-38) for various values of M.

By expanding $[k+1+(\dot{\theta}^2 b_0/b_2)]^{-3/2}$ in negative powers of $\dot{\theta}$ and retaining only the first nonzero term, Davis obtained the asymptotic form of Eq. (5.4-38) as $\dot{\theta}$ approaches infinity,[19]

$$p_S(\dot{\theta}) \sim \frac{2M!}{\sqrt{b_2/b_0}\ 2^{2M}(M-1)!} \left[\frac{b_2}{b_0\dot{\theta}^2} \right]^{M+1/2}. \tag{5.4-39}$$

The mean square value of $\dot{\theta}$ is given by

$$\langle \dot{\theta}^2 \rangle = \int_{-\infty}^{\infty} \dot{\theta}^2 p_S(\dot{\theta})\, d\dot{\theta}. \tag{5.4-40}$$

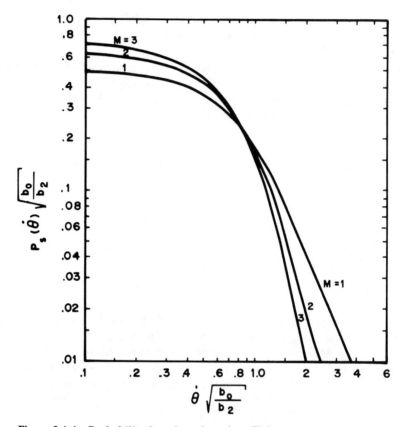

Figure 5.4-4   Probability function of random FM.

If we substitute for $p_S(\dot{\theta})$ from Eq. (5.4-38) and attempt to integrate term by term, each integral diverges since each term is $O(\dot{\theta}^{-3})$ as $\dot{\theta}$ tends to infinity. Davis avoided this difficulty by replacing the infinite limits in Eq. (5.4-40) by $\pm X$, integrating term by term, and then letting $X \to \infty$ to obtain[19]

$$\langle \dot{\theta}^2 \rangle = \frac{Nb_2}{2b_0} \sum_{k=0}^{M-1} \binom{M-1}{k} (-1)^{k+1} \log(k+1), \qquad M \geqslant 2. \quad (5.4\text{-}41)$$

Table 3 shows some values computed using Eq. (5.4-41), and we see that the random FM decreases as the number of diversity branches increases.

Table 3 *Mean Square Random*
FM with Selection Diversity

M	$\langle \dot{\theta}^2 \rangle b_0 / b_2$
1	∞
2	$\log(2) = 0.693$
3	$1.5 \log(4/3) = 0.432$
4	$2 \log(32/27) = 0.340$

Davis has derived the following expressions for the autocorrelation of the output random FM of a selection diversity system[19]:

$$R_M(\tau) = M \left[\frac{(b^2 + ad)}{2a^2} \right] f_M(a), \qquad (5.4\text{-}42)$$

where

$$f_1(a) = -\log(1 - a^2), \qquad (5.4\text{-}43)$$

$$f_2(a) = \log \left[\frac{2}{1 + \sqrt{1 - a^2}} \right], \qquad (5.4\text{-}44)$$

and

$$a \triangleq \frac{g(\tau)}{b_0}, \quad b \triangleq \frac{g'(\tau)}{b_0}, \quad \text{and} \quad d \triangleq -\frac{g''(\tau)}{b_0}. \qquad (5.4\text{-}45)$$

Davis[19] also evaluated $f_3(a)$ by a lengthy series expansion as shown in Figure 5.4-5, along with an empirical approximation,

$$f_3(a) \doteq \log \tfrac{4}{3} - 0.374\sqrt{1 - a^2} + 0.102(1 - a^2), \qquad (5.4\text{-}46)$$

and obtained the following properties:

$$f_M(a) \sim \frac{a^2}{M^2} \quad \text{for small} \quad a,$$

and

$$f_M(1) = \sum_{k=0}^{M-1} \binom{M-1}{k} (-1)^k \log(k+1), \qquad M \neq 1. \qquad (5.4\text{-}47)$$

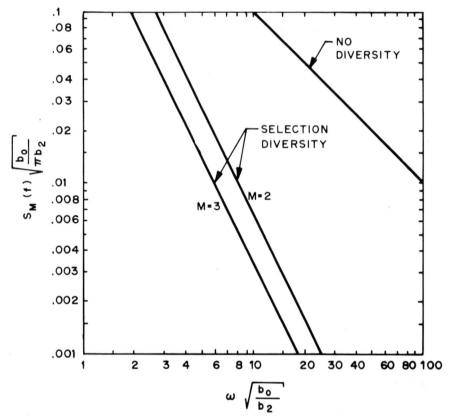

Figure 5.4-5 Asymptotic spectra for random FM.

The problem of determining the power spectrum, $S_{\dot\theta}(f)$, of the random FM involves taking the Fourier transform of the autocorrelation function [cf. Eq. (1.4-8)]. As mentioned above, a closed-form expression is not obtained for $S_{\dot\theta}(f)$ even for $M=1$. If the spectrum is required, then numerical integration is necessary, as in Section 1.4, or possibly an infinite series may be obtainable for specific correlation functions. For example, Rice[20] obtains an infinite series for $S_{\dot\theta}(f)$ for a single branch with $a = g(\tau) = e^{-\mu\tau^2}$.

In many cases of microwave mobile radio systems it is sufficient to know the asymptotic form of $S_{\dot\theta}(f)$ as $f\to\infty$ because the baseband frequencies of 300 Hz to 3 kHz are greater than the spread of the Doppler spectrum. For a correlation function that is infinitely smooth and absolutely integrable over an infinite range, $S(f)$ decreases more rapidly than any negative power of f as $f \to \infty$ [21]. Where step discontinuities or logarithmic infinities occur in the correlation function or its derivatives, the asymptotic spectrum as $f \to \infty$ decreases as some negative power of f.

The asymptotic form of the single branch $(M = 1)$ random FM power spectrum has already been derived and is given in Eq. (5.4-36), above. For $M > 1$ we need $g(\tau)$ to substitute in Eq. (5.4-42). From Eqs. (1.3-2) and (1.3-4) for small τ,

$$g(\tau) \doteq b_0 \left(1 - \frac{b_2}{2b_0} \tau^2 \right).$$

(5.4-48)

Then from Eqs. (5.4-42), (5.4-44), and (5.4-48) we see that for small τ and symmetric Doppler spectra, $R_2(\tau)$ is given by

$$R_2(\tau) \doteq \frac{b_2}{b_0} \log 2 - \left(\frac{b_2}{b_0} \right)^{3/2} |\tau|.$$

(5.4-49)

Thus at $\tau = 0$ there is a discontinuity of $2(b_2/b_0)^{3/2}$ in the first derivative of $R_2(\tau)$. The asymptotic two-sided spectrum is therefore

$$S_2(f) \sim 2 \left(\frac{b_2}{b_0} \right)^{3/2} (2\pi f)^{-2} \quad \text{as} \quad f \to \pm \infty.$$

(5.4-50)

Similarly, from Eqs. (5.4-42), (5.4-46), and (5.4-48), we have for three-branch selection diversity

$$R_3(\tau) \doteq \frac{3}{2} \frac{b_2}{b_0} \log \frac{4}{3} - (0.374) \frac{3}{2} \left(\frac{b_2}{b_0} \right)^{3/2} |\tau|.$$

(5.4-51)

Since the discontinuity in the first derivative is $3(0.374)(b_2/b_0)^{3/2}$ at $\tau = 0$, the asymptotic spectrum is

$$S_3(f) \sim \frac{3(0.374)(b_2/b_0)^{3/2}}{(2\pi f)^2} \quad \text{as} \quad f \to \pm \infty,$$

(5.4-52)

which constitutes an improvement of 2.5 dB for three-branch over two-branch diversity.

The three asymptotic spectra, S_1, S_2, and S_3 are compared in Figure 5.4-5. To give some quantitative feel for the reduction in random FM, the output random FM in the baseband frequencies $0.1\,W$ to W is given by

$$N_{\mathrm{RFM}} = 2 \int_{0.1\,W}^{W} S_{\dot{\theta}}(f)\,df = \begin{cases} \dfrac{b_2}{b_0} \log 10, & M = 1 \\[2ex] \left(\dfrac{b_2}{b_0}\right)^{3/2} \dfrac{9}{\pi^2 W}, & M = 2 \\[2ex] \left(\dfrac{b_2}{b_0}\right)^{3/2} \dfrac{5.05}{\pi^2 W}, & M = 3 \end{cases} \quad (5.4\text{-}53)$$

Typical values[19] for $\sqrt{b_2/b_0}$ and W in a microwave mobile radio system are 333 rad/sec and 3 kHz, respectively. Using these values in Eq. (5.4-53) indicates that two-branch selection diversity enjoys a 13.6-dB reduction in output random FM relative to a single branch, and three-branch selection diversity a 16.1-dB reduction relative to that of a single branch, for this typical case.

Thus selection diversity provides a sizeable reduction in random FM if the Doppler frequency is small compared to the maximum baseband frequency.

5.4.3 Squelch

Since the noise from an FM discriminator rises rapidly as the IF carrier-to-noise ratio falls, it may be desirable to mute the discriminator output if the carrier-to-noise falls below a certain value. Of course, not only is the noise removed during such times but also the signal modulation. The effect of this muting is to increase the noise due to signal suppression. In fact, this increase may, in many cases, be greater than the benefit obtained by eliminating the below-threshold noise of the discriminator.

Davis[17] has computed the effect of squelch on the average output signal to average output noise ratio by using the quasistatic approximation of averaging over the IF carrier-to-noise ratio, ρ. We will summarize his results here.

The squelch technique is assumed to reduce the total output voltage to zero whenever the IF carrier-to-noise ratio at the FM detector input drops below the muting level, ρ_1. Thus, the average quadratic noise and threshold noise are still computed from Eqs. (5.4-4) and (5.4-5), respectively, except that the lower limits of integration are ρ_1 instead of zero. Davis [17] has computed these integrals for an M-branch maximal ratio combiner to be

$$\langle N_2 \rangle_{MR} = \int_{\rho_1}^{\infty} N_2(\rho) p_{MR}(\rho) \, d\rho$$

$$= \frac{a}{(M-1)\rho_0} \left\{ A_{M-1}(r) - \frac{2}{(1+\rho_0)^{M-1}} A_{M-1}[r(1+\rho_0)] \right.$$

$$\left. + \frac{A_{M-1}[r(1+2\rho_0)]}{(1+2\rho_0)^{M-1}} \right\} \tag{5.4-54}$$

and

$$\langle N_D \rangle_{MR} = \int_{\rho_1}^{\infty} N_D(\rho) p_{MR}(\rho) \, d\rho = \frac{8\pi BW}{\sqrt{2}\,(M-1)! \rho_0^M} I_M(\rho_1), \tag{5.4-55}$$

where a is given in Eq. (5.4-2), B is the IF bandwidth, W is the maximum baseband frequency, ρ_0 is the mean IF CNR on each branch, and

$$r \triangleq \frac{\rho_1}{\rho_0}, \tag{5.4-56}$$

$$A_M(x) \triangleq 1 - \frac{\Gamma(M,x)}{\Gamma(M,\infty)} = e^{-x} \sum_{k=0}^{M-1} \frac{x^k}{k!}. \tag{5.4-57}$$

$\Gamma(M,x)$ is the incomplete gamma function.
For $M \geqslant 2$,

$$I_{M+1}(\rho) = \frac{\rho^{M-1}\sqrt{\gamma+\rho}}{\beta} e^{-\beta\rho} + \left(\frac{2M-1}{2\beta} - \gamma \right) I_M(\rho) + (M-1)\frac{\gamma}{\beta} I_{M-1}(\rho),$$

$$\tag{5.4-58}$$

where

$$I_1(\rho) = \sqrt{\frac{\pi}{\beta}} \, e^{\gamma\beta} \operatorname{erfc} \sqrt{\beta(\gamma+\rho)} \,, \tag{5.4-59}$$

$$I_2(\rho) = \frac{\sqrt{\gamma+\rho}}{\beta} e^{-\beta\rho} + \left(\frac{1}{2\beta} - \gamma \right) I_1(\rho), \, \cdot \tag{5.4-60}$$

and

$$\gamma = 2.35, \qquad \beta = 1 + \frac{1}{\rho_0}.$$

The average signal suppression noise and signal output are computed, since $v_0(t) = 0$ when $\rho < \rho_1$, by replacing the lower limit in the integral of Eq. (5.4-8) with ρ_1. Let

$$f(\rho) \overset{\Delta}{=} 1 - e^{-\rho}. \tag{5.4-61}$$

Then

$$\bar{v}_0(t) = v(t) \overline{f_{\mathrm{MR}}} (\rho_1), \tag{5.4-62}$$

where

$$\overline{f_{\mathrm{MR}}} (\rho_1) = \int_{\rho_1}^{\infty} (1 - e^{-\rho}) p_{\mathrm{MR}}(\rho) \, d\rho. \tag{5.4-63}$$

From Eq. (5.4-6), the average signal suppression noise is

$$\langle N_s \rangle_{\mathrm{MR}} = S \left\{ \overline{f_{\mathrm{MR}}^2} (\rho_1) \int_0^{\rho_1} p_{\mathrm{MR}}(\rho) \, d\rho + \int_{\rho_1}^{\infty} p_{\mathrm{MR}}(\rho) [f(\rho) - \overline{f_{\mathrm{MR}}} (\rho_1)]^2 \, d\rho \right\}$$

$$= S \left(\overline{f_{\mathrm{MR}}^2} (\rho_1) - \overline{f_{\mathrm{MR}}} (\rho_1)^2 \right), \tag{5.4-64}$$

where

$$\overline{f_{\mathrm{MR}}^2} (\rho_1) = \int_{\rho_1}^{\infty} (1 - e^{-\rho})^2 p_{\mathrm{MR}}(\rho) \, d\rho. \tag{5.4-65}$$

The average output signal power is given by Eqs. (5.4-10) and (5.4-62):

$$\langle S \rangle = \langle \bar{v}_0(t)^2 \rangle = S \overline{f_{\mathrm{MR}}}^2 (\rho_1). \tag{5.4-66}$$

Davis[17] has shown that

$$\overline{f_{\mathrm{MR}}} (\rho_1) = A_M(r) - \frac{A_M[r(1+\rho_0)]}{(1+\rho_0)^M} \tag{5.4-67}$$

and

$$\overline{f_{\mathrm{MR}}^2} (\rho_1) = A_M(r) - 2 \frac{A_M[r(1+\rho_0)]}{(1+\rho_0)M} + \frac{A_M[r(1+2\rho_0)]}{(1+2\rho_0)M}. \tag{5.4-68}$$

Due to the fact that $\overline{f^2_{MR}}(\rho_1)$ and $\overline{f_{MR}}^2(\rho_1)$ both approach unity as ρ_0 becomes large, whereas their difference may be small, computing $\langle N_s \rangle_{MR}$ by Eq. (5.4-64) may be subject to considerable error. This difficulty may be overcome by using the function

$$B_M(x) = 1 - A_M(x) = e^{-x} \sum_{k=M}^{\infty} \frac{x^k}{k!}. \qquad (5.4\text{-}69)$$

We obtain

$$\frac{1}{S}\langle N_s \rangle_{MR} = B_M(r) + A_M[r(1+2\rho_0)]$$

$$- \{ B_M(r) + A_M[r(1+\rho_0)] \}^2, \qquad (5.4\text{-}70)$$

which avoids taking the small difference of large quantities.

Optimum Muting Level

Since the signal suppression is small for most CNRs of interest, the output $\langle SNR \rangle$ is maximized approximately when the total noise, $\langle N_2 \rangle_{MR} + \langle N_D \rangle_{MR} + \langle N_s \rangle_{MR}$, is minimized. The level at which muting is applied, ρ_1, such that the total noise is minimum, is the optimum level, computed by Davis[17] as follows:

$$\langle N_2 \rangle_{MR} + \langle N_D \rangle_{MR} + \langle N_s \rangle_{MR} \triangleq \int_0^{\infty} [N_2(\rho) + N_D(\rho) + N_s(\rho)] P_{MR}(\rho) \, d\rho,$$

$$(5.4\text{-}71)$$

where

$$N_2(\rho) = \begin{cases} \dfrac{a}{\rho}(1 - e^{-\rho})^2, & \rho > \rho_1 \\ 0, & \rho < \rho_1 \end{cases}, \qquad (5.4\text{-}72)$$

$$N_D(\rho) = \begin{cases} b(\rho + \gamma)^{-1/2} e^{-\rho}, & \rho > \rho_1 \\ 0, & \rho < \rho_1 \end{cases}, \qquad (5.4\text{-}73)$$

$$N_s(\rho) = \begin{cases} S(f - \bar{f})^2, & \rho > \rho_1 \\ S_m \bar{f}^2, & \rho < \rho_1 \end{cases}, \qquad (5.4\text{-}74)$$

where a is given by Eq. (5.4-2),

$$b = \frac{8\pi BW}{\sqrt{2}}, \quad \gamma = 2.35, \quad f = f(\rho) = 1 - e^{-\rho},$$

and

$$\overline{f_{MR}} = \overline{f_{MR}}(\rho_1) = \int_{\rho_1}^{\infty} f(\rho) p_{MR}(\rho) \, d\rho.$$

To find the optimum ρ_1, we differentiate Eq. (5.4-71) with respect to ρ_1 and set the result equal to zero:

$$S\overline{f_{MR}}^2 = \frac{a}{\rho_1}(1 - e^{-\rho_1})^2 + be^{-\rho_1}[\rho_1 + \gamma]^{-1/2} + S(f - \overline{f_{MR}})^2; \quad (5.4\text{-}75)$$

here, $\overline{f_{MR}}$ is a function of ρ_1. However, in most cases of interest, ρ_1 is small compared to ρ_0 and ρ_0 is large compared to unity. In this case $\bar{f}$ is accurately approximated by unity and Eq. (5.4-75) is approximately given as

$$S \doteq \frac{a}{\rho_1}(1 - e^{-\rho_1})^2 + be^{-\rho_1}(\rho_1 + \gamma)^{-1/2} + S[f(\rho_1) - 1]^2. \quad (5.4\text{-}76)$$

Thus the optimum muting level is approximately independent of ρ_0 and the number of branches. Using Eq. (5.4-24), the solutions to (5.4-76) were obtained numerically and are presented for various IF to baseband bandwidth ratios, $B/2W$, in Table 4.[17]

Table 4   *Optimum Muting Levels versus Bandwidth Ratio*

$B/2W$	2	3	4	6	8	10	15	20
ρ_1, optimum (dB)	3.9	1.6	0.1	-2.0	-3.4	-4.5	-6.4	-7.7

It can be seen that in general these levels are too low to be achieved in practice. The benefits obtained by muting at the optimum level are not great, whereas muting at FM threshold will usually degrade the SNR compared with no muting. However, muting does have the advantage of reducing the output noise in the absence of modulation. For more than two branches of diversity, optimum muting results in very little improvement in SNR. The improvements for no diversity and two-branch maximal ratio are shown in Figure 5.4-6.

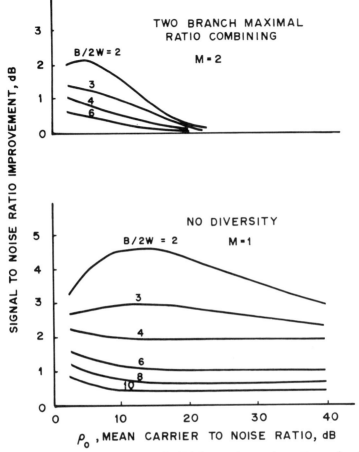

Figure 5.4-6   Improvement in ⟨SNR⟩ by muting at the optimum level.

Davis[17] has computed explicit expressions with M-branch selection diversity for the noise components $\langle N_2 \rangle_S$, $\langle N_D \rangle_S$, $\langle N_s \rangle_S$, and the signal $\langle S \rangle_S$ with squelch. The results again show that maximal ratio calculations of ⟨SNR⟩ may be applied to selection diversity by shifting the IF CNR by the amounts shown in Table 1.

The similarity in selection and maximal ratio IF CNR probability distributions allows the maximal ratio ⟨SNR⟩ results to be applied to selection diversity. As described above, this similarity also exists between equal-gain and maximal-ratio combining. Thus the effects of muting on output ⟨SNR⟩ with equal-gain combining can be obtained from those for maximal ratio combining by shifting the IF CNR as given in Table 2.

5.4.4 Cochannel Interference

In a high-capacity mobile radio system, the reduction of cochannel interference can be the most important advantage of diversity. A diversity combiner changes the probability distribution of the ratio of the desired signal and interfering signal power presented to the FM detector. The distribution of the IF signal-to-interference ratio can be converted to that of the baseband signal-to-interference ratio by computing the average output signal, average output interference, and signal suppression noise.

As shown in Section 4.1.5, the baseband interference is dominated by the occasions that the FM detector is captured by the interferer. As a result, suppression of interference by increasing the modulation index is not successful in the presence of Rayleigh fading without diversity. With diversity, not only is the IF signal-to-interference ratio improved, but one can achieve some interference suppression with FM demodulation by increasing the modulation index. The amount of "index cubed" baseband signal-to-interference ratio improvement that can be achieved in this way increases with the number of diversity branches.

As mentioned in Section 4.1.5, in the case where there are many cochannel interferers, each with different modulation signals, the central limit theorem may be applied. The sum of interferers is then approximated by Gaussian noise with power equal to the sum of average interferer powers. With this approximation, the analysis used in Section 5.4.1 applies to provide the average signal to average interference ratio in the baseband output.

Throughout the remainder of this section, however, we will consider only the case of a single interferer.

Selection Diversity

When the selection diversity combiner described in Section 5.2.1 is subjected to cochannel interference, the selection can be based on one of several decision algorithms. The "total-power algorithm," which selects the branch with the largest total IF received power (desired signal plus interferer), is probably the easiest to instrument in practice. Other decision algorithms can be proposed; for example, the signals and interferers could be identified by different pilots transmitted with each. The combiner then selects the branch with the largest desired signal power ("desired-signal-power algorithm").

Let the desired signal phasor on the $k\underline{th}$ branch be denoted g_k and the interferer phasor h_k. As described in Chapter 1, for a mobile radio environment with Rayleigh fading and with sufficient spacing of the diversity array to provide independent fading, the set of phasors $\{g_k\}$ and

$\{h_k\}$ are all independent complex Gaussian random variables, with $\langle g_k g_k^* \rangle = 2\sigma_g^2$ and $\langle h_k h_k^* \rangle = 2\sigma_h^2$ for all k.

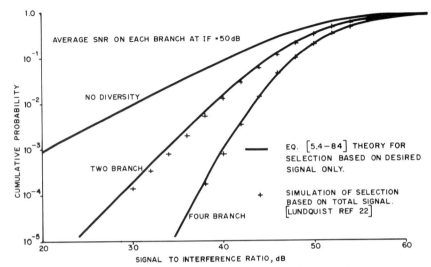

Figure 5.4-7 Cumulative probability distribution of signal to interference ratio with selection diversity.

Lundquist[22] has generated random $\{g_k\}$ and $\{h_k\}$ by Monte Carlo techniques with a computer and simulated the performance of a selection combiner operating with the "total-power algorithm." Figure 5.4-7 shows the results of his simulation for the cumulative probability distribution of the IF signal to interference power ratio at the combiner output. The average signal to interference power ratio, Γ, on each branch is assumed to be 50 dB. The figure shows a clear advantage of diversity in suppressing cochannel interference.

A closed-form solution for the probability distribution of the output IF signal-to-interference ratio may be obtained in the "desired-signal-power algorithm case." The probability density, $p(\gamma)$, of the selected branch signal to interference ratio γ is M times the joint probability density, $p_n(\gamma, n)$ of the signal-to-interference ratio on the $n\underline{th}$ branch and the event that the $n\underline{th}$ branch is the selected branch,

$$p(\gamma) = M p_n(\gamma, n), \qquad (5.4\text{-}77)$$

where M is the number of branches.

The joint probability density, $p_n(\gamma, n)$, is the integral over the desired signal amplitude on the $n\,\underline{th}$ branch, r_{gn}, of the product of the probability that all other $M - 1$ branches have desired signal amplitudes less than that of the $n\,\underline{th}$ branch and the joint probability density $p_n(\gamma, r_{gn})$ of the signal-to-interference ratio on the $n\,\underline{th}$ branch and r_{gn}:

$$p_n(\gamma, n) = \int_0^\infty dr_{gn} \left[\prod_{k \neq n} \int_0^{r_{gn}} p_k(r_{gk}) \, dr_{gk} \right] p_n(\gamma, r_{gn}). \qquad (5.4\text{-}78)$$

The probability that all other $M - 1$ branches have amplitudes r_{gk} less than r_{gn} is found by integrating over their independent Rayleigh densities,

$$\prod_{k \neq 1} \int_0^{r_{gn}} p_k(r_{gk}) \, dr_{gk} = \left[\int_0^{r_{gn}} \frac{r}{\sigma_g^2} e^{-r^2/2\sigma_g^2} \, dr \right]^{M-1} = (1 - e^{-r_{gn}^2/2\sigma_g^2})^{M-1}$$

$$= \sum_{k=0}^{M-1} \binom{M-1}{k} (-1)^k e^{-kr_{gn}^2/2\sigma_g^2}, \qquad (5.4\text{-}79)$$

where $\binom{M-1}{k}$ is the binomial coefficient.

The joint density of the desired signal amplitude r_{gn} and the independent interferer amplitude r_{hn} on the $n\,\underline{th}$ branch is

$$p_n(r_{gn}, r_{hn}) = \frac{r_{gn}}{\sigma_g^2} e^{-r_{gn}^2/2\sigma_g^2} \frac{r_{hn}}{\sigma_h^2} e^{-r_{hn}^2/2\sigma_h^2}. \qquad (5.4\text{-}80)$$

Since

$$\gamma = \left(\frac{r_{gn}}{r_{hn}} \right)^2, \qquad (5.4\text{-}81)$$

we may apply a coordinate transformation to Eq. (5.4-80) to obtain

$$p_n(\gamma, r_{gn}) = \frac{r_{gn}^3}{2\gamma^2 \sigma_g^2 \sigma_h^2} \exp\left\{ -\frac{r_{gn}^2}{2} (\sigma_g^{-2} + \sigma_h^{-2}\gamma^{-1}) \right\}. \qquad (5.4\text{-}82)$$

Substituting Eqs. (5.4-79) and (5.4-82) into Eq. (5.4-78) and integrating gives

$$p(\gamma) = M\Gamma \sum_{k=0}^{M-1} \binom{M-1}{k} (-1)^k [\gamma(k+1) + \Gamma]^{-2}, \qquad (5.4\text{-}83)$$

where $\Gamma = \sigma_g^2 / \sigma_h^2$.

The cumulative probability distribution of the output IF signal-to-interference ratio is obtained by integrating Eq. (5.4-83):

$$P(\gamma) = \int_0^\gamma p(x)\,dx = M\gamma \sum_{k=0}^{M-1} \binom{M-1}{k}(-1)^k \frac{1}{(k+1)\gamma + \Gamma}. \quad (5.4\text{-}84)$$

Equation (5.4-84) is also plotted in Figure 5.4-7. As seen in the figure, the "desired-signal-power algorithm" and the "total-power algorithm" provide nearly equal performance. This is to be expected since the total power on a branch is mainly determined by the desired signal power on that branch when the average desired signal power is much greater than the average interferer power.

Maximal Ratio Combining

When subjected to cochannel interference the performance of a maximal-ratio combiner depends on the means by which the branch gains are determined. Several methods of instrumenting a maximal-ratio combiner are described in Chapter 6. We will analyze two of them here: "perfect pilot" and "separate pilot" maximal-ratio combiners.

Perfect Pilot Maximal Ratio

The "perfect pilot" combiner adjusts the branch gains, $\{p_k\}$, to be equal to the complex conjugate of the phasors, $\{g_k\}$, of the desired signal only on each respective branch; that is, for "perfect pilot,"

$$p_k = g_k^*. \quad (5.4\text{-}85)$$

The interferer phasors, $\{h_k\}$, are assumed to have no effect on the branch gains. The "perfect pilot" might be obtained by transmitting different pilots with each signal so that the pilot of the desired signal only may be selected by the receiver by filtering. The Granlund maximal-ratio combiner, described in Chapter 6, is a good approximation to the "perfect pilot" combiner since the signal fed back from the combiner output to generate the branch pilots (pseudolocal oscillators) usually contains little contamination from the interferer.

The output signal-to-interference IF power ratio for an M-branch "perfect pilot" maximal-ratio combiner is

$$\gamma = \frac{\left| \sum_{k=1}^{M} p_k g_k \right|^2}{\left| \sum_{k=1}^{M} p_k h_k \right|^2} = \frac{\left(\sum_{k=1}^{M} |g_k|^2 \right)^2}{\left| \sum_{k=1}^{M} g_k^* h_k \right|^2}. \quad (5.4\text{-}86)$$

As mentioned above, the phasors $\{g_k\}$ and $\{h_k\}$ are independent symmetric complex Gaussian random variables with quadrature component variances σ_g^2 and σ_h^2, respectively.

Lundquist[22] has simulated the "perfect pilot" maximal-ratio combiner performance against cochannel interference, using the Monte Carlo technique with a computer. His results for the cumulative probability distribution of the output signal to interference IF power ratio are shown in Figure 5.4-8 for an average signal to interference power ratio, Γ, on each branch of 50 dB.

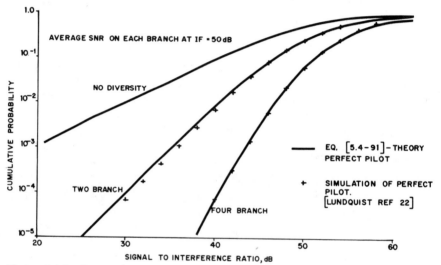

Figure 5.4-8 Cumulative probability distribution of signal to interference ratio with perfect pilot maximal ratio diversity.

A closed-form expression for the probability distribution of the output signal-to-interference IF power ratio for the "perfect pilot" maximal-ratio combiner may be derived as follows:

Define

$$x \stackrel{\Delta}{=} \mathrm{Re} \left\{ \frac{\displaystyle\sum_{k=1}^{M} g_k^* h_k}{\left[\displaystyle\sum_{k=1}^{M} |g_k|^2\right]^{1/2}} \right\} \qquad (5.4\text{-}87)$$

and

$$y \overset{\Delta}{=} \text{Im} \left\{ \frac{\sum\limits_{k=1}^{M} g_k^* h_k}{\left[\sum\limits_{k=1}^{M} |g_k|^2 \right]^{1/2}} \right\}. \tag{5.4-88}$$

As shown in Ref. 9, the conditional joint probability densities of x and y, given $\{g_k\}$, are symmetric Gaussian with a variance independent of whatever values are assigned to each of the g_k. Thus since the conditional distributions are identical, the unconditional distribution is the same; that is, the joint probability density of x and y is symmetric Gaussian, independent of g_k and with quadrature component variances:

$$\sigma_x^2 = \sigma_y^2 = \sigma_h^2. \tag{5.4-89}$$

Substituting Eqs. (5.4-87) and 5.4-88) into Eq. (5.4-86) gives

$$\gamma = \frac{\sum\limits_{k=1}^{M} |g_k|^2}{x^2 + y^2}. \tag{5.4-90}$$

Equation (5.4-90) is recognized as the ratio of two independent chi-squared variables, the numerator with $2M$ degrees of freedom (each with variance σ_y^2) and the denominator with two degrees of freedom (each with variance σ_h^2). The cumulative distribution, $P(\gamma)$, is therefore the F distribution of indices $2M$ and 2 as given in Ref. 23:

$$P(\gamma) = \left(\frac{\gamma}{\Gamma + \gamma} \right)^M \tag{5.4-91}$$

where $\Gamma = \sigma_g^2 / \sigma_h^2$. Equation (5.4-91) is plotted in Figure 5.4-8 as the exact result.

The simplicity of Eq. (5.4-91) facilitates the conversion of IF signal-to-interference ratios to baseband ratios of average signal to average interference and signal suppression noise by the formulas of Section 4.1.5. From Eqs. (4.1-67) and (5.4-91) the average baseband signal, $\bar{\varphi}_0(t)$, is related to the transmitted signal modulation, $\varphi(t)$, by

$$\bar{\varphi}_0(t) = \varphi(t)[1 - P(1)] = \varphi(t)\left[1 - (1+\Gamma)^{-M} \right]. \tag{5.4-92}$$

The signal suppression noise, $n_\varphi(t)$ is given by Eq. (4.1-68):

$$n_\varphi(t) = \begin{cases} -\left[1-(1+\Gamma)^{-M}\right]\varphi(t) & \text{if } \gamma < 1 \\ (1+\Gamma)^{-M}\varphi(t) & \text{if } \gamma > 1 \end{cases} \tag{5.4-93}$$

The average output signal power is

$$S = \Phi^2\left[1-(1+\Gamma)^{-M}\right]^2, \tag{5.4-94}$$

where Φ^2 is the mean square phase modulation of the transmitted signal. From Eq. (4.1-70), the sum of the average output interference and signal suppression noise is, for $M > 1$,

$$\overline{N}_\varphi = \int_1^\infty \frac{dP(\gamma)}{d\gamma} \left\{ \frac{1}{\gamma}\sqrt{\frac{3}{8\pi(\Phi^2+\Phi_i^2)}} + (1+\Gamma)^{-2M}\Phi^2 \right\} d\gamma$$

$$+ \int_0^1 \frac{dP(\gamma)}{d\gamma} \left\{ \Phi_i^2 + \left[1-(1+\Gamma)^{-M}\right]^2\Phi^2 \right\} d\gamma,$$

$$\overline{N}_\varphi = \sqrt{\frac{3}{8\pi(\Phi^2+\Phi_i^2)}} \left\{ \frac{M}{\Gamma(M-1)}\left[1-(1+\Gamma)^{1-M}\right] - \frac{1}{\Gamma}\left[1-(1+\Gamma)^{-M}\right] \right\}$$

$$+ (1+\Gamma)^{-M}\left\{ \left[1-(1+\Gamma)^{-M}\right]\Phi^2 + \Phi_i^2 \right\}, \tag{5.4-95}$$

where Φ_i^2 is the mean square phase modulation of the interferer.

Equations (5.4-94) and (5.4-95) have been used to compute the ratio of average baseband signal to average baseband interference versus modulation index (assuming $\Phi^2 = \Phi_i^2$) and the number of diversity branches, as plotted in Figure 5.4-9 through 5.4-12. As seen in the figures, diversity can allow the "index cubed" improvement with increasing modulation index to be regained. As pointed out in Section 4.1.5, the "index cubed" advantage was lost for a single branch, due to Rayleigh fading.

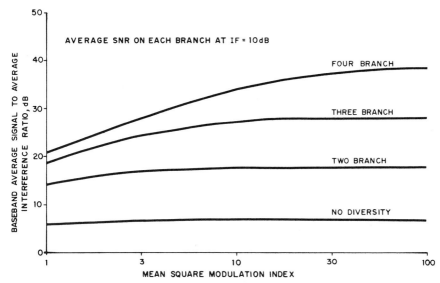

Figure 5.4-9 Baseband signal to interference ratio versus modulation index with perfect pilot maximal ratio diversity.

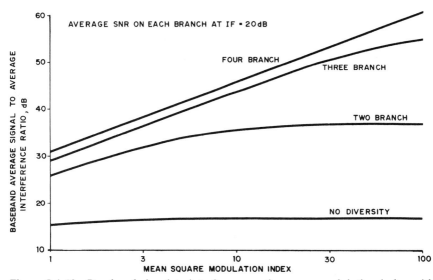

Figure 5.4-10 Baseband signal to interference ratio versus modulation index with perfect pilot maximal ratio diversity.

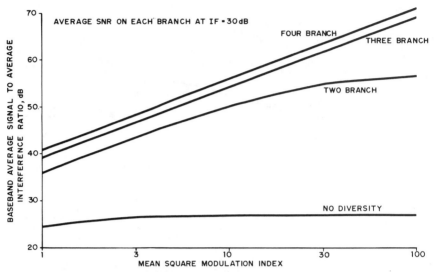

Figure 5.4-11 Baseband signal to interference ratio versus modulation index with perfect pilot maximal ratio diversity.

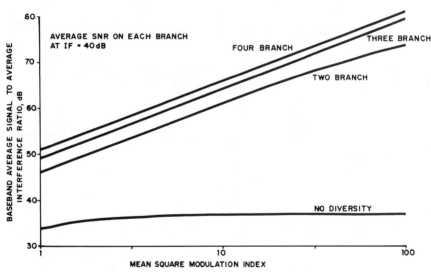

Figure 5.4-12 Baseband signal to interference ratio versus modulation index with perfect pilot maximal ratio diversity.

Separate Pilot Maximal Ratio

With "separate pilot" maximal-ratio combiners, described in Chapter 6, the branch gains do not differentiate between signal and interferer. Rather, the complex conjugate of the sum of the signal pilot phasor and the interferer pilot phasor is taken as the branch gain; that is, for "separate pilot,"

$$p_k = g_k^* + h_k^*. \tag{5.4-96}$$

The output signal-to-interference IF power ratio for an M-branch "separate pilot" maximal-ratio combiner is

$$\gamma = \frac{\left| \sum_{k=1}^{M} p_k g_k \right|^2}{\left| \sum_{k=1}^{M} p_k h_k \right|^2} = \frac{\left| \sum_{k=1}^{M} (|g_k|^2 + h_k^* g_k) \right|^2}{\left| \sum_{k=1}^{M} (g_k^* h_k + |h_k|^2) \right|^2}. \tag{5.4-97}$$

Using Monte Carlo techniques on a computer, Lundquist[24] has simulated the "separate pilot" maximal-ratio combiner performance against cochannel interference. His results for the cumulative probability distribution of the output signal to interference IF power ratio are shown in Figure 5.4-13, for an average signal-to-interference power ratio on each branch of 20 dB.

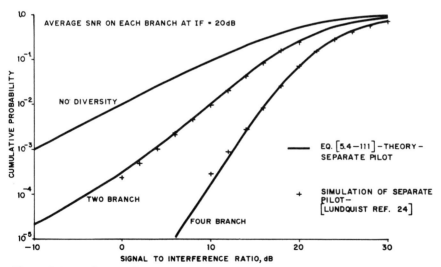

Figure 5.4-13 Cumulative probability distribution of signal to interference ratio with separate pilot maximal ratio diversity.

To solve for the probability distribution of γ in Eq. (5.4-97) we introduce the change of variables used in Ref 9:

$$r_k \overset{\Delta}{=} \sqrt{\frac{\Gamma}{\Gamma+1}} \ |p_k| \qquad \theta_k \overset{\Delta}{=} \arctan(p_k) \tag{5.4-98}$$

$$u_g \overset{\Delta}{=} \frac{\displaystyle\sum_{k=1}^{M} r_k \left(\mathrm{Re}\{ g_k e^{j\theta_k} \} - \sqrt{\frac{\Gamma}{\Gamma+1}} \ r_k \right)}{\left(\displaystyle\sum_{k=1}^{M} r_k^2 \right)^{1/2}}, \tag{5.4-99}$$

$$v_g \overset{\Delta}{=} \frac{\displaystyle\sum_{k=1}^{M} r_k (\mathrm{Im}\{ g_k e^{j\theta_k} \})}{\left(\displaystyle\sum_{k=1}^{M} r_k^2 \right)^{1/2}}, \tag{5.4-100}$$

and

$$U \overset{\Delta}{=} \sqrt{\frac{\Gamma}{\Gamma+1}} \left(\sum_{k=1}^{M} r_k^2 \right)^{1/2}.$$

Equation (5.4-97) then becomes

$$\gamma = \frac{(U+u_g)^2 + v_g^2}{[(1/\Gamma)U - u_g]^2 + v_g^2}, \tag{5.4-101}$$

where u_g, v_g, and U are independent random variables. u_g and v_g are zero-mean Gaussian variables with variance $\sigma_g^2/(\Gamma+1)$. U has a chi distribution of $2M$ degrees of freedom.[23] Each degree of freedom has a variance

$$\sigma^2 = \sigma_g^2 \frac{\Gamma}{\Gamma+1}, \tag{5.4-102}$$

so that the probability density of U is given by

$$p(U) = \frac{2 U^{2M-1} e^{-u^2/2\sigma^2}}{2^M \sigma^{2M} (M-1)!}.$$

If the signal-to-interference ratio is less than γ, then Eq. (5.4-101) implies that

$$(u_g^2 + v_g^2)(1-\gamma) + 2u_g U\left(1 + \frac{\gamma}{\Gamma}\right) + U^2\left(1 - \frac{\gamma}{\Gamma^2}\right) < 0. \quad (5.4\text{-}104)$$

For $\gamma = 1$, inequality (5.4-104) simplifies to

$$u_g < -U\left(\frac{\Gamma-1}{2\Gamma}\right) \quad (5.4\text{-}105)$$

As pointed out by Langseth[25], the $\gamma=1$ case is related to the F distribution[23] for the ratio of a chi square variable with one degree of freedom, u_g^2, to one with $2M$ degrees of freedom, U^2:

$$P(1) \triangleq \text{Prob}\{\gamma \leqslant 1\} = (1+\Gamma)^{-M} \sum_{k=0}^{M-1} \binom{k+M-1}{k}\left(\frac{\Gamma}{1+\Gamma}\right)^k, \quad (5.4\text{-}106)$$

where $\binom{k+M-1}{k}$ is the binomial coefficient.

If $\gamma > 1$, we define

$$c = \frac{\Gamma+\gamma}{\Gamma(\gamma-1)}, \quad (5.4\text{-}107)$$

and we write inequality (5.4-104) as

$$(u_g - cU)^2 + v_g^2 > U^2\gamma\left[\frac{\Gamma+1}{\Gamma(\gamma-1)}\right]^2. \quad (5.4\text{-}108)$$

Given a value of U as a condition, the probability that inequality (5.4-108) is satisfied, $P(\gamma|U)$, is recognized as the noncentral chi square distribution[23] with two degrees of freedom, variance equal to $\sigma_g^2/(\Gamma+1)$, and noncentrality parameter c^2U^2:

$$P(\gamma|U) = 1 - \int_0^d \frac{U^2}{2b_g}\exp\left[-\frac{U^2}{2b_g}(s+c^2)\right]I_0\left(\frac{U^2c\sqrt{s}}{b_g}\right)ds, \quad (5.4\text{-}109)$$

where we have defined

$$d \triangleq \left[\frac{\Gamma+1}{\Gamma(\gamma-1)} \right]^2 \gamma, \qquad s = \frac{(u_g - cU)^2 + v_g^2}{U^2},$$

$$b_g = \frac{\sigma_g^2}{1+\Gamma},$$

(5.4-110)

and $I_0(x)$ is the modified Bessel function of the first kind and zero order.

Multiplying Eq. (5.4-109) by the probability density of Eq. (5.4-103) and integrating (p. 712 of Ref. 26) over all values of U, one obtains the unconditional distribution of the output IF signal to interference ratio $(\gamma > 1)$:

$$P\{\gamma\} = 1 - \int_0^d \frac{(-1)^M ds}{(M-1)!\Gamma^M} \frac{\partial^M}{\partial p^M} (p^2 - \lambda^2)^{-1/2}, \qquad (5.4-111)$$

where we have defined

$$p \triangleq \frac{1}{2b_g} \left\{ s + \left[\frac{\Gamma+\gamma}{\Gamma(\gamma-1)} \right]^2 + \frac{1}{\Gamma} \right\},$$

$$\lambda^2 \triangleq \left[\frac{2(\Gamma+\gamma)}{\Gamma(\gamma-1)} \right]^2 s.$$

(5.4-112)

The integrals involved in Eq. (5.4-111) are all closed-form expressions (p. 83, Ref. 26). Following the same method for $\gamma < 1$ we obtain

$$P\{\gamma\} = \int_0^d \frac{(-1)^M ds}{(M-1)!\Gamma^M} \frac{\partial^M}{\partial p^M} (p^2 - \lambda^2)^{-1/2}. \qquad (5.4-113)$$

For $M=1$ Eqs. (5.4-111) and (5.4-113) reduce to the "no diversity" result,

$$P_1\{\gamma\} = \frac{\gamma}{\gamma+\Gamma}, \qquad (5.4-114)$$

and for $M=2$ they yield the result given by Langseth[25]:

$$P_2\{\gamma\} = \frac{\gamma^2(1+\Gamma)+2\Gamma\gamma}{(1+\Gamma)(\gamma+\Gamma)^2}.$$ (5.4-115)

As pointed out by Langseth,[25] the algebra becomes quite involved when one proceeds beyond two branches of "separate pilot" maximal-ratio diversity. As an example we present here the results of Eqs. (5.4-112) and 5.4-113) for four branches $(M=4)$:

$$P_4\{\gamma\} = \begin{cases} A(d)-A(0) & \text{if } \gamma<1 \\ 1-A(d)+A(0) & \text{if } \gamma>1 \end{cases},$$ (5.4-116)

where

$$A(s) \overset{\Delta}{=} \frac{s-b}{R}\left(\frac{\Gamma a_1}{2c^2}+\frac{\Gamma^2 a_2}{6c^4}+\frac{\Gamma^3 a_5}{20c^6}\right)+\frac{s-b}{R^3}\left(a_1+\frac{\Gamma a_2}{3c^2}+\frac{\Gamma^2 a_5}{10c^4}\right)$$

$$+\frac{1}{R^5}\left[(s-b)a_2-a_3+\frac{3\Gamma(s-b)}{10c^2}\right]+\frac{1}{R^7}[a_5(s-b)-a_4], \quad (5.4-117)$$

and

$$b \overset{\Delta}{=} c^2-\frac{1}{\Gamma} \qquad R \overset{\Delta}{=} \sqrt{s^2-2bs+h^2} \qquad h \overset{\Delta}{=} c^2+\frac{1}{\Gamma},$$

$$a_1 \overset{\Delta}{=} \frac{1}{3\Gamma^3 c^2} \qquad a_2 \overset{\Delta}{=} \frac{\Gamma}{} \frac{(b+h)(11+7h)}{2\Gamma^3 c^2} \qquad a_3 \overset{\Delta}{=} \frac{8(b+h)}{\Gamma^4}, \quad (5.4-118)$$

$$a_4 \overset{\Delta}{=} \frac{20b(b+h)^2}{\Gamma^4} \qquad a_5 \overset{\Delta}{=} \frac{5(b+h)^2(2b^2-h^2)}{2\Gamma^3 c^2}.$$

As a comparison to the simulation results, the values of Eq. (5.4-111) are also plotted as solid curves in Figure 5.4-13 for the case where the average signal-to-interference ratio on each branch is 20 dB.

Equal-Gain Combining

As in the case of diversity improvement against noise, the diversity improvement by equal-gain combining against cochannel interference closely follows that obtained by maximal-ratio combining. As an example, we will compute the performance of "perfect pilot" equal-gain combining. The branch gains of a "perfect pilot" equal-gain combiner all have unity gain with phase shift equal to the negative of the phase of the desired signal on the respective branches. The amplitude of the output desired signal is thus the sum of the amplitudes of the desired signal on each branch. The output interference is the sum of the branch interference phasors added with independent phase shifts. The output interference is thus a complex Gaussian phasor with M times the variance of that for a single branch. Thus, the distribution of the output signal-to-interference ratio is given by the distribution of the ratio:

$$\gamma = \frac{\left(\sum_{k=1}^{M} |g_k|\right)^2}{M|h_k|^2}, \tag{5.4-119}$$

where any one of the k branch interference phasors, h_k, may be used in the denominator of (5.4-119).

As shown on p. 457 of Ref. 1, the probability distribution of

$$\frac{1}{M}\left(\sum_{k=1}^{M} |g_k|\right)^2$$

is approximately the same as that of

$$\left(\sum_{k=1}^{M} |g_k|^2\right)/\nu(M),$$

where

$$\nu(M) = \frac{M}{2}\left[(M-\tfrac{1}{2})!/\pi^{-1/2}\right]^{-1/M}, \tag{5.4-120}$$

so

$$\nu(1) = 1, \quad \nu(2) = \sqrt{\tfrac{4}{3}}, \quad \nu(3) = \sqrt[3]{\tfrac{9}{5}} \quad \nu(4) = \sqrt[4]{\frac{256}{105}}.$$

By comparing Eqs. (5.4-119) and (5.4-90), and including the effect of $\nu(M)$, we have from Eq. (5.4-91) an approximation for the cumulative probability distribution of γ:

$$P_{EG}(\gamma) \cong \left[\frac{\gamma}{\gamma + \Gamma/\nu(M)}\right]^M. \tag{5.4-121}$$

Equation (5.4-121) is plotted in Figure 5.4-14 for $M = 1$, 2, and 4 with $\Gamma = 50$ dB average signal-to-interference ratio at the input to each branch. By comparing Figures 5.4-7, 5.4-8, and 5.4-14, it is seen that just as with diversity performance against thermal noise, the performance of equal-gain combining against cochannel interference is better than selection diversity but not quite as good as maximal-ratio combining.

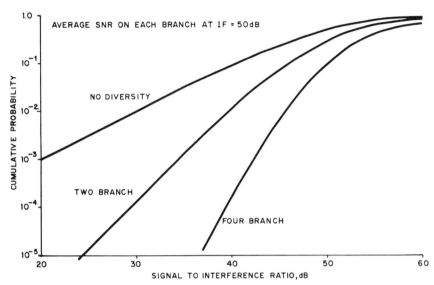

Figure 5.4-14 Cumulative probability distribution of signal-to-interference ratio with perfect pilot equal-gain diversity.

5.5 DIVERSITY AGAINST SHADOWING

5.5.1 Principles

As pointed out in Chapter 2, the other main cause of signal level variation in the mobile radio environment is shadowing, in which large obstructions in the radio path cause reduced local mean levels in areas tens or hundreds of feet in extent. Separation of antennas at mobile or base (by inches or by a few feet) clearly does not provide independent paths with respect to such obstructions. In typical systems using a single base-station site, shadowing leads to "holes" in the coverage area (sections of street in which the signal level is continuously unacceptable). If such a situation is deemed sufficiently serious, the hole may be filled in through use of a satellite base station, which would be located so as to have a less obstructed path to the problem area, as in Figure 5.5-1. Such a technique, however, is not desirable in small-cell systems, where one satellite per cell can raise site costs to prohibitive levels. The placement of base sites on alternate corners of the hexagonal cells as in Figure 5.5-2 provides three widely spaced sites for diversity against shadowing within each cell.[27] Except for edge effects, this plan requires only one base site per cell, since, while each cell is served by three sites, each site also serves three cells. The use of directive antennas at these sites can increase signal levels within the cell and reduce cochannel interference radiated in other directions. The shadow diversity created by selecting the best site will further increase the overall mean signal and, more importantly, will decrease the standard deviation of the local mean signal.

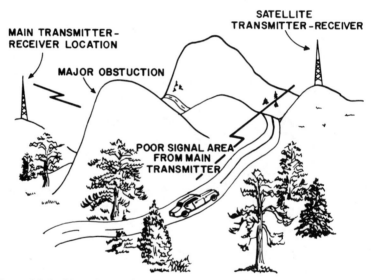

Figure 5.5-1   Use of a satellite base station to fill an area of poor reception.

To the extent that signals reaching the mobile from different directions are statistically independent, the diversity is completely effective. With certain assumptions regarding the statistical behavior of the mean signal variations, the diversity advantage can be calculated.

5.5.2 Multiple Base-Station Diversity to Improve Average SNR

In Chapter 1 it was shown that the local signal amplitude probability function in a scattering environment (such as a mobile vehicle encounters)

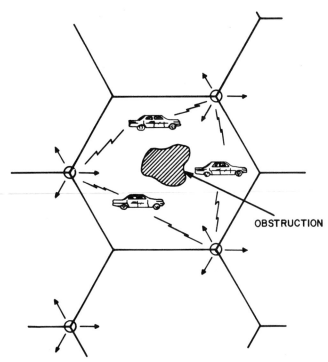

Figure 5.5-2 Use of inward-directed antennas at alternate cell corners to achieve diversity with respect to large obstructions.

is Rayleigh distributed. If a mobile diversity system is applied so that short-term fading, that is, fading due to multipath, is greatly reduced, then the signal out of a diversity combiner is a good representation of the average signal level. For example, if either equal-gain or maximal-ratio combining with eight diversity branches is used, 95% of the time the signal will be within 3 dB of its mean value. On the other hand, a Rayleigh envelope will vary from 16 dB below to 7 dB above its mean value within the same 95% time period. Therefore, as the vehicle moves about, the signal level from a diversity combiner will vary primarily with the long-term fading, that is, fading caused by shadows.

Even if a diversity receiver is not employed, it is possible to evaluate the output performance of an FM mobile receiver in the presence of both shadow-produced and Rayleigh fading while the vehicle is in motion. The calculation assumes that the mean received signal level remains constant for a time long enough to average over the Rayleigh fading. In general, significant changes in mean level do not occur for hundreds of wavelengths (at carrier frequencies around 1 GHz), so this approach should be valid in most instances. In Chapter 2 it was found that the long-term fading statistics (i.e., the local mean signals) are well described by a log-normal distribution with standard deviations in the range of 5–10 dB. The fact that the instantaneous amplitude fluctuates causes a degradation of the baseband signal, that can be calculated when the vehicle is in motion (Section 1.4). However, outage probabilities, which are important when the vehicle is stopped, cannot be calculated by assuming that the mean remains constant, but require the true distribution of the received signal strength.

Generally speaking, one could not simultaneously use multiple base stations with overlapping coverage areas and expect a conventional mobile receiver to give a diversity advantage. Although the RMS signal strength would be improved, multipath fading would still occur. At microwave frequencies, base-station frequency instabilities also give rise to beat frequencies similar to Doppler fading; however, this problem could be eliminated if a diversity receiver that corrects short-term fading were used. A diversity receiver such as equal gain or maximal ratio adds the signals at each antenna in phase; thus the resultant signal from the combiner would be proportional to the sum of the average signal powers from each base station. To analytically determine the statistics of a system using multiple base stations and employing a diversity type of receiver appears to be a difficult problem, however.

In the most general case the summed average signals come from individual base stations, each producing its own distribution of signal levels, yet with the mean signal from any two base stations having some degree of correlation that depends upon the terrain and location of the vehicle. If the joint distribution of the random variables describing the average signal strength is known, then the probability density for their sums can, in theory, be given. If the random variables are independent, then the probability of their sums becomes a convolution. Unfortunately, the convolution of two arbitrary, or as a matter of fact identical, probability functions cannot be performed analytically except in rather special cases. Furthermore, M-fold convolutions are required to determine the probability of the sum of M random variables representing the base stations, which permits even fewer examples to be integrated exactly. In particular the log-normal distribution, which best describes shadow fading, appears to be difficult to convolve.

Instead of using the base stations simultaneously, another approach is to employ selection diversity among base stations. In such a system a number of transmitters would be dedicated to serve a mobile but only one would actually transmit at any given time. This could be achieved by letting the base station that receives the strongest average signal from the mobile be the one to transmit back to it. The performance of such a system would be very similar to one using equal-gain combining, yet it is far easier to determine its statistical behavior analytically. Considering only long-term fading, the probability that the mean signal, s, is less than some value x is given by

$$P(s < x) = \sum_{k=1}^{M} P(s_k < x), \qquad (5.5\text{-}1)$$

where s_k is the average signal strength from the kth base station and there are M possible base stations to choose from.

This probability is very easy to calculate when the average received signal strengths are independent log-normals. The probability distribution function for the mean received signal from one base station assuming a log-normal distribution of mean signal strength is expressed as

$$P(s) = \tfrac{1}{2} + \tfrac{1}{2} \operatorname{erf} \frac{s - m_s}{\sqrt{2}\,\sigma}, \qquad (5.5\text{-}2)$$

where m_s and σ are expressed in decibel values. Substituting $\log_{10} x = s$, one obtains the log-normal distribution function with a density function:

$$p(x) = \frac{1}{x\sigma\sqrt{\pi}} \exp\left\{ -\frac{1}{2\sigma^2}(\log_{10} x - m_s)^2 \right\}, \qquad x > 0$$

$$= 0, \qquad x < 0. \qquad (5.5\text{-}3)$$

For selection diversity among base stations only the distribution function is needed. If M base stations surround the vehicle, the diversity advantage can be determined from Eq. (5.5-1), assuming that the received signals are independent. The improvement is plotted in Figure 5.5-3 assuming that all signals have the same mean (i.e., the mobile is roughly equidistant from all base stations) and a standard deviation of 8 dB. If one had hoped to cover 90% of the points at some fixed radius from one base station, one could reduce the transmitter power by 8, 11.5, or 14 dB by selecting the best of two, three, or four equidistant base stations.

If the diversity base stations are not equidistant from the mobile, the system performance will be degraded somewhat. For example, if the best of four base stations is chosen with two base stations having mean levels 6 dB below the other two, then the 98% coverage point is improved by 11 dB, compared to 14 dB in the previous example. A standard deviation of 8 dB for the distribution was chosen as typical. Figure 5.5-4 shows the diversity advantage for standard deviations of 10, 8, and 6 dB when four

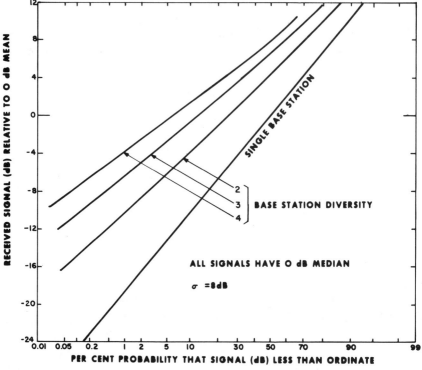

Figure 5.5-3   Base station diversity improvement.

base stations are used in a diversity configuration. As illustrated in Figure 5.5-4, for a 2% outage the improvement is 17, 14, or 10 dB for the three assumed standard deviations.

5.5.3 Cochannel Interference Reduction by Base-Station Diversity

In order to determine the base-station separations necessary to keep the probability of cochannel interference below some specified amount, one must take into account the following:

1. Multipath fading.
2. Receiver capture.
3. Average signal strength versus distance to the transmitter relationship.
4. Shadow-produced fading for both desired and interfering signals.

In a nonfading environment the interference at baseband is suppressed according to the cube of the FM index (Chapter 4). With large indices it is

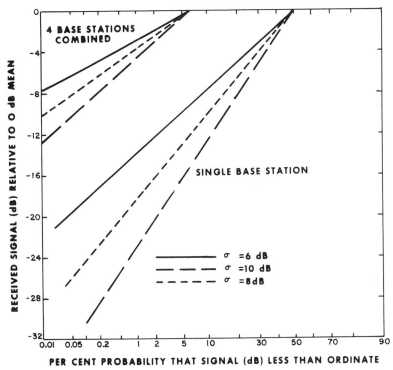

Figure 5.5-4 Comparison of four base station diversity to one base station for three standard deviations.

possible to achieve very high S/I at baseband when the signal level is only 1 dB greater than the interference level (hence the so-called capture effect). When Rayleigh fading is included in the analysis, most of the interference is caused when the receiver captures on the interfering stations, even if it is for a very short time. Systems employing diversity to protect against Rayleigh fading have a much higher baseband S/I. As shown in Chapter 4, two branches of diversity can improve the baseband performance by 10 dB or more. For simplicity, let us assume that the receiver is sensitive only to the mean signal levels from the desired and undesired stations. If we further assume that large index FM is transmitted and that the receiver has a good limiter, then we can define interference as occuring when the average signal strength from an undesired base station exceeds that from a desired base station. This will give an optimistic estimate, since it is the best one could hope to obtain. We will present the results in terms of the difference in average signal power necessary to stay below some given probability of interference, thus avoiding the propagation law.

Assume that the mobile is communicating with M base stations located at a radius R away. Let there be another set of N base stations located a further distance away that cause interference. The probability of interference assuming 0 dB capture ratio is expressed as $P(I) = P(s < i)$, where $P(I)$ denotes the probability of interference, s is the mean signal strength from the desired base stations, and i is the mean received signal strength from the interfering base stations. Assume that the distant base stations still have the same log-normal distribution as those nearby, and that the signals from all base stations are independent. Then the probability of interference may be written as

$$P(I) = \int_{-\infty}^{-\infty} di \int_{-\infty}^{i} ds \, p(s) p(i), \qquad (5.5\text{-}4)$$

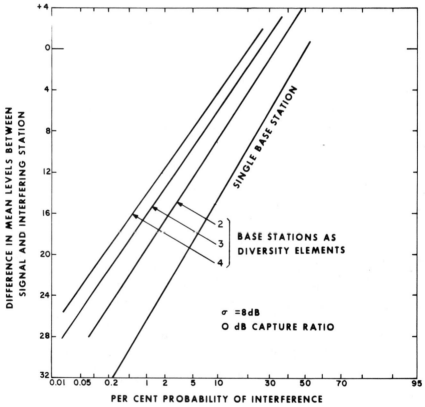

Figure 5.5-5 Probability of cochannel interference assuming only one interfering base station.

where $p(s)$ and $p(i)$ are the density functions of desired and interfering stations whose distributions are

$$P(s) = \left\{ \tfrac{1}{2} + \tfrac{1}{2} \operatorname{erf} \left[\frac{s - m_s}{\sqrt{2}\,\sigma} \right] \right\}^M, \qquad (5.5\text{-}5)$$

$$P(i) = \left\{ \tfrac{1}{2} + \tfrac{1}{2} \operatorname{erf} \left[\frac{s - m_i}{\sqrt{2}\,\sigma} \right] \right\}^N, \qquad (5.5\text{-}6)$$

where all values in the two equations are expressed in decibels.

An optimistic assumption would be that there is only one other base station or set of base stations that could cause interference to the mobile. Under these circumstances to obtain a 1% probability of cochannel interference, the difference in mean levels between the desired and the undesired base stations must be 27 dB for a standard deviation of 8 dB. Using two, three, and four base station diversity would allow the reduction

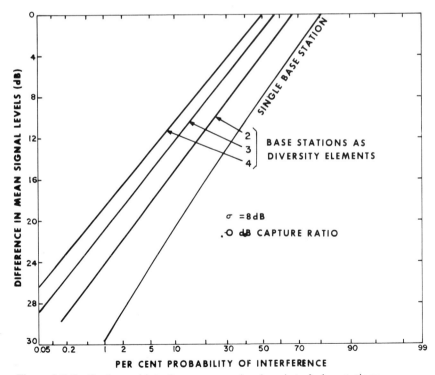

Figure 5.5-6 Cochannel interference assuming four interfering stations.

of the separation for reuse until the mean levels were only 20, 17, and 15 dB, respectively, for the same 1% chance of interference. This is illustrated in the curves presented in Figure 5.5-5.

More realistically, one might expect perhaps four sets of base stations separated by roughly equal distances as potential interferers. Under this assumption the separation of mean difference levels required for a 1% probability of interference is 24, 21, or 19 dB for systems employing two, three, or four base stations as diversity elements, compared to 21 dB for the single base station as shown in Figure 5.5-6.

If no diversity system to overcome short-term fading were used, an extra margin of about 20 dB would be required to prevent cochannel interference in the cases where even though the mean level from the primary base station is higher, the instantaneous level is lower (due to multipath) than the signal received from the interfering base station. It is evident that by using multiple base stations one can significantly reduce the power required to cover the mobile traffice in a given area. Furthermore, base-station diversity provides added protection against cochannel interference allowing closer reuse intervals.

REFERENCES

1. M. Schwartz, W. R. Bennett, and S. Stein, *Communication Systems and Techniques*, McGraw-Hill, New York, 1965, Chapters 10 and 11.

2. E. D. Sunde, *Communication Systems Engineering Theory*, Wiley, New York, 1969, Chapter 9.

3. A. Vigants, "Space Diversity Performance as a Function of Antenna Separation," *IEEE Trans. Comm. Tech.*, **16**, No. 6, December 1968, pp. 831–836.

4. W. T. Barnett, "Microwave Line-of-Sight Propagation with and without Frequency Diversity," *Bell System Tech. J.* **49**, No. 8, October 1970, pp. 1827–1871.

5. W. Y. S. Chen, "Estimated Outage in Long-Haul Radio Systems with Protection Switching," *Bell System Tech. J.* **50**, No. 4, April 1971, pp. 1455–1486.

6. A. J. Giger and F. K. Low, "The Automatic Protection Switching System of TH Radio," *Bell System Tech. J.* **40**, No. 6, November 1961, pp. 1665–1716.

7. D. G. Brennan, "Linear Diversity Combining Techniques," *Proc. IRE*, **47**, June 1959, pp. 1075–1102.

8. L. R. Kahn, "Ratio Squarer," *Proc. IRE (Corresp.)*, **42**, November 1954, p. 1704.

9. M. J. Gans, "The Effect of Gaussian Error in Maximal Ratio Combiners," *IEEE Trans. Comm. Tech.*, **COM-19**, No. 4, August 1971, pp. 492–500.

10. P. A. Bello and B. D. Nelin, "Predetection Diversity with Selectively Fading Channels," *IRE Trans. Comm. Systems*, **CS-10**, March 1962, pp. 32–42.

11. W. C. Y. Lee, "Effect of Mutual Coupling on a Mobile-Radio Maximum Ratio Diversity Combiner with a Large Number of Branches," *IEEE Trans. Comm.*, **COM-20**, December 1972, pp. 1188–1193.

12. W. C. Y. Lee, "Mutual Coupling Effect on Maximum-Ratio Diversity Combiners and Applications to Mobile Radio," *IEEE Ttrans. Comm. Tech.*, **COM-18**, No. 6, December 1970, pp. 779–791.

13. W. C. Y. Lee, "A Study of the Antenna Array Configuration of an M-Branch Diversity Combining Mobile Radio Receiver," *IEEE Trans. Veh. Tech.* **VT-20**, No. 4, November 1971, pp. 93–104.

14. W. C. Y. Lee, "Level Crossing Rates of an Equal-Gain Predetection Diversity Combiner," *IEEE Trans. Comm. Tech.*, **COM-18**, No. 4, August 1970, pp. 417–426.

15. H. A. Hans and R. B. Adler, *Circuit Theory of Linear Noise Network*, Wiley, New York, 1959.

16. C. T. Tai, "Coupled Antenna," *Proc. IRE*, **36**, April 1948, pp. 487–500.

17. B. R. Davis, "FM Noise with Fading Channels and Diversity," *IEEE Trans. Comm.*, **COM-19**, No. 6, December 1971, pp. 1189–1200.

18. W. C. Mason, M. Ginsburg, and D. G. Brennan, "Tables of the distribution function of sums of Rayleigh Variables," M.I.T. Lincoln Lab., Lexington, Mass., March 1960.

19. B. R. Davis, "Random FM in Mobile Radio with Diversity," *IEEE Trans. Comm. Tech.*, **COM-19**, No. 6, December 1971, pp. 1259–1267.

20. S. O. Rice, "Statistical Properties of a Sine Wave Plus Random Noise," *Bell System Tech. J.* 27, January, 1948, pp. 109–157.

21. J. Arsac, *Fourier Transforms and the Theory of Distributions*, Prentice-Hall, Englewood Cliffs, N. J., 1966, Chap. 2.

22. L. Lundquist, unpublished work.

23. M. Abramowitz and I. A. Stegun, *Handbook of Mathematical Functions* (Applied Mathematics Series 55), Washington, D. C.; NBS, 1964, p. 946.

24. L. Lundquist and M. M. Peritsky, "Co-channel Interference Rejection in a Mobile Radio Space Diversity System," *IEEE Trans. Veh. Tech., VT-20*, August 1971, pp. 68–75.

25. R. E. Langseth, "Some Aspects of Capture Phenomena in a Mobile Radio Diversity System Using a Separate Pilot," to be published in *IEEE Trans. Veh. Tech.* .

26. I. S. Gradshteyn and I. M. Ryzhik, *Table of Integrals, Series and Products*, Acedemic, New York, 1965.

27. Bell Laboratories, "High-Capacity Mobile Telephone System Technical Report," submitted to FCC in December, 1971.

chapter 6
diversity techniques

D. O. Reudink, Y. S. Yeh, and
W. C. Jakes

SYNOPSIS OF CHAPTER

This chapter deals with diversity techniques that were developed specifically for microwave mobile radio. Postdetection diversity is discussed in Section 6.1. In this section weighting factors for outputs of the individual branches are discussed, an experimental postdetection diversity receiver is described and some results from actual field measurements are presented.

True selection diversity, described in Chapter 5, may be impractical for mobile radio; however, a closely related diversity technique called switch diversity works well below 1 GHz and is easy to implement. Section 6.2 begins with a description of the technique and derives theoretical expressions for its performance. Because ˙switching is done at RF, switching transient noise is introduced. These transients can be reduced by audio blanking. This problem is discussed in Section 6.2.2. It is possible to put all of the diversity at the base station if there is a feedback path for the mobile to relay signal strength information back to the base station. This technique, which is discussed in Section 6.2.4, limits the vehicle range but works below certain Doppler shifts.

Coherent combining using carrier recovery involves bringing all branches to a common phase. Several techniques are discussed in Section 6.3. Some of these methods work for wideband FM where no carrier is actually transmitted.

Another method of cophasing diversity branches is to transmit a pilot signal close in frequency to the information signal. The phase information on the pilot is used to subtract the random phase from the signal. Section 6.4 discusses the performance of a particular receiver that achieves coherent combining by simply squaring the input signal.

When adaptive diversity arrays are used at both the mobile and base station the questions of "locking-up" and the optimum division of diversity

389

branches between base and mobile must be considered. These questions are answered in Section 6.5.1 and also the performance of the less complicated phase conjugate (compared to complex conjugate) retransmission systems is obtained.

Retransmission diversity is a technique where the base station (or mobile) has all of the diversity elements. When the mobile transmits a signal, each base-station antenna transmits the conjugate phase that it received so that the signals from all the base-station antennas arrive back at the mobile antenna in phase. This can be done by time separating the base and mobile station transmissions as discussed in Section 6.5.2 or in a continuous manner as discussed in Sections 6.5.3 and 6.5.4.

Multicarrier AM diversity is discussed in Section 6.6. It is characterized by a multiplicity of antennas at the transmitting site and one antenna at the receiver. Although AM transmission has drawbacks because of its vulnerability to cochannel interference, this technique may have application in special situations.

Section 6.7 discusses digital modulation diversity systems. The three major factors that increase the average bit error rate in digital systems in a mobile environment are (1) envelope fading, which degrades the carrier-to-noise ratio, (2) random FM, which causes errors in angle modulation systems, and (3) time delay spreads, which produce intersymbol interference. This section shows how diversity combining can cause orders of magnitude reduction in bit error rates.

This chapter concludes with a section that compares the performance of various diversity methods in the mobile radio environment. System comparisons are made on the basis of the transmitter power required to achieve an acceptable SNR with transmission bandwidth as a parameter. The comparisons are illustrated with numerical examples for an urban system operating at UHF.

6.1. POSTDETECTION DIVERSITY

Multipath fading is not unique to mobile radio. Diversity to combat fading has been suggested for early high-frequency and short-wave radio,[1] beyond-the-horizon UHF links, [2-4] and line-of-sight microwave transmission systems.[5], [6] The earliest diversity systems were of the postdetection type where an operator manually selected the receiver that sounded best. Methods of automatically selecting the receiver having the strongest signal were discussed by Bohn[7] in the late 1920s.

As discussed in Chapter 5, diversity receivers can be classified into two basic types, predetection and postdetection, each using one of four combining methods. A receiver may combine the separate received signals on a basis of scanning, selection, equal gain, or maximal ratio. In postdetection systems the equal-gain method is probably the simplest of all diversity combining methods. Two or more separately received signals are simply added together to produce the combined output. If exponential modulation with large index is used with this scheme, the output signal-to-noise ratio will be reduced drastically if a signal in one of the branches of the combiner falls below the noise-quieting threshold of that branch. When this occurs, the faded branch contributes only noise with a level that is approximately equal to the maximum signal output from the discriminator.

Maximal ratio combining is approximated by weighting the contribution of each branch to the combined output by a measure of its signal-to-noise ratio. Postdetection maximal ratio combiners thus require a gain control following the detector to adjust the output signal level proportional to the receiver branch input signal-to-noise ratio. When exponential modulation with large indices is used, a measure of the receiver branch signal-to-noise ratio may be obtained by sampling out-of-band noise from the discriminator output. This noise sample can be envelope detected and used to control the amplitude of the output signal of the discriminator.[8] When the receiver branch noise can be assumed constant, a measure of the signal-to-noise ratio may be obtained by envelope detecting the signal and using this to control the amplitude of the discriminator output before combining.

6.1.1 Postdetection Maximal-Ratio Combining

In a system using exponential modulation the demodulated output signal level from the discriminator is a function of the deviation only if the receiver input signal level is above threshold. The demodulated output noise level will vary inversely with the input signal-to-noise ratio down to threshold and then will increase nonlinearly below threshold. When the receiver input signal-to-noise ratio is about 10 dB (approximately at threshold), the output signal-to-noise ratio is about 37 dB for a modulation index of five. If the receiver input signal is above threshold, there is little gained in diversity combiner performance by weighting the output of each contributing branch by a measure of its signal-to-noise ratio. When the receiver input signal-to-noise ratio falls a few decibels below threshold, the discriminator output must be attenuated by 30 dB or greater, depending on modulation index. Brennan[9] has shown that it makes little difference in the performance of a postdetection diversity receiver that utilizes exponential modulation whether the weighting factor follows the output signal-to-noise ratio from the discriminator exactly or just "squelches" or discards the output when the receiver branch input falls below threshold.

When exponential modulation is used, the required weighting factor in each receiver branch can be obtained by using either a measure of the amplitude of the received signal envelope before detection or a measure of the out-of-band noise from the discriminator output. The first method provides an indication of the receiver input signal-to-noise ratio only if the receiver thermal noise is constant. The second method will provide a good indication of the receiver signal-to-noise ratio even if the receiver input thermal noise changes.

The weighting factor of the diversity receiver branch can be considered a function that modulates the message output from the discriminator. In the ideal case the weighting factor would be a continuously varying voltage proportional to the signal-to-noise ratio of the message output. In a practical case, this voltage will also have a randomly varying Gaussian component due to thermal noise from the source of the weighting factor. The thermal noise components can be minimized by band-limiting the weighting factor to the maximum fading rate of the receiver branch input signal. The weighting factor in a UHF or microwave mobile radio transmission system diversity receiver branch must be able to follow rapid fading at rates up to perhaps 2 kHz.

The signal-to-noise ratio of the weighting factor derived from detecting the received signal envelope will be a function of the receiver input signal-to-noise ratio, the IF bandwidth, and the postenvelope detector bandwidth. If the envelope detector is a square-law device, then the weighting factor will be a linear function of the input signal power. As the receiver input signal-to-noise ratio falls, the weighting factor signal-to-noise ratio will fall at about the same rate. When the input signal-to-noise ratio approaches unity, the output signal-to-noise ratio from the square-law detector falls at a faster rate due to the "noise times noise" components.

The weighting factor signal-to-noise ratio can be calculated for any value of receiver branch input signal-to-noise ratio. If, for example, the input signal-to-noise ratio within the IF bandwidth of 50 kHz of the receiver branch is 6 dB, the output signal to noise of the envelope detector is approximately the same as that of the input,[10] or 6 dB. If the weighting factor is band limited to about 5 kHz, the weighting factor signal to noise will be increased by the ratio of the IF bandwidth to the postenvelope detector bandwidth.

Let

$$B_{IF} = 50 \text{ kHz}, \qquad B_{wf} = 5 \text{ kHz};$$

then

$$10 \log_{10} \frac{B_{IF}}{B_{wf}} = 10 \log_{10} \frac{50}{5} = 10 \text{ dB}. \qquad (6.1\text{-}1)$$

Therefore, the weighting factor signal-to-noise ratio after the postenvelope detector filter will be about 16 dB for a branch input signal-to-noise ratio of 6 dB.

If the weighting factor is obtained by using the out-of-band noise sample from the discriminator output, the signal-to-noise ratio of the weighting factor can also be found. The weighting factor is obtained by taking a sample of noise from the discriminator output in a band of frequencies higher than the highest message-channel signal frequency, passing it through a square-law device, and extracting the dc component. The total bandwidth of the available out-of-band noise from the discriminator is a function of the receiver branch IF bandwidth and the modulation index of the exponential modulation. The noise output amplitude is a function of the receiver branch input signal-to-noise ratio and will increase rapidly as the signal-to-noise ratio approaches unity. The output from the square-law device is composed of a dc component and a randomly varying ac component. This can be shown by considering an input of bandlimited Gaussian noise corresponding to the out-of-band noise sample used to estimate the signal-to-noise ratio to a square-law detector and finding the output spectrum.

The required weighting factor for the receiver branch output is the impulsive component of the output spectrum. The randomly varying components in the square-law detector output can be considered as an undesired noise component. The weighting factor signal-to-noise ratio will be the ratio of the impulsive component to the randomly varying noise components.

Using the same bandwidths as in the envelope detection case, it can be shown[11] that as the diversity branch input signal-to-noise ratio decreases, the discriminator noise output increases with a spectral peak centered at a frequency equal to approximately 0.4 B_{IF} or 20 kHz. A further decrease in signal-to-noise ratio increases the lower-frequency noise output spectra amplitude, and this peak is obscured. A noise sample bandwidth B, of 10 kHz, centered at 20 kHz might be used to provide the input to a square-law detector. A weighting factor signal-to-noise ratio at the output of a low-pass filter following the square-law detector can be calculated.

Let

$$\left(\frac{S}{N}\right)_{wf} = \frac{S_y(f)_\delta}{S_y(f)_r}, \qquad (6.1\text{-}2)$$

where $S_y(f)_\delta$ is the impulsive component of the spectral density and $S_y(f)_r$

is the random component of the spectral density.

Referring to Figure 6.1-1, if the weighting factor is band limited to 2.5 kHz, then the weighting factor signal-to-noise ratio can be calculated from the relative areas of the spectral densities:

$$\left(\frac{S}{N}\right)_{wf} = \frac{16a^2A^2B}{7a^2A^2B} = \frac{16}{7} = 2.285,$$

$$10\log_{10}2.285 = 3.6\,dB.$$

(6.1-3)

This low weighting factor signal-to-noise ratio might be expected with

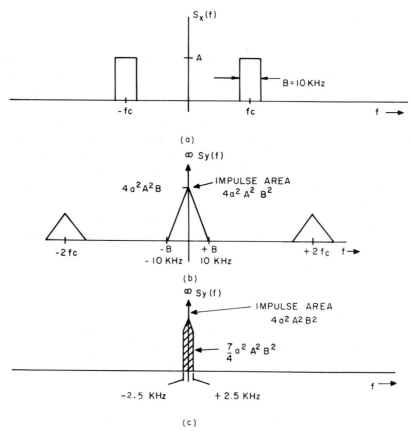

Figure 6.1-1 Input and output noise spectrum components for a square-law device.

the rapid fading requirement and limited noise sample bandwidth. This weighting factor signal-to-noise ratio would be improved if the post-noise detector bandwidth is reduced; however, the minimum bandwidth is limited by the maximum expected fading rate of the received signal.

If the weighting factor must follow rapid fading such as encountered in a mobile ratio transmission system operating at UHF or microwave frequencies, then the signal envelope detection technique appears to be more attractive because of its inherently better signal-to-noise ratio when the receiver branch input signal level is near the exponential modulation threshold.

For the previously given system parameters, the out-of-band noise detection technique will not provide a weighting factor signal-to-noise ratio of better than 3.6 dB as shown, whereas the envelope-detector technique at an input signal level slightly below the modulation threshold can provide a weighting factor signal-to-noise ratio of at least 16 dB.

6.1.2 Receiver Design and Field Measurements

A mobile radio postdetection combining receiver has been designed, built, and tested by Rustako.[12] A block diagram of the receiver is shown in Figure 6.1-2. It is composed of separate identical receivers or branches with their discriminator outputs summed at a common output. Each branch of the receiver uses a separate antenna. The discriminator audio output is fed in turn into an amplifier whose gain can be varied over a large range by a small control voltage. This gain control voltage, which is the receiver branch output signal weighting factor, is obtained by taking a sample of the received signal envelope from the integrated circuit limiter, amplifying it with an integrated circuit amplifier, and envelope detecting it. This detected output is a dc or slowly varying ac signal that is proportional in level to the input signal level as the signal level falls below limiting. The limiting threshold is set a few decibels below the exponential modulation threshold. This does not drastically degrade the overall performance of the receiver since it is assumed that the mean received signal level in either branch will be above the modulation threshold. As the fading received signal falls below the limiting threshold, the output from the envelope detector will change. This change is inverted in polarity and adjusted in level to provide the control voltage required to reduce the gain of the amplifier following the discriminator output.

The two-branch post detection combining space-diversity receiver was installed in a mobile vehicle to evaluate its performance in a typical scattering area. The transmitter and antenna were located on top of Crawford Hill in Holmdel Township, New Jersey, at an elevation of 360 ft

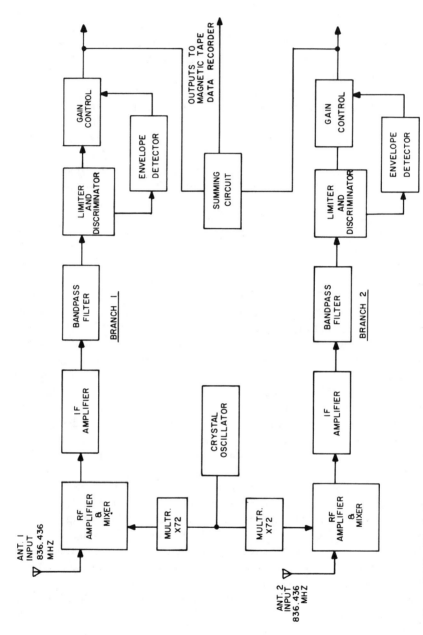

Figure 6.1-2 Block diagram of two-branch postdetection combining receiver.

a.m.s.* The path between the transmitting antenna and the area of the mobile measurements was obscured by a large group of trees approximately 80 ft high at a distance of about 500 ft from the transmitting antenna.

The mobile measurements were made on Annapolis Drive in Raritan Township, New Jersey at a distance of about $2\frac{1}{2}$ miles from the transmitting antenna. Figure 6.1-3 shows a map of the area of the measurements. The street is approximately orthogonal to a radial line of direct propagation between the transmitter and the mobile receiver. It is near the center of an area of closely spaced houses of uniform shape where typical roof heights are about 30 ft from the ground level. The street is at an elevation of 20 ft a.m.s. Figure 6.1-4 shows a diagram of the relative heights of obstruction between the transmitter and receiver.

Figure 6.1-5 shows the results of the test data. The probability that the output signal-to-noise ratio is greater than -10 dB (referred to 0 dB maximum output) is plotted as a function of the transmitted power. Comparing the curves for the single branch and the combined branches, it can be seen that for the same outage rate, the receiver input signal-to-noise ratio can be reduced significantly by combining two branches to obtain the same minimum output signal-to-noise ratio.

The cumulative probability distributions for a single-branch Rayleigh and for two-branch maximal ratio combining are plotted in Figure 5.2-7. For the combined case, the experimental data cannot be expected to fit the theoretical curve exactly, because of the detection and squelchinng characteristic of the receiver. However, the diversity gain at an outage rate of about 1% should be about the same. The theoretical curves show a diversity gain of 10 dB compared to the experimental receiver, which realized a diversity gain of about 9.5 dB.

The general complexity of a postdetection combining receiver is not sufficiently less than most predetection combining schemes to make its use attractive in a mobile radio transmission system in view of its overall performance. The postdetection combining receiver will reduce the outage time and consequently improve the average output signal-to-noise ratio of the message signal by a few decibels. By comparison, the predetection combining receiver can effectively increase the input signal-to-noise ratio before detection. If exponential modulation is used in the radio transmission system, then an extension in threshold is essentially provided that would result in a higher average output signal-to-noise ratio of the message signal than in the postdetection case.

*Above mean sea level.

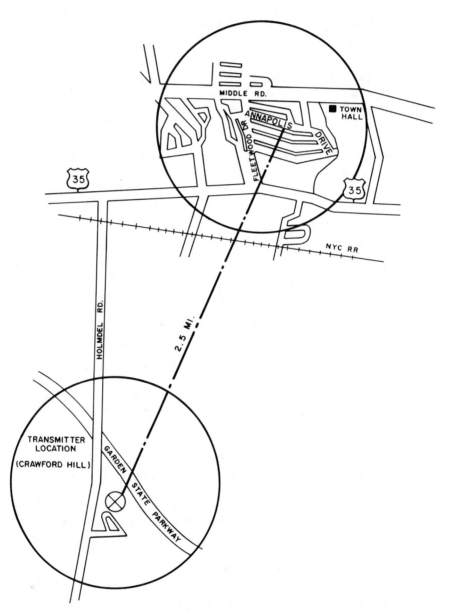

Figure 6.1-3 Map showing mobile test area.

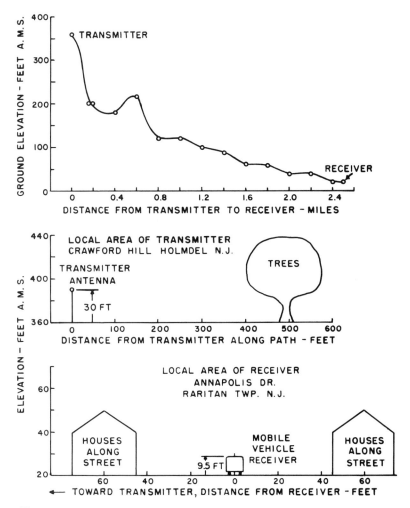

Figure 6.1-4 Path profile between transmitter and receiver.

6.2   SWITCHED DIVERSITY

A simplified block diagram of a predetection switch diversity system is shown in Figure 6.2-1. Two diversity branches are provided while using a single receiver front-end mixer and IF amplifiers. The instantaneous envelope of the received signal is monitored; if it falls below a predetermined threshold, the antenna switch is activated, thus selecting the second branch. If the second branch is above the threshold, switching ceases. If the second branch is also in a fade we can either revert to the first branch

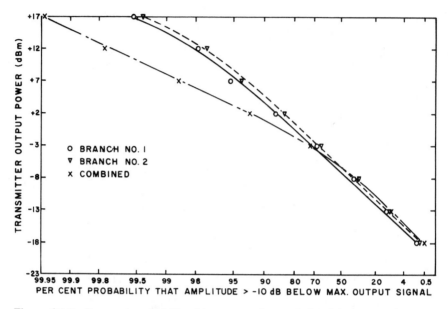

Figure 6.1-5 Per cent probability that output detected signal is greater than − 10 dB below maximum output signal Annapolis Drive, Raritan Twp.

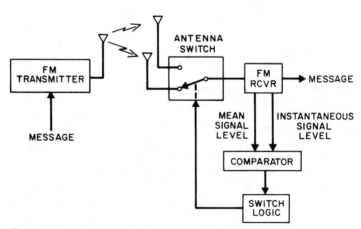

Figure 6.2-1 Switch diversity system block diagram.

400

and tolerate rapid switching between the two antennas until one of them emerges above threshold (switch and examine strategy), or we may stay with the second antenna (switch and stay strategy). The switch and stay strategy avoids excessive switching when both antennas are in simultaneous fades and is thus preferable. The switch and stay strategy can also be considered as switching at the instant the envelope crosses the threshold in the negative direction. The system performance is controlled by the choice of the switching threshold and the time delay in which the fading can be detected and the switching activated. These factors will be discussed in Section 6.2.1.

Common to all predetection switching diversity systems is the switching transient caused by the phase or amplitude difference between the two antennas. After FM detection, these transients will occur in the baseband as pulses whose width is proportional to the reciprocal of the IF bandwidth, and hence will degrade the signal message. An estimate of the transient noise and the means to control it are examined in Section 6.2.2.

Both theoretical and experimental results of baseband SNR as a function of vehicle speed, switching threshold, and average carrier-to-noise ratio are presented in Section 6.2.3. An optimum threshold for switching diversity is established.

By examining Figure 6.2-1, we notice that the system can be modified if instead of switching antennas at the receiver, we can provide a feedback link to switch transmitting antennas at the remote station within a limited amount of time delay. This technique, shown in Figure 6.2-2 is called feedback diversity. The command station, after the detection of a fade, adds a burst tone to message modulation. The responder station separates the tone from the message through baseband filtering and switches antennas whenever it detects the presence of burst tone. The command station thus requires only one antenna. The merit of this concept is that mobiles can be equipped with single antennas and only the base stations need be equipped with dual antennas. The system considerations in feedback diversity and some experimental results are presented in Section 6.2.4.

6.2.1 The Cumulative Probability Distribution (CPD) of the Carrier Envelope Using Switch and Stay Strategy

The density function of a Rayleigh stochastic process is, from Eq. (1.1-14),

$$p_r(A) = \begin{cases} \dfrac{A}{b_0} e^{-A^2/2b_0}, & A \geqslant 0 \\ 0, & A < 0 \end{cases} \tag{6.2-1}$$

where $\langle r^2 \rangle = 2b_0$.

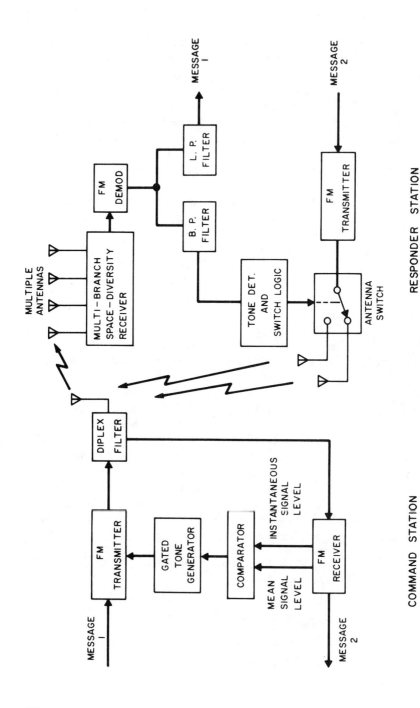

Figure 6.2-2 Feedback diversity system block diagram.

402

The cumulative distribution function of r is

$$P_r(A) = 1 - e^{-A^2/2b_0}. \tag{6.2-2}$$

The average duration that r fades below A is [Eq. (1.3-44)]

$$\tau_q = \frac{\sqrt{b_0}}{\sqrt{\pi}} \frac{e^{A^2/2b_0} - 1}{Af_m}. \tag{6.2-3}$$

In switching diversity let us assume that there are two independent Rayleigh processes $r_1(t)$ and $r_2(t)$. The mean and the autocorrelation functions of r_1 and r_2 are identical. Using switch and stay strategy, the resultant carrier envelope $R(t)$ is again a stochastic process and is illustrated in Figure 6.2-3. $R(t)$ consists of two portions derived from $r_1(t)$ and $r_2(t)$, respectively. Let the switching threshold be A, and define

$$q \overset{\Delta}{=} P\{r \leqslant A\} = 1 - e^{-A^2/2b_0}, \tag{6.2-4}$$

$$p \overset{\Delta}{=} P\{r > A\} = e^{-A^2/2b_0}. \tag{6.2-5}$$

The average duration that r is above A is, from Eq. (6.2-3),

$$\tau_p = \frac{p}{q}\tau_q = \frac{\sqrt{b_0}}{\sqrt{\pi}} \frac{1}{Af_m}. \tag{6.2-6}$$

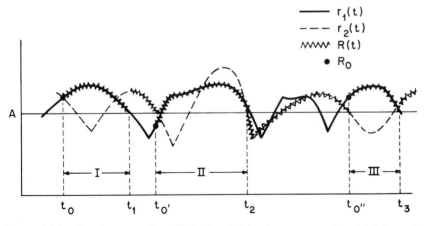

Figure 6.2-3 Resultant envelope $R(t)$ of switching between two Rayleigh branches using switch and stay strategy.

Referring to Figure 6.2-3 we note that R switches from r_1 to r_2 at $t = t_1$ and reverts back to r_1 at $t = t_0'$. The average duration between t_1 and t_0' is $\tau_p/2$. The normalized autocorrelation coefficient of $r_1(t_1)$ and $r_1(t + \tau_p/2)$ is

$$\rho = \frac{\langle r_1(t) r_1(t + \tau_p/2)\rangle - \langle r_1(t)\rangle^2}{\langle r_1^2(t)\rangle - \langle r_1(t)\rangle^2}$$

$$= \frac{R_r(\tau_p/2) - R_r(\infty)}{R_r(0) - R_r(\infty)}$$

$$\cong \rho^2(\tau_p/2) = J_0^2\left(\frac{\sqrt{\pi b_0}}{A}\right) \tag{6.2-7}$$

where

$$R_r(\tau) \cong \frac{\pi b_0}{2}\left(1 + \frac{\rho^2}{4}\right)$$

is given by Eq. (1.3-16). In most cases switching will be performed at small values of $A^2/2b_0$. Therefore ρ is extremely small and we may assume independence between $r_1(t_1)$ and $r_1(t_0')$. In other words, the assumption is that when we switch back from r_2 to r_1, the statistics of r_1 are already independent of that portion of r_1 that we switched out initially.

Referring to Figure 6.2-3, since r_1 and r_2 are indistinguishable, we have

$$P\{R \leqslant B\} = P\{R \leqslant B | R = r_1\}$$

$$= P\{R \leqslant B | = r_2\}, \tag{6.2-8}$$

where $\{R = r_1\}$ is the portion of R that belongs to r_1, $P\{C|D\}$ is the probability of C under the condition D, and B is any arbitrary level.

The set $\{R = r_1\}$ includes segments I, II, III,...marked on Figure 6.2-3. We note that the starting value R_0 of each segment can be either larger or smaller than A, the switching threshold level. Let us define

$$X = \{\text{portion of } \{R = r_1\} \text{ with } R_0 > A\}, \tag{6.2-9}$$

$$Y = \{\text{portion of } \{R = r_1\} \text{ with } R_0 \leqslant A\}. \tag{6.2-10}$$

It then follows that

$$\{R = r_1\} = \{R = X + Y\}. \tag{6.2-11}$$

Equation (6.2-8) can be written in terms of X and Y:[1]

$$P\{R \leqslant B | R = r_1\} = P\{R \leqslant B | R = X\} P\{R = X | R = X + Y\}$$

$$+ P\{R \leqslant B | R = Y\} P\{R = Y | R = X + Y\}. \qquad (6.2\text{-}12)$$

Since X and Y are mutually exclusive we have

$$P\{R = X + Y\} = P\{R = X\} + P\{R = Y\}. \qquad (6.2\text{-}13)$$

The typical segments of X are shown in Figure 6.2-4. Because of the assumption that the starting point t_0 is random, that is, it can be anywhere between the two level crossing points t_{-1} and t_1, the average length of the segment is

$$\tau_x = \frac{\tau_p}{2}. \qquad (6.2\text{-}14)$$

The typical Y segment is shown in Figure 6.2-5. Here, similarly, we obtain the average length of the segment to be

$$\tau_y = \frac{\tau_q}{2} + \tau_p. \qquad (6.2\text{-}15)$$

$R(t)$ would end up on X or Y only through a switch from the other branch. The probability that after each switch it would be on X is p. The probability that $R(t)$ would switch to Y is q. Therefore the ratio of

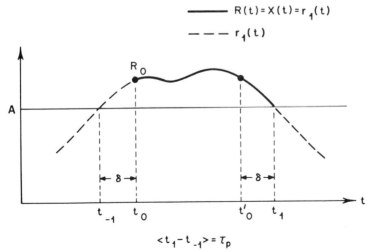

Figure 6.2-4   Segments of **R** with starting point $R_0 > A$.

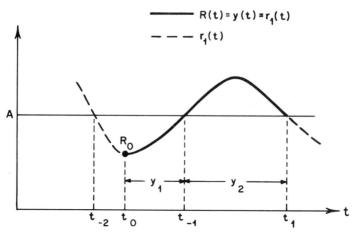

Figure 6.2-5   Segments of **R** with starting point $R_0 < A$.

$P\{R = X\}/P\{R = Y\}$ is

$$\frac{P\{R = X\}}{P\{R = Y\}} = \frac{p\tau_x}{q\tau_y}. \tag{6.2-16}$$

Substituting Eqs. (6.2-13)–(6.2-16) in (6.2-12) we have

$$P\{R \leqslant B \mid R = r_1\} = P\{R \leqslant B \mid R = X\} \cdot (1 - q)^2$$

$$+ P\{R \leqslant B \mid R = Y\} \cdot (2q - q^2). \tag{6.2-17}$$

Now let us go back to Figure 6.2-4. We know that t_0 is equally likely to be anywhere between t_{-1} and t_1. Given t_0, which is δ away from t_{-1}, then this segment of X traces the Rayleigh curve from t_0 to t_1. On the other hand, we should also have a t_0' that is δ away from t_1 and traces the portion of the Rayleigh curve from t_0' to t_1. However, the statistical distribution of the portion $t_0' \rightarrow t_1$ should not be different from the portion $t_{-1} \rightarrow t_0$. Therefore, the pair of segments with starting points at t_0 and t_0' traces the complete Rayleigh curve between t_{-1} and t_1. The conditional distribution can thus be written as

$$P\{R \leqslant B \mid R = X\} = P\{r \leqslant B \mid r \geqslant A\}$$

$$= \begin{cases} \dfrac{P_r(B) - q}{1 - q} & B > A \\ 0 & B \leqslant A \end{cases}, \tag{6.2-18}$$

where $P_r(B)$ is given by Eq. (6.2-2).

The typical segment of $\{R = Y\}$ is given by Figure 6.2-5. Here we notice that Y can be further divided into two mutually exclusive parts, that is, Y_1 and Y_2. And we have

$$P\{R \leqslant B | R = Y\} = \frac{P\{R \leqslant B | R = Y_1\} P\{R = Y_1\}}{P\{R = Y_1 + Y_2\}}$$

$$+ \frac{P\{R \leqslant B | R = Y_2\} P\{R = Y_2\}}{P\{R = Y_1 + Y_2\}}. \qquad (6.2\text{-}19)$$

It is simple to show that

$$\frac{P\{R = Y_1\}}{P\{R = Y_2\}} = \frac{\tau_{q/2}}{\tau_p} \qquad (6.2\text{-}20)$$

and

$$P\{R = Y_1 + Y_2\} = P\{R = Y_1\} + P\{R = Y_2\}. \qquad (6.2\text{-}21)$$

The conditional distributions are given by

$$P\{R \leqslant B | R = Y_2\} = P\{r \leqslant B | r \geqslant A\}, \qquad (6.2\text{-}22)$$

and

$$P\{R \leqslant B | R = Y_1\} = P\{r \leqslant B | r \leqslant A\}$$

$$= \begin{cases} 1 & B \geqslant A \\ \dfrac{P_r(B)}{q} & B < A \end{cases}. \qquad (6.2\text{-}23)$$

Substituting Eqs. (6.2-20)–(6.2-23) into (6.2-19) we obtain

$$P\{R \leqslant B | R = Y\} = \begin{cases} \dfrac{2P_r(B) - q}{2 - q} & B > A \\ \dfrac{P_r(B)}{2 - q} & B \leqslant A \end{cases}. \qquad (6.2\text{-}24)$$

Substituting Eqs. (6.2-18) and (6.2-24) into (6.2-17) we have the distribution of R:

$$P\{R \leqslant B\} = \begin{cases} P_r(B) - q + qP_r(B) & B > A \\ qP_r(B) & B \leqslant A \end{cases}, \qquad (6.2\text{-}25)$$

where

$$P_r(B) = 1 - e^{-B^2/2b_0},$$

$$q = P_r(A).$$

The density function of R is

$$p_R(B) = \begin{cases} (1+q)p_r(B) & B > A \\ qp_r(B) & B \leq A \end{cases}. \qquad (6.2\text{-}26)$$

Thus it can be seen that the density of R consists of two Rayleigh densities but with different weighting factors. The distribution of R for several values of switching threshold, A, is shown in Figure 6.2-6. For comparison, the envelope distribution of an ideal two branch selection diversity is also plotted. In Figure 6.2-6, the CPD of the switch and stay case shows a breaking point at the switching threshold. Below the threshold, the CPD is

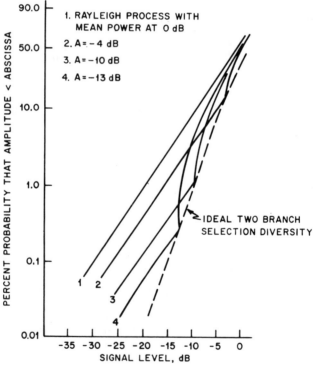

Figure 6.2-6 Cumulative probability distribution curves of two-branch switching diversity systems using switch and stay strategy.

of Rayleigh slope but is displaced by A dB from the CPD of the single Rayleigh. Above the threshold, it quickly merges with the single Rayleigh again.

If there is a fixed time delay T between the threshold crossing and the switching, the distribution of R will be degraded. The resultant envelope $R(t)$ with time delay is illustrated in Figure 6.2-7. The subsets X' and Y', corresponding to X and Y in the no-delay case, are presented in Figures 6.2-8 and 6.2-9. First of all we note that the average durations of X' and Y' are given by

$$\tau'_x = \tau_x + T = \frac{\tau_p}{2} + T, \qquad (6.2\text{-}27)$$

$$\tau'_y = \tau_y + T = \frac{1}{2}\tau_q + \tau_p + T. \qquad (6.2\text{-}28)$$

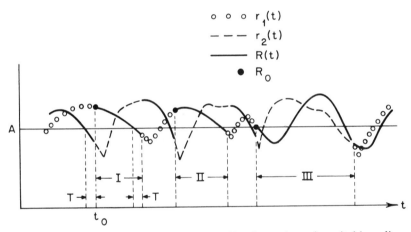

Figure 6.2-7 Resultant signal envelope **R**(t) of two-branch switching diversity systems with decision time delay T.

Therefore Eq. (6.2-17) can be modified to

$$P\{R \leqslant B\} = P\{R \leqslant B \mid R = X'\}\frac{(1-q)^2\tau_q + 2q(1-q)T}{\tau_q + 2qT}$$

$$+ P\{R \leqslant B \mid R = Y'\}\frac{(2q-q^2)\tau_q + 2q^2T}{\tau_q + 2qT} \qquad (6.2\text{-}29)$$

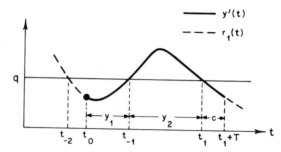

Figure 6.2-8   Segments of **R** with starting point $R_0 > A$ and time delay T.

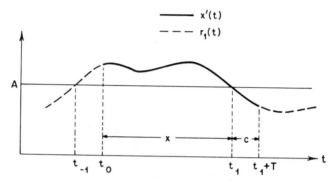

Figure 6.2-9   Segments of **R** with starting point $R_0 < A$ and time delay T.

Typical segments of X' (Figure 6.2-8) consist of X and C. The distribution of X is given by Eq. (6.2-18). The distribution of C can be approximated if we allow T to be small in comparison to τ_q, the average duration of fades below A. In this case we have

$$R(t) \cong A + \dot{r}(t_1)t, \qquad t_1 \leqslant t \leqslant t_1 + T. \tag{6.2-30}$$

Knowing the joint distribution of r and $\dot{r}$ [Eq. (1.3-34)] the distribution of C is calculated to be

$$P_C(B) = \begin{cases} 1, & B > A \\ 1 - \mathrm{erf}\left(\dfrac{B-A}{\omega_m T \sqrt{b_0}}\right) + \dfrac{B-A}{\sqrt{\pi b_0}\,\omega_m T} E_i\left(-\dfrac{(B-A)^2}{\omega_m^2 T^2 b_0}\right), & B \leqslant A \end{cases}$$

$$\tag{6.2-31}$$

where

$$\text{erf}(x) = \frac{2}{\sqrt{\pi}} \int_0^x e^{-t^2} dt,$$

$$E_i(x) = \int_{-\infty}^x \frac{e^t}{t} dt.$$

A typical segment of Y' is shown in Figure 6.2-9. Again it consists of sets Y_1, Y_2, and C with known distributions. Through a tedious operation of breaking X' and Y' into their subsets, Eq. (6.2-29) can be transformed to

$$P\{R \leqslant B\} = \frac{\tau_q}{\tau_q + 2qT} [P_r(B)q + P_r(B) - q] u(B - A)$$

$$+ \frac{\tau_q}{\tau_q + 2qT} P_r(B) qu(A - B)$$

$$+ \frac{2qT}{\tau_q + 2qT} P_C(B), \qquad (6.2\text{-}32)$$

where

$$U(x) = \begin{cases} 1 & x > 0 \\ 0 & x \leqslant 0 \end{cases}$$

Equation (6.2-32) is presented in Figure 6.2-10. These curves illustrate the degradation of the CPD of R as a function of T/τ_q. It is observed that up to $T/\tau_q = 0.288$ we might still expect some diversity advantages. For $T/\tau_q < 0.144$ the delayed case is almost identical to the no-delay case.

The tolerable time delay is a function of switching threshold and the Doppler frequency. For the case $f_m = 75$ Hz ($f_c = 840$ MHz, $v = 60$ mi/hr) the tolerable time delays are presented in Figure 6.2-11. This puts an upper limit on the allowable time delay in any switching diversity systems.

6.2.2 Switching Transient Noise

In predetection switching diversity systems the carrier may experience abrupt changes both in amplitude and phase after each switching. The subsequent transient responses of a CW carrier are shown in Figure 6.2-12. After the IF filter of bandwidth B, we have both amplitude and phase transients. The limiter removes the amplitude variation, but the phase

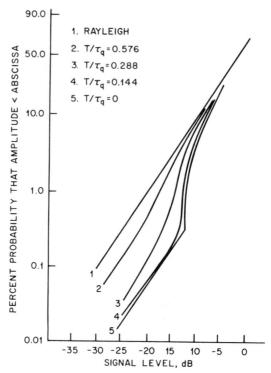

Figure 6.2-10 Cumulative probability distribution curves of two-branch switching diversity systems using switch and stay strategy with − 13 dB switching threshold and time delay T.

transients get through and appear as clicks in the discriminator output. It is worthwhile to note that the duration of the click is of the order of $1/B$.

To analyze the effect of the phase transient we will consider the transition between two unmodulated carriers with a relative phase difference of α radians. We will assume that during the switching time one envelope decreases linearly to zero and the other rises from zero to maximum linearly at the same time.

Let the original phasor be z_1 and the new phasor z_2. Suppose the switching takes place in a time τ, so that during this time we have a phasor $z(t)$ given by

$$z(t) = \frac{1}{2}\left\{ z_1\left(1 - \frac{2t}{\tau}\right) + z_2\left(1 + \frac{2t}{\tau}\right)\right\}, \qquad -\frac{\tau}{2} \leqslant t \leqslant \frac{\tau}{2}. \quad (6.2\text{-}33)$$

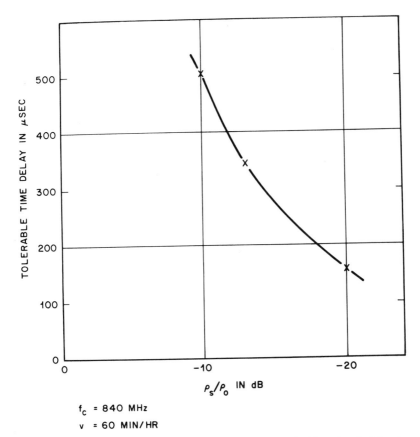

f$_c$ = 840 MHz

v = 60 MIN/HR

Figure 6.2-11   Tolerable time delay as function of switching threshold ρ_s/ρ_0.

Thus the instantaneous frequency is given by

$$\dot{\theta}(t) = \mathrm{Im}\{\dot{z}(t)/z(t)\}$$

$$= \mathrm{Im}\left\{t + \frac{\tau}{2}\frac{z_2 + z_1}{z_2 - z_1}\right\}^{-1}. \tag{6.2-34}$$

For the case where z_1 and z_2 have equal magnitudes, this becomes

$$\dot{\theta}(t) = \frac{c}{t^2 + c^2}, \tag{6.2-35}$$

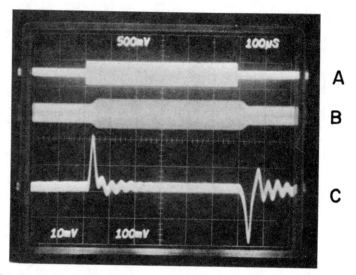

Figure 6.2-12   Typical transient responses at different states of an FM receiver. (a) Receiver RF input with amplitude and phase step. (b) IF signal after 50 KHz bandpass filter before limiter. (c) Discriminator output.

where

$$c = \frac{\tau}{2}\cot\frac{\alpha}{2}.$$

We now wish to determine how much of the energy of this pulse lies within the baseband. Consider a pulse $p(t)$ and its Fourier transform $P(f)$. The energy spectrum is given by $|P(f)|^2$. Thus the energy in a baseband extending to W Hz is given by

$$E = \int_{-W}^{W} |P(f)|^2 df. \tag{6.2-36}$$

If we substitute the following expression for $P(f)$,

$$P(f) = \int_{-\tau/2}^{\tau/2} p(t) e^{-j2\pi ft} dt,$$

into (6.2-36) we obtain

$$E = \int_{-W}^{W} df \int_{-\tau/2}^{\tau/2} ds \int_{-\tau/2}^{\tau/2} dt\, p(t) p(s) e^{j2\pi f(s-t)}.$$

Performing the integration on f first, then integrating with respect to $(s-t)$, we obtain

$$E=2W\int_{-\tau}^{\tau}I(x,\alpha)\frac{\sin 2\pi Wx}{2\pi Wx}\,dx, \qquad (6.2\text{-}37)$$

where

$$I(x,\alpha)=\frac{2c^2}{x(x^2+4c^2)}\log\left\{\frac{\tau^2+4c^2}{(2x-\tau)^2+4c^2}\right\}$$

$$+\frac{2c}{x^2+4c^2}\left\{\frac{\alpha}{2}-\tan^{-1}\frac{2x-\tau}{2c}\right\}$$

$$=I(-x,\alpha), \qquad 0\le x\le\tau.$$

Finally, if we assume the phase transients are uniformly distributed from $-\pi$ to $+\pi$ and that they occur at random at a rate ν per second, the noise power in the baseband is

$$N=\frac{\nu W}{\pi}\int_{-\pi}^{\pi}d\alpha\int_{-\tau}^{\tau}dx\,I(x,\alpha)\frac{\sin(2\pi Wx)}{2\pi Wx}. \qquad (6.2\text{-}38)$$

We note that if τ is small, $p(t)$ is essentially an impulse of area α, $I(x,\alpha)$ is an impulse of area α^2, and the integration (6.2-38) reduces to $N=2\pi^2\nu W/3$.

The integration (6.2-38) was performed numerically for $W\tau=0.4$, 0.8, 1.6, and 3.2 corresponding to delays of 133, 267, 533, and 1067 μsec with a baseband bandwidth of 3 KHz. The result is shown in Table 1.

From Table 1 it appears that unless we allow "soft" switching up to around 500 μsec, the advantage gained by soft switching is rather minimal.

If we consider an RF bandwidth of B and assume Gaussian modulation of rms deviation σ Hz such that $B=2(W+\sigma\sqrt{10})$, then the signal power at the frequency discriminator output is

$$S=4\pi^2\sigma^2. \qquad (6.2\text{-}39)$$

The limiting signal to noise ratio is given by the ratio of (6.2-39) and (6.2-38). In the case where τ is small we note that the result is

$$\frac{S}{N}=0.6\frac{W}{\nu}\left(\frac{B}{2W}-1\right)^2. \qquad (6.2\text{-}40)$$

For arbitrary τ, the noise reduction shown in Table 1 should be applied to Eq. (6.2-40), giving a larger signal-to-noise ratio.

Table 1   *Reduction of Switching Transient by Soft Switching*

$W\tau$	0	0.4	0.8	1.6	3.2
$\dfrac{N}{(2\pi^2\nu W/3)}$	0 dB	-0.5 dB	-1.2 dB	-2.8 dB	-4.8 dB

The switching rate is therefore of great interest. Experimental results have indicated that the switching rate with switching diversity systems is around 1.3 times that of the threshold level crossing rate. In UHF mobile radio ($f_c = 840$ MHz) let us take the extreme situation in which $v = 60$ mi/hr. The level crossing rate at $A^2/2b_0 = 0.1$ (-10 dB) is 53 Hz. Therefore $\nu = 70$ Hz. The limiting baseband signal-to-noise ratio is presented in Table 2.

Table 2   *Limiting Signal-to-Noise Ratio Due to Switching*

	$B/2w$	2	3	4	6	8	10	15	20
Limiting	No delay	13.8	19.8	23.3	27.8	30.5	32.9	36.7	39.4
SNR									
(dB)	100 μsec delay	14.1	20.1	23.6	28.1	30.8	33.2	37.0	39.7

For systems that require baseband SNR around 20 dB, the switching transient noise can apparently be tolerated. However, if we are interested in a high-quality system having SNR on the order of 30 dB, means must be provided to reduce the switching transients.

In Figure 6.2-13 the discriminator output before the baseband filter is shown as a combination of signal modulation, FM noise, and transient spikes. It is therefore plausible after examining Figure 6.2-13 to suggest that we either blank or sample and hold the discriminator output for a duration of T that is about the length of the transient spikes. Since both blanking and sample and hold will distort signal modulation, the reduction of transient noise is obtained at the price of additional signal distortion noise. The amount of distortion can be estimated similarly to the way we estimate the switching transient. For example, each time we blank the discriminator output we introduce a pulse of the form $-s(t)$ with duration T. The average energy spectrum of the pulse is

$$E(f) = \left\langle \left| \int_0^T s(t) e^{-j2\pi ft} \, dt \right|^2 \right\rangle$$

$$= 2 \int_0^T R_s(\tau)(T - \tau) \cos 2\pi f\tau \, d\tau, \qquad (6.2\text{-}41)$$

where

$$R_s(\tau) = \langle s(t)s(t + \tau) \rangle.$$

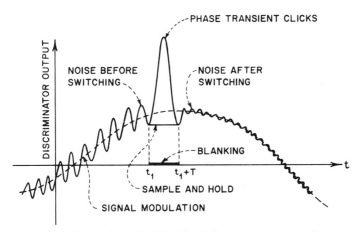

Figure 6.2-13 An illustration of FM discriminator output in the presence of switching transient and thermal noise.

The switching rate ν times the energy in baseband gives us the signal distortion noise due to blanking,

$$N_b = \nu \int_{-W}^{W} E(f) \, df. \qquad (6.2\text{-}42)$$

For the case $W = 3$ kHz and $s(t)$ a Gaussian distribution with a flat power spectrum from 0 to 3 kHz we obtain the following signal-to-distortion ratio [for $WT \ll 1$]:

$$\frac{S}{N_b} = \frac{1 - \frac{1}{18}(2\pi WT)^2 + \cdots}{2\nu WT^2} . \tag{6.2-43}$$

Similar calculations can be applied to sample and hold to obtain

$$\frac{S}{N_{\mathrm{SH}}} = \frac{3}{2\nu W^2 \pi^2 T^4 \left[1 - 0.0528(2\pi WT)^2 + \cdots\right]} . \tag{6.2-44}$$

Equations (6.2-43) and (6.2-44) are plotted in Figure 6.2-14 as a function of T with a switching rate $\nu = 70$ Hz.

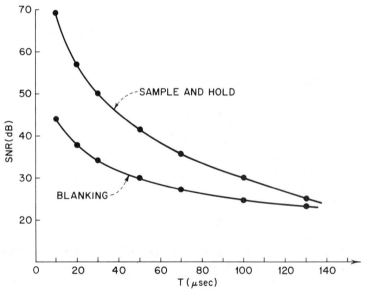

Figure 6.2-14 Baseband signal-to-distortion ratio caused by blanking or sample and hold at an average rate of 70 Hz.

6.2.3   Baseband Signal-to-Noise Ratio and Optimum Threshold

The ultimate judgment of the system performance is the improvement in baseband SNR over that of a single-branch system. The baseband noise in a switch diversity system consists of the ordinary FM noise and switching transient noise and is highly dependent on the switching threshold.

A laboratory test was performed with $B = 32$ KHz and baseband filter bandwidth 300 Hz to 3 KHz. The signal modulation was a 2 KHz sine wave tone with peak deviation 5.8 KHz such that the signal power equals $4\pi^2\sigma^2$, as given by Eq. (6.2-39). Two independent simulated Rayleigh fading carriers at $f_c = 840$ MHz were used. The fading rate was 100 Hz (equivalent to $v = 80$ mi/hr). The baseband SNR was measured by first measuring the received baseband signal power and then the noise power with signal modulation turned off.

The measurement results are presented in Figure 6.2-15 as a function of ρ_s (the carrier-to-noise ratio at switching threshold) and ρ_0 (the average carrier-to-noise ratio of each branch). The theoretical baseband SNR is

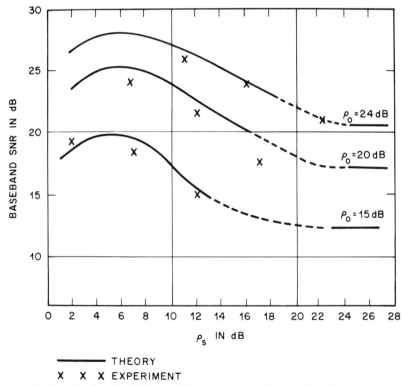

Figure 6.2-15 Baseband SNR at different ρ_0 (CNR) as a function of switching threshold ρ_s. ($B_{if} = 32$ kHz, $f_m = 100$ Hz, baseband 0.3-3 kHz).

calculated by integrating the baseband noise (Section 5.4) over the density function [Eq. (6.2-26)]. The switching transient noise [Eq. (6.2-40)] is also included with $\nu = 1.3 \ N\rho_s$ ($N\rho_s$ is the level crossing rate of ρ_s by Rayleigh fading).

The comparison between the theoretical and experimental results indicated close agreement. Let us look at the case $\rho_0 = 24$ dB first. We note that the baseband SNR reaches a peak at $\rho_s = 7$ dB, and there is a considerable range where variation of ρ_s makes little variations in SNR. As ρ_s steadily increases, the baseband SNR decreases until it finally reaches the baseband performance of a single Rayleigh branch. The best improvement in baseband SNR is about 7.5 dB. Similar behavior is also demonstrated in the curves with $\rho_0 = 20$ dB and 15 dB. From these curves it appears that an optimum threshold is around 6 dB carrier-to-noise ratio (i.e., 3 dB below FM threshold), and within a broad range (± 4 dB) the baseband SNR is quite close to its maximum value.

To demonstrate the system limitations due to switching transient noise, we have plotted in Figure 6.2-16 the results of a two-branch switching diversity system with $\rho_s/\rho_0 = -13$ dB. We note that as ρ_0 increases, baseband SNR fails to increase indefinitely, indicating the dominance of switching transient noise at high CNR. The ν for $\rho_s/\rho_0 = -13$ dB is 68 Hz. Evaluating Eq. (6.2-40) we find that the SNR due to switching noise alone is 27 dB, which checks well with measured limiting behavior.

By including circuitry that blanks the discriminator output for a duration of 50 μsec after each transition the SNR was improved, as shown in Figure 6.2-16. For comparison the theoretical SNR (neglecting switching transient) is also presented. It is observed that blanking has an effect only when the baseband SNR is greater than 25 dB. For example, at $\rho_0 = 25$ dB blanking yields a baseband improvement of 4 dB. At $\rho_0 = 30$ dB blanking improves baseband SNR by 6 dB.

The reduction of transmitter power by the use of switching diversity versus nondiversity can now be assessed from Figure 6.2-16. The comparison is based on the measured performance. For example, if 20-dB baseband SNR is required, two-branch switching diversity would yield a 7.5-dB reduction in transmitter power. The transmitter power reduction at 30-dB SNR is close to 14 dB. Although this comparison is taken at $f_m = 100$ Hz, that is, at a vehicle speed of 80 mi/hr at 840 MHz, the performance at lower vehicle speeds should not be much different because with blanking the baseband SNR for ρ_0 up to 30 dB is quite close to that for the case $f_m = 0$ Hz.

As reported before, ideally the optimum threshold would have to be around a carrier-to-noise ratio of 7 dB. To implement the system would require a fixed reference threshold that is sensitive to gain variations (Section 6.1). An alternative method might be that of using a threshold

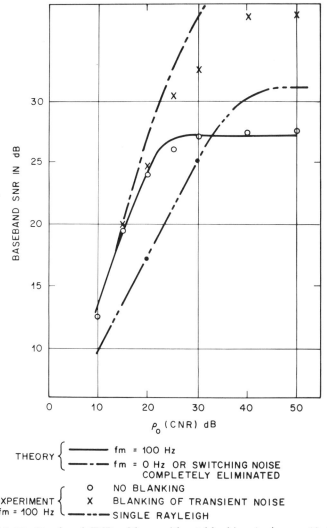

Figure 6.2-16 Baseband SNR with or without blanking ($\rho_s/\rho_0 = -13$ dB $B_{if} = 32$ kHz baseband $0.3 - 3$ kHz).

that is a fixed fraction of the average power of the Rayleigh fading envelope. Since the range of optimum threshold is broad, we can pick, say, $\rho_s/\rho_0 = 0.05$. Referring to Figure 6.2-15, we note that this amounts to switching at $\rho_s = 2$, 7, and 11 dB for $\rho_0 = 15$, 20, and 24 dB, respectively. The system is still near optimum in this range with baseband SNR ranging from 20 to 30 dB.

6.2.4 Feedback Diversity

The system configuration is shown in Figure 6.2-2. An out-of-band tone burst is added to the baseband message whenever the command station requires a switch of antenna.

As discussed in Sections 6.2.1 and 5.3.4, the major consideration for feedback diversity is the time delay involved. The other considerations are the failure to detect the tone at the responder station or the false triggering of the antenna switch when the tone is not present. These in turn depend on the peak frequency deviation of the tone burst, the responder station tone filter bandwidth, and also the signal-to-noise ratio at the discriminator output. The power spectrum of the discriminator output at the responder station is sketched in Figure 6.2-17. It consists of parabolic noise, flat click noise, signal modulation, and the tone burst.

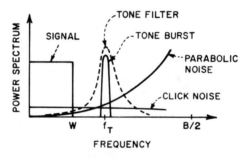

W = 3 KHz BASEBAND BAND WIDTH
f_T = TONE BURST FREQUENCY
B = I.F. BANDWIDTH

Figure 6.2-17 The power spectrum at the discriminator output at the responder station.

The tone burst filter bandwidth must be a compromise between the time delay and the amount of noise allowed within its bandwidth. This in turn depends on the expected discriminator output noise at the responder station. As an example let us assume that the responder station uses a two-branch predetection equal-gain combining system with $B = 50$ KHz and $W = 3$ KHz. A baseband SNR of 25 dB can be obtained with an average carrier-to-noise ratio of 15 dB (Section 5.4). The choice of a two-pole Chebychev filter with $f_T = 7$ KHz and $B_T = 2$ KHz will yield a tone-to-noise ratio around 24 dB for a 7-KHz tone with modulation index equal to unity and a time delay around 240 μsec.

The RF spectrum broadening due to the 7-kHz tone burst of index unity in a 25-kHz peak deviation system is small; thus the signaling method does not require much additional RF spectrum.

The other delays involved in the system are the round-trip propagation delay of 40 μsec for the 4-mile round trip and two IF filter delays of 20 μsec each for a 50 kHz IF bandwidth.

From the above considerations it appears that the total time delays are still within the tolerable range even at $v = 60$ miles/hr and a switching threshold of -13 dB with respect to the mean carrier power (Figure 6.2-11). The measured baseband SNR of the feedback diversity system is shown in Figure 6.2-18. Here the measurement is taken under the same condition as Figure 6.2-13 but with a time delay of 270 μsec (equivalent to 340 μsec at $v = 60$ miles/hr) inserted in the decision loop. Comparing with Figure 6.2-16 we note that the diversity advantage has been reduced by approximately 2 dB. This is due to the degradation of the distribution of R as illustrated by curve 3 of Figure 6.2-10.

6.3 COHERENT COMBINING USING CARRIER RECOVERY

Systems that coherently combine independent signals from spatially separated antennas have better carrier statistics and reduce random FM better than switch-combining systems. More important, perhaps, is that coherent combining systems do not suffer degradation from phase transients that are inherent in antenna-switching systems. The obvious disadvantage of coherent combining is cost. In its simplest form a switch-diversity system requires a means of selecting one of several antennas and logic (for example, circuits that compare the instantaneous received signal level to a predetermined threshold as in Section 6.2) to control the switch. On the other hand, coherent combining is almost always performed at IF, requiring individual down converters and IF amplifiers plus the cophasing circuitry.

There are several methods of coherently combining signals that have been developed over the years for a number of applications. As far back as 1925, for example, Rice patented[14] such a radio receiving system. One straightforward method of achieving phase control is shown in Figure 6.3-1. The signal from one antenna passes through a variable phase shifter controlled by a circuit that compares the relative phase of the signals from each branch. Several variations on this basic theme have been employed. In 1965 Altman and Sichak[2] discussed a coherent combiner for beyond-

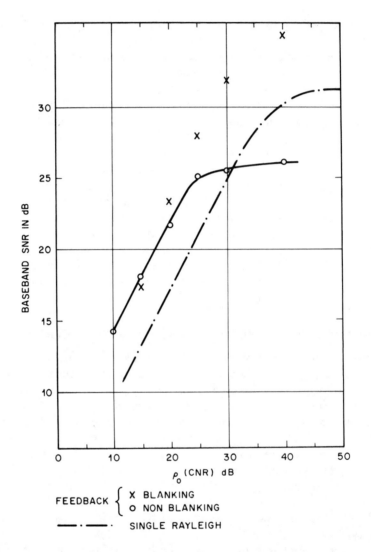

Figure 6.2-18 Measured baseband SNR with and without blanking in feedback system. ($\rho_s/\rho_0 = -13$ dB, fm = 100 Hz, time delay = 270 μsec, $B_{if} = 32$ kHz, baseband 3.0–3 kHz).

the-horizon links. In this type of combiner, shown in Figure 6.3-2, cophasing is achieved by varying the frequency of a local oscillator. A similar receiver was patented by Adams[15] and is discussed in a paper coauthored with D. M. Mindes.[16] In the following sections we shall look in detail at the operation of three coherent combining schemes that have been considered for use in microwave mobile radio receivers.

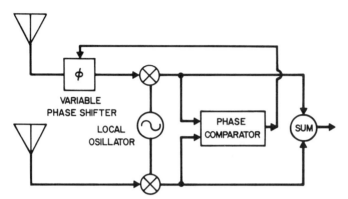

Figure 6.3-1 Cophasing circuit using a variable phase shifter.

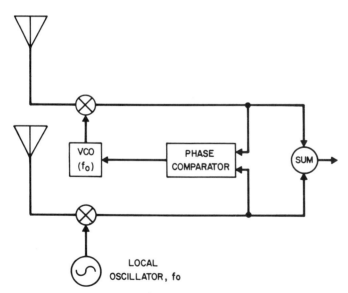

Figure 6.3-2 Cophasing circuit using a variable-frequency local oscillator.

6.3.1 Use of the Signal as a Pilot

The first published work describing a diversity receiver for mobile radio applications using a coherent combiner appeared in 1966 in the *IEEE Transactions on Vehicular Technology*. In this journal Black, Kopel, and Novy[17] describe a dual diversity predetection combining mobile radio receiver that was based upon an invention of Cutler, Kompfner, and Tillotson[18] for use in satellite repeaters.

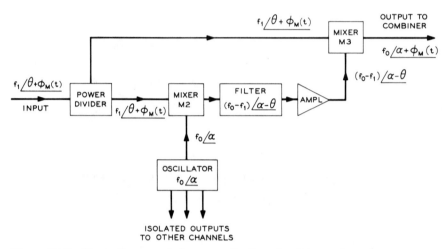

Figure 6.3-3 Block diagram of phase-equalization circuit in one branch.

A four-branch diversity receiver using the cophasing method of Cutler et al. was constructed by Rustako[19] and tested in a mobile radio environment. Four identical branches provided four inputs to a linear combiner. Figure 6.3-3 shows a simplified block diagram of the phase equalization circuit used in one receiver branch. The received signal with a frequency of $f_1 \angle \theta$ is split and fed into two mixers M_1 and M_2. The output of a local oscillator common to all channels at a frequency of $f_0 > f_1$ and arbitrary phase angle α is mixed with part of $f_1 \angle \beta$ in mixer M_1. The lower mixing product at a frequency of $(f_0 - f_1) \angle (\theta - \alpha)$ is passed through a narrow bandpass filter to eliminate modulation sidebands and reduce the noise in this part of the circuit, and then is amplified and limited. This signal at a frequency and phase $(f_0 - f_1) \angle (\alpha - \beta)$ has nearly constant amplitude with a fading received input signal down to a threshold determined by the receiver noise level. The output of mixer M_2 is tuned to the upper sideband

and is given by

$$(f_1+f_0-f_1)\angle(\theta+\alpha-\theta)=f_0\angle\alpha. \qquad (6.3\text{-}1)$$

This process is identical for all four branches resulting in four outputs each at a frequency of $f_0\angle\alpha$ with four inputs at f_1 of any arbitrary phase angles.

Figure 6.3-4 shows a complete block diagram of one branch of the diversity receiver actually tested. The incoming UHF signal was mixed with a signal from a local oscillator common to all channels to produce an IF signal, $f_1\angle\beta$. This signal was amplified and split in a power divider; one output from the divider was mixed with an oscillator at $f_0\angle\alpha$ producing a lower sideband $(f_0-f_1)\angle(\alpha-\beta)$. This signal was passed through a narrow-band filter, amplified, limited, and passed to a second mixer, M_2, where it was mixed with the other output from the power divider at $f_1\angle\theta$ to produce the final IF frequency, $f_0\angle\alpha$. This signal now contained the information carried by the signal $f_1\angle\theta$, including the amplitude modulation of the envelope produced by the fading and the frequency modulation carrying the intelligence. The output from the combiner and four other outputs from the single-branch power dividers were fed into five IF amplifiers from which the received signal levels and modulation could be recovered. At the option of the user the receiver could be arranged for two-, three-, or four-branch diversity operation.

Figure 6.3-5 shows a plot of the IF output level from the four single branches and the combined branches as a function of the input level. As the input level is decreased, the noise threshold is reached in the single branches at a higher signal level than in the combined channel, since in the latter the signals add coherently. The curves for the individual branches show a gradual increase in level below the noise threshold. This is caused by the increase in noise power output from the narrow-band recovery filter amplifier as its limiting threshold is reached. Under these conditions the receiver threshold improvement is close to the theoretical value of 6 dB.

The diversity receiver was installed in a mobile radio test van to evaluate its performance in a multipath environment. The measurements were made on three typical suburban residential streets in New Jersey. The mobile antennas were short $(\lambda/4)$ vertical whips mounted on the roof of the test van, and separated by $\frac{5}{4}\lambda$ along a line perpendicular to the direction of travel. Some examples of the test results are shown in Figures 6.3-6 to 6.3-8. Figure 6.3-6 shows the cumulative probability distribution for a single branch and for two, three, and four branches combined for a test run on a typical street. A Rayleigh distribution is plotted for comparison. The single branch is nearly Rayleigh distributed except at very low

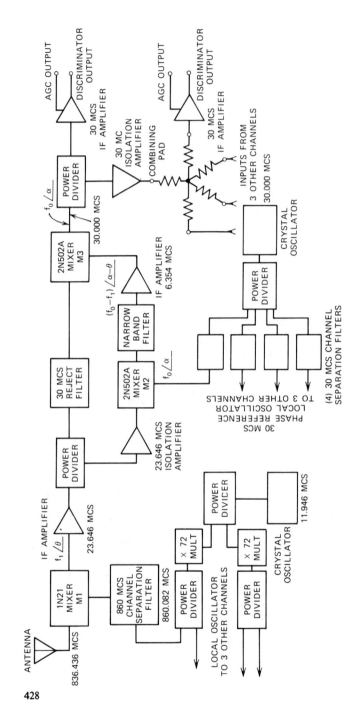

Figure 6.3-4 Block diagram of one branch of an experimental diversity receiver.

428

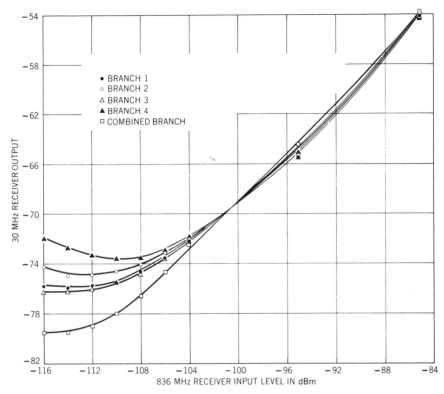

Figure 6.3-5   IF output as a function of 836 MHz input level.

probabilities where it deviates, possibly due to direct transmission components. As more branches are combined, the reduction in fading is evident.

The level crossing rates and power spectra (see Chapter 1) of the received signal variations were also calculated from the data. Figure 6.3-7 shows the power spectra for one of the test runs. It shows the spreading of energy of the transmitted signal over a band of energy about the carrier frequency by the random scattering media between the transmitter and the mobile receiver. These power spectra were calculated for the same sample of data as the cumulative probability distribution shown in Figure 6.3-6. These spectra are typical of those calculated from all of the data. The solid line in Figure 6.3-7 shows the power spectrum for a single channel. This spectrum shows trends similar to those in the power spectrum calculated in Chapter 1; namely, the gradual rise towards zero frequency and the sharp falloff near 40 Hz or twice the maximum Doppler frequency shift at this

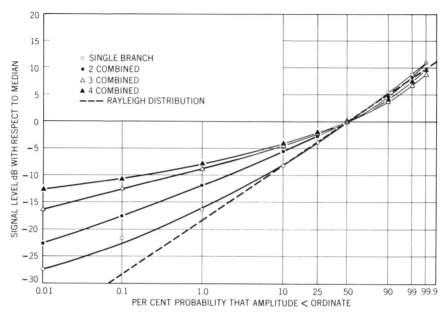

Figure 6.3-6 Cumulative probability distributions, Sherwood drive.

carrier frequency and vehicle speed. As additional branches are combined, the relative power at the higher fading frequencies is reduced.

Figure 6.3-8 shows the level crossing rates of the received signals, again for the same sections of data. These curves show the number of times per second the received signal crosses some level with respect to the rms value of the fading signal. As additional branches are combined, the number of times the signal crosses at some level below the rms level is reduced. At a signal level 10 dB below the rms values, the fading rate with four branches is less than $\frac{1}{10}$ that of a single channel.

Random FM

The signal received by a mobile radio vehicle moving through a scattered field experiences random frequency modulation (see Section 1.4) as well as amplitude modulation, due to the interference pattern in the field. Diversity techniques serve to minimize this random frequency modulation, but it is significant that a coherent adaptive system can *completely eliminate* this random FM, even when only one antenna element is used.

In order to show how this system eliminates the random FM, consider the phase correction circuit in Figure 6.3-9. The pilot and signal frequencies are assumed to be close enough so that the phase modulation, θ_m, introduced on each signal due to the motion through standing waves is the

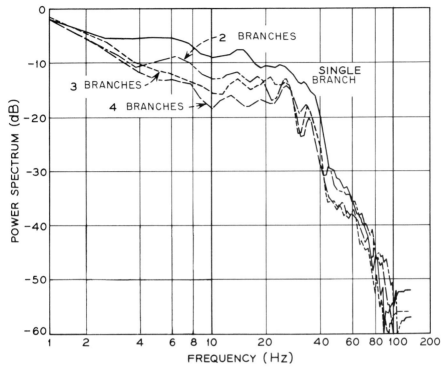

Figure 6.3-7 Power spectra of fading envelope, Sherwood Drive.

same (see Section 1.5). Thus the phase of the incoming signal is $\theta_m(t) + \phi(t)$, where $\phi(t)$ is the phase modulation associated with the information placed on the carrier. The pilot is separated from the signal by filter F_1 after being shifted in frequency by mixer M_1 and then amplified so as to serve as a pseudo-local-oscillator signal to mixer M_2. The filter F_1 must reject all signals out of mixer M_1 due to the signal and pass only the proper sideband due to the pilot, including the phase modulation θ_m on the pilot.

For example, if the signal is FM and the pilot is the FM carrier (as assumed in Figure 6.3-3), the filter F_1 must reject the information modulation so that the filter output behaves as if only the carrier had been transmitted. The phase modulation due to motion through the standing waves, $\theta_m(t)$, must be left undistorted. In order to recover the FM carrier in this manner there are two requirements: (a) the minimum frequency in the information modulation baseband must be greater than the maximum Doppler frequency, and (b) the FM must be small index. (Large-index FM may be reduced to small-index FM for filtering purposes by frequency division.)

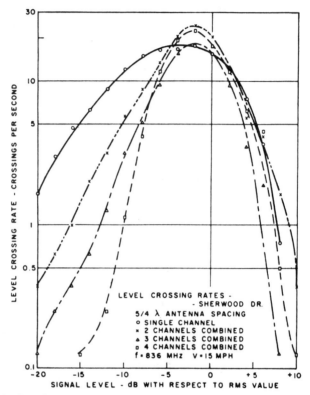

Figure 6.3-8 Level crossing rates as a function of number of branches combined, Sherwood Drive.

If the filter F_1 and the amplifier A_1 have a phase shift that increases rapidly with frequency in their passband, then an equal time delay must be inserted at point A so that the phase correlation between the pilot and signal is retained.

The pseudo-local-oscillator signal then has a phase of $\alpha - \theta_m$, where α is the phase of the local oscillator at frequency f_0 and we have selected the lower sideband out of mixer M_1, just as an example. By selecting the proper upper sideband output of mixer M_2 with filter F_2, the phase of the output is seen to be $\alpha + \phi$; that is, the phase modulation due to motion through standing waves, $\theta_m(t)$, has been *eliminated* and the information phase modulation ϕ has been left undistorted.

It was pointed out above that F_1 will introduce a time delay in its channel that should be compensated by an equal time delay in the upper signal path to obtain cancellation of the random FM. The remainder of

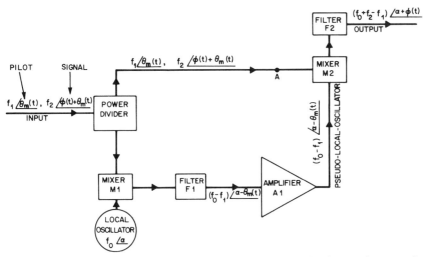

Figure 6.3-9 Block diagram of the phase correction circuit of one element of a coherent combiner using a pilot.

this discussion will be devoted to a study of the properties of the undesired random FM output of a single branch when such delay compensation is not made.* The results will help to determine the bandwidth of F_1 for an acceptable reduction in the random FM.

First, it will be instructive to consider the effect of this channel on an FM signal with one sinusoidal modulation component, assuming the entire FM signal is passed through F_1:

$$e_{in}(t) = \cos[\omega_c t + \phi(t)],$$
$$\phi(t) = \beta \sin \omega_0 t. \tag{6.3-2}$$

Here ω_c is the carrier frequency and ω_0 the modulation frequency. After filtering by F_1 the signal is

$$e(t) = \cos[(\omega_s - \omega_c)(t - \tau_0) - \phi(t - \tau_0)]. \tag{6.3-3}$$

M_2 multiplies e_{in} and e, then the upper sideband is selected by F_2, giving

$$e_{out}(t) = \cos[\omega_s t - (\omega_s - \omega_c)\tau_0 + \phi(t) - \phi(t - \tau_0)]. \tag{6.3-4}$$

Let $\omega(t)$ be the departure of the instantaneous frequency from the new

*This material is based on unpublished work of W. C. Jakes.

carrier, ω_s:

$$\omega(t) = \frac{d}{dt} \text{Arg}[e_{\text{out}}(t)] - \omega_s = \dot\phi(t) - \dot\phi(t - \tau_0)$$

$$= \beta\omega_0[\cos\omega_0 t - \cos\omega_0(t - \tau_0)].$$

This may be put in the form

$$\omega(t) = 2\beta\omega_0 \sin\frac{\omega_0\tau_0}{2} \sin\left[\omega_0\left(t - \frac{\tau_0}{2}\right)\right]. \tag{6.3-5}$$

The instantaneous phase is given by

$$\psi(t) = \int \omega(t)\,dt$$

$$= 2\beta \sin\frac{\omega_0\tau_0}{2} \cos\left[\omega_0\left(t - \frac{\tau_0}{2}\right)\right], \tag{6.3-6}$$

which represents an FM signal with index

$$\beta' = 2\beta\sin\frac{\omega_0\tau_0}{2}. \tag{6.3-7}$$

For $\tau_0 \leqslant (6f_0)^{-1}$, $\beta' \leqslant \beta$, and there will be a reduction in the modulation index. Roughly speaking, the delay τ_0 is comparable to the reciprocal of the filter bandwidth, so we see that this bandwidth must be at least six times the modulation frequency in order to achieve any reduction in the FM index (providing F_1 passes the entire FM signal, as stated earlier).

In what follows we shall calculate in more detail the effect of the filter on the actual random FM in a simple mobile radio transmission path. The power spectrum of the output random FM is discussed in the next section and following that the probability density function is obtained.

Power Spectrum of the Random FM

The one-sided power spectrum of the random frequency deviations of $e_{\text{in}}(t)$ was derived in Chapter 1, Section 1.4 by taking the Fourier transform of the autocorrelation function of the instantaneous frequency deviations, $\dot\phi(t)$:

$$W(f) = 4\int_0^\infty R(\tau)\cos\omega\tau\,d\tau, \tag{6.3-8}$$

where

$$R(\tau) \equiv \overline{\dot\phi(t)\dot\phi(t - \tau)}. \tag{6.3-9}$$

Applying the same signal to our system with delay τ_0, the autocorrelation function of the random frequency deviations from the carrier ω_s is

$$R_D(\tau) = \overline{[\dot{\phi}(t) - \dot{\phi}(t - \tau_0)][\dot{\phi}(t - \tau) - \dot{\phi}(t - \tau - \tau_0)]}$$

$$= \overline{\dot{\phi}(t)\dot{\phi}(t - \tau)} + \overline{\dot{\phi}(t - \tau_0)\dot{\phi}(t - \tau - \tau_0)}$$

$$- \overline{\dot{\phi}(t - \tau_0)\dot{\phi}(t - \tau)} - \overline{\dot{\phi}(t)\dot{\phi}(t - \tau - \tau_0)}$$

$$= 2R(\tau) - R(\tau - \tau_0) - R(\tau + \tau_0), \qquad (6.3\text{-}10)$$

where we have used the assumption that the mobile radio statistics are stationary. The power spectrum is then

$$S_D(f) = \int_{-\infty}^{\infty} R_0(\tau) e^{-i\omega\tau} \, d\tau$$

$$= 2 \int_{-\infty}^{\infty} R(\tau) e^{-i\omega\tau} \, d\tau - \int_{-\infty}^{\infty} R(\tau - \tau_0) e^{-i\omega\tau} \, d\tau$$

$$- \int_{-\infty}^{\infty} R(\tau + \tau_0) e^{-i\omega\tau} \, d\tau. \qquad (6.3\text{-}11)$$

But

$$\int_{-\infty}^{\infty} R(\tau \pm \tau_0) e^{-i\omega\tau} \, d\tau = e^{\pm i\omega\tau_0} \int_{-\infty}^{\infty} R(\tau) e^{-i\omega\tau} \, d\tau; \qquad (6.3\text{-}12)$$

thus

$$S_D(f) = (2 - e^{-i\omega\tau_0} - e^{i\omega\tau_0}) \int_{-\infty}^{\infty} R(\tau) e^{-i\omega\tau} \, d\tau$$

$$= 2 \sin^2 \frac{\omega\tau_0}{2} \int_{-\infty}^{\infty} R(\tau) e^{-i\omega\tau} \, d\tau.$$

Since $R(-\tau) = R(\tau)$, the one-sided spectrum is

$$W_D(f) = 4 \sin^2 \frac{\omega\tau_0}{2} W(f). \qquad (6.3\text{-}13)$$

Note that this is a fairly general result, requiring only that the autocorrelation function be stationary.

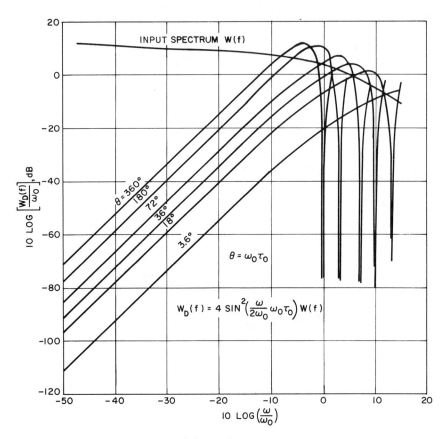

Figure 6.3-10   Power spectrum of the random FM.

Figure 6.3-10 shows a plot of $W_D(f)$ (valid to 4%) and curves for $W_D(f)$ for a range of values of $\theta = \omega_0\tau$ using Eq. (6.3-13). If the filter bandwidth is at least $5f_0$, the random frequency deviations have a power spectrum that equals the value for the input spectrum at $\omega = \omega_0$ and falls off rapidly for lower frequencies, approaching a slope of 6 dB per octave. This value thus seems to be a reasonable criterion for the filter (F_1) bandwidth and agrees with that for a single sinusoidal modulation component.

Probability Density of the Random FM

As in the previous paragraphs, we let the input signal be a narrow-band Gaussian process:

$$e_{in}(t) = R(t)\cos[\omega_c t + \phi(t)], \tag{6.3-14}$$

where the envelope $R(t)$ of the signal is Rayleigh distributed, and all values of $\phi(t)$ from 0 to 2π are equally probable. After mixing by M_1 and delay by τ_0, we write the signal

$$e(t + \tau_0) = R(t + \tau_0)\cos[(\omega_s - \omega_c)(t + \tau_0) - \phi(t + \tau_0)], \tag{6.3-15}$$

where we now use $t + \tau_0$ instead of $t - \tau_0$ in order to conform to the notation used by others. (For the type of signal assumed it does not matter which signal branch is delayed.) After mixing by M_2 and selection of the upper sideband by F_2, the output frequency deviations from ω_s are

$$\omega(t, \tau_0) = \dot{\phi}(t) - \dot{\phi}(t + \tau_0). \tag{6.3-16}$$

We seek the probability density function (pdf) of ω, $p[\omega(t, \tau_0)]$. The straightforward approach would be to obtain the eightfold joint pdf of the required variables:

$$p[R(t), R(t + \tau_0), \dot{R}(t), \dot{R}(t + \tau_0), \phi(t), \phi(t + \tau_0), \dot{\phi}(t), \dot{\phi}(t + \tau_0)], \tag{6.3-17}$$

and integrate over all values of all variables except $\dot{\phi}(t)$, $\dot{\phi}(t + \tau_0)$, yielding the joint pdf of these two:

$$p[\dot{\phi}(t), \dot{\phi}(t + \tau_0)].$$

Since our desired variable, $\omega(t, \tau_0)$, is a linear combination of $\dot{\phi}(t), \dot{\phi}(t + \tau_0)$, we could get $p(\omega)$ by a final integration:

$$p[\omega(t, \tau_0)] = \int_{-\infty}^{\infty} p[\dot{\phi}(t), \dot{\phi}(t) - \omega(t, \tau_0)] d[\dot{\phi}(t)]. \tag{6.3-18}$$

However, this method contains formidable difficulties of integration, and hence a Monte Carlo computer simulation was used instead.

In the case where the delay is very large ($\tau_0 \to \infty$) the input signal and its delayed counterpart are independent and uncorrelated. Their joint pdf is then simply the product of their individual pdfs:

$$p(R,\phi,\dot{R},\dot{\phi},R_1,\phi_1,\dot{R}_1,\dot{\phi}_1)=p(R,\phi,\dot{R},\dot{\phi})p(R_1,\phi_1,\dot{R}_1,\dot{\phi}_1), \quad (6.3\text{-}19)$$

where the unsubscripted variables refer to $e_{in}(t)$ and subscript 1 to $e'_{in}(t+\tau_0)$. We thus obtain

$$p(R,\ldots,\dot{\phi}_1)$$

$$=\frac{R^2 R_1^2}{4\pi^4 \omega_0^4}\exp\left\{-\frac{1}{2}\left[R^2+R_1^2+\frac{2}{\omega_0^2}(\dot{R}^2+\dot{R}_1^2+R^2\dot{\phi}^2+R_1^2\dot{\phi}_1^2)\right]\right\}. \quad (6.3\text{-}20)$$

The integrations on $R,R_1,\dot{R},\dot{R}_1,\phi,\phi_1$ may be easily carried out, obtaining

$$p(\dot{\phi},\dot{\phi}_1)=\frac{1}{2\omega_0^2}\left[\left(1+2\frac{\dot{\phi}^2}{\omega_0^2}\right)\left(1+2\frac{\dot{\phi}_1^2}{\omega_0^2}\right)\right]^{-3/2}. \quad (6.3\text{-}21)$$

Finally, $p(\omega)$, where $\omega=\dot{\phi}-\dot{\phi}_1$, is given by

$$p(\omega)=\frac{1}{2\omega_0^2}\int_{-\infty}^{\infty}\left\{\left[1+2\left(\frac{\dot{\phi}}{\omega_0}\right)^2\right]\left[1+2\left(\frac{\dot{\phi}-\omega}{\omega_0}\right)^2\right]\right\}^{-3/2}d\dot{\phi}. \quad (6.3\text{-}22)$$

This integral can be evaluated exactly in terms of elliptic integrals by substituting

$$u=\frac{\sqrt{2}\,\dot{\phi}}{\omega_0}-\frac{\omega}{\omega_0\sqrt{2}}.$$

Thus

$$p(\omega)=\frac{1}{\omega_0\sqrt{2}}\frac{(1-M)^{5/2}}{4M}\left[K(k)+\frac{2M-1}{1-M}E(k)\right], \quad (6.3\text{-}23)$$

where

$$k = \sqrt{M} = \frac{\omega}{\omega_0} \left[2 + \left(\frac{\omega}{\omega_0} \right)^2 \right]^{1/2},$$

and $K(k)$, $E(k)$ are the complete elliptic integrals.

Figure 6.3-11 shows the pdfs for $\theta = \omega_0 \tau_0 = 3.6°$, $18°$, $36°$, $72°$, $180°$, $360°$, and ∞. It is evident that the curves computed for larger values of $\theta(180°, 360°)$ are approaching that for $\theta = \infty$ with reasonable accuracy.

The power spectra and probability density functions are useful in revealing various aspects of the random FM resulting from the processing system studied here. Both imply that the filter bandwidth must be at least five or six times the maximum Doppler shift, v/λ, to appreciably reduce the FM. At 11.2 GHz with $v = 60$ mi/hr this means a bandwidth of at least 5 kHz or a delay less than 200 μsec. For $\tau_0 = 0$ the pdf is a delta function, and the trend toward this is obvious in Figure 6.3-11.

6.3.2 Delayed Signal as a Pilot

An alternative to the combining scheme is to use the complete signal but delayed in time by a separate circuit at the receiver, as the pilot. The time-delayed pilot diversity combiner was first proposed by Earp.[20] The phase-correction loop of this combiner (see Figure 6.3-12) has been used in other systems besides combiners, such as FM suppression[21] and airborne clutter cancellation.[22] If the time delay τ were small enough relative to the rate of change of $\phi(t)$, then $\phi(t) - \phi(t - \tau)$ would be negligible for all branches and the resulting branch outputs (at point 7) would all be in phase and could be added directly to provide the desired diversity advantage.

Although the term $s(t) - s(t - \tau)$ in the output phase indicates a reduction in FM index (when the delay τ is short compared to fluctuations in the signal modulation), it will be shown later that signal detection is not necessarily degraded, but may, in some cases, be improved.

There are four major considerations with the use of the delayed signal as a pilot. First, the signal should not be delayed so long that the relative phase shifts between antennas have changed so much as to be unrelated to the relative phase shift between pilots. This would prevent the various branches from adding in phase. The maximum allowable delay is $\sim 1/B_c$, where B_c is the coherence bandwidth discussed in Section 1.5.

Second, the longer the delay the less the pilot will cancel the random

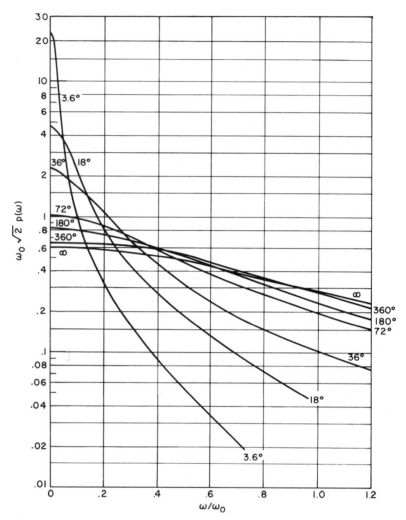

Figure 6.3-11   Computed curves for $p(\omega)$.

FM. The extent to which the loop cancels the random FM depends on how short the time delay is relative to the period of the maximum Doppler frequency.

Third, using the delayed signal as a pilot tends to cancel the information modulation on the signal (e.g., the effect of very short time delays on an FM signal is to reduce the modulation index and convert to phase modulation). To prevent this, the time delay should not be too much smaller than the period of the highest modulation frequency. If one wishes to use the phase-correction loop to suppress random FM relative to the information modulation, the bulk of the information modulation spectrum should be higher in frequency than the major portion of the spectrum of the random FM.

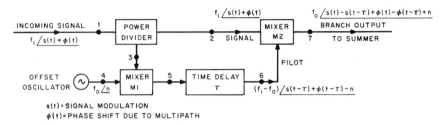

Figure 6.3-12   Block diagram of phase-correction loop.

Fourth, since the pilot bandwidth is now as wide as that of the signal, the level of the pilot relative to noise is no better than that of the signal. Therefore, the resulting signal-to-noise ratio after the phase-correction loop will be lower. However, we will show later in this section that this does not degrade the baseband signal-to-noise ratio when the signal is above threshold and can, in some cases, result in an improved threshold for FM.

The following analyses* provide formulas for determining the extent to which the above considerations affect the performance of a mobile radio system that uses the delayed signal as a pilot.

Phase Error

We now derive the rms phase difference in the relative phase between antennas compared at two times separated by a time delay, τ. The mean square phase difference in the relative phase shift between any two antennas in the array compared at two times separated by a time delay τ is just twice the mean square phase difference in $\phi(t)$ and $\phi(t - \tau)$ at one antenna; that is, for antennas a and b,

*The material in this section is based on unpublished work of M. J. Gans.

$$\overline{\{[\phi_a(t)-\phi_b(t)]-[\phi_a(t-\tau)-\phi_b(t-\tau)]\}^2}$$

$$= \overline{[\phi_a(t)-\phi_a(t-\tau)]^2} + \overline{[\phi_b(t)-\phi_b(t-\tau)]^2}$$

$$+ \overline{\phi_a(t)\ \phi_b(t-\tau)} + \overline{\phi_b(t)\ \phi_a(t-\tau)}$$

$$= \overline{2[\phi(t)-\phi(t-\tau)]^2} \qquad [\phi(t) \text{ has a zero mean}]$$

where the subscript has been dropped since the result is the same for all antennas. Thus, we need only consider the phase difference $\phi(t)-\phi(t-\tau)$ at one antenna to find the difference in relative phase shift between antennas. The joint probability density of the phase at t and at $t-\tau$ is given in Ref. 23, page 164, and is the same form as $p(\theta_1,\theta_2)$ derived in Chapter 1:

$$p[\phi(t),\phi(t-\tau)] = \frac{|\Lambda|^{1/2}}{4\pi^2\sigma_x^4} \left[\frac{(1-\beta^2)^{1/2}+\beta(\pi-\cos^{-1}\beta)}{(1-\beta^2)^{3/2}} \right]$$

$$\text{if} \quad 0<\phi(t), \quad \phi(t-\tau)<2\pi$$

$$=0 \qquad \text{otherwise,} \qquad\qquad (6.3\text{-}24)$$

where

$$\beta = \frac{R_c(\tau)}{\sigma_x^2}\cos\Delta\phi + \frac{R_{cs}(\tau)}{\sigma_x^2}\sin\Delta\phi,$$

$$R_c(\tau) = \int_0^\infty S(f)\cos 2\pi(f-f_c)\tau\, df,$$

$$R_{cs}(\tau) = \int_0^\infty S(f)\sin 2\pi(f-f_c)\tau\, df,$$

$$|\Lambda|^{1/2} = \sigma_x^4 - R_c^2(\tau) - R_{cs}^2(\tau),$$

$$\sigma_x^2 = \int_0^\infty S(f)\, df,$$

$$\Delta\phi = \phi(t)-\phi(t-\tau).$$

A probability density of the type shown in Eq. (6.3-24) can be integrated, as shown in Section 1.5, to obtain the probability density of $\Delta\phi$, the phase error due to assuming $\phi(t) = \phi(t - \tau)$:

$$p(\Delta\phi) = \frac{|\Lambda|^{1/2}}{4\pi^2\sigma_x^2} \left[\frac{(1-\beta^2)^{1/2} + \beta(\pi - \cos^{-1}\beta)}{(1-\beta^2)^{3/2}} \right] (2\pi - |\Delta\phi|)$$

$$\text{if} \quad -2\pi < \Delta\phi < 2\pi$$

$$= 0 \quad \text{otherwise.} \tag{6.3-25}$$

If one assumes an omnidirectional antenna, then

$$R_c(\tau) = \frac{b_0}{2\pi^2} \int_{f_c - f_m}^{f_c + f_m} \frac{\cos[2\pi(f - f_c)\tau]}{\sqrt{f_m^2 - (f - f_c)^2}} df$$

$$= \frac{b_0}{2\pi} J_0(2\pi f_m \tau), \tag{6.3-26}$$

while

$$R_{cs}(\tau) = 0,$$

$$\sigma_x^2 = \tfrac{1}{2} V^2(t) = \frac{b_0}{2\pi}$$

$$= \text{average power level,}$$

$$|\Lambda|^{1/2} = \frac{b_0^2}{4\pi^2} [1 - J_0^2(2\pi f_m \tau)],$$

$$\beta = J_0(2\pi f_m \tau) \cos\Delta\phi.$$

By comparing Eq. (6.3-26) with Eqs. (1.5-28) and (1.5-20), one notices that all of the statistics are identical to those of the separate frequency pilot case by making the transformation

$$J_0(2\pi f_m \tau) \leftrightarrow e^{-\phi^2/2}.$$

Using this transformation the results derived previously can be used to give the following:

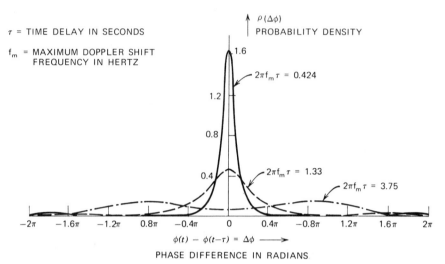

Figure 6.3-13   The probability density of phase difference versus time delay.

Figure 6.3-13:   the probability density of the difference phase on one antenna at t and at $t - \tau$, that is, $\phi(t) - \phi(t - \tau)$.

Figure 6.3-14:   the probability density of the difference in relative phase shift between two antennas at t and $t - \tau$.

Figure 6.3-15:   the normalized correlation coefficient of the phase at one antenna at t and at $t - \tau$.

Figure 6.3-16:   the root mean square error in relative phasing between antennas by using the delayed signal as a pilot.

Figure 6.3-17:   the average signal-to-noise improvement that can be obtained by combining pilot corrected branches relative to that of one pilot corrected branch. Two types of combiners are compared: (a) the maximal ratio combiner, in which the gain of a branch is proportional to the signal level of the branch (ambient noise in all branches assumed equal), and (b) the linear combiner in which the gain of all branches are equal.

Figures 6.3-13 to 6.3-17 point up a difference between using a separate frequency pilot (Section 6.4) and using a delayed signal as a pilot. From Figure 6.3-15, it is seen that the correlation of pilot phase to signal phase is oscillatory with time delay, varying between positive and negative correla-

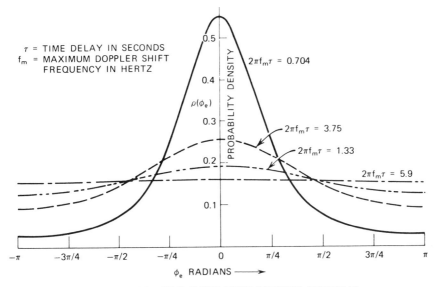

Figure 6.3-14 Probability density of the error ın antenna phasing by using a delayed pilot.

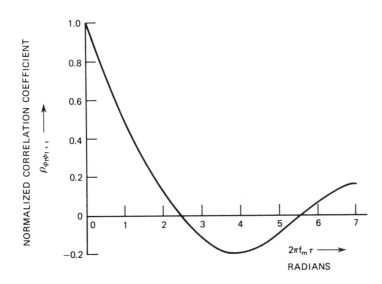

PRODUCT OF MAXIMUM DOPPLER SHIFT AND TIME DELAY

Figure 6.3-15 Normalized correlation of the phase at t and the phase at $t + \tau$.

445

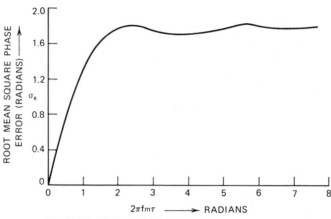

Figure 6.3-16 Root mean square error in relative phasing between antennas by using the delayed signal as a pilot.

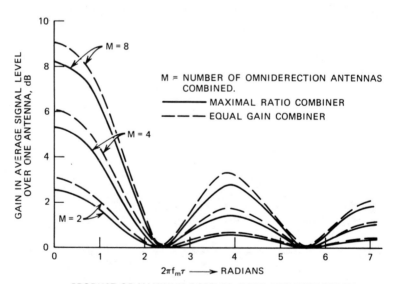

Figure 6.3-17 Average improvement in signal level relative to background noise by combining M antennas.

tion, for the case of the delayed signal pilot, whereas the phase correlation between pilot and signal monotonically decreases with increasing frequency separation for the case of the separate frequency pilot. The negative correlation arises because of the phase reversals that occur roughly every half wavelength. The omnidirectional antenna in a uniformly scattered field has singularities in its power spectrum at $f_c + f_m$ and $f_c - f_m$. These singularities interfere to give a sort of standing-wave pattern that has rapid phase changes of 180° at the nulls of the pattern.

However, the negative correlation of phase between pilot and signal still gives a signal-to-noise improvement by combining because the difference between pilot and signal phase is peaked in probability at $\approx \pi$ radians (see Figure 6.3-13, curve for $2\pi f_m \tau = 3.75$) instead of 0 radians. Thus as long as all branches cluster around the same phase they add coherently, independent of whether that phase is 0° or 180°. (Note that the probability of relative phase between antennas is still slightly peaked near zero degrees in Figure 6.3-14 for $2\pi f_m \tau = 3.75$ and the rms error in relative phase between antennas is at a local minimum in Figure 6.3-16 for $2\pi f_m \tau = 3.75$.)

From Figure 6.3-17 is seen that good combiner improvement can be obtained with fairly long time delays. For example, at 11 GHz and 60 mi/hr, the maximum Doppler shift is 1 kHz $= f_m$. Thus a time delay for the pilot of 0.1 msec results in only about $\frac{1}{2}$ dB reduction in combiner improvement relative to a perfectly correlated pilot.

The use of directional antennas instead of omnidirectional antennas in each branch of the combiner allows larger time delays to be used for the pilot without reducing the combiner improvement. The allowable time delay, before appreciable degradation occurs in combiner signal-to-noise improvement, increases roughly as the inverse square of the antenna beamwidth.*

Random FM

By using the delayed signal as a pilot, the random FM on the pilot is not exactly the same as that on the signal due to the time difference. Thus the longer the delay, the less the pilot will cancel the random FM by the phase subtraction that occurs when the pilot and signal are mixed. In fact, if the delay is large, the pilot can cause the random FM to increase by a factor of 2.

An estimate of the cancellation of random FM by using the delayed signal as a pilot can be obtained from the calculations presented in Section 6.3.2. Since the random FM in a Rayleigh fading channel does not have

---

*M. J. Gans, private communication.

finite rms value (Section 1.5), it is necessary to assume the baseband output circuitry is limited in bandwidth and/or in amplitude, if one wishes to compare the reduction in rms random FM for various time delays of the pilot relative to the signal.

The baseband power spectrum of the random FM for various time delays in the pilot was computed in Eq. (6.3-13). If one assumes a limited bandwidth of $W/10$ to W Hz, for the baseband output, then the mean square frequency excursion indicated by the filter output is obtained by integrating Eq. (6.3-13) from $W/10$ to W Hz. This integration was performed numerically, and the resulting ratio of rms frequency excursion to maximum Doppler frequency is plotted versus baseband bandwidth relative to maximum Doppler frequency, for several values of time delay in Figure 6.3-18. Here it is seen that, in order to reduce the random FM by 20 dB from that obtained with no pilot, the time delay should be less than

$$\tau < \frac{1}{36W}.$$

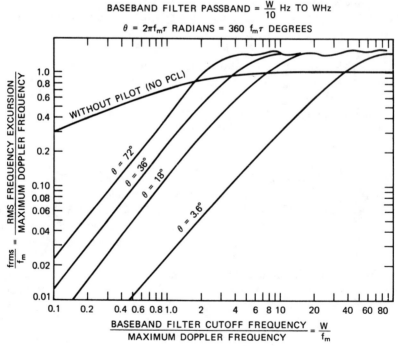

Figure 6.3-18   Random FM through a baseband filter for several pilot delays.

If one assumes the transmitted signal to be frequency modulated with a sine wave of frequency f_a to give a peak frequency excursion of Δf, then the signal-to-random FM ratio at the receiver baseband output is found from Eq. (6.3-13):

$$\frac{S}{R} \equiv \frac{\frac{1}{2}(2\sin\pi f_a\tau)^2\Delta f^2}{f_{\text{rms}}^2}. \qquad (6.3\text{-}27)$$

Receiver Noise

The phase-correction loop, shown in Figure 6.3-12, resembles the frequency feedback demodulator in that the effect of mixing the pilot with the signal in mixer M_2 is to equally reduce both the signal index and the quadrature noise. Thus one might expect threshold extension in FM detection to result from the phase-correction loop similar to that obtained with frequency feedback demodulation.

The incoming signal at point 1 in Figure 6.3-12 is $A\cos\omega_c t$, while the noise may be expressed by

$$n_1(t) = X_c(t)\cos\omega_c t - X_s(t)\sin\omega_c t, \qquad (6.3\text{-}28)$$

where the subscript numeral denotes circuit location in Figure 6.3-12, and where $X_c(t)$ and $X_s(t)$ are the in-phase and quadrature noise components, respectively, and are slowly varying with respect to the carrier frequency ω_c. The power spectrum at the input is shown in Figure 6.3-19(a); note that the carrier frequency is assumed centered in the noise band, assumed rectangular with bandwidth B. In the following analyses, B will refer to this bandwidth at the input to the phase-correction loop. In computing the performance of the phase-correction loop with respect to thermal noise, we will neglect the effects of fading and will assume the signal is not modulated. w_0 is the power spectral density of the input noise spectrum in watts per Hertz, so the input signal-to-noise ratio is $\gamma_1 = A^2/(2w_0B)$. Figure 6.3-19 shows the power spectrum of the input in-phase and quadrature components.

Except for a multiplicative constant, which we shall neglect, the signal and noise at points 2 and 3 are the same as at the input, point 1. The offset oscillator signal, $\cos(\omega_0 t + \eta)$, at point 4 is mixed with the input signal and noise in mixer M_1 and the lower sideband is taken at point 5:

$$v_5(t) = C\{[A + X_c(t)]\cos[(\omega_c - \omega_0)t - \eta]$$

$$- X_s(t)\sin[(\omega_c - \omega_0)t - \eta]\},$$

where C represents the gain of the pilot portion of the loop and η is a constant phase associated with the offset oscillator. After being delayed by the time delay, τ seconds, the desired pilot is produced at point 6:

$$v_6(t) = C\{[A + X_c(t-\tau)]\cos[(\omega_c - \omega_0)(t-\tau) - \eta]$$
$$- X_s(t-\tau)\sin[(\omega_c - \omega_0)(t-\tau) - \eta]\}.$$

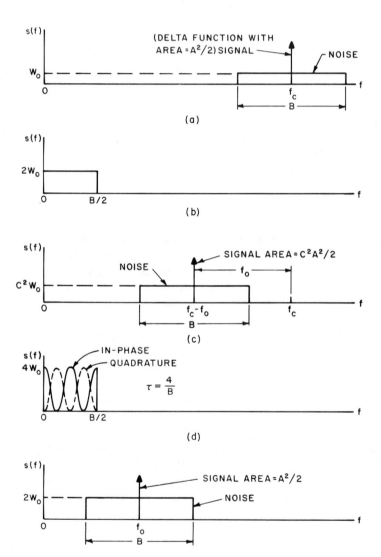

Figure 6.3-19 (a) Input spectrum, (b) Spectrum of input in-phase and quadrature components, (c) Pilot spectrum, (d) Output in-phase and quadrature noise spectra, (e) Output spectrum.

The power spectrum of v_6 is shown in Figure 6.3-19(c). The pilot and input are mixed in mixer M_2 and the difference frequency component is taken at point 7, the output of the phase correction loop, which is to be added with similar signals from the other branches of the combiner. We assume sufficient signal-to-noise ratio so that noise-times-noise components may be neglected:

$$v_7(t) = DCA \{ A \cos (\omega_0 t + v) + Y_c(t) \cos (\omega_0 t + v)$$

$$- Y_s(t) \sin (\omega_0 t + v) \}, \tag{6.3-29}$$

where $v = (\omega_c - \omega_0) \tau + \eta$ is a constant phase shift, D is the gain of the output mixer M_2, and the output in-phase and quadrature noise components are given, in terms of those at the input, by

$$Y_c(t) = X_c(t) + X_c(t - \tau), \tag{6.3-30}$$

$$Y_s(t) = X_s(t) - X_s(t - \tau). \tag{6.3-31}$$

Since it does not enter into signal-to-noise ratios, we will neglect the common multiplicative factor DCA in Eq. (6.3-29).

By the same reasoning that led to Eq. (6.3-27), it is seen that (6.3-30) and (6.3-31) imply that each component of frequency, $\omega/2\pi$, of the input in-phase and quadrature noise power spectra are multiplied by $4 \cos^2 (\omega \tau / 2)$ and $4 \sin^2 (\omega \tau / 2)$, respectively, to obtain the output noise spectra shown in Figure 6.3-19(d) for an example $\tau = 4/B$. Since the input noise components, X_c and X_s, are independent, the output noise components, Y_c and Y_s, are independent. Thus, since $\sin^2 (\omega \tau / 2) + \cos^2 (\omega \tau / 2) = 1$, the noise spectrum of v_7 is again rectangular (but with height $2w_0$) and centered at f_0 as shown in Figure 6.3-19(e).

The signal-to-noise ratio at point 7 is reduced by 3 dB from that at the input; however, for high input signal-to-noise ratios the baseband signal-to-noise ratio is not degraded. The phase noise, ψ, when $A^2/2 \gg w_0 B$ (high input signal-to-noise) at point 1 is

$$\psi_1(t) = \arctan \frac{X_s(t)}{A + X_c(t)}$$

$$\cong \frac{X_s(t)}{A}, \tag{6.3-32}$$

and at point 7 is

$$\psi_7 \cong \frac{Y_s(t)}{A} = \frac{1}{A} [X_s(t) - X_s(t - \tau)]. \tag{6.3-33}$$

Thus the phase noise spectral density at the output differs from that at the input by the multiplicative factor $4\sin^2(\omega\tau/2)$, which is the same factor applied to the baseband signal spectrum due to the phase-correction loop [see Eq. (6.3-27)]. This is the basis of the above statement that both the signal index and quadrature noise are reduced equally by the loop.

To compute the threshold characteristics, Rice's "clicks" technique discussed in Chapter 4 can be applied to Eq. (6.3-29). The expected number of clicks per second, N_c, is given by

$$N_c = \int_{-\infty}^{-A} dY_c \int_{-\infty}^{\infty} d\dot{Y}_s p_{Y_c, Y_s, \dot{Y}_s}(Y_c, 0, \dot{Y}_s)|\dot{Y}_s|, \qquad (6.3\text{-}34)$$

where $\dot{Y}_s$ is the time derivative of Y_s, and $p_{Y_c, Y_s, \dot{Y}_s}(Y_c, 0, \dot{Y}_s)$ is the joint probability density of Y_c, Y_s, and $\dot{Y}_s$ evaluated at $Y_s = 0$. The integration yields (see Chapter 4)

$$N_c = \frac{B}{\sqrt{12}} \left[\frac{\kappa^3 + (6 - 3\kappa^2)\sin\kappa - 6\kappa\cos\kappa}{\kappa^3 - \kappa^2\sin\kappa} \right]^{1/2}$$

$$\times \operatorname{erfc}\sqrt{\frac{\gamma\kappa/2}{\kappa + \sin\kappa}}, \qquad (6.3\text{-}35)$$

where $\kappa \equiv \pi B\tau$ and $\gamma = A^2/2w_0 B$ is the input signal-to-noise ratio. Note as $\kappa \to \infty$, Eq. (6.3-35) approaches Rice's standard click formula[24] for a signal-to-noise ratio 3 dB lower than that at the input:

$$\lim_{\kappa \to \infty} N_c = \frac{B}{\sqrt{12}} \operatorname{erfc}\sqrt{\frac{\gamma}{2}}. \qquad (6.3\text{-}36)$$

This is as expected, since the noise on the pilot becomes uncorrelated with that on the signal for an infinite time delay, and the loop does nothing but reduce the signal-to-noise ratio 3 dB.

At the other extreme of $\tau \to 0$, Eq. (6.3-35) approaches

$$\lim_{\kappa \to 0} N_c = \sqrt{\tfrac{3}{20}}\, B \operatorname{erfc}\sqrt{\frac{\gamma}{4}}. \qquad (6.3\text{-}37)$$

The total baseband noise is the sum of the click noise plus quadrature noise in the baseband bandwidth, which, to allow comparison with Rice, we will assume extends from dc to W Hz. The click noise spectrum is flat and the quadrature noise spectrum is given by Eq. (6.3-33), remembering that the spectrum of the instantaneous frequency is ω^2 times that of instantaneous phase.[12] Thus the baseband noise spectrum is given by

$$W(f) \cong 8\pi^2 N_c + 4^2 A^{-2} \sin^2\left(\frac{\omega\tau}{2}\right) S_{x_s}(f), \tag{6.3-38}$$

where N_c is given by Eq. (6.3-35) and $S_{x_s}(f)$ is the power spectral density of the input quadrature noise, which from Figure 6.3-19(b) is

$$S_{x_s}(f) = \begin{cases} 2w_0 & 0 < f < \dfrac{B}{2} \\ 0 & \text{otherwise} \end{cases} \tag{6.3-39}$$

An important feature of the time-delay phase-correction loop is that its above-described index reduction of $2\sin(\omega\tau/2)$ allows a narrower bandwidth filter to be used at the output. For example, if the signal modulation is assumed to be a Gaussian random variable whose spectrum is flat from 0 to W Hz, then the rms frequency excursion, δ, is reduced by the phase-correction loop by the factor m:

$$m = \frac{\delta_7}{\delta_1} = \sqrt{\frac{4}{W} \int_0^W \sin^2 \frac{\omega\tau}{2} \, df}$$

$$= \sqrt{\frac{2\pi W\tau - \sin 2\pi W\tau}{\pi W\tau}} \ . \tag{6.3-40}$$

We also assume the peak to average power ratio is 10 dB for the audio signal (representative of moderately clipped or compressed speech[25]). Then using Carson's rule, as derived in Chapter 4, for FM bandwidth,

$$B_7 = 2\left(\sqrt{10}\,\delta_7 + W\right) = 2m\sqrt{10}\,\delta_1 + 2W$$

$$= \left[\sqrt{\frac{2\pi W\tau - \sin 2\pi W\tau}{\pi W\tau}}\left(\frac{B}{2W} - 1\right) + 1\right] 2W. \tag{6.3-41}$$

Since the output in-phase and quadrature noise components are linearly related to those at the input [see Eqs. (6.3-30) and (6.3-31)], the effect of further filtering at the output (point 7, Figure 6.3-12) is the same as if the input noise had been restricted by the same filter. Thus the above analysis of click rate still applies if we use B_7 for B.

The total baseband noise N_{BB} is found by integrating Eq. (6.3-38) from 0 to W Hz:

$$N_{BB} = 8\pi^2 N_c W + \frac{16\pi^2 w_0 W^3}{A^2}$$

$$\times \left\{ \frac{1}{3} - \left[1 - \frac{2}{(2\pi W\tau)^2} \right] \frac{\sin 2\pi W\tau}{2\pi W\tau} - \frac{2\cos 2\pi W\tau}{(2\pi W\tau)^2} \right\}. \quad (6.3\text{-}42)$$

If we had demodulated the signal directly at the input, the baseband noise would have been[12]

$$N_{BB} = \frac{4\pi^2}{\sqrt{3}} WB \operatorname{erfc} \sqrt{\gamma_1} + \frac{4\pi^2 W^3}{3\gamma_1 B}. \quad (6.3\text{-}43)$$

Let us compare the threshold before and after the loop. We define threshold as the IF signal-to-noise ratio at which the baseband noise is $\frac{1}{2}$ dB higher than that predicted by quadrature noise alone. The input signal-to-noise ratio at threshold was determined from Eqs. (6.3-42) and (6.3-43) by computer and is plotted in Figure 6.3-20 versus the ratio W/B; that is, baseband cutoff frequency divided by input bandwidth. The parameter $W\tau$ is the product of baseband cutoff frequency and time delay. Also shown on the abscissa is the rms frequency excursion of the input FM signal, δ_1, divided by the highest baseband frequency, W. This ratio, δ_1/W, may be considered as the rms index of the input. Since the index must be greater than zero, W/B must be less than $\frac{1}{2}$. If one assumes that the smallest rms index that is practical to demodulate (due to noise generated in the audio circuit, for example) is $\delta_7/W = 10^{-2}$ radians, then the curves of Figure 6.3-20 are applicable only to the left of the hash marks. Values of $W\tau$ greater than $\frac{1}{2}$ are not considered because of the resultant nonlinear distortion of the signal frequency modulation.

As shown by the curves in Figure 6.3-20, the use of the time-delayed pilot can improve the threshold over that of the case with no phase-correction loop, if W/B is less than about 0.05 and $W\tau$ is less than about 0.1. This threshold reduction is a result of the reduction of index provided by the loop. There is a limit of threshold reduction possible that is not shown by Figure 6.3-20. This limit is due to the mixer M_2 (Figure 6.3-12), which exhibits a "square-law detector" threshold (Ref. 22, page 266) when the input signal-to-noise ratio is less than about -3dB. Thus, although the phase-correction loop can continue to provide improved threshold as the input bandwidth is increased, its absolute threshold is limited to about -3 dB.

In summary, the use of the delayed signal as a pilot does not degrade the baseband signal-to-noise ratio above threshold. Also for high index FM,

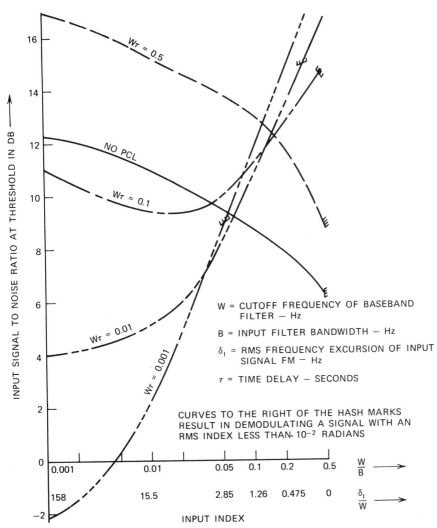

W = CUTOFF FREQUENCY OF BASEBAND
    FILTER — Hz

B = INPUT FILTER BANDWIDTH — Hz

δ_I = RMS FREQUENCY EXCURSION OF INPUT
    SIGNAL FM — Hz

τ = TIME DELAY — SECONDS

CURVES TO THE RIGHT OF THE HASH MARKS
RESULT IN DEMODULATING A SIGNAL WITH AN
RMS INDEX LESS THAN 10^{-2} RADIANS

Figure 6.3-20   Threshold level versus index with time delay as a parameter.

where the input IF bandwidth exceeds about 20 times the maximum baseband frequency, the phase-correction loop can provide threshold improvement over a conventional demodulator, providing the time delay is less than one-tenth the reciprocal of the maximum baseband frequency. Such high index FM transmissions may be preferable for mobile radio in the microwave range.

6.3.3 The Signal Feedback Combiner

Granlund[26] in 1956 proposed a predetection diversity combiner employing a combination of feedback and feedforward. This same technique was apparently discovered independently about 10 years later by Bickford et al.,[27] and a telemetry signal processor operating on the Granlund principle saw commercial use.

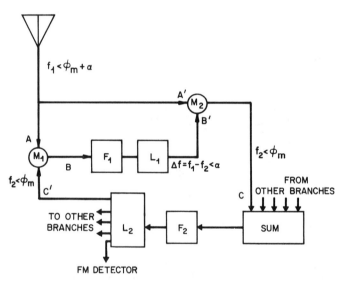

Figure 6.3-21 Block diagram of an equal-gain combiner.

A single branch of the receiver is shown in Figure 6.3-21. Let us assume that the loop is locked, then we can see in a simplified way how the combiner operates. Let the input signal (at points A and A' of Figure 6.3-21) be denoted by

$$a \cos (\omega_1 t + \phi_m + \alpha), \tag{6.3-44}$$

where ω_1 is the carrier frequency, ϕ_m is the audio phase modulation, α is the uniformly distributed random carrier phase, and a is the Rayleigh distributed random carrier amplitude.

Assume that in some mysterious way the feedback signal at point C' contains the audio phase modulation, that is, at C' the local oscillator for mixer M_1 is

$$e_{c'} = b \cos(\omega_2 t + \phi_m), \qquad b = \text{constant}. \qquad (6.3\text{-}45)$$

The result of the mixing process is to strip the input signal of the audio information leaving only the random carrier phase and amplitude information at the output of the F_1 difference-frequency narrow-bandpass filter. (This filter has to be only wide enough to pass the spectrum of the unmodulated Rayleigh fading carrier without introducing excessive time delay.) By passing this signal through a limiting amplifier L_1, the amplitude variations are removed leaving only the random phase information, $(f_1 - f_2) \angle \alpha$. This resultant signal is then mixed in the second mixer, M_2, with the original input signal. Here the input signal is stripped of its random phase leaving only the audio modulation $f_2 \angle \phi_m$. As a result, signals from each channel are at the same reference phase and can be added directly in a power combiner. The combined signal that is produced will have the statistics of a signal from an equal-gain diversity combiner. By passing this combined signal from the difference-frequency filter F_2 into a limiting amplifier L_2, the final output signal, which is also the reference signal for the input, is derived. (The mystery is solved!) This is then fed to a frequency discriminator, where the audio information is detected.

Detailed Analysis (Noise Free) for Two Branches*

We shall show that this signal feedback combiner has *unique* operating frequencies in the loops that depend only on the carrier frequency and the filter time delays [Eq. (6.3-67)]. With reference to the circuit shown in Figure 6.3-22 let the input to each branch be $e_{1a}(t)$ and $e_{1b}(t)$, respectively, where (using complex notation)

$$e_{1a}(t) = A(t)e^{j\phi_{1a}(t)},$$

$$e_{1b}(t) = B(t)e^{j\phi_{1b}(t)}, \qquad (6.3\text{-}46)$$

and

$$\phi_{1a}(t) = \omega_1 t + \phi_m(t) + \alpha(t),$$

$$\phi_{1b}(t) = \omega_1 t + \phi_m(t) + \beta(t). \qquad (6.3\text{-}47)$$

In the above equations, ω_1 is the carrier center frequency, $\phi_m(t)$ is the signal modulation, $A(t)$ and $B(t)$ are normalized independent Rayleigh

*Based on work of S. W. Halpern.[28]

distributed processes, and $\alpha(t)$ and $\beta(t)$ are independent uniformly distributed random phase modulations caused by multipath propagation.

It will be assumed that the mixers take the product of their input signals and the filters pass only the resulting lower sidebands. These filters will be represented as ideal bandpass with constant delay over the band. The limiters are also assumed ideal with constant unity amplitude output and zero phase shift.

With the above simplifications, the equations of the loop can now be written.

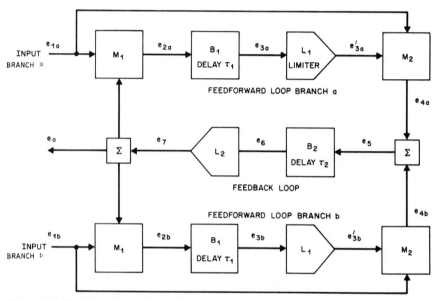

Figure 6.3-22   Two-channel equal-gain predetection diversity combiner.

Let the signal at the output of limiter L_2 be given by

$$e_0(t) = e^{j\phi_0(t)}, \tag{6.3-48}$$

where $\phi_0(t)$ is its phase angle. It is reasonable to assume that $\phi_0(t)$ will consist of a linear term containing the output carrier frequency, $\omega_0 t + \theta$, a modulation term $\phi_m(t - \tau_2)$ that is delayed from the input by the delay in filter F_2, and a general term $\gamma(t)$, which, along with the output carrier frequency, is to be determined.

With $\Delta\omega = \omega_1 - \omega_0$ and $\Delta\phi_m(t) = \phi_m(t) - \phi_m(t - \tau_2)$, the phase angles of the signals at the outputs of the first mixers are

$$\phi_{2a}(t) = \Delta\omega t + \Delta\phi_m(t) - \gamma(t) + \alpha(t) \tag{6.3-49}$$

and

$$\phi_{2b}(t) = \Delta\omega t + \Delta\phi_m(t) - \gamma(t) + \beta(t),$$

and θ was set to zero without loss of generality.

Now we assume that the spectral components of the modulation lie outside the band of the F_1 filters. This is reasonable if $\Delta\phi_m(t)$ is low index and its frequency content is greater than one-half the bandwidth of F_1.

Writing the phase modulation as a sum of Fourier components,

$$\phi_m(t) = \sum_{m=1}^{N} C_m \cos(\omega_m t + \theta_m), \qquad (6.3\text{-}50)$$

then

$$\Delta\phi_m(t) = \sum_{m=1}^{N} C_m[\cos(\omega_m t + \theta_m) - \cos(\omega_m t + \theta_m - \omega_m \tau_2)]. \quad (6.3\text{-}51)$$

With $\omega_m \tau_2 \ll 1$ for all $m = 1, \dots, N$, then

$$\Delta\phi_m(t) \approx - \sum_{m=1}^{N} C_m \omega_m \tau_2 \sin(\omega_m t + \theta_m)$$

$$= \tau_2 \frac{d\phi(t)}{dt}, \qquad (6.3\text{-}52)$$

which one recognizes as a preemphasized modulation signal. The index of the signal at this point is reduced by the weighting factor $\omega_m \tau_2 \ll 1$ for each component, and we have small index modulation as previously assumed.

Since the modulation signal, $\Delta\phi_m(t)$, cannot pass through F_1, one is left with signals at the input to the L_1 limiters with phase angles given by

$$\phi_{3a}(t) = \Delta\omega t - \gamma(t - \tau_1) - \Delta\omega\tau_1 + \alpha(t - \tau_1),$$
$$\phi_{3b}(t) = \Delta\omega t - \gamma(t - \tau_1) - \Delta\omega\tau_1 + \beta(t - \tau_1). \qquad (6.3\text{-}53)$$

The limiters strip the signal of any amplitude variations and thus deliver constant-amplitude signals as local oscillators to the second mixers. Here, these local oscillator signals, with phase angles ϕ_{3a} and ϕ_{3b}, respectively, are mixed with the original incoming signals. Out of these mixers, the lower sideband signals are

$$e_{4a}(t) = A(t)e^{j\phi_{4a}(t)},$$
$$e_{4b}(t) = B(t)e^{j\phi_{4b}(t)}, \qquad (6.3\text{-}54)$$

where

$$\phi_{4a}(t) = \phi_{1a}(t) - \phi_{3a}(t)$$

$$= \omega_0 t + \phi_m(t) + \gamma(t - \tau_1) + \Delta\omega\tau_1 + \alpha(t) - \alpha(t - \tau_1).$$

Similarly,

$$\phi_{4b}(t) = \omega_0 t + \phi_m(t) + \gamma(t - \tau_1) + \Delta\omega\tau_1 + \beta(t) - \beta(t - \tau_1).$$

Let

$$\Psi(t) = \omega_0 t + \phi_m(t) + \gamma(t - \tau_1) + (\omega_1 - \omega_0)\tau_1,$$

$$\Delta\alpha(t) = \alpha(t) - \alpha(t - \tau_1),$$

and

$$\Delta\beta(t) = \beta(t) - \beta(t - \tau_1).$$

Then

$$e_{4a}(t) = A(t)e^{j\Delta\alpha(t)}e^{j\Psi(t)},$$

and

$$e_{4b}(t) = B(t)e^{j\Delta\beta(t)}e^{j\Psi(t)}.$$

These two signals are added together in the power combiner. The resulting signal is

$$e_5(t) = e_{4a}(t) + e_{4b}(t)$$

$$= e^{j\Psi(t)}\left[A(t)e^{j\Delta\alpha(t)} + B(t)e^{j\Delta\beta(t)}\right]. \qquad (6.3\text{-}55)$$

With $\alpha(t)$ and $\beta(t)$ slowly varying with time, and with τ_1 small enough so that $|\Delta\alpha(t)| \ll 1$ and $|\Delta\beta(t)| \ll 1$, then

$$e_5(t) \approx e^{j\Psi(t)}\left\{[A(t) + B(t)] + j[A(t)\Delta\alpha(t) + B(t)\Delta\beta(t)]\right\}$$

$$= R(t)e^{j\xi(t)}e^{j\Psi(t)}, \qquad (6.3\text{-}56)$$

where

$$R(t) = \sqrt{[A(t) + B(t)]^2 + [A(t)\Delta\alpha(t) + B(t)\Delta\beta(t)]^2}$$

$$\approx A(t) + B(t),$$

and

$$\xi(t) = \arctan\left[\frac{A(t)\Delta\alpha(t) + B(t)\Delta\beta(t)}{A(t) + B(t)}\right]$$

$$\approx \frac{A(t)\Delta\alpha(t) + B(t)\Delta\beta(t)}{A(t) + B(t)},$$

since

$$\frac{A|\Delta\alpha|}{A+B} \leqslant |\Delta\alpha| \ll 1 \quad \text{and} \quad \frac{B|\Delta\beta|}{A+B} \leqslant |\Delta\beta| \ll 1.$$

With

$$e_5(t) = R(t)e^{j[\Psi(t)+\xi(t)]}$$

applied to the lower sideband filter F_2, it is delayed by τ_2, and one obtains

$$e_6(t) = e_5(t-\tau_2) = R(t-\tau_2)e^{j[\Psi(t-\tau_2)+\xi(t-\tau_2)]}, \qquad (6.3\text{-}57)$$

which is then limited by L_2. The output of this limiter strips the amplitude variations and leaves

$$e_7(t) = e^{j[\Psi(t-\tau_2)+\xi(t-\tau_2)]}, \qquad (6.3\text{-}58)$$

which must be equal to

$$e_0(t) = e^{j[\Psi_0(t)+2k\pi]}.$$

Therefore,

$$\Psi_0(t) + 2k\pi = \Psi(t-\tau_2) + \xi(t-\tau_2).$$

Substituting in the above equation (6.3-58), one obtains

$$\omega_0 t + \psi_m(t-\tau_2) + \gamma(t) + 2k\pi$$

$$= \omega_0 t + \psi_m(t-\tau_2) + \gamma(t-\tau_1-\tau_2) + \xi(t-\tau_2) - D \qquad (6.3\text{-}59)$$

where

$$D = \omega_0(\tau_1 + \tau_2) - \omega_1\tau_1 = \text{constant}.$$

After reducing the above, one is left with

$$\gamma(t-\tau_1-\tau_2)-\gamma(t)$$

$$= D+2k\pi - \frac{A(t-\tau_2)\Delta\alpha(t-\tau_2)+B(t-\tau_2)\Delta\beta(t-\tau_2)}{A(t-\tau_2)+B(t-\tau_2)}. \qquad (6.3\text{-}60)$$

Filter F_2 is much wider than filter F_1 since it has to pass the incoming FM spectrum; therefore, $\tau_2 \ll \tau_1$. This being the case,

$$A(t-\tau_2)\approx A(t), \qquad B(t-\tau_2)\approx B(t), \qquad (6.3\text{-}61)$$

$$\alpha(t-\tau_2)\approx\alpha(t), \qquad \beta(t-\tau_2)\approx\beta(t), \qquad (6.3\text{-}62)$$

and

$$\Delta\alpha(t-\tau_2)\approx\alpha(t)-\alpha(t-\tau_1)=\Delta\alpha(t),$$

$$\Delta\beta(t-\tau_2)\approx\beta(t)-\beta(t-\tau_1)=\Delta\beta(t),$$

$$\Delta\gamma(t)=\gamma(t-\tau_3)-\gamma(t),$$

where

$$\tau_3=\tau_1+\tau_2\approx\tau_1. \qquad (6.3\text{-}63)$$

One now has the following equation:

$$\Delta\gamma(t)=D+2k\pi - \frac{A(t)\Delta\alpha(t)}{A(t)+B(t)} - \frac{B(t)\Delta\beta(t)}{A(t)+B(t)}. \qquad (6.3\text{-}64)$$

If A, B, α, and β are constant (the case when there is no fading), the above equation becomes

$$\Delta\gamma(t)=D+2k\pi=\text{constant}, \qquad (6.3\text{-}65)$$

since $\Delta\alpha=\Delta\beta=0$. One possible solution is a periodic solution $\gamma(t) = g(t-n\tau_3)$. However, this solution cannot be accepted since it would imply signals with components outside the bandwidths of the filters in the system. Another solution to this equation is $\gamma(t)=\omega t+\theta$. This form of solution, with $\theta\equiv 0$, was part of the assumption for

$$\phi_0(t)=[\omega_0 t+\theta]+\gamma(t)+\phi_m(t-\tau_2),$$

therefore, $\gamma(t)=0$ when $\Delta\alpha=\Delta\beta=0$, which requires $\Delta\gamma(t)=0=D+2k\pi$. Substituting for D,

$$\omega_0(\tau_1+\tau_2)-\omega_1\tau_1=-2k\pi,$$

or

$$\omega_0 = \frac{\omega_1 \tau_1 - 2k\pi}{\tau_1 + \tau_2}, \qquad (6.3\text{-}66)$$

which is the desired outout carrier frequency. The frequency in the feed-forward loop is then

$$\omega_1 - \omega_0 = \Delta\omega = \frac{\omega_1 \tau_2 + 2k\pi}{\tau_1 + \tau_2}.$$

Or,

$$\Delta f = \frac{f_1 \tau_2 + k}{\tau_1 + \tau_2}. \qquad (6.3\text{-}67)$$

There exists an integer, k, such that Δf is within the bandwidth of filter F_1.

The above equation for Δf determines uniquely the operating frequencies in the loops in terms of the fixed quantities: f_1, τ_1, and τ_2. For $\tau_2 \ll \tau_1$ $\Delta f \approx (f_1 \tau_2 + k)/\tau_1$; therefore, a change in incoming carrier frequency of δf_1 would cause a corresponding change in the Δf loop frequency of $\delta f_1 \tau_2 / \tau_1$ and a change in the f_0 loop frequency of

$$\delta f_1 - \frac{\delta f_1 \tau_2}{\tau_1} = \delta f_1 \left(\frac{1 - \tau_2}{\tau_1} \right) \approx \delta f_1.$$

In the above analysis, it was assumed that there was no phase shift in the loops except for that caused by time delay in the filters. It should be mentioned that by adding phase shift, the operating frequency of the loop can be adjusted so that Δf will fall in the center of the F_1 filter band.

The above results were established from the special case $\Delta\alpha = \Delta\beta = 0$. It was then shown that the constant term $D + 2k\pi = 0$. Generally, $\Delta\alpha(t)$ and $\Delta\beta(t)$ are not zero, with the result that

$$\Delta\gamma(t) = -\frac{A(t)\Delta\alpha(t) + B(t)\Delta\beta(t)}{A(t) + B(t)}. \qquad (6.3\text{-}68)$$

It is seen that

$$\Delta\alpha(t) \approx \frac{\tau_1 d\alpha(t)}{dt},$$

$$\Delta\beta(t) \approx \frac{\tau_1 d\beta(t)}{dt},$$

$$\Delta\gamma(t) \approx \frac{-\tau_3 d\gamma(t)}{dt},$$

by taking the first two terms of the Taylor expansion of $\alpha(t) - \alpha(t - \tau_1)$, $\beta(t) - (t - \tau_1)$, and $\gamma(t - \tau_3) - \gamma(t)$. From this, one obtains

$$-\tau_3 \frac{d\gamma(t)}{dt} = \frac{-\tau_1}{A(t) + B(t)} \left[A(t) \frac{d\alpha(t)}{dt} + B(t) \frac{d\beta(t)}{dt} \right]. \quad (6.3\text{-}69)$$

Thus, the output random FM is

$$\frac{d\gamma(t)}{dt} = \frac{\tau_1 / (\tau_1 + \tau_2)}{A(t) + B(t)} \left[A(t) \frac{d\alpha(t)}{dt} + B(t) \frac{d\beta(t)}{dt} \right]$$

$$\approx \frac{A(t)}{A(t) + B(t)} \frac{d\alpha(t)}{dt} + \frac{B(t)}{A(t) + B(t)} \frac{d\beta(t)}{dt}, \quad (6.3\text{-}70)$$

which is a weighted sum of the random FM from each channel.

Halpern[28] derived an upper limit on the rms value of the additional random FM caused by any phase-shift error between the loops within the diversity circuit. The expression is

$$\sqrt{\overline{(d\gamma_\delta(t)/dt)^2}} \leqslant \frac{|\delta|}{\tau_1 + \tau_2} \quad \text{rad/sec}, \quad (6.3\text{-}71)$$

where $|\delta|$ is the magnitude of the phase-shift error between the two loops and $d\gamma_\delta(t)/dt$ is the additional random FM caused by phase-shift error between the loops.

An analysis of the noise performance of this receiver is beyond the scope of this book. An approximate analysis of a maximal-ratio version of this combiner has been published by Tsao et al.,[29] and measured data show that this type of combiner performs according to the theoretical predictions of an ideal N-branch combiner. One can argue intuitively that this should be true as long as the bandwidth of filter F_1 in Figure 6.3-21 is narrow compared to F_2. In this case the difference frequency signal $f_1 - f_2$ is relatively noise free; thus the major noise contribution is the front-end noise accompanying the input signal. Since the signals all add in phase at the summing point and the noises add incoherently, the predetection coherent combining statistics should be realized.

6.4 COHERENT COMBINING USING A SEPARATE PILOT

Another method of coherently combining diversity branches is to have the transmitter send a CW pilot wave along with the modulated signal. The receiver mixer stage beats the signal against the received pilot (instead of against a locally generated tone). Doppler distortion, which affects the signal and pilot in nearly the same way, cancels out during mixing. The diversity system with N antennas adds the outputs of N such mixers providing predetection maximal ratio combining (Section 3) and demodulates the sum by means of an ordinary AM or FM detector.

The receiver obtains a signal-to-noise advantage by adding signal components from the N mixers in phase while adding most interference terms powerwise. To obtain this advantage under multipath propagation conditions, the receiver IF (that is, the difference f between the signal and pilot frequencies) must be chosen small enough. It suffices to make f so small that the propagation times along the different paths all agree to within a small fraction of $1/f$, as will be shown.

We shall present an analysis of this receiver based upon work published by Gilbert.[30]

The effectiveness of this receiver is most clearly seen by examining the signal and noise levels at their outputs. Here the noise in question may be either random noise or an unwanted beat from an interfering station. Several kinds of signal-to-noise ratios can be defined because the signal and noise levels fluctuate as the receiver moves. The ratio of output signal power to output noise power depends on the receiver's position. Here snr is regarded as a random variable and its probability distribution function will be derived. A simpler ratio called SNR, is obtained by dividing the mean output signal power by the mean output noise power. SNR is simply a fixed number but it gives less information about receiver failure than the distribution of snr. In the diversity receiver an interfering station produces three noise signals having different properties. These are called $2PS'$, $2P'S$, $2P'S'$, the letters denoting the components that beat to produce the noise. Thus $2P'S$ is a beat between an interfering pilot and the desired signal.

The transmitter sends a pilot tone $A\cos 2\pi Ft$ along with the modulated signal $AB\cos[2\pi(F+f)t+\theta]$. Here f is an intermediate frequency, small compared with F but large enough so that the signal spectrum does not overlap the pilot. B and θ are an amplitude and a phase, either one of which may be varied slowly to represent the modulating signal. The receiver (see the block diagram, Figure 6.4-1) contains "squarer" elements that square received antenna voltages. Each squared voltage contains a component at frequency f that results from a beat between the pilot and the modulated signal. This component contains the modulation, AM or FM, of the original transmission. The N squares are added and the sum is filtered to remove other components at frequencies far from f. The filtered

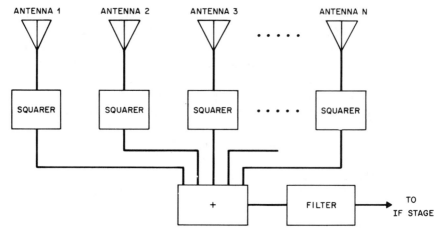

Figure 6.4-1   Diversity receiver.

sum is an IF signal to be demodulated in the usual way.

In effect the transmitted pilot tone replaces the local oscillator tone that a conventional receiver generates internally. The advantage is that any Doppler distortion affects the pilot as well as the modulated signal in the same way. As a result, the circuit of Figure 6.4-1 tends to add IF components in-phase if f is small. This may be seen as follows.

Figure 6.4-2 shows N antennas receiving a signal that arrives from the direction indicated by the arrow. Suppose for the moment that this is the only incident signal (no multipath effects). Now consider two typical antennas, say 1 and 2. Let the difference between the lengths of the paths from 1 and 2 to the transmitter be called s.

If the voltage in antenna 1 is

$$A \cos(2\pi Ft + \phi) + AB \cos[2\pi(F+f)t + \psi], \tag{6.4-1}$$

then the voltage in antenna 2 is

$$A \cos\left[2\pi F\left(t - \frac{s}{c}\right) + \phi\right] + AB \cos\left[2\pi(F+f)\left(t - \frac{s}{c}\right) + \psi\right], \tag{6.4-2}$$

where c is the velocity of light. After squaring, the IF components are $\frac{1}{2}A^2B \cos(2\pi ft + \psi - \phi)$ from antenna 1, and $\frac{1}{2}A^2B \cos(2\pi ft + \psi - \phi - 2\pi fs/c)$ from antenna 2.

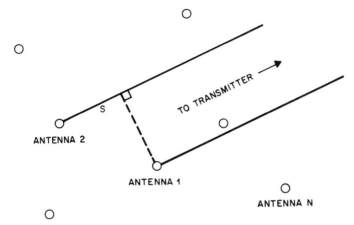

Figure 6.4-2 Reception by N antennas.

These two components differ in phase by $2\pi f s/c$ radians. To keep this angle small, s must be a small fraction of c/f, the wavelength at IF. For instance, if the IF is $f \leqslant 1$ MHz and if no two antennas are more than 10 feet apart, then s is less than 0.01 wavelength and the N beat frequency components are in phase to within $3.6°$.

Under multipath conditions cross beats occur between pilots and modulated signals received via different paths. We shall now derive a more stringent sufficient condition for in-phase addition. The lengths of all major propagation paths from transmitter to receiving antennas must agree within a small fraction of the IF wavelength. For example, if the IF were 100 kHz, the wavelength in question would be 3000 meters. Path differences of hundreds of feet would still permit nearly in-phase addition.

The voltages in antennas 1 and 2 of Figure 6.4-2 are now sums of voltages received over different paths. The kth path contributes terms like (1) and (2) but with parameters A_k, ϕ_k, ψ_k, and s_k that depend on k. Suppose the kth path has length L_k. Then ϕ_k is a sum of phase shifts at reflections plus a propagation term $-2\pi F L_k/c$. Likewise ψ_k is a sum of the same phase shifts at reflections, a propagation term $-2\pi(F+f)L_k/c$, and the modulation angle θ. Then

$$\psi_k = \phi_k + \theta - \frac{2\pi f L_k}{c}.$$

At antenna 2 the kth pilot is

$$P_k = A_k \cos\left(2\pi F t + \phi_k - \frac{2\pi F s_k}{c}\right)$$

and the kth modulated signal is

$$S_k = A_k B \cos \left[2\pi(F+f)t + \theta + \phi_k - \frac{2\pi f L_k}{c} - \frac{2\pi(F+f)s_k}{c} \right].$$

At antenna 1 the kth path produces voltages of the same form but with $s_k = 0$.

When the antenna 2 voltage is squared, cross beats between the ith and kth paths occur. The IF part of $P_k S_i$ is

$$P_k S_i: \quad \tfrac{1}{2} A_k A_i B \cos \left[2\pi f t + \theta + \phi_i - \phi_k \right.$$

$$\left. - \frac{2\pi f L_i}{c} - \frac{2\pi(F+f)s_i}{c} + \frac{2\pi F s_k}{c} \right].$$

There is also a $P_i S_k$ beat, and the sum of the two beats contains the IF component

$$P_k S_i + P_i S_k: \quad A_k A_i B \cos \left[2\pi f t + \theta - \frac{\pi f(L_k + L_i + s_k + s_i)}{c} \right]$$

$$\times \cos \left[\phi_i - \phi_k - \frac{\pi f(L_i - L_k + s_i - s_k)}{c} - \frac{2\pi F(s_i - s_k)}{c} \right].$$

The same expression gives the IF component of $P_k S_i + P_i S_k$ at antenna 1 when s_i and s_k are replaced by zero. In this expression the first cosine contains the time dependence while the second cosine is purely an amplitude factor.

Now suppose, that $s_1, s_2, \ldots,$ are all so small that the terms $\pi f s_k / c$ are small angles. Then the first cosine in the $P_k S_i + P_i S_k$ contribution is nearly the same at antenna 2 as it is at antenna 1. However the second cosine contains the large angle $2\pi F(s_i - s_k)/c$ at antenna 2 only. Indeed one can construct numerical examples to show that further assumptions are needed to make the total IF outputs of the two squarers be inphase. It will suffice to assume that the path lengths $L_1, L_2, \ldots,$ are nearly equal, differing from one another by only a small fraction of c/f. Under this extra condition, the first cosine factor is approximately $\cos(2\pi f t + \theta - 2\pi L_1/c)$ for all k, i

and at both antennas. For a given k, i the second cosine factor can still have opposite signs at the two antennas. However, when all beats are combined, the amplitude at antenna 2 is approximately

$$\frac{1}{2} \sum_{k,i} A_k A_i B \cos\left[\phi_i - \phi_k - \frac{2\pi F(s_i - s_k)}{c}\right]$$

$$= \frac{1}{2} B \operatorname{Re}\left[\sum_{k,i} A_k A_i \exp j\left[\phi_i - \phi_k - \frac{2\pi F(s_i - s_k)}{c}\right]\right]$$

$$= \frac{1}{2} B \operatorname{Re}\left[\sum_{i} A_i \exp j\left[\phi_i - \frac{2\pi F s_i}{c}\right]\right]^2,$$

which is positive. The same argument with $s_i = 0$ gives a positive amplitude at antenna 1; the two sums are in phase.

6.4.1 Noise Performance

This section considers the effect of random noise on diversity reception and gives expressions for output noise spectra. Multipath fading effects make the output signal-to-noise ratio depend on the position of the receiver. A single mathematically convenient figure of merit is the ratio of expected signal power to expected noise power. This ratio is called SNR here.

When making SNR comparisons one must also recognize qualitative differences between the output noises from different receivers. The conventional receiver has a steady noise output resulting from input noise beating against the steady local oscillator signal. In the diversity system the output noise results largely from input noise beating against fluctuating pilot and modulated signals. During fades the output noise from the diversity receiver also fades while the noise from the conventional receiver does not. Thus, the diversity receiver has acceptable SNR more often than a conventional receiver with the same output SNR.

The mathematical treatment will begin with the case $N = 1$; the extension to more antennas will be easy. The input to the squarer is the sum of three voltages:

Pilot $\qquad P(t) = A \cos(2\pi F t + \phi),$ $\qquad$ (6.4-3)

Signal $\qquad S(t) = AB \cos[2\pi(F+f)t + \psi],$ $\qquad$ (6.4-4)

Noise $\qquad n(t) = \sum n_i \cos(2\pi f t + \xi_i),$ $\qquad$ (6.4-5) where the multipath notation of Section 6.4 has been omitted. Here the noise is represented, as by Rice,[31] as a sum of sinusoids with random phases ξ_i and amplitudes n_i. Rice studied the effect of squaring a random noise; this section adapts his work to the present problem.

The received pilot power is $\frac{1}{2}A^2$ (into a 1 ohm load); likewise the signal power $\frac{1}{2}A^2B^2$. The noise has a one-sided power spectrum function $w(\nu)$ such that

$$w(\nu)\Delta\nu = \frac{1}{2} \sum_{\nu < f_i < \nu + \Delta\nu} n_i^2$$

represents the noise power in the frequency band from ν to $\nu + \Delta\nu$. The shape of the function $w(\nu)$ is determined by the tuned circuits (not shown in Figure 6.4-1) that filter the antenna signal before squaring. Figure 6.4-3 shows a typical case

$$w(\nu) = \begin{cases} N_0, & F - b \leqslant \nu \leqslant F + f + a \\ 0, & \text{otherwise} \end{cases}, \tag{6.4-6}$$

which uses a bandpass filter slightly wider than necessary to pass the pilot and signal at frequencies F and $F + f$.

Squaring $P + S + n$ produces six terms: P^2, S^2, n^2, $2PS$, $2Pn8$ $2Sn$. P^2 and S^2 contribute nothing to the output after the output filter removes components remote from frequency f. The other contributions are

$$A^2B\cos(2\pi ft + \psi - \phi) \qquad \text{from } 2PS, \tag{6.4-7}$$

$$A\sum_i n_i \cos[2\pi(f_i - F)t + \xi_i - \phi] \qquad \text{from } 2Pn, \tag{6.4-8}$$

$$AB\sum_i n_i \cos[2(f_i - F - f)t + \xi_i - \psi] \qquad \text{from } 2Sn, \tag{6.4-9}$$

$$\sum_{i<j} n_i n_j \cos[2\pi(f_i - f_j)t + \xi_i - \xi_j] \qquad \text{from } n^2. \tag{6.4-10}$$

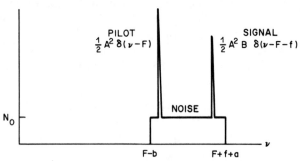

Figure 6.4-3 Power spectra at the input to a squarer.

The $2PS$ contribution is the desired output; its power is $\frac{1}{2}A^4B^2$. The spectra of the other contributions appear in Figure 6.4-4. The spectral density functions are

$$A^2w(\nu + F) \qquad \text{from } 2Pn, \qquad (6.4\text{-}11)$$

$$A^2B^2w(F+f+\nu) \qquad \text{from } 2Sn, \qquad (6.4\text{-}12)$$

$$2\int_0^\infty w(x)w(\nu + x)\,dx \quad \text{from } n^2. \qquad (6.4\text{-}13)$$

Functions (6.4-11), (6.4-12), and (6.4-13) assign some power to negative values of ν; these are to be aliased to positive frequencies. This aliasing accounts for the peculiar discontinuities in the spectra at low frequencies. The dotted lines in Figure 6.4-4 show functions (6.4-11), (6.4-12), and (6.4-13) before aliasing. The values of a and b will be assumed smaller than f so that, as in Figure 6.4-4, the noise power densities at frequency f are A^2N_0 for Pn noise and $A^2B^2N_0$ for Sn noise.

In the case of Gaussian noise, the phases ξ_i in functions (6.4-8), (6.4-9), and (6.4-10) are independent. It then follows that the three kinds of output

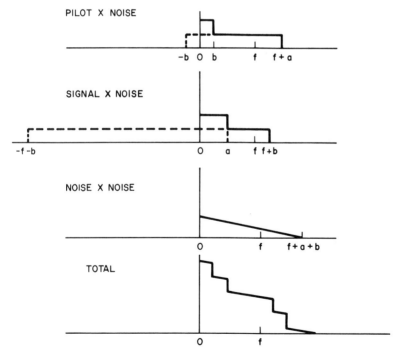

Figure 6.4-4 Output noise spectra.

noise components at a given frequency ν are uncorrelated. Then these noises add powerwise and the total noise spectral density is the sum of functions (6.4-11), (6.4-12), and (6.4-13).

In a diversity system the same kind of analysis applies for each of N antennas. The amplitudes and phases would now be written as A_k, n_{ik}, ψ_k, ϕ_k, and ξ_{ik}, where the subscript k $(k=1,\ldots,N)$ specifies the antenna. All these random variables are independent of one another except for ψ_k and ϕ_k, which satisfy $\psi_1-\phi_1=\psi_2-\phi_2=\cdots=\psi_N-\phi_N=\theta$ because the $2PS$ terms have a common phase angle θ. Thus the N signal components add voltagewise and the expected signal power at the output is*

$$\tfrac{1}{2}B^2E(\Sigma A_k^2)^2=\tfrac{1}{2}B^2E\left\{NE(A^4)+N(N-1)[E(A^2)]^2\right\}.$$

Let k_0 denote the (dimensionless) ratio

$$k_0=\frac{E(A^4)}{[E(A^2)]^2}. \tag{6.4-14}$$

For Rayleigh fading, $k_0=2$. For no fading, $k_0=1$. The expected output signal power is

$$E(\text{signal power out})=\tfrac{1}{2}N(N+k_0-1)[E(A^2)]^2B^2$$

$$=\tfrac{1}{2}N(N+1)[E(A^2)]^2B^2 \quad\text{(Rayleigh)}. \tag{6.4-15}$$

According to Equations (6.4-3) and (6.4-4), $\tfrac{1}{2}E(A^2)$ and $\tfrac{1}{2}B^2E(A^2)$ are the expected received powers of the pilot and signal. With fixed total transmitter power equal to $\tfrac{1}{2}(1+B^2)E(A^2)=P_0$, the output signal, Eq. (6.4-15), is maximized when the transmitted power is divided equally between pilot and signal $[B=1, E(A^2)=P_0]$. Then Eq. (6.4-15) becomes

$$E(\text{signal power out})=\tfrac{1}{2}N(N+k_0-1)P_0^2.$$

The noise terms (6.4-11), (6.4-12), and (6.4-13) for the N antennas add powerwise, and the expected output noise power spectrum is a sum of three terms

$$NE(A^2)w(\nu+F) \quad\text{from }2Pn, \tag{6.4-16}$$

$$NB^2E(A^2)w(F+f+\nu) \quad\text{from }2Sn, \tag{6.4-17}$$

*E stands for expected value.

and

$$2N \int_0^\infty w(x)w(\nu + x)dx \qquad \text{from } n^2. \qquad (6.4\text{-}18)$$

For a typical case, suppose $w(\nu)$ is the function (6.4-6) with $a < f$ and $b < f$. Suppose also that the output filter in Figure 6.4-1 has a narrow rectangular transfer function with bandwidth Δf about frequency f. Then the expected output noise power is

$$E(\text{noise power out}) = 2NN_0[P_0 + N_0(a + b)]\Delta f \qquad (6.4\text{-}19)$$

where again $P_0 = \frac{1}{2}(1 + B^2)E(A^2)$ is the total expected power that an antenna receives from pilot and signal. In this case the output noise power does not depend on B, and the choice $B = 1$ maximizes not only the output signal power but the output signal-to-noise ratio as well. With $B = 1$, Eqs. (6.4-15) and (6.4-19) combine to give

$$\text{SNR} = \frac{N + k_0 - 1}{4} \frac{P_0/(N_0\Delta f)}{1 + (a + b)N_0/P_0}, \qquad (6.4\text{-}20)$$

where k_0 is given by Eq. (6.4-14) ($k_0 = 2$ for Rayleigh fading).

If the input noise spectrum is not flat, the output noise contributions (6.4-16) and (6.4-17) do not combine into the term $2NN_0P_0\Delta f$ that appears in Eq. (6.4-19). In that case the value of B that gives the best output SNR may not be unity but will depend on the input noise power densities at F and $F + f$.

Equations (6.4-11), (6.4-12), and (6.4-13) also apply, with slight reinterpretations, to the conventional receiver without diversity. $\frac{1}{2}A^2$ is the power of a local oscillator and $\frac{1}{2}A^2B^2$ is the received signal power. Then A has a well-determined value, but B is a random variable having perhaps a Rayleigh distribution. Now P_0 is $E(\frac{1}{2}A^2B^2)$. The desired output signal has amplitude A^2B and so has expected power $E(\text{signal power out}) = \frac{1}{2}A^4E(B^2) = A^2P_0$. The local oscillator is deliberately made much stronger than the incoming signal or noise; then the output noise components (6.4-12) and (6.4-13) are negligible compared with Eq. (6.4-11). For the output filter of bandwidth Δf, $E(\text{noise power out}) = A^2w(F + f)\Delta f$. When $w(\nu)$ is the function (6.4-6) again,

$$\text{SNR} = \frac{P_0}{N_0\Delta f}. \qquad (6.4\text{-}21)$$

The output signal-to-noise ratios in Eqs. (6.4-20) and (6.4-21) differ by a factor

$$\frac{(\text{SNR})_{\text{conventional}}}{(\text{SNR})_{\text{diversity}}} = \frac{4[1+(a+b)N_0/P_0]}{N+k_0-1}. \tag{6.4-22}$$

The term $(a+b)N_0/P_0$ represents that part of the input noise-to-signal ratio that results from noise arriving outside the band $F \leqslant \nu \leqslant F+f$. Then this term will be small in any useful case. The remaining factor $4/(N+k_0-1)$ gives the conventional system the advantage unless $N \geqslant 5-k_0$. When Rayleigh fading holds, a three-antenna diversity system has the same output SNR as the conventional system. This result may appear somewhat misleading because it considers only the mean signal and noise powers. More useful results in terms of probability distributions are discussed next.

The signal and noise levels of conventional and diversity receivers fluctuate differently as the receiver moves. In the case of Rayleigh fading one can obtain the probability distribution functions for snr=(signal power out/noise power out) for the two receivers. Again take the simple input noise spectrum of Eq. (6.4-6) with small values of a and b.

Expressions (6.4-7), (6.4-11), and (6.4-12) show that

$$\text{snr} = (4N_0\Delta f)^{-1}\Sigma A_k^2$$

for the diversity receiver $(B=1)$. Each Rayleigh amplitude A_k may be expressed in terms of independent Gaussian variables x_k, y_k of mean zero and unit variance by means of $A_k^2 = \frac{1}{2}P_0(x_k^2+y_k^2)$.

Then

$$\text{snr} = \left(\frac{X_{2N}^2}{8}\right)\left(\frac{P_0}{N_0\Delta f}\right),$$

where $X_{2N}^2 = X_1^2 + \cdots + X_N^2 + Y_1^2 + \cdots + Y_N^2$ has the chi-squared probability distribution with $2N$ degrees of freedom. The same result would be obtained by interpreting the receiver as a maximal-ratio combiner[32] (see Section 6.8).

Expressions (6.4-7), (6.4-11), and (6.4-12) also apply to the conventional receiver if, as explained above, A is a fixed number while B is a small Rayleigh variable. Only $2Pn$ noise need be considered; then

$$\text{snr} = (2N_0\Delta f)^{-1}A^2B^2 = \frac{1}{2}X_2^2\left(\frac{P_0}{N_0\Delta f}\right),$$

where again X_2^2 has the chi-squared distribution, now with two degrees of freedom.

Suppose the system fails when snr is below some known critical value. Suppose such failure can be tolerated only a small fraction Q of the time. The given value of Q is reached at some X^2 value which can be read from probability tables. To achieve the desired small failure probability the ratio $P_0/(N_0\Delta f)$ (a kind of input SNR) must be

$$\frac{P_0}{(N_0\Delta f)} = \begin{cases} \left(\dfrac{8}{X_{2N}^2}\right)\text{snr} & \text{(diversity)} \\[2ex] \dfrac{1}{4}\left(\dfrac{8}{X_2^2}\right)\text{snr} & \text{(conventional)} \end{cases}$$

Table 1 gives $10\log(8/X_{2N}^2)$ as a function of Q. Thus for a 0.01 probability of failure, $P_0/(N_0\Delta f)$ must exceed the critical snr by 26.0, 14.3, 6.87, and 1.39 dB for diversity systems of one, two, four, and eight antennas, respectively. The conventional receiver requires $26.0-6.0=20.0$ dB and so is intermediate between diversity systems with $N=1$ and 2.

Table 1   *Values of $10\log(8/X_{2N}^2)$ for which the Probability of Failure Equals Q*

Number of Antennas	Q					
	0.001	0.005	0.01	0.025	0.05	0.1
1	36.0	29.0	26.0	22.0	18.9	15.8
2	19.5	15.9	14.3	12.1	10.5	8.78
3	13.3	10.7	9.62	8.01	6.89	5.60
4	9.74	7.75	6.57	5.65	4.67	3.60
5	7.36	5.70	4.95	3.92	3.08	2.16
6	5.61	4.15	3.50	2.59	1.85	1.03
7	4.23	2.96	2.35	1.53	0.86	0.12
8	3.15	1.92	1.39	0.64	0.02	-0.63

6.4.2   Cochannel Interference Performance

Suppose a diversity system tries to receive a desired signal while another station uses the same channel. The pilots and modulated signals of the two stations produce a variety of beat components, three of which cause interference at IF [functions (6.4-23), (6.4-24), and (6.4-25) below]. Two components sound like Doppler-distorted versions of the modulated signals from the desired station and its competitor. The third component is an undistorted copy of the modulated signal from the competing station.

Under multipath conditions, the two Doppler-distorted beats have phases that are uncorrelated from antenna to antenna. The output SNRs for these noises grow linearly with the number N of antennas, Eq. (6.4-26). The third components from the separate squarers add in phase. Then the SNR for this interference is not reduced by increasing N, Eq. (6.4-27). However, increasing N reduces the variability of the power levels of the output signal and noise. Thus, if the desired station is a few decibels stronger than the competing station, increasing N reduces the chance that multipath fading will allow the competing station to override the desired station (Table 2.)

Table 2   *Values of* $10\log F$ *Such That the Probability of Failure Equals Q (2P'S'Noise)*

Number of Antennas	Q							
	0.001	0.005	0.01	0.025	0.05	0.1	0.25	0.50
1	30.0	23.0	20.0	15.9	12.8	9.54	4.77	0
2	17.3	13.6	12.0	9.82	8.05	6.14	3.14	0
3	13.0	10.4	9.27	7.64	6.31	4.84	2.50	0
4	10.8	8.74	7.80	6.46	5.36	4.13	2.14	0
5	9.56	7.70	6.88	5.71	4.79	3.67	1.90	0
6	8.35	6.92	6.19	5.16	4.29	3.32	1.73	0
7	7.73	6.33	5.70	4.75	3.94	3.04	1.58	0
8	7.17	5.86	5.29	4.42	3.69	2.88	1.49	0

A single antenna again receives a pilot, Eq. (6.4-3), modulated signal, Eq. (6.4-4), and a noise that is a special case of Eq. (6.4-5). The noise now has only two components. One is a pilot $P'(t)$ of frequency F, phase ϕ', and amplitude A'. The other is a modulated signal $S'(t)$ of frequency $F+f$, phase ψ', and amplitude $A'B'$.

Squaring produces IF components that are obtainable from functions (6.4-7), (6.4-8), (6.4-9), and (6.4-10). The desired signal component is function (6.4-7) again. The $2Pn$ component, function (6.4-8), has two parts, one of which $[P(t)$ beating against $P'(t)]$ contributes nothing. The remaining IF contribution from function (6.4-8) is

$$AA'B'\cos(2\pi ft + \psi' - \phi) \qquad \text{from } 2PS'. \qquad (6.4\text{-}23)$$

Likewise functions (6.4-9) and (6.4-10) contribute only

$$AA'B\cos(2\pi ft + \psi - \phi') \qquad \text{from } 2SP' \qquad (6.4\text{-}24)$$

and

$$A'^2B'\cos(2\pi ft + \psi' - \phi') \qquad \text{from } 2P'S'; \qquad (6.4\text{-}25)$$

the $2SS'$ and S'^2 terms do not contribute at IF.

The three interference terms (6.4-23), (6.4-24), and (6.4-25) have different characteristics. The $2PS'$ and $2P'S'$ components carry the modulation (AM or FM) of $S'(t)$ and act like interfering stations at IF. Likewise the $2SP'$ term sounds like a station with the desired modulation of $S(t)$. As the receiver moves, the two angles, ψ' and ϕ, undergo different Doppler shifts. Then the $2PS'$ component contains a residual Doppler distortion. Likewise the $2SP'$ component is Doppler distorted and so will be considered a noise. By contrast, as in the $2PS$ term, the Doppler shifts in the $2P'S'$ term cancel out leaving an undistorted interfering signal.

Because the $2P'S$ component has both the desired modulation and Doppler distortion, it is not clear whether it should be treated as a signal term or as a noise term. If it were counted as part of the signal, the $2P'S$ term would be a source of fluctuation of the output signal level (it differs in phase from the $2PS$ term by a random amount). To call the $2P'S$ term a kind of noise is probably overconservative if the system uses FM of index high enough to make the Doppler distortion unimportant. It turns out that the power levels of the $2PS'$ and $2P'S$ terms have the same probability distribution. Thus, whenever other interference terms are small, it does not matter much whether $2P'S$ components are treated as signal or as noise.

One can compute an SNR, defined as $E(\text{signal power out})/E(\text{noise power out})$, for each of the three interferences. Again multipath fading conditions will be assumed so that pilot amplitudes and phases from the N antennas are independent variables. The conditions

$$\psi_1 - \phi_1 = \psi_2 - \phi_2 = \cdots = \psi_N - \phi_N = \theta$$

and

$$\psi_1' - \phi_1' = \psi_2' - \phi_2' = \cdots = \psi_N' - \phi_N' = \theta'$$

relate the signal phases to the pilot phases.

The expected signal output power is given by Eq. (6.4-15) as before. The expected power from the N terms of type $2PS'$ is

$$E\left(\sum \tfrac{1}{2}A_k^2 A_k'^2 B'^2\right) = \tfrac{1}{2}NB'^2 E(A^2)E(A'^2).$$

Likewise the $2SP'$ power has expected value

$$\tfrac{1}{2}NB^2E(A^2)E(A'^2).$$

The SNRs are

$$\text{SNR} = (N + k_0 - 1)\left(\frac{B}{B'}\right)^2 \frac{E(A^2)}{E(A'^2)}$$

for $2SP'$ interference and

$$\text{SNR} = (N + k_0 - 1)\frac{E(A^2)}{E(A'^2)}$$

for $2SP'$ interference [recall the definition of k_0 given by Eq. (6.4-14)]. When $B = B' = 1$ and the expected received power from the two stations are P_0 and P_0', both interferences have

$$\text{SNR} = \frac{(N + k_0 - 1)P_0}{P_0'}. \tag{6.4-26}$$

The expected power of $2P'S'$ interference is given by Eq. (6.4-15) with A' and B' replacing A and B. Then, if $B' = B$, the SNR for $2P'S'$ is

$$\text{SNR} = \left(\frac{P_0}{P'_0}\right)^2. \tag{6.4-27}$$

The two expressions (6.4-26) and (6.4-27) have interesting differences. They depend on N in different ways because the $2SP'$ and $2S'P$ components from separate antennas add with random phases, while the $2S'P'$ components add in phase. The input signal-to-noise ratio P_0/P_0' appears with different exponents in Eqs. (6.4-26) and (6.4-27) because Eq. (6.4-26) relates to beats between the desired station and the interfering one, while Eq. (6.4-27) relates to beats of the interfering station itself.

Because of these differences, either kind of output noise can be the more serious one, depending on the situation. For a given number of antennas, the $2S'P$ and $2SP'$ noises are stronger than the $2S'P'$ noise when P_0/P_0' is large. As P_0/P_0' becomes smaller, all noises increase and, at $P_0/P_0' = N + k_0 - 1$, they have equal powers. When P_0/P_0' is still smaller, the $2S'P'$ noise (undistorted copy of the interfering signal) predominates. With Rayleigh fading and $N = 4$ antennas, the $2S'P'$ noise predominates at input signal-to-noise ratios of 7 dB or less.

In conventional systems, an interfering station produces only one output noise component. It has

$$\text{SNR} = \frac{P_0}{P_0'} . \tag{6.4-28}$$

None of the noise components of the diversity system is as bad as this unless the interfering station is stronger than the desired one.

Equation (6.4-27) shows that adding more antennas does not improve the SNR for $2P'S'$ noise. However, diversity helps by reducing the chance that a large fluctuation of the interfering signal level will cause the system to fail. To study this effect let $A_1, \dots, A_N$ be signal amplitudes, as in expressions (6.4-7) and (6.4-8), received by the N antennas. Likewise, let these antennas receive $A_1', \dots, A_N'$ from the interfering station. Under severe multipath conditions these $2N$ amplitudes may be regarded as independent random variables. Again take $B = B' = 1$ so that $E(A_k^2) = P_0$, $E(A_k'^2) = P_0'$. The desired and interfering stations produce output signals with amplitudes ΣA_k^2 and $\Sigma A_k'^2$. Then

$$\text{snr} = \left(\frac{\Sigma A_k^2}{\Sigma A_k'^2} \right)^2 \tag{6.4-29}$$

is the random variable that must be studied.

The probability distribution function for snr can be obtained easily in the case of Rayleigh fading. Each Rayleigh amplitude A may be represented by the formula $A^2 = X^2 + Y^2$, where X and Y are independent Gaussian variables with variance $E(X^2) = E(Y^2) = \frac{1}{2}P_0$.

In these terms, the quantity

$$F = \frac{(X_1'^2 + Y_1'^2 + X_2'^2 + \cdots + Y_N'^2)/(\frac{1}{2}P_0')}{(X_1^2 + Y_1^2 + X_2^2 \cdots + Y_N^2)/(\frac{1}{2}P_0)} \tag{6.4-30}$$

$$F = \left(\frac{P_0}{P_0'} \right) \text{snr}^{-1/2}$$

is the ratio of two sums of $2N$ independent squares of Gaussian variables of unit variance. Statisticians use such ratios frequently and have tabulated their probability distributions. Abramowitz and Stegun give such a table.[33] In their notation the cumulative probability function for F is $P(F|2N, 2N)$, a special case of their $P(F|\nu_1, \nu_2)$. Their Table 26.9 gives $Q(F|\nu_1, \nu_2)$

$= 1 - P(F | \nu_1, \nu_2)$, so that snr has the distribution function

$$\text{Prob}\left\{ \text{snr} \leqslant \left(\frac{P_0}{P_0'}\right)^2 F^{-2} \right\} = Q(F | 2N, 2N). \qquad (6.4\text{-}31)$$

Table 2 reproduces part of Abramowitz and Stegun's table after convert-ing F values to decibels. The numbers tabulated are values $10 \log_{10} F$ that are needed to make the probability of Eq. (6.4-31) a small value $Q = 0.001$, 0.005, 0.01, 0.025, 0.05, 0.1, 0.25, or 0.5. To use Table 2 one must first know how small snr can become before the systems will fail; one also decides on an acceptable probability Q of failure. The table gives a corresponding value of F and the conditions for not failing are met as long as the input signal to noise ratio P_0 / P_0' satisfies

$$F \text{snr}^{1/2} \leqslant \frac{P_0}{P_0'} \qquad (6.4\text{-}32)$$

For example, suppose the system fails if snr becomes as small as 3 dB. Suppose failure can be tolerated only 1% of the time. The tabulated values of F for $Q = 0.01$ and $N = 1, 4, 8$ are 20.0, 7.80, and 5.29 dB. Then inequality (6.4-32) requires the input signal-to-noise ratio to be

$$20.0 + 1.50 = 21.5 \text{ dB} \qquad \text{for one antenna,}$$

$$7.80 + 1.50 = 9.30 \text{ dB} \qquad \text{for four antennas,}$$

$$5.29 + 1.50 = 6.79 \text{ dB} \qquad \text{for eight antennas.}$$

In the case of one and four antennas at these signal levels, Eqs. (6.4-26) and (6.4-27) show that the other noise components $2SP'$ and $2PS'$ are stronger than the $2P'S'$ components. Thus the snr for $2SP'$ and $2PS'$ noises must be considered later.

To show the advantage of diversity over a conventional system, one may examine the probability distribution function for the conventional snr. This function is not just Eq. (6.4-31) with $N = 1$. A conventional system has

$$\text{snr} = \frac{(AB)^2}{(A'B')^2} = \frac{X^2 + Y^2}{X'^2 + Y'^2}$$

instead of Eq. (6.4-29). To get a ratio of sums of squares of Gaussian variables with unit variance, one must now define $F = (P_0 / P_0') / \text{snr}$ instead of Eq. (6.4-30). The value of F for a given failure probability Q is again obtained from Table 2 with $N = 1$. The input signal-to-noise ratio P_0/P_0' must then satisfy

$$Fsnr \leqslant \frac{P_0}{P_0'} \qquad (6.4\text{-}33)$$

instead of inequality (32). To have snr as low as 3 dB for only a fraction $Q = 0.01$ of the time, the input signal-to-noise ratio SNR must now be 23 dB or more.

The SNR calculation showed that $2SP'$ and $2PS'$ components are apt to be the strongest noises when N is small. The distribution functions for their snr may also be derived. Again Rayleigh fading is assumed and $B' = B = 1$. The latter assumption makes the $2PS'$ and $2SP'$ components have the same snr distribution [compare expressions (6.4-23) and (6.4-24)].

It is convenient to rewrite the $2PS'$ component (6.4-23) in terms of cosine and sine amplitudes:

$$X' = A' \cos(\psi' - \phi), \qquad Y' = -A' \sin(\psi' - \phi). \qquad (6.4\text{-}34)$$

Then expression (6.4-23) becomes $AX' \cos 2\pi ft + AY' \sin 2\pi ft$. Now X' and Y' are independent Gaussian random variables with mean zero and variance $\frac{1}{2} P_0'$. When there are N antennas, Eqs. (6.4-34) give amplitudes X_k' and Y_k' for the kth antenna. The kind of argument that produced Eqs. (6.4-28) and (6.4-29) now leads to

$$snr = \frac{\left(\Sigma A_k^2\right)^2}{\left(\Sigma A_k X_k'\right)^2 + \left(\Sigma A_k Y_k'\right)^2} \qquad (6.4\text{-}35)$$

It is possible to transform Eq. (6.4-35) into a form to which an F distribution again applies. As a first step, introduce two new random variables

$$x' = \frac{\Sigma A_k X_k'}{\left(\frac{1}{2} P_0' \Sigma A_i^2\right)^{1/2}}, \qquad y' = \frac{\Sigma A_k Y_k'}{\left(\frac{1}{2} P_0' \Sigma A_i^2\right)^{1/2}}.$$

For any $A_1, \ldots, A_N$, x' and y' are independent Gaussian variables of mean zero and variance 1. Now Eq. (6.4-35) becomes

$$snr = \frac{2\Sigma A_k^2}{P_0'(x'^2 + y'^2)}. \qquad (6.4\text{-}36)$$

Next one can express the pilot $P(t)$ in terms of cosine and sine amplitudes. In this way one obtains $A_k^2 = \frac{1}{2} P_0(x_k^2 + y_k^2)$, where x_k and y_k are independent Gaussian random variables of mean zero and variance 1. Finally Eq. (6.4-36) becomes

$$snr = (P_0/P_0')/G, \qquad (6.4\text{-}37)$$

where $G = (x'^2 + y'^2)/\Sigma(x_k^2 + y_k^2)$.

Again the snr involves a ratio G of sums of squares of Gaussian variables and formulas for a suitable F distribution are applicable. This time the numerator and denominator of the ratio contain unequal numbers of terms; the appropriate definition of F is $F = NG$. In the notation of Abramowitz and Stegun,[33] the cumulative probability function for F is $1 - Q(F|2, 2N)$. From their Table I, we obtain Table 3 giving values of 10 log G that may be used with Eq. (6.4-37). Thus if a given output snr must be maintained for all but a fraction Q of the time, Table 3 determines G. Then Eq. (6.4-37) determines the input signal-to-noise ratio $P_0/P_0' = G$ snr.

Table 3 *Values of* 10 log *G Such That Probability of Failure Equals Q* (2*SP'* and 2*PS'* Noises)

Number of Antennas	Q							
	0.001	0.005	0.01	0.025	0.05	0.1	0.25	0.50
1	30.0	23.0	20.0	15.9	12.8	9.54	4.77	0
2	14.9	11.2	9.54	7.26	5.40	3.34	0	−3.83
3	9.54	6.86	5.61	3.84	2.33	0.61	−2.32	−5.85
4	6.64	4.41	3.34	1.79	0.45	−1.09	−3.82	−7.24
5	4.74	2.76	1.79	0.37	−0.86	−2.34	−4.95	−8.28
6	3.34	1.52	0.61	−0.70	−1.88	−3.31	−5.85	−9.12
7	2.25	0.53	−0.32	−1.59	−2.72	−4.09	−6.64	−9.83
8	1.37	−0.27	−1.08	−2.32	−3.43	−4.76	−7.24	−10.5

Continuing the earlier example with snr = 3 dB, and $Q = 1\%$, Table 3 gives G values of 20.0, 3.34, and −1.08 dB for one, four, and eight antennas. The required input signal-to-noise ratios are

$$20.0 + 3.0 = 23.0 \text{ dB} \qquad \text{for one antenna,}$$

$$3.34 + 3.00 = 6.34 \text{ dB} \qquad \text{for four antennas,}$$

$$-1.08 + 3.00 = 1.92 \text{ dB} \qquad \text{for eight antennas.}$$

6.4.3 Baseband Performance

We have just seen how coherently combining several diversity branches using a separate pilot can reduce fading and provide significantly increased performance over a single-branch system. The resultant signal derived using a separate pilot is the input to a detector and has some peculiar properties that we shall now consider in more detail.

Let us assume a system where the input power spectra to the squaring circuit appears as shown in Figure 6.4-3. Let there be equal power in both the pilot and signal frequencies and assume that the signal can be frequency modulated. As we have shown in Section 6.4.1 the output noise spectra, Figure 6.4-4 is unsymmetrical about the signal. From Eqs. (6.4-8) and (6.4-9) we can see that portions of this noise output periodically go to zero.

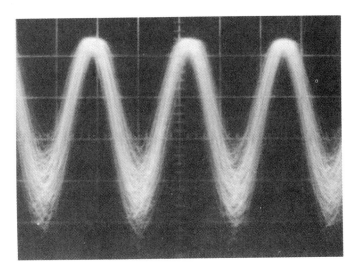

1/f

Figure 6.4-5 Nonstationary noise plus signal in difference frequency band. IF = 90 kHz.

This nonstationary noise is shown in the photograph in Figure 6.4-5 for the case of an unmodulated signal. The fact that the noise, as seen in the photograph, does not completely go to zero is a consequence of the components of the noise mixing with other noise terms, Eq. (6.4-10). For very low signal-to-noise ratios ($S/N < 0$ dB), one cannot neglect these components; however, noise-times-noise terms will be neglected in this discussion.

It is evident that in the arrangement shown in Figure 6.4-3, the deviation of the carrier when modulation is applied can be almost equal to the separation between the pilot frequency and the carrier center frequency. The resulting difference frequency signal, which is centered at $f/2$, would have sideband components from almost zero frequency to f. Thus, the bandwidth of the FM signal and nonstationary noise is on the order of twice the center frequency. Photographs of the unmodulated signal and noise spectra for different values of f are shown in Figures 6.4-6*a* through 6.4-6*c*. These photographs depict the same spectra as sketched in Figure 6.4-4 and agree well with the theory.

The important question is the performance of a frequency demodulator when presented with a signal in a band of nonstationary noise. Figure 6.4-7 is a plot of the measured output noise as a function of the input signal-to-noise ratio for the case where the signal frequency is centered in the IF noise band. Plotted on this same graph are theoretical points from Rice's clicks analysis[24] giving the demodulator output as a function of input for stationary noise. These results indicate that the observed threshold for equal pilot carrier and that for a conventional FM signal are essentially the same. This might seem surprising since there is a high degree of nonstationarity in the noise at the demodulator input.

6.4.4   Design Factors and Field Tests

Systems which derive coherent combining by using a separate pilot require a slightly increased bandwidth and pay a noise penalty over techniques that do not employ a pilot. The major advantage of the pilot diversity system is that it achieves maximal ratio predetection combining with extremely simple circuitry, namely, a mixer.

It is possible to overcome to a large extent the noise penalty of the pilot diversity system by employing special filters and adjusting the level of the pilot. If one reduces the amplitude of the pilot signal [Figure 6.4-3], say to one-half the level of the signal, then most of the energy is carried by the signal. Now, at the receiver prior to the squaring operation one can insert a very narrow band filter centered on the pilot frequency. This filter can be made only as wide as the expected Doppler shifts, which for wideband FM can be 100 times smaller than the signal deviation. Thus, this filter serves to improve the signal-to-noise ratio of the pilot, possibly to a point where it is high compared to the signal component. This has the effect of reducing the pilot-times-noise term in the squaring operation Eq. (6.4-8), and the net effect approaches that of conventional superheterodyne operation.

Some experiments were carried out to compare the performance of a four-branch pilot diversity FM receiver of the type shown in Figure 6.4-1

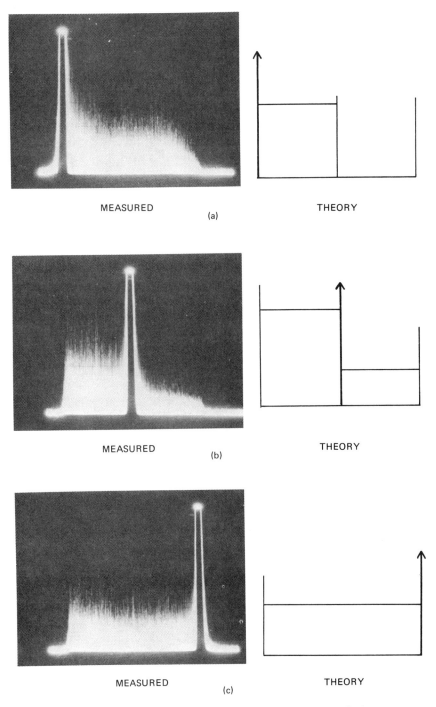

MEASURED   THEORY

(a)

MEASURED   THEORY

(b)

MEASURED   THEORY

(c)

Figure 6.4-6   Spectra of signal plus noise. (a) $f \approx 0$, (b) $f = 0.5$, (c) $f \approx 1$.

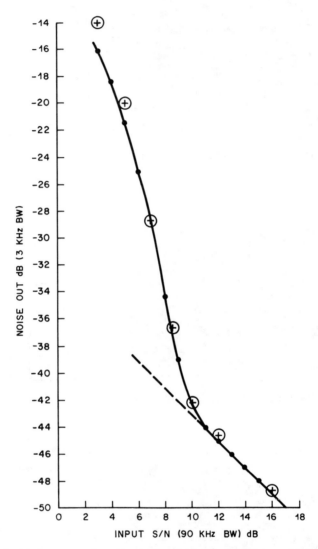

Figure 6.4-7 Noise out versus S/N in for $S = 0.5$ equal pilot carrier amplitude.

to a conventional receiver in an urban multipath area.[34] The tests were carried out at a frequency of 836 MHz by transmission from a 500-ft-high fixed station to a mobile unit that roved about the streets of Philadelphia.

In order to obtain a fair comparison between a four-branch predetection combining and a conventional FM receiver, the two were designed to have nearly identical threshold characteristics by keeping receiver noise figures, IF, and audio bandwidths similar. The combining scheme in the four-branch receiver requires the transmission of a CW pilot signal along with the modulated signal for use as a phase reference. Since this additional power would normally be available in the modulated signal for a conventional receiver, an adjustment was made in the four-branch receiver threshold by placing a 3-dB attenuator in each antenna lead.

The predetection combining receiver was composed of four identical branches with inputs from four physically separated antennas. The outputs from the four branches were combined in a resistive combining pad and fed into a common limiter and discriminator. Figure 6.4-8 shows a block

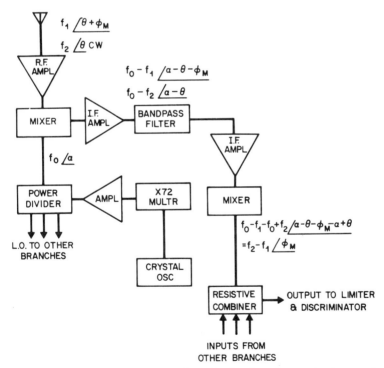

Figure 6.4-8 A block diagram of one branch of a diversity combining receiver.

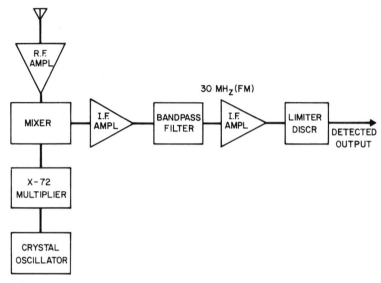

Figure 6.4-9 A block diagram of a conventional FM receiver.

diagram of one branch of the diversity combining receiver, and a block diagram of the conventional FM receiver is shown in Figure 6.4-9. The detected outputs from the discriminators were band limited to 3 kHz.

In a Rayleigh fading environment the comparative performance of these systems is readily calculated [see Chapter 5, Section 5.3]. There are subjective factors that appear to have an important effect on the quality of the detected audio. These are (1) average signal strength, (2) multipath scattering, and (3) vehicle speed. In the absence of multipath effects, that is, in line-of-sight areas, both the conventional FM receiver and the four-branch diversity receiver perform well until the average signal level approaches threshold; however, this occurs rather infrequently and cannot be considered important in the comparisons.

Vehicle speed plays an important role in the subjective evaluation of the conventional FM receiver. The signal fading on the conventional receiver is particularly annoying at speeds under 5 miles/hr even with average signal levels up to as high as 25 dB above the FM threshold. On the other hand, at 50 miles/hr the fade durations are so short that one does not seem to be particularly annoyed while listening to voice transmissions.

The amount by which a four-branch diversity system improves reception is directly dependent upon how much degradation in speech quality one wishes to allow. To illustrate the point, one might graph the subjective performance of the two receivers as below:

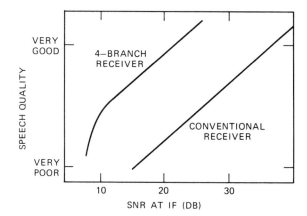

In an idealized situation where there is much multipath scattering but the rms level has no variation, a conventional FM receiver continually sounds worse as the rms signal level decreases, since as the level goes down, more fades go below the receiver threshold. On the other hand, four branches of diversity virtually eliminate the multiple signal interference problem, allowing this system to give good quality reception down to an average signal level near the FM threshold. Roughly speaking, the tests showed that the average received signal level in a conventional receiver ought to be 15–20 dB above FM threshold in order that the speech quality be tolerable, whereas a four-branch diversity receiver operates well near FM threshold.

ACKNOWLEDGMENTS

We thank Mr. E. N. Gilbert for allowing us to reproduce much of the material used in this section and Mr. S. W. Halpern for supplying the photographs.

6.5 RETRANSMISSION DIVERSITY

6.5.1 Analysis of Adaptive Diversity Array Systems

While the previous diversity combination schemes dealt with reception only, we are also interested in the transmitting and receiving systems as a whole. In particular, we want to know the optimal division of diversity branches between the base and the mobile, as well as the specific type of diversity combination scheme to be used. By optimal, we mean that both the simplicity of the system and the improvement in signal threshold are to be considered.

Adaptive antenna arrays have been the subject of numerous investigations.[35,36] In an adaptive transmitting array, the individual element is excited according to information derived from the incident pilot field. For example, in a complex conjugate system, the excitation currents are proportional to the complex conjugate of the incident voltages while the total power radiated is kept constant. In a phase conjugate system, the currents are kept constant while the phases are adjusted according to the conjugate phase of the incident voltages.

In a free-space environment, that is, plane wave incident from a particular direction, it is well known that phase reversal would steer the radiated beam toward the source antenna. Cutler and others [18] have shown how phase reversal can be achieved by frequency conversion of the pilot signal.

The role of adaptive retransmission in a multipath fading environment, for example, mobile radio, troposcatter communication, and so on, has received far less attention. Still unanswered is the question of whether the phase conjugate or the complex conjugate retransmission schemes could improve the communication link and reach a stable state. Morgan has shown that, in a stationary arbitrary environment, stable state and maximal power transfer can be achieved by complex conjugate retransmission.[36]

Yeh[37] has shown that the much simpler phase conjugate will also reach a stable state. Furthermore, assuming equal amplitude transmitting currents on the antenna elements, the summation of voltages received at one array is equal to that of the other array and is maximized. Consequently, the phase conjugate retransmission system will maximize the signal-to-noise ratio (S/N) of an equal-gain diversity reception system (Section 5.2).

In general, the fundamental differences of the two retransmission schemes are that the phase conjugate retransmission maximizes the sum of the amplitudes of the voltages received and the complex conjugate retransmission maximizes the total power received.

Where fading is slow in comparison to the time required to reach an equilibrium state, both systems could be used to improve the quality of a fading communication link.

Yeh[37] investigated the performance of these two systems in fading environments. In particular, he found how these two systems differ in average S/N, what the S/N probability distributions are, how much they improve fading statistics over a single branch system, and, finally, what the optimal division of number of antennas would be between the two antenna arrays. The remainder of this subsection is a summary of his results.

The configuration of the arrays is depicted in Figure 6.5-1. The open circuit voltages and the transmitting currents in each array are represented by column vectors with the time factor $e^{j\omega t}$ suppressed. The mutual couplings are neglected and the antennas in each array are assumed to be identical, with input resistance R during transmission and admittance G during reception.

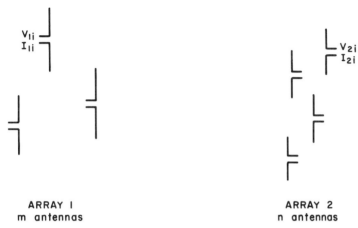

ARRAY 1
m antennas

ARRAY 2
n antennas

Figure 6.5-1   Arrays in adaptive retransmission system.

The transmitting current vector I_2 at array 2 produces the received voltage vector at array 1,

$$V_1 = C\Gamma I_2, \tag{6.5-1}$$

where Γ is an $m \times n$ matrix whose elements are proportional to the transmission between a particular pair of antennas. The real constant C stands for the average transmission loss.

By reciprocity, the received voltage at array 2 is

$$V_2 = C\Gamma^t I_1, \tag{6.5-2}$$

where the superscript t stands for the transpose of the Γ matrix.

Here according to our definition of phase conjugate retransmission, the elements of I_1 and I_2 are of unity amplitudes although their phases could be different. Multiplying Eq. (6.5-1) and (6.5-2) by I_1 and I_2, respectively, we obtain

$$\langle V_1, I_1 \rangle = C\langle \Gamma I_2, I_1 \rangle, \tag{6.5-3}$$

$$\langle V_2, I_2 \rangle = C\langle \Gamma^t I_1, I_2 \rangle, \tag{6.5-4}$$

where the angular brackets stand for inner product. Equations (6.5-3) and (6.5-4) are equal, and we obtain the following reciprocity relation:

$$\langle V_1, I_1 \rangle = \langle V_2, I_2 \rangle. \tag{6.5-5}$$

Let array 1 be excited initially with current I_1, which produces V_2 at array 2. And let array 2 be excited with I_2, which produces V_1 at array 1. Equation (6.5-5) holds and we have the following:

$$\sum_{i=1}^{m} V_{1i} I_{1i} = \sum_{i=1}^{n} V_{2i} I_{2i}, \tag{6.5-6}$$

where the subscript i stands for the ith element of the array.

Consider now the excitation at array 2. Since the I_{2i} are of unity amplitude, the quantity $\sum_{i=1}^{n} V_{2i} I_2 i$ can be maximized by choosing I'_{2i} to be phase conjugate to V_{2i}. We shall call this real maximum quantity λ. Let V'_1 be the voltage vector producted by I'_{2i}; then we have

$$\sum_{i=1}^{m} V'_{1i} I_{1i} = \sum_{i=1}^{n} V_{2i} I'_{2i} = \sum_{i=1}^{n} |V_{2i}| = \lambda. \tag{6.5-7}$$

Let us now consider the excitation of array 1. Obviously the quantity $\sum_{i=1}^{m} V'_{1i} I_{1i}$ can be maximized if we choose I'_{1i} to be the phase conjugate of V'_{1i}. It then follows that

$$\sum_{i=1}^{m} V'_{1i} I'_{1i} = \sum_{i=1}^{m} |V'_{1i}| = \lambda' \geq \lambda. \tag{6.5-8}$$

Let V'_{2i} be the voltages produced by I_{1i}. We obtain, by applying Eq. (6.5-6), the following:

$$\sum_{i=1}^{m} V'_{1i} I'_{1i} = \sum_{i=1}^{n} V'_{2i} I'_{2i} = \lambda' \geq \lambda. \tag{6.5-9}$$

Now I''_{2i} can again be chosen to be phase conjugate to V'_{2i} and we obtain

$$\sum_{i=1}^{m} V'_{2i} I''_{2i} = \sum_{i=1}^{n} |V'_{2i}| = \lambda'' \geq \lambda' \geq \lambda. \tag{6.5-10}$$

This process continues with each new choice of I representing the actual retransmission adjustment made by the antenna system. It is obvious that each retransmission yields a new value of λ that is real and bigger than or equal to the previous value. However, because of the finite number of antennas involved, λ cannot increase indefinitely. The iteration process must therefore finally settle down to a value λ_f that no longer changes. If this is so, we have

$$\sum_{i=1}^{m} V_{1i}^{f} I_{1i}^{f} = \sum_{i=1}^{n} V_{2i}^{f} I_{2i}^{f} = \lambda_f.$$ (6.5-11)

The fact that λ_f is real, and also that we cannot vary the phase of I_{2i}^{f} and I_{1i}^{f} to make λ_f larger, automatically guarantees that I_{1i}^{f} and I_{2i}^{f} are phase conjugate to V_{1i}^{f} and V_{2i}^{f}, respectively. In this case, our phase conjugate retransmission apparatus will no longer change the phases of I_{1i}^{f} and I_{2i}^{f} because they have already reached their proper value. Therefore, we have arrived at a stable state. In this case Eq. (6.5-11) can be further simplified to

$$\sum_{i=1}^{m} |V_{1i}^{f}| = \sum_{i=1}^{n} |V_{2i}^{f}| = \lambda_f.$$ (6.5-12)

So far we have demonstrated that each retransmission tends to increase λ and a stable state must finally be reached. It still remains to be shown that this stable state yields the absolute maximum λ. It is quite possible that several pairs of I_1 and I_2 exist such that they are phase conjugate to V_1 and V_2 but their corresponding λ_fs are different. This is similar to the existence of different eigenstates in matrix analysis. As is well known in matrix algebra, unless the initial vector is orthogonal to the maximum eigenstate, we would invariably obtain the maximum eigenstate through iterations.

Since the phase conjugate operation on V to produce I is a nonlinear operation, an analytical analysis along the above lines is extremely difficult, if not impossible. However, with computer simulation it was shown that the phase conjugate retransmission process converges rapidly and the probability of ending up in a nonmaximum state of λ_f is practically zero.[37]

Let V_{1i} be the voltage response at the ith elementary antenna. Furthermore, let η_{1i} be the corresponding noise voltage, which satisfies

$$\langle \eta_{1i} \eta_{1j} \rangle_{av} = \begin{array}{ll} N^2, & i=j \\ 0, & i \neq j \end{array}$$ (6.5-13)

where $\langle\ \rangle_{av}$ stands for time average.

The S/N of an m-branch diversity equal-gain system is

$$\frac{S}{N} = \frac{\left[\sum_{i=1}^{m} |V_{1i}|\right]^2}{mN^2} = \frac{\lambda_f^2}{mN^2}.$$

(6.5-14)

Recall that there are n elements in the other array, which radiates a total power nR; therefore the S/N of the received signal per unit power radiated is

$$\frac{S}{N} = \frac{\lambda_f^2}{nmN^2R}.$$

(6.5-15)

It is therefore obvious that the S/N ratios at both arrays are identical.

The excitation currents of a complex conjugate retransmission system are related to the incoming voltages by

$$I_2 = K_2 V_2^*,$$

(6.5-16)

$$I_1 = K_1 V_1^*,$$

(6.5-17)

where K_1 and K_2 are scalars to keep the total radiated power constant. For unity transmitter power, the received power at arrays 1 and 2 are maximized and are equal,[36]

$$P_{1R} = P_{2R} = \frac{G}{R} C^2 \lambda_m,$$

(6.5-18)

where λ_m is the maximum eigenvalue of the Hermitian matrix $\Gamma\Gamma^+$. The validity of Eq. (6.5-18) is subject to the constraint that when the adaptive retransmission array starts operation, its current vector should not be orthogonal to the maximum eigenvector of the $\Gamma\Gamma^+$ matrix. The S/N of a multibranch maximal-ratio reception system then is

$$\frac{S}{N} = \frac{C^2}{RN^2} \lambda_m.$$

(6.5-19)

It can be seen that the S/N at both arrays are equal.

The complexity of the quantities λ_m and λ_f makes a closed-form solution of the cumulative probability distribution extremely difficult, if not impossible. Therefore, we try instead the Monte Carlo method and aim at a numerical solution. The essence of the method is to choose for each element of the Γ matrix a random variable of the form $u + jv$. The variables u and v, according to our assumption of independent Rayleigh fading statistics, are normalized independent Gaussian variables. For a particular $m:n$ array system, we can therefore evaluate the maximum eigenvalue λ_m by repeated matrix multiplication. The value λ_f is evaluated by iterations according to the retransmission schemes defined earlier.

We look at the complex conjugate retransmission system first. Incorporated with maximal-ratio diversity reception, this system provides the best S/N performance obtainable from a particular $m:n$ array system.

The average S/N is presented in Figure 6.5-2. It is seen that for small numbers of n, there do exist appreciable improvements in average signal

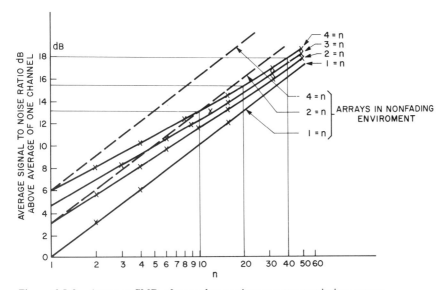

Figure 6.5-2 Average SNR of complex conjugate retransmission arrays.

level as m changes from 1 to 4. However, as n increases the advantage diminishes. For example, a 1:50 array has the same average signal level as 2:44, 3:39, and 4:35 arrays. This is in sharp contrast to the case of adaptive arrays with nonfading signals. In that case, plane-wave incidence is assumed and a $m:n$ array would have the same S/N as a $1:mn$ array (Figure 6.5-2).

A simple explanation of the difference between the fading and the nonfading arrays is the following: In both cases, the 1:mn adaptive retransmission system guarantees that the voltages produced by the mn elements at the single array add in phase. In the m:n system, the voltage components produced by the n antennas again add in phase at each antenna of the m array if plane wave incidence is assumed. Consequently, the power received is identical to that of the 1:mn array. However, in a random environment the n voltage components at each antenna element in the m array no longer add in phase; therefore, the m:n system receives less power than that of the 1:mn system.

For a single Rayleigh fading signal, the level is above the -20.6-dB point (with respect to its mean) 99% of the time. We can thus designate -20.6 dB as the 99% reliability level. Hence the difference in decibels values of two antenna systems for a particular reliability indicates their difference in signal threshold or their difference in the required transmitter power. The 99% reliability level is presented in Figure 6.5-3.

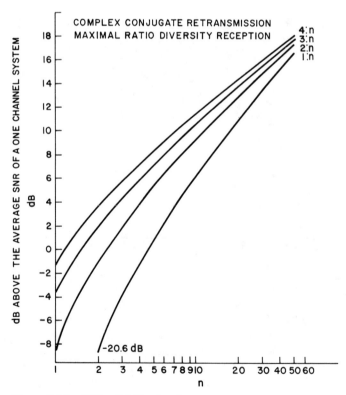

Figure 6.5-3   99% reliability level.

We discuss now results obtained from the phase conjugate retransmission system. In this system, the S/N of an equal-gain diversity reception system is maximized. It is observed that because of this maximization effect, the performance of the phase conjugate system is not much inferior to that of the complex conjugate system. For example, the cumulative probability distributions of the S/N for both systems in the case of a 2:4 array system are presented in Figure 6.5-4. The distributions of the two systems differ approximately by the average S/N difference. Therefore, the difference in average S/N of the two systems is also a good indication of their difference in percentile reliability levels.

The average S/N of the two systems is shown in Figure 6.5-5 for 2:n and 4:n array systems. It is seen that for the same m:n array, the difference of the two systems is small, that is, within a decibel or so.

The performance of these two systems differs little. Therefore the choice of a particular scheme should be based on practical considerations. For example, in the phase conjugate system, the total power is divided equally among all the antenna elements. On the other hand, the complex-conjugate retransmission system requires that the total power be distributed in a complicated fashion. In practice this means that each antenna-feeding apparatus must be equipped to handle power far exceeding that of the phase conjugate syystem.

In view of the simplicity of the phase conjugate retransmission compared to the complex conjugate retransmission (which must keep the total power transmitted constant), and only slightly inferior performance, the phase conjugate system appears to be more attractive.

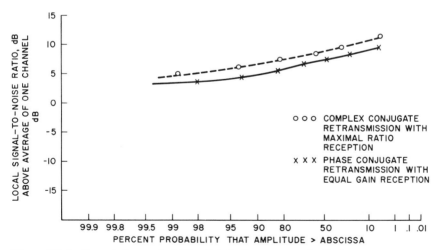

Figure 6.5-4 Cumulative probability distribution curves of a 2 : 4 array system.

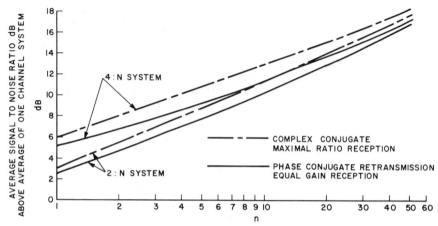

Figure 6.5-5 Average SNR of antenna systems $2:N$ and $4:N$.

As far as the division of diversity branches is concerned, it can be seen from Figure 6.5-2 that for small numbers of antennas, an $m:n$ array would have similar performance to an $1:mn$ array. However, as the number of elements involved becomes larger, this relation no longer holds. For example the performance of a $4:n$ array would approach a $1:n$ array as n increases indefinitely.

6.5.2 Adaptive Retransmission by Time Division

In this section we shall discuss a method of achieving adaptive retransmission proposed by Bitler, Hoffman, and Stevens[38] where transmission and reception are on exactly the same frequency, but at different times.

The mobile and base station timing sequence is shown in Figure 6.5-6. At time t_0 the mobile (master) station transmits an unmodulated pulse of RF energy, P_1, of duration τ_1 followed by a modulated pulse, P_2, of duration τ_2. After the propagation delay, τ_d, the base (slave) station receives and processes the two pulses from the mobile station. The base station then transmits a modulated pulse, P_3, of duration τ_3 back to the mobile station where, after another propagation delay, τ_d, it is received and processed. The entire sequence is initiated again by the mobile station at time $t_0 + T$.

The system transmits pulses of modulated RF carrier at a rate $f_s = 1/T$. The sampling rate, f_s, must be fast enough to satisfy the sampling theorem in order to reconstruct continuous signals at the base and mobile receiver. Two signals are involved. First, the speech must be reconstructed. Second, the phase distortion caused by the medium that is required for adaptive retransmission must be restored.

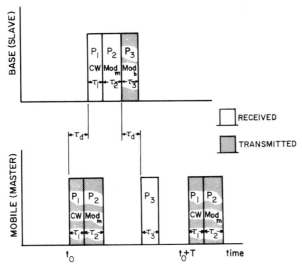

Figure 6.5-6   Pulse-timing sequence.

There are two methods which can be used to reconstruct the baseband signals from the RF pulses. The complete RF signal can be reconstructed and then demodulated. The sampling rate required is then greater than twice the RF bandwidth. Assume Carson's rule,

$$f_s \geq 4f_b(1+\beta), \qquad (6.5\text{-}20)$$

where f_b is the highest baseband frequency and β is the modulation index. Typical parameter values, $f_b = 3\text{kHz}$, $\beta = 4$, require the sampling frequency to be greater than 60 kHz. Sampling rates of this order of magnitude are not realistic due to range limitations, which we shall consider shortly.

A second method is to place the reconstruction filter after the demodulator. The sampling rate must now be $\geq 2f_b$, or $f_s \geq 6\text{kHz}$. Associated with the reduction in sampling rate is a decrease in signal-to-noise performance and an increase in distortion.

Distortion occurs when an FM signal is gated by a periodic pulse waveform, the result passed through a bandpass filter, demodulated, gated again, and passed through a low pass filter. The amount of distortion generated is a function of the ratio of the FM bandwidth to the sampling frequency. It can be shown that for less than 10% distortion the FM bandwidth must be at least five times greater than the sampling frequency.

Applying either Carson's rule or the distortion constraint

$$B = \begin{cases} 2f_b(1+\beta) & \beta \leqslant 4 \\ 5f_s & \beta < 4 \end{cases}.$$

For a given sampling rate, the maximum range is a function of propagation delays and the duty cycle of each pulse. From Figure 6.5-6,

$$2\tau_d + \tau_1 + \tau_2 + \tau_3 \leqslant T,$$

or

$$\text{Range} < \frac{[T - (\tau_1 + \tau_2 + \tau_3)]c}{2}. \qquad (6.5\text{-}21)$$

where c is the velocity of light. For voice transmission the maximum range is about 10 miles.

The receiver eliminates the phase distortion due to the medium and brings the phase of each branch to a common reference value. The signals from each branch can then be added coherently before the detection process. It is also the purpose of the cophaser to provide the conjugate of the phase distortion at each branch transmitter for adaptive retransmission.

Figure 6.5-7 shows a cophaser block diagram for a typical branch. Assume that two pulses are present at the cophaser input, the first, P_1, is unmodulated and of duration τ_1; the second, P_2, is modulated and of duration τ_2 (see Figure 6.5-6). Each pulse has the same phase distortion due to the medium.

The pulses are converted to a convenient IF frequency and amplified by a low noise preamplifier. The pulse detection network detects the leading edge of P_1. The timing circuit then produces outputs at the appropriate times that control three sets of gates. The blanking gate is normally closed, but it is open when the base station transmits its pulse. Its output is split into two paths. One path goes through an amplifier, $A2$, whose output is then mixed with a second local oscillator, $f_0 \angle \alpha$, where $f_0 > f_{\text{IF}}$. The output of mixer 2 now contains pulses modulated with the conjugate phase. A carrier recovery circuit uses these pulses to construct a continuous signal of constant amplitude that contains the conjugate of the phase distortion. The modulated pulse, P_2, is gated out of the carrier recovery circuit so that phase information due to modulation will not be present at its output. The CW output of the carrier recovery circuit is split. One signal goes to mixer 3, which mixes with P_1 and P_2. The mixing process cancels the phase distortion due to the medium, but does not affect the modulation. The output of mixer 3 is gated to pass only the cophased P_2 pulse to the

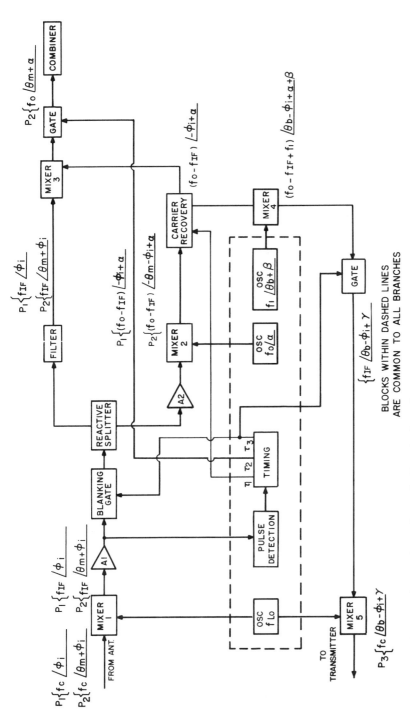

$P_2\{f_o\ \underline{/\theta_m+\alpha}$

$P_1\{f_{IF}\ \underline{/\phi_i}$

$P_2\{f_{IF}\ \underline{/\theta_m+\phi_i}$

$P_1\{(f_o-f_{IF})\ \underline{/-\phi_i+\alpha}$

$P_2\{(f_o-f_{IF})\ \underline{/\theta_m-\phi_i+\alpha}$

$(f_o-f_{IF})\ \underline{/-\phi_i+\alpha}$

$(f_o-f_{IF}+f_l)\ \underline{/\theta_b-\phi_i+\alpha+\beta}$

$\{f_{IF}\ \underline{/\theta_b-\phi_i+\gamma}$

BLOCKS WITHIN DASHED LINES
ARE COMMON TO ALL BRANCHES

$P_1\{f_{IF}\ \underline{/\phi_i}$

$P_1\{f_{IF}\ \underline{/\phi_i}$

$P_2\{f_{IF}\ \underline{/\theta_m+\phi_i}$

$P_1\{f_c\ \underline{/\phi_i}$

$P_2\{f_c\ \underline{/\theta_m+\phi_i}$

FROM ANT.

TO TRANSMITTER

$P_3\{f_c\ \underline{/\theta_b-\phi_i+\gamma}$

Figure 6.5-7 Base-station cophaser for each diversity branch.

501

combiner network. The other signal is mixed with an oscillator that contains the base-station modulation to be transmitted to the mobile station. The output of mixer 4 contains both the conjugate phase distortion due to the medium and the base-station modulation. It is gated into mixer 5 where it is heterodyned up to the carrier frequency, f_c, for transmission to the mobile station.

All elements within the dotted line are common to each branch; thus any phase shift they introduce is common to all branches.

An experimental program was set up to test the adaptive retransmission capabilities of the system in a suburban environment at a frequency near 1 GHz. A two-branch system was constructed and tested. Since the purpose of the test was to determine the retransmission capabilities, only the unmodulated pulses, P_1, were used. The envelope fading statistics at both the base and mobile stations were measured and compared with a theoretical two branch system. Table 1 lists the various frequencies and parameters used in this system.

Table 1   *Frequencies and Parameters of Experimental System*

Carrier frequency	$f_c = 836$ MHz
IF frequency	$f_{IF} = 23.646$ MHz
Local oscillator for Mixer 1	$f_0 = 30$ MHz
Local oscillator for Mixer 4	$f_1 = 17.292$ MHz
Output pulses	10 watts, 15 μsec, every 100 μs
Antenna gain	base 10 dB
	mobile 5.5 dB

Data runs were made on local, suburban streets located 2–3 miles from an elevated base station. The first run of each set used both branch transmitters operating from the base station. The second run used branch 1 transmitting only, and third, branch 2 only. Cumulative distribution curves were calculated for the same sections of street for the three conditions.

The calculated data for the base and mobile stations were compared to each other and to the theoretical performance of a two-branch equal-gain diversity system. The performance of the theoretical system is given by the solid lines on the cumulative distribution curves of Figures 6.5-8 and 6.5-9. The points marked branch 1 and branch 2 correspond to the same run at the base station. They correspond to different runs at the mobile station with either transmitter 2 or transmitter 1 turned off at the base station.

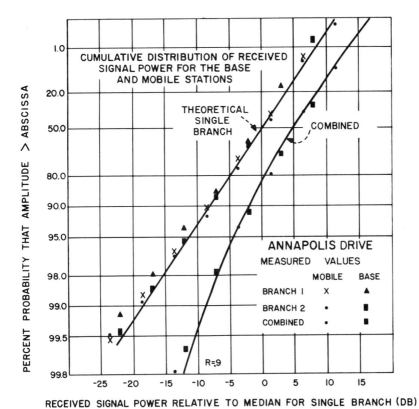

Figure 6.5-8 Cumulative distribution of received signal power for the base and mobile stations for Annapolis Drive.

The single-branch data taken on Annapolis Drive (Figure 6.5-8) fit a Rayleigh distribution very closely. The points for the combined branches fall very close to a theoretical curve for a 0.9 correlation between the base antennas which were about 25 ft apart.

Holmdel-Keyport Road data is shown on Figure 6.5-9. These data show a strong influence of a direct path in the curvature of the single-branch data to the left at high signal levels. However, deep fades were still present, as indicated by the small deviation from pure Rayleigh at low signal levels. The combined data lies generally between the theoretical curves assuming 0.5 and 0 correltation. The proof of the capability of· the system for adaptive retransmission lies in the fact that the statistics of the combined branches at both the base station and the mobile station were equal. Thus, the reduction of fading at the mobile receiver was virtually the same as that at the base receiver where all the signal processing was actually done.

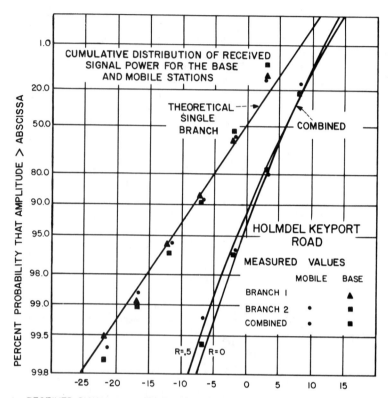

Figure 6.5-9 Cumulative distribution of received signal power for the base and mobile stations for Holmdel Keyport Road.

6.5.3 Continuous Adaptive Retransmission

In this section we shall illustrate some methods for transmitting back over a mobile radio path the complex conjugate excitation of the signal received at each antenna element.

In order to minimize the cross-talk between the incoming pilot and the retransmitted signal, they should be separated in frequency as widely as possible. However, the frequency separation should be small enough so that the pilot has a high phase correlation (after traversing a mobile radio path) with the original signal. Satisfactory phase correlation exists for frequency separations less than about 100–200 kHz (Section 1.5). This maximum limit on frequency separation makes radio-frequency filtering of the incoming pilot from the retransmission signal impractical at microwave frequencies.

To overcome this problem of separating the pilot from the signal to be transmitted, a scheme was discussed in the previous section where time diversity is employed. Alternatively, the methods described in this section propose to allow continuous reception of the pilot during continuous retransmission.

For coherent reception and retransmission it is necessary to derive not only the voice information from the signal, but also a signal that indicates the distortion introduced by the medium. The latter signal is called a pilot. Various methods of obtaining a pilot signal have been discussed in Chapter 5.

The basic idea for coherent adaptive retransmission is to receive a weak pilot signal and extract the phase information on this pilot, and then use this information to control the phase of the retransmitted signal in a continuous manner. The problem which must be overcome is to detect the weak incoming pilot signal in the presence of the strong retransmission signal. Probably the most direct method of attack is that of filtering the retransmission signal. We have in mind transmission frequencies in the UHF to X-band region. A reasonable estimate for the coherence bandwidth, that is, where high phase correlation exists between two frequencies, is approximately 100 kHz. Separating two frequencies this close together by RF filters is obviously impractical. However, filtering at IF might be possible, and a basic retransmission circuit which relies on IF filtering is shown in Figure 6.5-10.

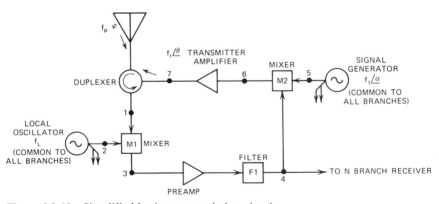

Figure 6.5-10 Simplified basic retransmission circuit.

At position (1) of Figure 6.5-10, the incoming pilot signal, $f_p \angle \phi$, and a fraction of the retransmitted signal, $f_t \angle \theta$, arrive at mixer $M1$. For simplicity, when describing systems, we will assume single signal transmissions.

Even with a good match at the antenna, a large amount of signal will be reflected into the mixer. We must assume that this power is not so large as to drive the mixer into saturation. Upon mixing to a convenient intermediate frequency (position 3), we want a filter to suppress as much as possible the signal $(f_t - f_L) \angle \theta$ while passing undisturbed the phasing signal $(f_p - f_L) \angle \phi$. Assume for the moment that the signal $(f_t - f_L)$ has been entirely suppressed at position 4; then, the desired conjugate phase for the retransmitted signal is obtained by mixing $(f_p - f_L) \angle \phi$ with a frequency modulated RF signal, $f_1 \angle \alpha$. Note that only the lower sideband from this product gives the proper phase for the retransmission signal.

When we equate frequencies and phases around the loop of Figure 6.5-10 we find:

$$f_t = f_1 - f_p + f_L$$

and

$$\theta = \alpha (\text{modulation}) - \phi.$$

A problem arises when $(f_t - f_L) \angle \theta$ is not suppressed sufficiently before mixer $M2$, because we see that the lower sideband after mixing also contains

$$f_1 - (f_t - f_L) = f_1 - (f_1 - f_p + f_L - f_L) = f_p,$$

which is the frequency we are trying to receive! The difference in level between f_t and f_p may be on the order of 100 dB at $M1$, so that in order to detect f_p while transmitting, we must suppress $f_t - f_L$ by 200 dB. Even though filters with 200 dB rejection 100 kHz off-center frequency might be realistic for the proper IF, a serious problem of preventing the leakage of this signal into mixer $M2$ remains. In the following paragraphs we show how the use of an offset oscillator can reduce the leakage problem.

We previously pointed out that if the IF frequency derived from the transmitted signal leaks back into the last RF mixer, a signal at the pilot frequency is generated. An offset oscillator (LO_2 in Figure 6.5-11) provides a means of shifting the IF for upconversion so that leakage of the strong signal, $(f_t - f_L)$, into the final mixer does not produce a signal at the pilot frequency.

As shown in Figure 6.5-11, the local oscillator signal f_L and the incoming pilot signal $f_p \angle \phi$ are then amplified and filtered to suppress the dominant mixer product from $M1$, $(f_t - f_L) \angle \theta$, due to $f_t \angle \theta$ leaking into $M1$. $(f_p - f_L) \angle \phi$ is then mixed with an offset local oscillator $f_0 \angle \beta$ in mixer $M2$. The upper product out of $M2$, $(f_0 + f_p - f_L) \angle (\beta + \phi)$, is amplified and

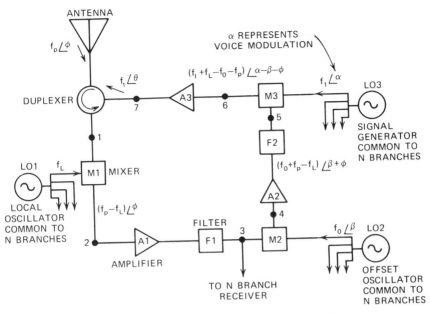

ANTENNA

$f_p \angle \phi$

α REPRESENTS
VOICE MODULATION

$f_t \angle \theta$

$(f_1 + f_L - f_0 - f_p) \angle \alpha - \beta - \phi$

$f_1 \angle \alpha$

LO3

DUPLEXER

A3

7

M3

6

5

SIGNAL
GENERATOR
COMMON TO
N BRANCHES

1

F2

LO1 f_L

M1 MIXER

$(f_0 + f_p - f_L) \angle \beta + \phi$

LOCAL
OSCILLATOR
COMMON TO
N BRANCHES

$(f_p - f_L) \angle \phi$

A2

4

2

A1

FILTER
F1

3

M2

$f_0 \angle \beta$ LO2

AMPLIFIER

TO N BRANCH
RECEIVER

OFFSET
OSCILLATOR
COMMON TO
N BRANCHES

Figure 6.5-11 Basic retransmission circuit with an offset oscillator.

filtered to eliminate the component $(f_0 + f_t - f_L) \angle (\beta + \theta)$. Mixer $M3$ mixes the phasing signal $(f_0 + f_p - f_L) \angle (\beta + \phi)$ with the voice modulated signal $f_1 \angle \alpha$ so that the lower mixer product $(f_1 + f_L - f_0 - f_p) \angle (\alpha - \beta - \phi) = f_t \angle \theta$ is the desired retransmission signal in that it contains minus the pilot phase, $-\phi$, and the voice modulation, α. Note, by using the offset oscillator, there is no longer a component at the pilot frequency.

This scheme is maximal ratio in power output; that is, each branch retransmits a signal proportional to the voltage generated on the antenna from the pilot signal. Hence, not only is the proper relative phase provided to each antenna, but also the optimum power level is retransmitted.

It was shown in Section 6.5.1 that if both mobile and base stations employ maximal ratio retransmission, then stability requires that the total power retransmitted be held constant. This may be accomplished by detecting the sum of the squares of the received pilot signals from all branches and making the output power of the signal generator (LO_3) inversely proportional to this sum. If one desires only phase conjugate retransmission, rather than maximal ratio, a limiter can be placed at position 5.

6.5.4   Retransmission Using Transmitted Signal as Local Oscillator

In the previous section, the retransmitted signal appeared as an undesired signal in the receiver, and special efforts were made to suppress it. In this section we present two solutions to the adaptive retransmission problem where the retransmitted signal is *necessary* to the circuit operation.[39] A particular result of these solutions is that the severe filtering requirements of the previous method are substantially reduced.

A Method Using Frequency Division with Feedback

The key to this retransmission scheme is to allow the transmitted signal to leak back into the receiving section of the retransmission circuit and serve as a local oscillator in detecting a weak pilot signal. As shown in Figure 6.5-12, the retransmitted signal $f_t \angle \theta$ and the incoming pilot signal $f_p \angle \phi$ are mixed together at $M1$. The lower mixer product, $(f_t - f_p) \angle (\theta - \phi)$, is then amplified and mixed at mixer $M2$ with an offset oscillator frequency, $f_0 \angle (\beta - \alpha)$, where the upper product is required, $(f_0 + f_t - f_p) \angle (\theta - \phi - \alpha + \beta)$. As mentioned, α is the information modulation of the retransmission signal. It is subtracted at this point to facilitate frequency division and filtering. β is a constant phase of the offset oscillator. The signal is amplified by a limiting amplifier and fed into a frequency divider of ratio $\frac{1}{2}$. The output from the frequency divider,

$$\frac{f_0 + f_t - f_p}{2} \angle \left(\frac{\theta}{2} - \frac{\phi}{2} - \frac{\alpha}{2} + \frac{\beta}{2} \right),$$

is now mixed with the signal to be transmitted in the transmitting mixer $M3$. Again the upper mixer product is taken which determines the actual transmitted frequency and phase:

$$\left(f_1 + \frac{f_0}{2} + \frac{f_t}{2} - \frac{f_p}{2} \right) \angle \left(\frac{\alpha}{2} + \frac{\theta}{2} - \frac{\phi}{2} + \frac{\beta}{2} \right).$$

Equating this expression to $f_t \angle \theta$ we find that

$$f_t = 2f_1 + f_0 - f_p$$

and the phase

$$\theta = \alpha + \beta - \phi.$$

Thus we see that the relative phasing between an array of antennas is automatically adjusted to the optimum value, since θ contains the conjugate phase $(-\phi)$ of the incoming signal. Note that the modulation index of the FM signal transmitted from the antenna will equal that of the generator at f_1.

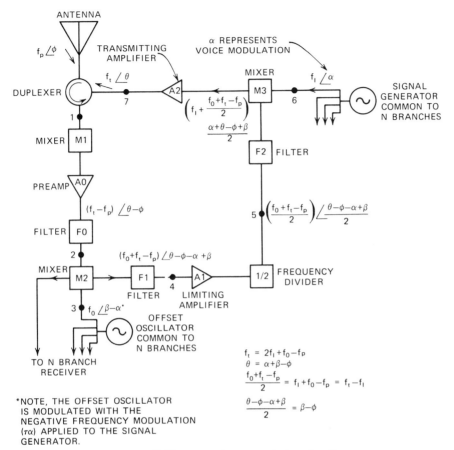

Figure 6.5-12   Frequency divider type of retransmission circuit.

Various aspects of the above system deserve further explanation. First, the use of the retransmission signal f_t as a local oscillator assumes that the amount of f_t that leaks into mixer $M1$ is limited to about 10 mW. Typically, a single branch of the retransmission system for mobile radio might transmit 1 W; thus the isolation of the antenna duplexer (circulator, directional coupler) must be better than 20 dB and the VSWR at the input to the antenna should be lower than 1.2.

The limiting amplifier is used to facilitate frequency division, so this retransmission system provides phase correction only. (One could perhaps devise a frequency division scheme which would not require limiting, however.)

The frequency division by 2 is necessary to provide the correct resultant phase for retransmission. This is because the signal out of mixer $M1$ contains the retransmission phase as well as the pilot phase, and direct subtraction of this phase at mixer $M3$ to obtain the retransmission signal would result in an indeterminate phase. In fact, it can be shown that the only factor of frequency division that will provide the correct retransmission phase is frequency division by 2.

The frequency translation by f_0 is required so that the unwanted (lower) sideband from mixer $M3$ will not be at the same frequency as the pilot. If this sideband were at the pilot frequency, it would be very difficult to suppress it sufficiently over the whole bandwidth of the loop to prevent self-oscillation.

This scheme is essentially a feedback scheme in that it uses a signal phase-dependent on the transmission signal phase to *control* the transmission signal phase. Thus there will be a time delay before the system reaches equilibrium. It is desirable that this time delay be minimized because the pilot signal phase becomes uncorrelated in a short time due to motion of the mobile.

A Method Using Double Mixing

An alternative scheme which also achieves the optimum antenna phasing is shown in Figure 6.5-13. This method eliminates the frequency divider by substituting a double mixing process. A local oscillator, f_L, is used to mix with both the transmitted signal, $f_t \angle \theta$, and the weak incoming pilot signal, $f_p \angle \phi$. The following lower mixer products will be observed after mixing at $M1$ (we assume $f_L < f_p < f_t$ and that $f_t - f_p$ is within the coherence bandwidth for the medium):

$$(f_p - f_L) \angle \phi, \qquad (f_t - f_p) \angle (\theta - \phi), \qquad (f_t - f_L) \angle \theta.$$

The first term is desired; however, if it is impractical to filter this signal from the other two frequencies, the following scheme offers an alternative. The term $(f_t - f_L) \angle \theta$ provides a strong steady local oscillator signal for mixer $M2$. The lower mixer product yields the desired term

$$(f_p - f_L) \angle \phi.$$

An offset oscillator will be used for the same reasons as discussed in Section 6.5.3. We require the lower sideband from mixer $M3$. The transmitted modulation may be provided by modulating the offset oscillator, $f_0 \angle \alpha$. The desired component is

$$(f_0 + f_L - f_p) \angle (-\phi + \alpha).$$

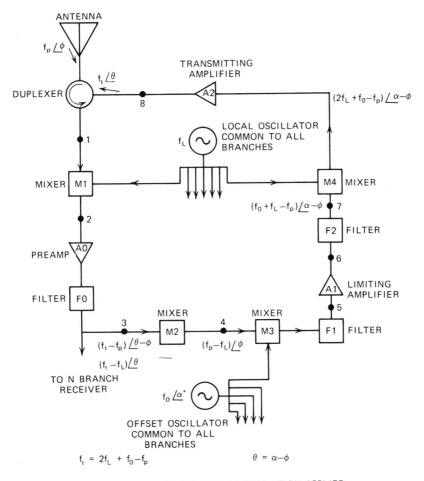

$f_t = 2f_L + f_0 - f_p$ $\theta = \alpha - \phi$

*α REPRESENTS THE VOICE MODULATION APPLIED
TO THE OFFSET OSCILLATOR

Figure 6.5-13 Double mixing type of retransmission circuit.

The transmitted frequency is derived at mixer $M4$ from the upper side-band, and we find that

$$f_t = 2f_L - f_p + f_0, \qquad \theta = -\phi + \alpha.$$

Frequencies f_t and f_p are chosen to lie within the coherence bandwidth (approximately 100 kHz), and they are centered at the particular RF frequency assigned to the channel (f_{RF}). Thus, the above relation determines f_0:

$$f_0 = 2(f_{RF} - f_L).$$

This scheme differs from that mentioned previously in that it is not a feedback system. Regardless of the initial phase of f_t, its phase is automatically corrected as soon as f_p is received.

6.6 MULTICARRIER AM DIVERSITY

A simple diversity technique suitable for AM systems that may have application at low microwave frequencies was proposed and tested in the early 1940s.[40] It is diversity of the "transmitter" type, characterized by a multiplicity of antennas at the transmitting site and one antenna at the receiver. It can be used in this way to mitigate the effects of multipath fading, or, by widely dispersing the antennas, to achieve an advantage with respect to "shadows" caused by gross terrain features. The use of AM for high-capacity systems characterized by frequency reuse and small cell coverage is generally discouraged, however, because of the greater vulnerability of AM receivers to cochannel interference. Nevertheless, this kind of diversity system may have application in some special situations.

6.6.1 Principles of operation

The basic idea is to transmit the same audio signal by conventional AM on several adjacent frequencies, each one on a separate antenna. A linear detector will then yield an output proportional to the sum of the modulation envelopes, equivalent to predetection, equal-gain combining. The signal envelopes at the different carrier frequencies are uncorrelated by virtue of the transmitting antenna spacing; the frequency separation between carriers would not be enough, in general, to decorrelate the signals (see Chapter 1).

The M different carrier frequencies must be separated from each other by at least twice the audio bandwidth, and with a separation precision of about $(\pm 150/M)$ in Hertz. The reason for this is to prevent beat products between harmonics of the carriers from falling back into the audio band. The receiver bandwidth must be wide enough to accept all of the modulated carriers, as shown in Figure 6.6-1.

The modulation applied to all transmitters must be very nearly identical in amplitude and phase to avoid distortion effects and to make most efficient use of the method. If the transmitters are at one site this can be easily accomplished. If they are widely separated, however, in applications where the objective is to overcome shadowing, there may be appreciable delay difference in getting the modulating signals to the transmitters. This delay difference should be kept well below 90° at the highest modulating frequency. Differential propagation delays from the various transmitters to the receivers can be neglected, however, for path length differences of 20 miles of less.

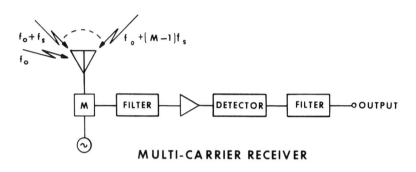

MULTI-CARRIER RECEIVER

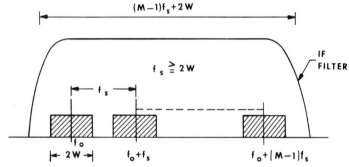

Figure 6.6-1 Spacing of carriers and relationship to receiver bandwidth in the multicarrier AM diversity scheme.

In the absence of fading the baseband signal to thermal noise ratio is linearly increased by adding more carriers (assuming equal amplitudes):

$$\left(\frac{S}{N}\right)_M = M\left(\frac{S}{N}\right)_1, \tag{6.6-1}$$

where $(S/N)_1$ is the S/N ratio for a conventional AM channel as given by Eq. (4.1-32). It is assumed that the receiver bandwidth is just enough to accept the M signals.

6.6.2 Effect of Rayleigh fading

Motion of the mobile receiver through the field strength maxima and minima produced by multipath propagation cause a degradation in the signal quality. The effect is to broaden every spectral line of the composite signal until it occupies a spectrum width of $2f_m = 2v/\lambda$ (cf. Section 1.2) where v is the velocity of the mobile and λ the transmitted wavelength. For values of $f_m \leqslant 150$ Hz the velocity modulation of the carriers may be blocked by a high-pass filter after the detector. Modulation of the sidebands, however, cannot be removed by filtering, and remains as motion-induced noise. The magnitude of this noise will now be calculated for various cases.

We assume that the kth antenna at the transmitter radiates a signal of the form

$$e_k(t) = E_k[1 + v(t)]\cos\omega_k t, \qquad (6.6\text{-}2)$$

where E_k is the carrier amplitude and ω_k its frequency, $v(t)$ is the modulating signal (identical for all transmitters). For M such signals applied to a linear envelope detector the baseband output voltage is, from Eq. (4.1-81),

$$a(t) = [1 + v(t)]\sum_{k=1}^{M} r_k(t), \qquad (6.6\text{-}3)$$

where $r_k(t)$ is the fading envelope amplitude of the signal from the kth transmitter. The noise due to the fading can be calculated as in Section 4.1.7. To get the baseband signal power we first ensemble average $a(t)$ over the Rayleigh fading:

$$\langle a(t)\rangle = [1 + v(t)]\sum_{k=1}^{M} \langle r_k(t)\rangle. \qquad (6.6\text{-}4)$$

Recalling that $\langle r_k(t)\rangle = \sqrt{(\pi/2)b_{0k}}$, from Eq. (1.3-48), we can write

$$\langle a(t)\rangle = dc + v(t)\sqrt{\frac{\pi}{2}}\sum_{k=1}^{M}\sqrt{b_{0k}} . \qquad (6.6\text{-}5)$$

The dc term does not contribute to the useful signal power, P_s; thus we assume

$$P_s = \langle v^2(t)\rangle\frac{\pi}{2}\left(\sum_{k=1}^{M}\sqrt{b_{0k}}\right)^2$$

$$= S\frac{\pi}{2}\left(\sum_{k=1}^{M}\sqrt{b_{0k}}\right)^2 \qquad (6.6\text{-}6)$$

where $S = \langle v^2(t) \rangle$ is the signal modulation power. (For a sinusoidal modulating signal of index m, $S = m^2/2 \leqslant 0.5$; generally $S \leqslant 0.1$ for voice modulation to avoid distortion).

The noise part of the output is taken to be $a(t) - \langle a(t) \rangle$, that is, the total output less the output averaged over the fading:

$$n(t) = [1 + v(t)] \sum_{k=1}^{M} \left[r_k(t) - \sqrt{\frac{\pi}{2} b_{0k}} \right].$$ (6.6-7)

The average noise power is then

$$P_n = \langle n^2(t) \rangle = (1 + S) \left\langle \left\{ \sum_{k=1}^{M} \left[r_k(t) - \sqrt{\frac{\pi}{2} b_{0k}} \right] \right\}^2 \right\rangle.$$ (6.6-8)

(We have assumed $v(t)$ is zero mean). Assuming further that the r_k are uncorrelated ($\langle r_i r_j \rangle = \langle r_i \rangle \langle r_j \rangle$, $i \neq j$), we can evaluate the average:

$$P_n = (1 + S) \sum_{k=1}^{M} \left[\langle r_k^2 \rangle - \frac{\pi}{2} b_{0k} \right].$$ (6.6-9)

From Eq. (1.3-14), $\langle r_k^2 \rangle = 2b_{0k}$; thus

$$P_n = (1 + S) \left(2 - \frac{\pi}{2} \sum_{k=1}^{M} b_{0k} \right)$$ (6.6-10)

and the signal-to-noise ratio is

$$\frac{P_s}{P_n} = \frac{S}{1 + S} \frac{\pi \left(\sum_{k=1}^{M} b_{0k} \right)^2}{(4 - \pi) \sum_{k=1}^{M} b_{0k}}.$$ (6.6-11)

If all of the signals are of equal average strength, then $b_{0k} = b_0$ and

$$\frac{P_s}{P_n} = \frac{S}{1 + S} \frac{\pi}{4 - \pi} M.$$ (6.6-12)

This expression implies a prohibitively large number of transmitters to achieve a reasonable signal-to-noise ratio, as shown in Figure 6.6-2, where it is plotted (lower curve) for $S = 0.1$.

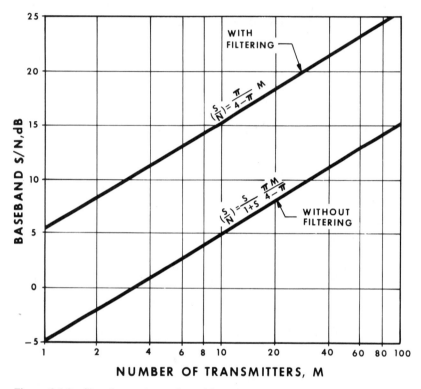

Figure 6.6-2 Signal-to-noise ratios with and without filtering, $s = 0.1$.

If most of the spectrum of $r_k(t)$ lies below the lowest baseband frequency ($f_m \leqslant 150$ Hz), then the noise may be considerably reduced by filtering. In this case the noise is associated with the broadening of the modulation spectrum; the component due to the carrier fading is suppressed, leaving as the residual noise

$$n_f(t) \doteq v(t) \sum_{k=1}^{M} \left[r_k(t) - \sqrt{\frac{\pi}{2}} b_{0k} \right]. \qquad (6.6\text{-}13)$$

The noise power $\langle n_f^2(t) \rangle$ may be evaluated as before; the signal power is unchanged, thus the signal-to-noise ratio with filtering becomes (assuming equal b_{0k}):

$$\left(\frac{P_s}{P_n}\right)_f = \frac{\pi}{4-\pi}M. \qquad (6.6\text{-}14)$$

The S/N ratio is thus improved over the unfiltered case by the factor $1+1/S=11$ of 10.4 dB for $S=0.1$. Nevertheless, a prohibitively large number of transmitters is still required, as shown in Figure 6.6-2.

A further improvement in the signal-to-noise ratio for slow fading may be obtained by the proper application of automatic gain control (AGC). Since there is only one IF amplifier, its gain would be controlled by an average of the signal strengths. The AGC bandwidth would have to be carefully adjusted to reach an optimum compromise between the desired objective of suppressing the low-frequency fading modulation while not affecting the desired signal modulation. As much as 20 dB of improvement might be obtained over the "filtered" case just treated. The number of transmitters required could thereby be brought within reason, with as few as three or four resulting in a 30-dB signal-to-noise ratio. The restriction on fading rate limits the maximum carrier frequency to about 1500 MHz, assuming a top vehicle speed of 60 mi/hr.

Alternatively, by providing a separate receiver for each transmitted carrier and adding the detector outputs, it would be possible to use "pilot" AGC (Section 4.1.7) and remove much of the higher frequency noise appearing in the audio band. The simplicity of one receiver has now been lost, however.

6.7 DIGITAL MODULATION-DIVERSITY SYSTEMS

There are three major factors which degrade the average bit error rate of digital transmission systems in a fading environment. The envelope fading degrades the carrier-to-noise ratio, the random FM causes errors in angle modulation systems, and the time delay spreads produce intersymbol interferences.

The bit error rate caused by envelope fading may be reduced by increasing transmitted power but increasing power is not at all effective against random FM and time delay spread. In this section we shall show that diversity combining can be used to obtain orders of magnitude reduction in bit error rate attributable to these three causes.

The bit error rate caused by envelope fading alone will be defined as P_1, the irreducible error rate due to random FM will be P_2 and the irreducible error rate caused by time delay spread will be P_3. In the discussions to follow, material from Section 4.2 is used freely; therefore the reader is advised to read Section 4.2 first.

6.7.1 Rayleigh Fading

The received digital signal after the front-end matched filter in the lth branch of an m-branch diversity system [Eq. (4.2-37)] is

$$y_l(t) = r_l(t) \sum_k \mathbf{a}_k h(t - kT) \cos[\omega_c t + \phi_l(t)]$$

$$+ n_c \cos \omega_c t - n_s \sin \omega_c t, \tag{6.7-1}$$

where

$$h(kT) = \begin{cases} 1 & k = 0 \\ 0 & k \neq 0, \end{cases}$$

and r_l is a Rayleigh fading envelope, ϕ_l is the random FM, n_c and n_s are in-phase and quadrature noises, $\langle n_c^2 \rangle = \langle n_s^2 \rangle = \sigma^2$,

$$\mathbf{a}_k = \begin{cases} \pm 1 & \text{binary AM (or PSK)} \\ 1, 0 & \text{on-off AM,} \end{cases}$$

T is the interval between pulses, and $f_s = 1/T$ is the bit rate.

The peak carrier-to-noise ratio $[t = kT]$ is

$$\rho_p = \frac{r_l^2}{2\sigma^2}. \tag{6.7-2}$$

In order to facilitate comparison of systems on the average transmitter power basis an average carrier-to-noise ratio, ρ can be defined as

$$\rho = \frac{r_l^2 P_T}{N_0 f_s}, \tag{6.7-3}$$

where P_T is the transmitter power, $r_l^2 P_T$ is the average carrier power before the receiving filter, and $N_0 f_s$ is the thermal noise power in bandwidth f_s.

The average CNR and peak CNR are related through Eq. (4.2-30) for systems with optimum filter distributions by

$$\rho = \begin{cases} \rho_p & \text{binary AM} \\ \frac{1}{2}\rho_p & \text{on-off AM} \end{cases}. \tag{6.7-4}$$

Since r is Rayleigh distributed, the probability density of ρ is

$$f(\rho) = \frac{1}{\rho_0} e^{-\rho/\rho_0}, \qquad \rho \geqslant 0, \tag{6.7-5}$$

and

$$\rho_0 = \langle \rho \rangle_{av} = \frac{\langle r_l^2 \rangle_{av} P_T}{N_0 f_s} \tag{6.7-6}$$

is the average carrier-to-noise ratio, averaged over the Rayleigh fading with noise bandwidth f_s.

We shall examine binary AM first, that is, $a_k = \pm 1$. If predetection diversity combining is used we certainly need means to bring each branch of y_l to the same RF phase. This may be accomplished by a pilot tone or other carrier phase recovery techniques. In the case of m-branch predetection maximal-ratio diversity combining, assuming independent fading on each branch, the carrier-to-noise ratio after combining is no longer Rayleigh distributed. It is given in Section 5.2 by

$$f_m(\rho) = \frac{1}{\rho_0} \left(\frac{\rho}{\rho_0} \right)^{m-1} \frac{1}{(m-1)!} e^{-\rho/\rho_0}, \qquad \rho \geqslant 0,$$

where ρ_0 is the average carrier-to-noise ratio of a single branch, defined in Eq. (6.7-6).

The bit error rate of a coherent detection system can be obtained by averaging over $f_m(\rho)$

$$P_1 = \int_0^\infty \tfrac{1}{2} \operatorname{erfc}(\rho^{1/2}) f_m(\rho) \, d\rho$$

$$\cong \frac{(2m-1)!}{m!(m-1)!} \frac{1}{(4\rho_0)m}, \tag{6.7-7}$$

where $\tfrac{1}{2} \operatorname{erfc}(\rho^{1/2})$ is the error rate of a coherent detection system at CNR ρ (Eq. 4.2-13).

Equation (6.7-7) is plotted in curve 1 of Figures 6.7-1 and 6.7-2 for two- and four-branch systems. These curves represent the best theoretical error rate performance.

The bit error rate of a differentially coherent detection system can also be calculated after recalling that the error rate under CNR ρ is $\tfrac{1}{2} e^{-\rho}$ [Eq. (4.2-19)]. The bit error rate with m-branch predetection maximal ratio combining is[32]

$$P_1 = \int_0^\infty \tfrac{1}{2} e^{-\rho} f_m(\rho)\, d\rho$$

$$\cong \frac{1}{2}\left(\frac{1}{\rho_0}\right)^m. \tag{6.7-8}$$

Equation (6.7-8) is presented in curve 2 of Figures 6.7-1 and 6.7-2.

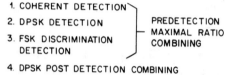

1. COHERENT DETECTION

2. DPSK DETECTION

3. FSK DISCRIMINATION DETECTION

PREDETECTION MAXIMAL RATIO COMBINING

4. DPSK POST DETECTION COMBINING

5. ON−OFF AM { A. OPTIMUM FIXED THRESHOLD B. OPTIMUM MOVING THRESHOLD

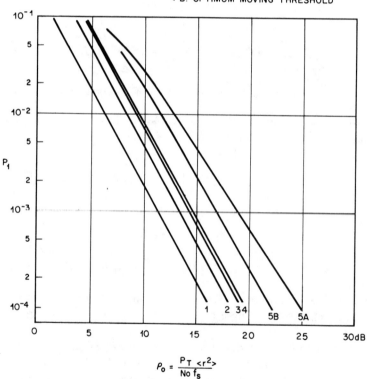

$$\rho_0 = \frac{P_T \langle r^2 \rangle}{No\, f_s}$$

Figure 6.7-1 Bit error rate due to envelope fading in two branch diversity combining systems.

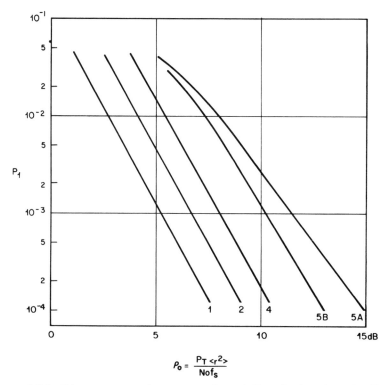

1. COHERENT DETECTION ⎫ PREDETECTION
                             ⎬ MAXIMAL RATIO
2. DPSK DETECTION   ⎭ COMBINING
4. DPSK POST DETECTION COMBINING
5. ON-OFF AM ⎰ A. OPTIMUM FIXED THRESHOLD
              ⎱ B. OPTIMUM MOVING THRESHOLD

$$P_0 = \frac{P_T <r^2>}{N_0 f_s}$$

Figure 6.7-2 Bit error rate due to envelope fading in four-branch diversity combining systems.

For FSK with discriminator detection, since there are yet no analytic expressions for the bit error rate under low CNR, integration over $f_m(\rho)$ cannot be performed. Nevertheless, after examining curve 5 of Figure 4.2-6 we observed that the experimental error rate curve follows the coherent detection curve closely except for about 3 dB shift in CNR. Thus we may approximate the error rate by $\frac{1}{2} \text{erfc}([\rho/2]^{1/2})$, and with this approximation the system error rate using maximal ratio predetection combining is

$$P_1 \cong \frac{(2m-1)!}{m!(m-1)!} \left(\frac{1}{2\rho_0}\right)^m . \tag{6.7-9}$$

Equation (6.7-9) is presented in curve 3 of Figure 6.7-1.

Predetection combining requires cophasing of the different branches at RF and hence complicated circuitries. In DPSK a simple post detection system can be readily implemented as shown in Figure 6.7-3. There the signal in each branch is multiplied with its one-bit delayed version and then combined at baseband with signals from other branches. To derive the error rates in such a system let us first define complex envelopes $\tilde{y}_l(t)$ for the real waveform $y_l(t)$ in Eq. (6.7-1):

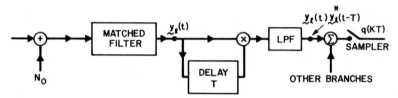

Figure 6.7-3 Postdetection diversity combining in DPSK systems.

$$y_l(t) = \mathrm{Re}\left\{ \tilde{y}_l(t) e^{j\omega_c t} \right\}. \tag{6.7-10}$$

Therefore

$$\tilde{y}_l(t) = \tilde{u}_l(t) \sum_k a_k h(t - kT) + \tilde{n}_l(t), \tag{6.7-11}$$

where

$$\tilde{u}_l(t) = r_l(t) e^{j\phi(t)}, \tag{6.7-12}$$

$$\tilde{n}_l(t) = n_c + jn_s. \tag{6.7-13}$$

The sampler output at $t = 0$ is

$$q = \sum_{l=1}^m y_l(0) y_l^*(-T)$$

$$= \sum_{l=1}^m [a_0 u_l(0) + \tilde{n}_l(0)][a_0 u_l^*(-T) + \tilde{n}_l^*(-T)], \tag{6.7-14}$$

where the asterisk indicates complex conjugate. The decision rules are

$$a_0 a_{-1} = 1 \quad \text{if} \quad q > 0 \tag{6.7-15}$$

$$a_0 a_{-1} = -1 \quad \text{if} \quad q < 0.$$

Bello and Nelin[41] have shown that for complex Gaussian variables u_l and v_l having same mean and variance the following product,

$$q = \sum_{l=1}^{m} (u_l v_l^* + u_l^* v_l), \tag{6.7-16}$$

satisfies

$$P_r\{q>0\} = \left[\frac{1+\gamma}{2+\gamma}\right]^m \sum_{k=0}^{m-1} \left(\frac{1}{2+\gamma}\right)^k \binom{m+k-1}{k}, \tag{6.7-17}$$

$$P_r\{q<0\} = \frac{1}{(2+\gamma)^m} \sum_{k=0}^{m-1} \left(\frac{1+\gamma}{2+\gamma}\right)^k \binom{m+k-1}{k}, \tag{6.7-18}$$

where

$$\binom{m}{n} = \frac{m!}{n!(m-n)!},$$

$$\gamma = \frac{2(m_{10}+m_{10}^*)}{\sqrt{(m_{10}+m_{10}^*)^2 + [m_{11}m_{00} - |m_{10}|^2]} - (m_{10}+m^*_{10})}, \tag{6.7-19}$$

$$m_{00} = \langle u_l u_l^* \rangle_{av}$$

$$m_{11} = \langle v_l v_l^* \rangle_{av}, \tag{6.7-20}$$

$$m_{10} = \langle u_l v_l^* \rangle_{av},$$

Referring to Eq. (6.7-15) we notice that an error would be committed if $q<0$ while $a_0 a_{-1}=1$ or $q>0$ while $a_0 a_1 = -1$. The error rate, assuming equal probability of $a_0 a_{-1} = \pm 1$, is

$$P_e = \tfrac{1}{2} P_r\{q>0|a_0 a_{-1} = -1\} + \tfrac{1}{2} P_r\{q<0|a_0 a_1 = 1\} \tag{6.7-21}$$

where $P_r\{A|B\}$ is the probability of A under condition B. Since $\tilde{y}_l(t)$ is a complex Gaussian random variable, Eqs. (6.7-17) and (6.7-18) can be applied. With the reasonable assumption that $\tilde{n}_0(0)$ and $\tilde{n}_0(T)$ are independent we obtain the following moments:

$$m_{00} = \langle r_l^2 \rangle_{av} + 2\sigma^2$$

$$m_{11} = \langle r_l^2 \rangle_{av} + 2\sigma^2 \tag{6.7-22}$$

$$m_{10} = a_0 a_{-1} \langle r_l^2 \rangle_{av} J_0(2\pi f_m T).$$

Therefore

$$\gamma = \frac{2a_0 a_{-1} \langle r_l^2 \rangle_{av} J_0(2\pi f_m T)}{\langle r_l^2 \rangle_{av} + 2\sigma^2 - a_0 a_{-1} \langle r_l^2 \rangle_{av} J_0(2\pi f_m T)} .$$

Recalling that $\rho_0 = \langle r_l^2 \rangle_{av} / 2\sigma^2$ we have

$$\gamma = \frac{2\rho_0 a_0 a_{-1} J_0(2\pi f_m T)}{1 + \rho_0 - a_0 a_{-1} J_0(2\rho f_m T)} . \qquad (6.7\text{-}23)$$

It is simple to show that the conditional probabilities in Eq. (6.7-21) are equal. Hence, we need only to consider one case. For $a_0 a_{-1} = 1$ we have

$$\gamma = \frac{2\rho_0 J_0(2\pi f_m T)}{1 + \rho_0 - J_0(2\pi f_m T)} . \qquad (6.7\text{-}24)$$

Substituting Eq. (6.7-24) into Eq. (6.7-18) yields P_e.

Letting $m = 1$ we obtain the P_e for Rayleigh fading of Section 4.2:

Single branch: $\qquad P_e = \dfrac{1 + \rho_0[1 - J_0(2\pi f_m T)]}{2(1 + \rho_0)} .\qquad (6.7\text{-}25)$

For two-branch diversity combining we have

$$P_e = \left[\frac{1 + \rho_0[1 - J_0(2\pi f_m T)]}{2(1 + \rho_0)} \right]^2 \left[\frac{2(\rho_0 + 1) + \rho_0 J_0(2\pi f_m T)}{\rho_0 + 1} \right]. \qquad (6.7\text{-}26)$$

Letting $\rho \to \infty$ (i.e., infinite carrier-to-noise ratio) we obtain the irreducible probability of error due to random FM:

Two branch: $\qquad P_2 \cong \frac{1}{4}[1 - J_0(2\pi f_m T)]^2 [2 + J_0(2\pi f_m T)]. \qquad (6.7\text{-}27)$

This is the irreducible error due to random FM.

Letting $f_m \to 0$, that is, extremely slow fading we obtain the error rate due to envelope fading alone:

Two branch: $\qquad P_1 \cong \dfrac{3}{4} \left(\dfrac{1}{\rho_0 + 1} \right)^2 . \qquad (6.7\text{-}28)$

For four-branch combining we obtain

$$P_1 \cong \frac{35}{16}\left(\frac{1}{1+\rho_0}\right)^4.$$ (6.7-29)

The results of Eq. (6.7-28) and (6.7-29) are presented in curve 4 of Figures 6.7-1 and 6.7-2, respectively. It is immediately apparent that the error rates of postdetection diversity combining of DPSK degrade only slightly in comparison to that of the predetection cases.

Postdetection diversity combining of FSK with discriminator detection can also be argued to perform only slightly inferior in comparison to predetection combining on the basis that it is inherently a small index system and hence the FM threshold effect is not pronounced. Therefore, diversity combining before or after discriminator detection would not be significantly different. In postdetection combining random FM would again produce irreducible errors. For example, in a maximal-ratio postdetection system with peak signal frequency deviation of f_d, the combiner output at the sampling instant is

$$v = \sum_{l=1}^{m} r_l^2 [2\pi f_d + \dot{\phi}_l],$$ (6.7-30)

where γ_l is the Rayleigh envelope at lth branch, and $\dot{\phi}_l$ is the random FM at lth branch.

For positive f_d, the irreducible error P_2 caused by random FM is

$$P_2 = P_r\{v < 0 | f_d > 0\}.$$ (6.7-31)

Assuming that the fading on different diversity branches is independent, Eq. (1.4-5) contains sufficient information for evaulation of Eq. (6.7-30). After some lengthy calculations, we obtain

$$P_2 = \frac{(2m-2)!}{(m-1)!^2} \frac{1}{(2Q^2+4)^m} \frac{2\sqrt{Q^2+2}}{Q+\sqrt{Q^2+2}},$$ (6.7-32)

where $Q = 2f_d/f_m$, and f_m is the Doppler frequency.

For large values of Q, we have

$$P_2 \cong \frac{(2m-2)!}{(m-1)!^2}\left[\frac{f_m^2}{8f_d^2}\right]^m.$$ (6.7-33)

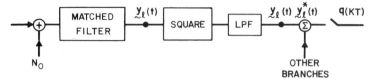

Figure 6.7-4   Diversity combining in on-off AM systems.

We shall now examine the on-off AM case, that is, $a_\kappa = 1$ or 0 in Eq. (6.7-1). A postdetection combining system is shown in Figure 6.7-4. Here again for simplicity we have adopted the complex notation defined in Eqs. (6.7-10)–(6.7-13). After squaring and combining at baseband the combiner output at the sampling instant $(t=0)$ is

$$y = \sum_{l=1}^{m} [a_0 \tilde{u}_l(0) + \tilde{n}_l(0)][a_0 \tilde{u}_l^*(0) + \tilde{n}_l^*(0)]. \qquad (6.7\text{-}34)$$

Since $a_0 \tilde{u}_l(0) + \tilde{n}_l(0)$ is a complex Gaussian random variable. y can be considered as the sum of n squared independent Rayleigh variables. The probability density of y is given by a gamma function,[43]

$$f(y) = \frac{y^{m-1}}{(m-1)! b^{2m}} e^{-y/b^2}, \qquad y \geqslant 0, \qquad (6.7\text{-}35)$$

where

$$b^2 = \langle [a_0 \tilde{u}_l(0) + \tilde{n}_l(0)][a_0 \tilde{u}_l^*(0) + \tilde{n}_l^*(0)] \rangle_{av}$$

$$= a_0^2 \langle r^2 \rangle + 2\sigma^2. \qquad (6.7\text{-}36)$$

when $a_0 = 1$ the density function is

$$f_1(y) = \frac{y^{m-1}}{(m-1)! [\langle r^2 \rangle_{av} + 2\sigma^2]^m} e^{-y/[\langle r^2 \rangle_{av} + 2\sigma^2]}, \qquad (6.7\text{-}37)$$

and when $a_0 = 0$ the density function is

$$f_0(y) = \frac{y^{m-1}}{(m-1)! [2\sigma^2]^m} e^{-y/2\sigma^2}. \qquad (6.7\text{-}38)$$

Let the decision threshold be q, that is,

$$\begin{aligned} \mathbf{a}_0 &= 1 \quad \text{if} \quad y > q \\ \mathbf{a}_0 &= 0 \quad \text{if} \quad y < q. \end{aligned} \tag{6.7-39}$$

Assuming equal probability of one or zero being transmitted, the bit error rate is

$$P_e = \frac{1}{2} \int_q^\infty f_0(y)\, dy + \frac{1}{2} \int_0^q f_1(y)\, dy. \tag{6.7-40}$$

Differentiating Eq. (6.7-40) with respect to q leads to the following optimum (minimum error-probability) threshold, q_0:

$$f_0(q_0) = f_1(q_0). \tag{6.7-41}$$

Using (6.7-37) and (6.7-38) we solve for q_0 to obtain

$$q_0 = 2\sigma^2 \frac{1 + \langle r^2 \rangle_{\text{av}}/2\sigma^2}{\langle r^2 \rangle_{\text{av}}/2\sigma^2} m \log\left(1 + \frac{\langle r^2 \rangle_{\text{av}}}{2\sigma^2}\right). \tag{6.7-42}$$

Since $\langle r^2 \rangle_{\text{av}}/2\sigma^2$ is the average peak carrier-to-noise in one branch, the threshold q_0 can be determined. In terms of average transmitter power required we may recall that $\rho_P = 2\rho$ [Eq. (4.2-32)] for on-off AM; therefore

$$\frac{\langle r^2 \rangle_{\text{av}}}{2\sigma^2} = \langle \rho_P \rangle_{\text{av}} = 2\rho_0, \tag{6.7-43}$$

where ρ_0 is defined by Eq. (6.7-6).Substituting Eq. (6.7-42) into (6.7-40) we obtain the following series expansion:

$$P_e = e^{-q_0/2\sigma^2} \sum_{l=0}^{m-1} \frac{(q_0/2\sigma^2)^l}{l!} + 1 - e^{-x} \sum_{l=0}^{m-1} \frac{x^l}{l!}, \tag{6.7-44}$$

where

$$x = \frac{q_0}{2\sigma^2(1 + 2\rho_0)}.$$

Evaluation[42] of Eq. (6.7-44) as a function of ρ_0 is presented in curve 5A of Figure 6.7-1 and 6.7-2.

In previous discussions of the on-off AM system we derived the optimum threshold based on the average fading envelope. The system error rate can be further improved if we change our threshold according to instantaneous combined carrier power. This is an optimum "moving threshold" strategy. Results due to Langseth[43] are presented in curve 5B of Figure 6.7-1 and 6.7-2.

We shall now examine the performance of various systems with Rayleigh fading and two branch diversity combining, shown in Figure 6.7-1. We immediately notice that the postdetection combining of DPSK is about only 1.5 dB inferior to predetection combining in required carrier-to-noise ratio. Predetection combining and coherent detection provide the best error rate performance but also the most complicated circuitry requirements. FSK with discriminator detector performs about the same as DPSK systems and probably is the simplest to implement. On-off AM with fixed optimum threshold is about 5 dB inferior to DPSK at an error rate of 10^{-4}. With optimum moving threshold it is about 25 dB inferior to DPSK in the required average transmitter power. If the transmitter is peak power limited the FSK and PSK systems would have an additional 3 dB power advantage over on-off AM because the latter transmits only half the time.

The bit error rate of various systems with four-branch diversity combining is shown in Figure 6.7-2. We notice that DPSK with postdetection combining requires only 10 dB average carrier-to-noise to produce a reasonable error rate of 10^{-4}. Here the noise power is referred to the noise in a bandwidth of bit rate f_s.

6.7.2   Frequency-Selective Fading

The irreducible intersymbol error rate P_3 caused by time delay spread for DPSK detection of binary AM with fully raised cosine shaping (Eq. 4.2-7) has been calculated by Bailey and Lindenlaub[44] for postdetection combining systems. The results are presented in Figure 6.7-5 as a function of $d = f_s \Delta$. Here Δ is the rms value of the time delay spread. For $\Delta = 2.5$ μsec (values encountered in downtown New York) and $f_s = 30$ kHz, we obtain an irreducible error rate due to delay spread of about 3×10^{-6}, which is quite small. This is in sharp contrast to the case where P_3 is about 10^{-3} in the single-branch Rayleigh fading case reported in Section 4.2.4.

Since P_3 is small with two-branch diversity combining, we may not need fully raised cosine baseband shaping for the control of intersymbol interferences. For example, a partial cosine roll-off filter may be used to save RF bandwidth while allowing somewhat higher values of P_3 than 3×10^{-6}.

For comparison, the ideal rectangular pulsed FSK system with peak

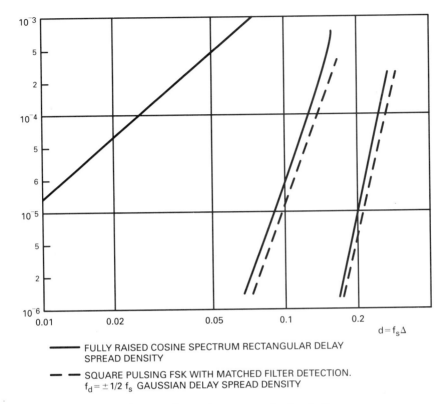

FULLY RAISED COSINE SPECTRUM RECTANGULAR DELAY
SPREAD DENSITY

SQUARE PULSING FSK WITH MATCHED FILTER DETECTION.
$f_d = \pm 1/2 f_s$ GAUSSIAN DELAY SPREAD DENSITY

Figure 6.7-5 Irreducible error due to frequency-selective fading.

deviation $f_s/2$ and incoherent matched filter detection[45] is also shown in
Figure 6.7-5. The results are quite similar to those of DPSK.

6.7.3 Examples

Recall that in Eq. (4.2-48) we have shown that $P_e \cong P_1 + P_2$ and in Eq.
(4.2-72) that $P_e \cong P_1 + P_3$. It is therefore a reasonable approximation to
include the effects of P_1 (Rayleigh fading), P_2 (random FM), and P_3
(frequency-selective fading) on system error rate by $P_e \stackrel{\Delta}{=} P_1 + P_2 + P_3$. On
this basis the system parameters can be determined for given environ-
ments. For example, the error rate of two-branch postdetection combining
of DPSK with fully raised cosine spectrum and optimum filter is presented
in Figure 6.7-6. The error rate P_e is shown as a function of f_s/f_m (the ratio
of signaling rate to Doppler frequency), $f_s\Delta$ (the product of signaling rate
and rms time delay spread), and ρ_0 (the average carrier-to-noise ratio

referred to a bandwidth of f_s). It is observed that $\rho_0 = 19$ dB would give an error rate $P_1 = 10^{-4}$. So long as $f_s/f_m > 30$ and $f_s\Delta < 0.09$ the contributions of P_2 and P_3 are negligible and P_e is dominated by P_1. As we decrease f_s/f_m or increase $f_s\Delta$, the influence of random FM and time delay spread starts to show up. As a numerical example, let $f_m = 100$ Hz, $\Delta = 2.5$ μsec, and $f_s = 30$ kHz; we thus have $f_s/f_m = 300$ and $f_s\Delta = 0.075$. From Figure 6.7-6 it is seen that P_e is predominantly determined by the envelope fading.

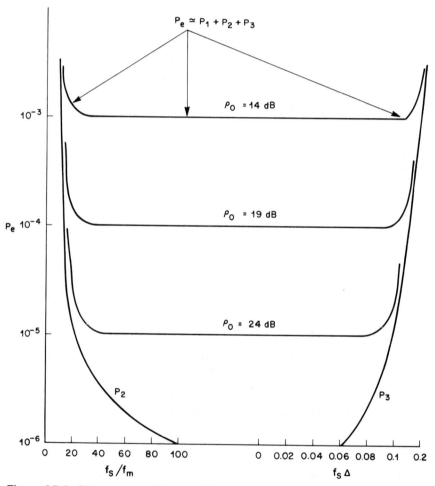

Figure 6.7-6 Bit error rate in a two-branch diversity combining DPSK system subject to envelope fading, random FM, and frequency-selective fading.

The required RF carrier-to-noise ratio for 10^{-4} error rate is 19 dB. In terms of required transmitter power, this is equivalent to that of an analog FM system with 30 kHz RF bandwidth and 19 dB average carrier-to-noise ratio at the receiver. The required RF bandwidth for the digital system may not need to be 60 kHz in view of the small P_3 cited above. With partial roll-off spectrum there are hopes that 45 kHz bandwidth may suffice, that is, 50% roll-off in baseband spectrum.

6.8 COMPARISON OF DIVERSITY SYSTEMS

In this section we will make comparisons of the performance of various diversity methods in the mobile radio environment.[46] Establishing a performance criterion for mobile systems is difficult since we are dealing with a randomly varying signal whose fading rate may change from zero (vehicle stationary) to a value equal approximately to the Doppler shift (75 Hz for 60 mi/hr and 840 MHz). Listening tests have indicated that above about 5 mi/hr the subjective effect of the fast fading is that of noise, and can be evaluated by calculating the baseband signal and noise averaged over the fading statistics for the specific system under study. In the following, system comparison will be made primarily on the basis of the transmitter power required to achieve a certain SNR (signal-to-noise ratio) computed in this manner, with transmission bandwidth as a parameter. A secondary measure of system performance is its reliability, defined here as the probability of obtaining an SNR at least as good as some specified value whenever the vehicle stops. Only the Rayleigh distribution will be considered; gross terrain effects (shadows due to hills, and the like) will be neglected. The comparison method will be illustrated in the following with numerical examples for an urban system operating at UHF. This simplifies the treatment somewhat, since the effects of random FM, which become important at the higher microwave frequencies, can then be neglected.

6.8.1   Reference System (No Diversity)

We assume the following parameters for a system in an urban environment:

frequency	840MHz
base antenna height	30 meters (100ft)
mobile antenna height	1.5 meters (5ft)
range d	3.2 km (2 mi)
base antenna gain G_B	9 dB
mobile antenna gain G_M	3 dB

The line-of-sight loss between these antennas is

$$L_0 = 20\log_{10}\frac{4\pi d}{\lambda} - (G_B + G_M) = 101 - 12 = 89 \text{ dB} \qquad (6.8\text{-}1)$$

Applying an additional loss of 41 dB to account for the mobile path characteristics (Section 2.5), the net loss is

$$L = 89 + 41 = 130 \text{ dB.} \qquad (6.8\text{-}2)$$

We will assume that FM is used with Gaussian modulation flat over the band W extending from 0 to 3 kHz. The baseband SNR is a function of the parameters α and ρ, where $\alpha = B/2W$, B is the RF bandwidth, and ρ is the carrier-to-noise ratio (CNR). Now ρ represents the Rayleigh fading signal at the receiver input. To obtain the effect of the fading, the signal is averaged over the fading distribution as described in Section 4.1, and the resulting baseband SNR calculated as a function of ρ_0 (the mean value of ρ) and the parameter α. Using these results the required value of ρ_0 to obtain a given SNR is determined as a function of α, with the results shown in Table 1. In this and subsequent systems we will choose a baseband SNR of 30 dB as a reference.

Table 1 *Values of ρ_0 Required to Obtain 30 dB SNR without Diversity*

α	B (kHz)	ρ_0 (dB)
2	12	42.0
3	18	37.4
4	24	35.0
6	36	32.7
8	48	31.5
10	60	30.6
15	90	29.7
20	120	29.0

Knowing ρ_0 and the receiver noise figure, assumed to be 3 dB, the transmitter power required may then be calculated:

$$\begin{aligned} P_T &= \rho_0 + NF + 10\log_{10}kTB + L \\ &= \rho_0 + 3 - 174 + 10\log_{10}B + L \\ &= \rho_0 + 10\log_{10}\alpha - 3.2, \text{ dBm,} \end{aligned} \qquad (6.8\text{-}3)$$

where

$$kT = (1.38 \times 10^{-20})(290°) = 4 \times 10^{-18}, \text{ mW/Hz.}$$

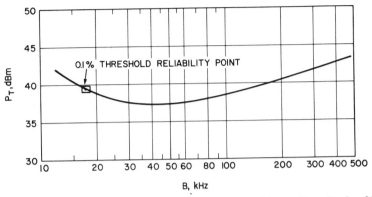

Figure 6.8-1 Transmitter power required at 836 MHz with no diversity for 30 dB SNR at baseband with Gaussian FM modulation. B is RF bandwidth, vehicle speed is 60 mi/hr, range is 2 mi in city, base antenna height is 30 meters, total antenna gain is 12 dB.

NF is the noise figure and L is the net loss of 130 dB, previously computed. The variation of required power with bandwidth B is shown in Figure 6.8-1, and we see that the power falls from 16 to 5.5 W as B increases from 12 to 40 kHz. For $B \geqslant 40$ kHz, the power slowly increases; thus $B = 40$ kHz would probably be the maximum value considered.

The reliability of this system may be obtained from the Rayleigh distribution curve ($M = 1$) of Figure 5.2-2. Suppose we require the probability of stopping in a location where $\rho \geqslant \rho_t$, the FM click threshold (Chapter 4) given by the condition that the click noise, N_c, be 7.5 dB less than the quadrature noise, $N_{\theta'}$:

$$\rho_t \, \text{erfc} \, \sqrt{\rho_t} = \frac{0.75\sqrt{2\pi}}{(10\alpha)^2}. \tag{6.8-4}$$

Values of ρ_t versus α are listed below.

α	2	3	4	6	8	10	15	20
ρ_t (dB)	7.5	8.2	8.6	9.1	9.4	9.6	10.0	10.3

The Rayleigh curve is then entered at a signal level of ρ_t/ρ_0 for the various values of α and ρ_0 already given, and the corresponding probability is then obtained. The results are plotted in Figure 6.8-2 for $B \leqslant 400$ kHz.

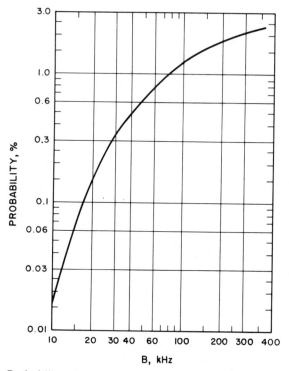

Figure 6.8-2   Probability of stopping in a location where signal level is at or below click threshold in reference system (no-diversity).

In order to select the system design parameters, the relative importance of the SNR experienced while moving compared with the chance of stopping at an unfavorable signal level will have to be examined from a subjective viewpoint. We will arbitrarily assume that enjoying a 30 dB SNR while moving is just as important as stopping at the click threshold only one time in a thousand, that is, a reliability of 99.9% with respect to threshold. This point corresponds to a transmitted power of 39.4 dBm and a bandwidth of 17.5 kHz, as marked in Figure 6.8-1. These numbers characterize the reference system, and will be compared with equivalent values for diversity systems.

The reader is reminded that random FM has been neglected in the analysis here. At 60 mi/hr and 840 MHz, requiring a 30 dB SNR, this assumption is valid for $\alpha > 4$ ($B > 24$ kHz), or, alternatively, for $B > 12$ kHz if the speed is limited to 15 mi/hr.

6.8.2 Selection Diversity

The selection diversity system shown in Figure 5.2-1 is a receiver diversity type that can be used either at the base or mobile, the only difference being the larger antenna separation required at the base station.

We proceed as above to determine the required transmitter power and realibility. Davis[47] gives the required value of ρ_0, the mean CNR per diversity branch, to obtain a desired baseband SNR in the presence of fading for various numbers of diversity branches M. We again assume 30 dB baseband SNR. The transmitted power for this system is then

$$P_T = \rho_0 + 10\log_{10}\alpha - 3.2, \quad \text{dBm.} \tag{6.8-5}$$

Table 2 lists the values of ρ_0 in decibels and P_T in dBm. The values of required power are plotted in Figure 6.8-3, and behavior similar to the no-diversity case is seen except for a substantial reduction in required power.

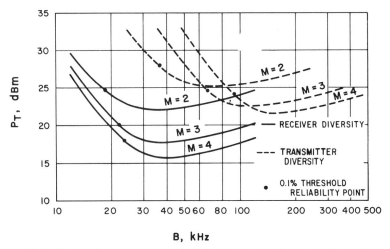

Figure 6.8-3 Transmitter power required for selection diversity with 2, 3, and 4 branches to obtain 30 dB SNR.

For the transmitter diversity arrangement of Figure 5.3-4 the required total power and RF bandwidth are simply M times the values for the receiver diversity scheme, that is

$$P_T = \rho_0 + 10\log_{10}\alpha - 3.2 + 10\log_{10}M, \quad \text{dBm.} \tag{6.8-6}$$

Table 2 *Selection Diversity*

α	B (kHz)	$M=2$ ρ_0(dB)	P_T(dBm)	$M=3$ ρ_0	P_T	$M=4$ ρ_0	P_T
2	12	30.0	29.8	28.1	27.9	27.2	27.0
3	18	23.3	24.9	20.6	22.2	19.6	21.2
4	24	20.0	22.8	16.4	19.2	15.0	17.8
6	36	17.7	22.3	13.2	17.8	11.1	15.7
8	48	16.5	22.3	12.2	18.0	10.0	15.8
10	60	16.0	22.8	11.6	18.4	9.6	16.4
15	90	15.1	23.7	10.9	19.5	8.9	17.5
20	120	14.8	24.6	10.5	20.3	8.5	18.3

These values are shown in Figure 6.8-3 as dotted curves.

The 99.9% threshold reliability points are also marked on the curves of Figure 6.8-3 and tabulated in Table 3. They are obtained as previously described for the no-diversity case: enter the distribution curves of Figure 5.2-2 for signal levels of ρ_t/ρ_0 for the various values of α and ρ_0 already given and the corresponding probability is obtained. As an example, let $M=2$ and $\alpha=4$; then $\rho_0=20.0$ dB, $\rho_t(4)=8.6$ dB, and $\rho_t/\rho_0=11.4$ dB. The curve for $M=2$ in Figure 5.2-2 gives a probability of 99.5%. Plotting the probability versus α enables us to find the point for 99.9%. The results are given in Table 3.

Table 3 *99.9% Reliability Points*

M	Receiver Diversity P_T (dBm)	B (kHz)	Transmitter Diversity P_T (dBm)	B (kHz)
2	24.8	18.5	27.8	37.0
3	20.0	22.0	24.8	66.0
4	18.0	23.5	24.0	94.0

In the foregoing treatment the effects of random FM have again been neglected. Any kind of diversity helps to reduce random FM; this is due to the fact that the phase of the RF signal changes more and more rapidly as it drops deeper into a fade. Thus the worst random FM excursions occur during the fading instants of the signal. Selection diversity utilizes the signal during times when the fading is less; thus it discriminates against the more severe bursts of random FM.[48] For this reason it is felt that the preceding results may be applied for vehicle speeds up to 60 mi/hr at 840 MHz without restrictions on bandwidth, contrary to the case with no diversity.

6.8.3 Maximal-Ratio Combining

The diversity scheme of Figure 5.2-5 that uses a separate unmodulated carrier (the "pilot") transmitted along with a modulated carrier provides predetection maximal ratio combining. As noted earlier, there is an overall penalty of 6 dB in CNR: 3 dB because half the power is in the pilot and 3 dB because both pilot and carrier heterodyne noise into the baseband. The total required transmitter power for our 2-mile path is given by

$$P_T = \rho_0 + 10\log_{10}\alpha + 2.8, \quad \text{dBm} \tag{6.8-7}$$

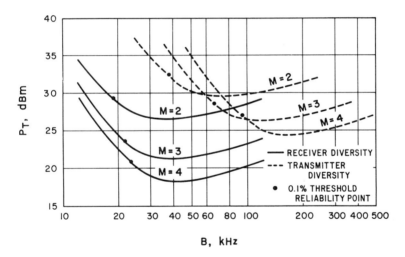

Figure 6.8-4 Transmitter power required for pilot-carrier maximal ratio diversity with 2, 3, and 4 branches to obtain 30 dB SNR.

where ρ_0 is the CNR per branch to obtain a 30-dB SNR while moving, given by Davis,[48] for maximal ratio combining, as shown in Table 4. The 6-dB penalty has also been applied. The values of required power are plotted in Figure 6.8-4. The curves are similar to selection diversity except for an additional 4–6 dB of transmitter power.

Table 4   *Maximal Ratio Combining*

α	B (kHz)	$M=2$		$M=3$		$M=4$	
		ρ_0(dB)	P_T(dBm)	ρ_0	P_T	ρ_0	P_T
2	12	28.6	34.4	25.6	31.4	23.9	29.9
3	18	21.9	29.5	18.1	25.7	16.3	23.9
4	24	18.6	27.4	13.9	22.7	11.7	20.5
6	36	16.3	26.9	10.7	21.3	7.8	18.4
8	48	15.1	26.9	9.7	21.5	6.7	18.5
10	60	14.6	27.4	9.1	21.9	6.3	19.1
15	90	13.7	28.3	8.4	23.0	5.6	20.2
20	120	13.4	29.2	8.0	23.8	5.2	21.0

For the transmitter diversity scheme of Figure 5.2-4 the required total power and RF bandwidth are M times the values for receiver diversity, and the performance is shown by the dotted curves of Figure 6.8-4. Following the method previously described, we can obtain the 99.9% threshold reliability points. They are marked on the curves shown in Figure 6.8-4 and are listed in Table 5.

Table 5   *9.9.9% Reliability Points*

M	Receiver Diversity		Transmitter Diversity	
	P_T (dBm)	B (kHz)	P_T (dBm)	B (kHz)
2	29.3	19.0	32.3	38.0
3	23.5	21.8	28.3	65.4
4	20.8	23.0	26.8	92.0

The performance characteristics of the various diversity systems have been presented in terms of total power required at the transmitter versus bandwidth for a typical mobile transmission path in an urban environment. This gives one a feel for the order of magnitude of the powers involved. To make comparisons between systems, the ratio of the total required transmitter power for the various diversity systems to that of the reference system with no diversity is of interest. The curves of Figures 6.8-5 and 6.8-6 show these ratios expressed in decibels as a function of total RF

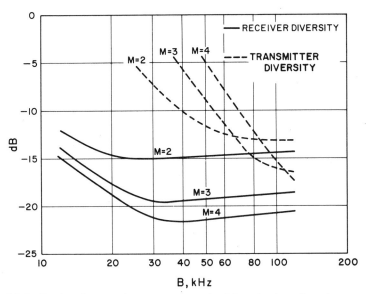

Figure 6.8-5 Ratio of transmitter power required in selection diversity system to that of no-diversity system.

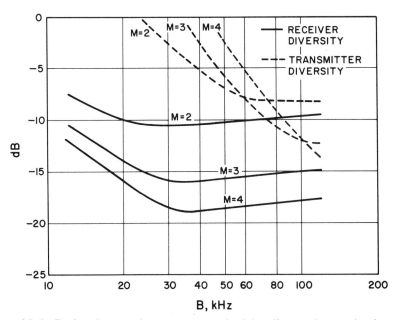

Figure 6.8-6 Ratio of transmitter power required in pilot-carrier maximal ratio diversity combiner to that of no-diversity system.

bandwidth for selection diversity and maximal-ratio combining, respectively. The solid curves compare receiver diversity with the no-diversity case, and the dotted curves are for transmitter diversity. Considering the receiver diversity cases, substantial reductions of transmitter power of 10–20 dB result from either selection or maximal ratio combining of two to four branches, with selection somewhat better by about one order of diversity. That is, two-branch selection diversity is about equivalent to three-branch maximal-ratio combining. This is due mainly to the 6-dB noise penalty incurred by the pilot carrier scheme. The advantages of transmitter diversity appear for larger bandwidths and generally are not as great.

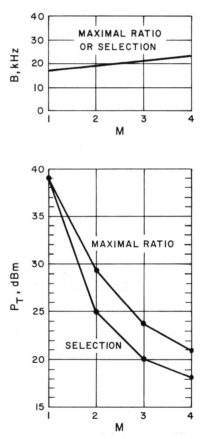

Figure 6.8-7 Bandwidth and required transmitter power versus diversity order for 30 dB SNR in motion and 99.9% threshold reliability when stopped.

Another interesting comparison uses the design criteria that the transmitted power and bandwidth are chosen to provide 30-dB SNR while moving and 99.9% threshold reliability when stopped. These values have been given earlier for the various techniques and are plotted in Figure 6.8-7 as a function of diversity order. Figure 6.8-8 shows the transmitter power saving for these design points, which again is in the range of 10–20 dB. These savings are substantial; for example, with no diversity the design point requires 8 W of transmitter power, whereas with two-branch selection receiver diversity only 300 mW are needed.

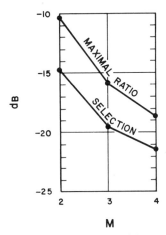

M

Figure 6.8-8 Ratio of transmitter power required in receiver diversity systems to that of no-diversity system versus diversity order for 30 dB SNR in motion and 99.9% threshold reliability when stopped.

It was pointed out that the large-scale shadowing caused by gross terrain features, such as large hills or especially high buildings, has been neglected. This shadowing causes variations in mean signal level that may be described by a log-normal distribution. Thus there may be times when the mean signal is so low that diversity schemes addressed to the problem of reducing multipath interference of the Rayleigh type will still not provide a usable signal. The mean signal must be increased in these cases, but the use of diversity makes possible much smaller an increase by the aforementioned amounts.

REFERENCES

1. H. H. Beverage and H. O. Peterson, "Diversity Receiving System of RCA for Radiotelegraphy," *Proc. IRE*, **19**, April 1931, pp. 531–584.

2. F. J. Altman and W. Sichak, "A Simplified Diversity Communication System for Beyond-The-Horizon Links," *IRE Trans. Comm. Systems*, **CS-4**, March 1956, pp.50–55.

3. P. Bello and B. Nelin, "Predetection Diversity Combining with Selectively Fading Channels," *IRE Trans. Comm. Systems*, **CS-9**, March 1963, pp. 32–42.

4. J. W. Boyhan, "A New Forward Acting Predetection Combiner," *IEEE Trans. Comm. Tech.*, **COM-15**, October 1967, pp. 41–47.

5. R. F. White, "Space Diversity on Line-of-Sight Microwave Systems," *IEEE Trans. Comm. Tech.*, **Com-16**, No. 1, February 1968, pp. 119–133.

6. H. Makino and K. Morita, "Design of Space Diversity Receiving and Transmitting Systems for Line-of-Sight Microwave Links," *IEEE Trans. Comm. Tech.*, **COM-15**, August 1967, pp. 603–614.

7. W. C. Bohn, "Automatic Selection of Receiving Channels," U. S. Patent No. 1,747,218, February 18, 1930.

8. C. L. Mack, "Diversity Reception in UHF Long-Range Communications," *Proc. IRE*, **43**, October 1955, pp. 1281–1289.

9. D. G. Brennan, "Linear Diversity Combining Techniques," *Proc. IRE*, **47**, June 1959, pp. 1075–1102.

10. P. F. Panter, *Modulation, Noise, and Spectral Analysis*, McGraw-Hill, New York 1965, pp. 228–235.

11. J. L. Lawson and G. E. Uhlenbeck, *Threshold Signals*, McGraw-Hill, New York, 1950, pp. 369–383.

12. A. J. Rustako, Jr., "Performance of a Two Branch Post-Detection Combining UHF Mobile Radio Space Diversity Receiver," MS Thesis, Newark College of Engineering, Newark, N. J., 1969.

13. A. Papoulis, *Probability, Random Variables, and Stochastic Processes*, McGraw-Hill, New York, 1965.

14. C. W. Rice, "Radio Receiving System," U. S. Patent No. 1,562,056, November 17, 1925.

15. R. T. Adams, "Combining System for Diversity Communication Systems," U. S. Patent No. 2,975,275, March 14, 1961.

16. R. T. Adams and B. M. Mindes, "Evaluation of Intermediate Frequency in Baseband Diversity Combining Receivers," *Elec. Comm.*, **36**, No. 2, 1960.

17. D. M. Black, P. S. Kopel, and R. J. Novy, "An Experimental UHF Dual-Diversity Receiver Using a Predetection Combining System," *IEEE Trans. Veh. Comm.*, **VC-15**, No. 2, October 1966, p. 43.

18. C. C. Cutler, R. Kompfner, and L. C. Tillotson, "A Self-Steering Array Repeater," *Bell System Tech. J.* **42**, September 1963, pp. 2013–2032.

19. A. J. Rustako, Jr., "Evaluation of a Mobile Radio Multiple Channel Diversity Receiver Using Pre-Detection Combining," *IEEE Trans. Veh. Tech.* **VT-16**, October 1967, pp. 46–57.

20. C. W. Earp, "Radio Diversity Receiving System," U. S. Patent No. 2,683,213,

July 6, 1954. (Priority application to Great Britain, Feb. 14, 1950.)

21. F. R. Shirley, "FM Suppression in a Mix-On-Self-Loop," *IEEE Trans. Comm. Tech.*, **COM-13**, No. 4, December 1965, pp. 471–475.

22. D. A. McKee, "An FM MTI Cancellation System," Lincoln Laboratory, MIT Tech. Report No. 171, January 8, 1958.

23. W. B. Davenport, Jr. and W. L. Root, *An Introduction to the Theory of Random Signals and Noise*, McGraw-Hill, New York, 1958, p. 158.

24. S. O. Rice, "Noise in FM Receivers," Chapter 25, pp. 395–422, in *Proceedings of the Symposium of Time Series Analysis*, M. Rosenblatt (ed.), Wiley, New York, 1963.

25. S. R. Lines, "Frequency-Division Multiplex Communications for the Land Mobile Radio Service," Report No. T-6801, Federal Communications Commision, Office of the Chief Engineer, June 20, 1968, p. 38.

26. J. Granlund, "Topics in the Design of Antennas for Scatter," Technical Report 135, Lincoln Laboratory, MIT, November 1956.

27. W. J. Bickford, R. G. Cease, D. B. Cooper, and H. J. Rowland, "Study of a Signal Processor Employing a Synthetic Phase Isolator," Raytheon Company, CADPO, Contract NAS 12-82, Tech. Report CDP-TR-2, October 1966.

28. S. W. Halpern, "The Theory of Operation of an Equal-Gain Predetection Regenerative Diversity Combiner with Rayleigh Fading Channels," *IEEE Trans. on Communication*, **COM-22**, August, 1974.

29. C. K. H. Tsao, R. G. Cease, W. J. Bickford, and H. J. Rowland, "Analysis of a Signal Processor for an Antenna Array," *IEEE Trans. Aerospace and Electronic Systems*, **AES-6**, January 1970, p 79.

30. E. N. Gilbert, "Mobile Radio Reception," *Bell System Tech. J.* **48**, September 1969, pp. 2473–2492.

31. S. O. Rice, "Mathematical Analysis of Random Noise," *Bell System Tech. J.* **23**, July 1944, pp. 282–322; and **24**, January 1945, pp. 46–156.

32. M. Schwartz, W. R. Bennett, and S. Stein, *Communication Systems and Techniques*, McGraw-Hill, New York, 1966.

33. M. Abramowitz and I. A. Stegun, *Handbook of Mathematical Functions*, National Bureau of Standards, AMS55 (1964).

34. D. O. Reudink and A. J. Rustako, Jr., "Mobile Signal Strength Measurements at 900 MHz in an Urban Area," IEEE Int. Conf. Comm., University of Colorado, June 9–11, 1969.

35. Special Issue on Active and Adaptive Arrays, *IEEE Trans. Ant. Prop.*, **AP-12**, March 1964, pp. 140–233.

36. S. P. Morgan, "Interaction of Adaptive Antenna Arrays in an Arbitrary Environment," *Bell System Tech. J.* **44**, January 1965, pp. 23–47.

37. Y. S. Yeh, "An Analysis of Adaptive Retransmission Arrays in a Fading Environment," *Bell System Tech. J.* **49**, October 1970, pp. 1811–1824.

38. J. S. Bitler, H. H. Hoffman, and C. O. Stevens, "A Mobile Radio Single

Frequency Two-Way Diversity System Using Adaptive Retransmission from the Base," *Joint IEEE Comm. Soc–Veh. Tech. Group Special Trans. Mobile Radio Comm.*, November 1973, pp. 1241–1247.

39. M. J. Gans and D. O. Reudink, "Retransmission System," U. S. Patent 3,631,494, December 28, 1971.

40. J. R. Brinkely, "A Method of Increasing the Range of VHF Communication Systems by Multi-Carrier Amplitude Modulation," *J. IEE*, **93**, Part III, May 1946, pp. 159–176.

41. P. A. Bello and B. D. Nelin, "The Influence of Fading Spectrum on the Binary Error Probabilities of Incoherent and Differentially Coherent Matched Filter Receivers," *IRE Trans. Comm. Systems*, June 1962, pp. 160–168.

42. H. B. Voelker, "Phase-Shift Keying in Fading Channels," *Proc. IEEE*, **107**, Part B, January 1960, p. 31.

43. R. E. Langseth, "On the Use of Space Diversity in the Square-Law Reception of Fading On-Off Keyed Signals," *IEEE Trans. Veh. Tech.* August 1973.

44. C. C. Bailey and T. C. Lindenlaub, "Further Results Concerning the Effect of Frequency Selective Fading on Differentially Coherent Filtered Receivers," *IEEE Trans. Comm. Tech.*, **COM-16**, October 1968, p 749.

45. P. A. Bello and B. D. Nelin, "Effect of Frequency Selective Fading on the Binary Error Probability of Incoherent and Differentially Coherent Matched Filter Receivers," *IEEE Trans. Comm. Systems*, **CS-11**, June 1963, pp. 170–186.

46. W. C. Jakes, Jr., "A Comparison of Specific Space Diversity Techniques for Reduction of Fast Fading in UHF Mobile Radio Systems," *IEEE Trans. Veh. Tech.* **VT-20**, November 1971, pp. 81–92.

47. B. R. Davis, "FM Noise with Fading Channels and Diversity," *IEEE Trans. Comm. Tech.*, **COM-19**, December 1971, pp. 1189–1971.

48. B. R. Davis, "Random FM in Mobile Radio with Diversity," *IEEE Trans. Comm. Tech.*, **COM-19**, December 1971, pp. 1259–1267.

chapter 7

layout and control of high-capacity systems

D. C. Cox and D. O. Reudink

SYNOPSIS OF CHAPTER

The geographical layout of base stations and the assignment of radio channels and base stations to serve calls are system requirements that must take into account the total aggregate of channels available to the radio system as well as the constraints of propagation, modulation techniques, noise, and so on. A number of base stations operating on many radio channels must be assembled into an economically viable system that makes efficient use of the radio spectrum while providing the number of channels per unit area required to meet the traffic offered by the customers. This chapter considers techniques for laying out small contiguous coverage cell systems and for assigning base stations and radio channels to serve calls in such systems.

Section 7.1 discusses conventional large radio coverage area systems with emphasis on traffic considerations and channel requirements. These systems are important both historically and because they will continue to be used to serve small relatively isolated cities. A simple model for estimating traffic in a single cell from geometry and vehicle density estimates is developed first. Then telephone traffic formulas are derived for different queueing disciplines. These formulas relate demand in a single coverage cell to call blocking and the number of channels available. The traffic formulas are applicable both to large cell systems, and to individual small cells of a small cell system using fixed-channel-assignment techniques.

Channel reuse and ways of contiguously covering planar regions with small cells are covered in Section 7.2. Base-station assignment and vehicle locating techniques for small cell systems are discussed in Section 7.3.

Section 7.4 is devoted to the performance of several different channel assignment methods. Performance of dynamic channel assignment techniques that do not assume a fixed relationship between channels and cells is compared with performance of a fixed channel assignment technique that reserves specific sets of channels for use in specific cells. Because of the complex interaction between cells and channels in dynamic channel assignment systems, performance data can be obtained only by computer simulation. Brief descriptions of three simulations are included in Sections 7.4.3, 7.4.4, and 7.4.6 along with performance data relating traffic offered to systems, call blocking, and traffic carried by systems. Data illustrating effects produced by vehicles crossing coverage cell boundaries also are presented. One-dimensional systems of cells laid out in a line are considered first in Section 7.4.3 and then two-dimensional systems of square cells in Section 7.4.4. Section 7.4.6 considers a system consisting of a small set of hexagonal cells. The improvement in the efficiency of channel usage resulting from mixing fixed and dynamic channel assignment techniques and reassigning channels to calls in progress is illustrated in Section 7.4.5.

7.1 LARGE RADIO COVERAGE AREA SYSTEMS

Historically, mobile radio engineers have sought the highest mountain or building within a proposed service area for their antenna location. On this they have erected the tallest practical structure to support the base-station antenna for the system and have transmitted very high power in an attempt to "cover" the largest area possible. Since it was not practical to transmit the same power from the mobile units, additional receivers eventually had to be distributed within the area to service the mobile-to-base link in low-signal regions.[1]

Because of large obstructions like hills or tunnels, holes in the radio coverage from the high base-station antenna sometimes existed within the primary coverage area. Attempts have been made to fill in some of the coverage holes by installing secondary transmitters and antennas. This solution creates problems in the coverage overlap because of frequency beats between the signals from the different transmitters.[2] In addition, careful equalization of delay in the baseband circuits to the different transmitters is required to keep distortion at acceptable levels.[2]

Large radio coverage areas are often separated by distances that put them beyond the earth's horizon from one another as illustrated in Figure 7.1-1. As discussed in Chapter 2, signal attenuation increases rapidly for this condition; therefore the areas are usually isolated sufficiently so that

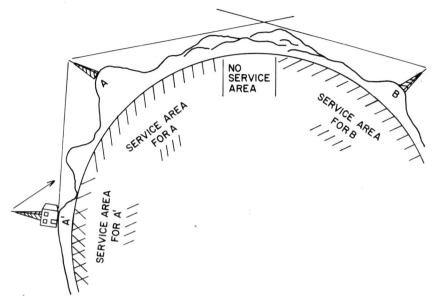

Figure 7.1-1 Large area mobile radio coverage.

the same radio channel or set of channels can be used relatively free of interference.

The number of channels allocated for mobile radio services has been grossly inadequate. For example, cities of over a million population often have fewer than ten channels available for mobile radio-telephone service. The radius of coverage for some of these systems in over 30 miles; thus the number of circuits (channels) per square mile may be less than 0.003, a number that is infinitesimal compared to the density of vehicles.

Because of the scarcity of channels available, dispatch services often have installed several base stations serving several unrelated and un-coordinated activities on the same channel within one overall coverage area. Channel useage is theoretically time shared, but often in practice simultaneous use on the same channel is attempted and much interference results.

7.1.1 A Model for the Call Attempt Rate*

It seems reasonable that an upper bound on the mobile telephone call attempt rate may be obtained by calculating the expected number of cars on all the streets within a circle of radius R. The streets are assumed to form a uniform grid of squares of side length w, as shown in Figure 7.1-2.

*This model was developed by W. C. Jakes, Jr., in a private communication.

It is further assumed that all streets are two-way, with a single lane in each direction. From Figure 7.1-2 it is simple to calculate the total length of the streets within the circle:

$$L = 8R \left[\sum_{k=1}^{N} \sqrt{1 - \left(\frac{kw}{R} \right)^2} + \frac{1}{2} \right], \qquad N = \frac{R}{w}. \qquad (7.1\text{-}1)$$

It turns out that L can be approximated by $2\pi R^2/w$ for the regions of interest (street spacings from 250 to 1000 ft, $R > 0.5$ mile), and this approximation will be used henceforth.

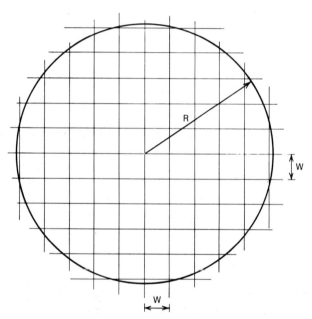

Figure 7.1-2 Coverage area bounded by a circle of radius R with a uniform grid of streets spaced w apart.

Assuming the cars are randomly spaced with a mean separation S, the number of cars in the circle is

$$C = \frac{2L}{S} = \frac{4\pi R^2}{wS}. \qquad (7.1\text{-}2)$$

We now assume that call attempts associated with each car, A_v, occur at random and independent from those associated with other cars. The overall attempt rate, A_1, within the assumed coverage circle is then

$$A_1 = A_v C = 4\pi \frac{R^2}{wS} A_v. \tag{7.1-3}$$

The actual attempt rate, A, on the radio system will be reduced because some cars wishing service will not receive radio coverage. Thus $A = A_1 F_u$, where F_u is the fraction of cars in the coverage circle that actually receive radio coverage as determined by the log-normal distribution of the received mean signal power (see Chapter 2). In an actual coverage area we can obtain C from vehicle density studies, but at the present time we can only guess at A_v and therefore also A_1. However, F_u was determined with some degree of precision in Chapter 2 and is plotted as a function of propagation parameters in Figure 2.5-1. The fraction of cars receiving radio coverage is the same as the fraction of the area within a coverage circle receiving usable signal if the car density is uniform. Otherwise an integration over the spatial car density function is required.

It is worth noting that cars not receiving radio coverage but wanting to make calls are "blocked" just as effectively as though all the channels were tied up. An "effective" system blocking probability can thus be defined that takes into account the channel blocking, P, and radio coverage blocking, F_u:

$$P_s = 1 - F_u(1 - P). \tag{7.1-4}$$

If P_s is specified as a system requirement, a trade-off between F_u and P is thus established.

7.1.2 Traffic Considerations*

The study of traffic is a well-established discipline in telephone engineering. Because of this the word traffic will not be used in this chapter to refer to the motion of vehicles, that is, it will not be applied to vehicular "traffic."

Telephone calls are made by individual customers as they fit into their living habit or into the conduct of their business. The aggregate of customers' calls follows a varying pattern throughout the day, and facilities must be sufficient in quantity to care satisfactorily for the period of maximum demand, usually termed the busy hour. The basic factors involved in the provision of facilities are the call attempt rate, call holding

*Most of the material in this section was taken from a set of lecture notes provided by S. R. Neal of Bell Telephone Laboratories, Incorporated, Holmdel, New Jersey.

times, numbers of channels (trunks or facilities), and grade of service.

The product of the first two factors is the "offered traffic" or "offered load." It denotes the amount of time that a quantity of callers desires the use of facilities. A load that engages one channel (trunk) completely is known as an Erlang. Offered traffic is also expressed in terms of hundred call-seconds per hour (CCS) or call-minutes per hour (min). Since there are 3600 call-seconds in an hour, an Erlang is equal to 36 CCS or 60 call min.

It has been found that probability theory can be used to derive relationships among the three factors: offered traffic, number of channels, and grade of service. Formulas have been developed for the derivation of suitable capacity tables. The formulas used assume the characteristics of telephone calls as well as the physical relationship of call sources and communication channels. Among call characteristics are distribution in time of calls placed by customers, customer calling rate, call holding time variation, and customer or equipment behavior upon encountering busy facilities.

System call attempt rates are estimated from predictions based upon anticipated individual calling rates, vehicle densities, and propagation considerations as described previously. Sometimes estimates of offered traffic are made directly instead of estimating attempt rates and holding times separately. As we shall see later, a small change in the calling rate can sometimes produce a large change in grade of service.

The elapsed time during which a call occupies a channel is the call duration or holding time. This consists principally of the conversation time plus relatively small intervals required by the equipment to make completion and for the customer to answer. Holding times for local land-line calls within a base rate exchange area are found to vary according to an exponential law. Mobile radio calls are expected to vary similarly. This will be discussed later. Ideally, every mobile would have complete access to every channel available for completing a call. Economic and physical considerations in the design of equipment are such that this may not always be true. Any limitation of access adversely affects the relationship of the basic factors.

Grade of service may be described in terms of either the blocking rate (frequency with which the channels are found unavailable to offered calls) or the average delay encountered. This will be discussed in more detail in the following section.

General Assumptions in Regard to Offered Traffic

The distribution of calls placed by customers has an important bearing on the number of channels required for a given grade of service. It is customary to assume that each customer originates his calls at random during his idle time and independently of all other customers. This assumption cannot be strictly true because when two customers are talking there is clearly a restriction on their ability to originate calls independently. However, for large numbers of customers where each has a small probability of initiating a call, this restriction can be neglected. In the land network for periods as long as the busy hour, the random placement of calls usually seems to be quite well realized.

It is also assumed that each customer over a long period of time originates the same total Erlangs of load as every other customer. Under this assumption the probability that at any random moment any particular customer will be using his telephone is then a constant. Obviously, individual customer calling rates vary widely, but the average calling rate of customers is used in deriving the theory. It can be shown in most cases, however, that nonuniformity in offered traffic among customers results in a slightly better service than predicted by formulas based upon the uniform-load assumption. Holding times for the cases of delayed call situations are assumed either to be constant or to vary exponentially.

Our goal in this section is to derive formulas that interrelate the basic parameters of offered traffic, numbers of channels, and grade of service. These formulas require knowledge of the negative-exponential distribution, the Poisson distribution, and the birth-death process whose properties are presented in the next few pages.

Properties of the Negative-Exponential Distribution

The exponential law is used to describe the distribution of local call conversation times.[3, 4] Its formula may be expressed as

$$F(T) = 1 - e^{-\lambda T}, \tag{7.1-5}$$

where $F(T)$ is the probability that any randomly selected holding time will equal or exceed length T and $1/\lambda$ is the average holding time of all customer calls. The negative-exponential distribution plays a major role in telephone similar in nature to that portrayed by the Gaussian distribution in communication theory.

We shall investigate several of the important properties of the negative-exponential distribution before proceeding with the derivation of the telephone traffic formulas.

Given that an arrival (i.e., a call attempt) occurred at $t=0$ and no arrival in $(0,t)$, what is the probability of an arrival in $(t, t+\Delta t)$?

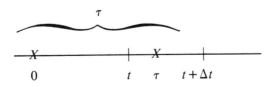

Assume that the interarrival times τ are independent and identically distributed according to $F(t)$. We are interested in

$$P_1 = P\{t < \tau \leqslant t + \Delta t | \tau > t\}$$

$$= \frac{F(t + \Delta t) - F(t)}{1 - F(t)} \tag{7.1-6}$$

The conditional arrival-rate is "roughly speaking" the probability of an arrival at t given an arrival at 0 and no arrival in $(0, t)$. When $F(t) = 1 - e^{-\lambda t}$, one can see that the conditional arrival rate is λ. That is to say

$$P\{t < \tau \leqslant t + \Delta t | \tau > t\} = \lambda \Delta t + 0(\Delta t)^* \tag{7.1-7}$$

is the probability of an arrival in $(t, t + \Delta t)$ given that "we've waited" t units of time since the last arrival. Notice that this probability is independent of t; that is, the negative-exponential process has no memory.

In a similar fashion the conditional probability of no arrival in $(t, t + \Delta t)$ is

$$P_0 = 1 - \lambda \Delta t + O(\Delta t), \tag{7.1-8}$$

and for two or more arrivals

$$1 - P_1 - P_0 = O(\Delta t). \tag{7.1-9}$$

We have just seen that properties (7.1-7), (7.1-8), and (7.1-9) follow from the (negative) exponential distribution. If the times between arrivals are independent and identically distributed according to a distribution function $A(t)$ that satisfies (7.1-7), (7.1-8), and (7.1-9), then $A(t) = 1 - e^{-\lambda t}$. *Proof:*

$$\frac{A(t + \Delta t) - A(t)}{1 - A(t)} = P\{t < \tau \leqslant t + \Delta t | \tau > t\}$$

$$= \lambda \Delta t + O(\Delta t)$$

Hence,

$$\frac{A'(t)}{1 - A(t)} = \lambda \quad \text{and} \quad A(t) = 1 - ce^{-\lambda t}.$$

Property (7.1-7) implies that $A(0) = 0$ so that $c = 1$.

*We define $O(\Delta t)$ by

$$\lim_{\Delta t \to 0} \frac{O(\Delta t)}{\Delta t} = O.$$

Poisson Distribution

We are concerned with the number $N(t)$ of calls in an arbitrary interval $(h, h+t)$ of length t. For negative-exponentially distributed times between arrivals (interarrival times), we know that it is immaterial whether or not an arrival occurred at h, so let

$$p_n(t) = P\{n \text{ arrivals occur in } (0,t)\}$$

$$= P\{N(t) = n\}.$$

where $N(t)$ is the number of calls. Now,

$$p_0(t) = P\{N(t) = 0\} = P\{\text{interarrival time exceeds } t\}$$

$$= F(t) = 1 - e^{-\lambda t}. \tag{7.1-10}$$

In order for n arrivals to occur in $(0,t)$, the first arrival can take place between x and $x + dx$ for any $x < t$ (that is, $x < \tau \leqslant x + dx$) and then $n-1$ arrivals must occur during (x,t):

$$\begin{array}{ccccc}
+ & + & \hspace{-0.5em}\times\hspace{-0.5em} & + & + \\
0 & x & & x+dx & t
\end{array}$$

Since interarrival times are independent, the probability of the joint occurrence is

$$(\lambda e^{-\lambda x}\, dx)p_{n-1}(t-x).$$

Thus, summing (integrating) over all possible x, $0 < x < t$, we have

$$p_n(t) = \lambda \int_0^t e^{-\lambda x} p_{n-1}(t-x)\, dx, \qquad n \geqslant 1. \tag{7.1-11}$$

Using mathematical induction it follows that

$$p_n(t) = \frac{(\lambda t)^n}{n!} e^{-\lambda t}. \tag{7.1-12}$$

We have seen that properties (7.1-7), (7.1-8) and (7.1-9) are essentially equivalent to the negative-exponential distribution. We also know that if the interarrivals are exponentially distributed, then $N(t)$ is Poisson. The converse is also true: If $N(t)$ is Poisson with intensity λ and the associated interarrivals τ are independent, then τ has a negative-exponential distribution $1 - e^{-\lambda t}$.

Proof:

$$p_0(\Delta t) = e^{-\lambda \Delta t} = 1 - \lambda \Delta t + \frac{(\lambda \Delta t)^2}{2!} + \frac{(\lambda \Delta t)^3}{3!} + \cdots$$

$$= 1 - \lambda \Delta t + O(\Delta t),$$

$$p_1(\Delta t) = \lambda \Delta t e^{-\lambda \Delta t} = \lambda \Delta t + O(\Delta t)$$

and for $n \geqslant 2$,

$$p_n(\Delta t) = \frac{(\lambda \Delta t)^n}{n!} e^{-\lambda \Delta t}$$

$$= O(\Delta t).$$

These properties imply the negative-exponential distribution.

Birth-Death Process

We shall consider mobile radio systems in which both the interarrival times and service times have negative-exponential distributions with parameters that depend on the number of customers in the system. Let $\delta n(t)$ denote the event "the system is in state n at time t." Then as $\Delta t \to 0$, it follows that for $n \geqslant 0$,

$$P\{\delta_{n+1}(t+\Delta t)|\delta_n(t)\} = \lambda_n \Delta t + O(\Delta t) \qquad \text{(Birth)} \qquad (7.1\text{-}13)$$

$$P\{\delta_n(t+\Delta t)|\delta_{n+1}(t)\} = \mu_{n+1} \Delta t + O(\Delta t), \qquad \text{(Death)} \qquad (7.1\text{-}14)$$

and for $k \geqslant 2$,

$$P\{\delta_{n \pm k}(t+\Delta t)|\delta_n(t)\} = O(\Delta t). \qquad (7.1\text{-}15)$$

Such a process is called a birth-death process with birth rate λ_n and death rate μ_n.

Let $p_n(t) = P\{\delta_n(t)\}$. We can see that

$$p_n(t+\Delta t) = (1 - \lambda_n \Delta t)(1 - \mu_n \Delta t)p_n(t)$$

$$+ \lambda_{n-1} \Delta t p_{n-1}(t) + \mu_{n+1} \Delta t p_{n+1}(t) + O(\Delta t)$$

as $\Delta t \to 0$.

It follows directly that for $n \geqslant 0$,

$$\frac{d}{dt} p_n(t) = -(\lambda_n + \mu_n)p_n(t) + \lambda_{n-1} p_{n-1}(t)$$

$$+ \mu_{n+1} P_{n+1}(t), \qquad (7.1\text{-}16)$$

where we define $p_{-1}(t) = 0$.

A solution to the birth-death Equation (7.1-16) is difficult to obtain except for very special cases of λ_n and μ_n. However, we can obtain results when the limiting distribution $\lim_{t\to\infty} p_n(t) = p_n$ exists. In such a case

$$\lim_{t\to\infty} \frac{d}{dt} p_n(t) = 0$$

and we have for $n \geqslant 0$,

$$(\lambda_n + \mu_n)p_n = \lambda_{n-1}p_{n-1} + \mu_{n+1}p_{n+1}. \qquad (7.1\text{-}17)$$

It is worthwhile to observe that Eq. (7.1-17) can be obtained directly from a conservation-of-flow argument. First of all, $(\lambda_n + \mu_n)p_n$ is the rate of flow out of the state n. Then, $\lambda_{n-1}p_{n-1}$ is the rate of flow into state n from state $n-1$ and $\lambda_{n+1}p_{n+1}$ is the rate of flow into state n from $n+1$. Since the flow into a state must be the same as the flow out, we have (7.1-17).

The flow concept is represented schematically by a state flow diagram[5] in Figure 7.1-3.

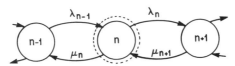

Figure 7.1-3   State flow diagram with surface enclosing node n.

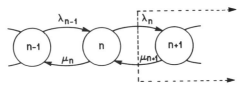

Figure 7.1-4   State flow diagram with surface enclosing all states above n.

Notice that Eq. (7.1-17) is obtained by considering the flow across a surface which completely encloses the node n. For the birth-death process, consider a surface that encloses all the states above n as shown in Figure 7.1-4. In this case, flow considerations lead to

$$\lambda_n p_n = \mu_{n+1}p_{n+1} \qquad \text{for} \quad n \geqslant 0. \qquad (7.1\text{-}18)$$

Of course, any solution to (7.1-18) is a solution for (7.1-17). In the last case,

$$p_{n+1} = \frac{\lambda_n}{\mu_{n+1}} p_n \qquad \text{for} \quad n \geqslant 0$$

$$= \frac{\lambda_n \lambda_{n-1} \cdots \lambda_0}{\mu_{n+1} \mu_n \cdots \mu_1} p_0, \qquad (7.1\text{-}19)$$

and p_0 is determined by the requirement $\sum\limits_{n=0}^{\infty} p_n = 1$.

Telephone Traffic Formulas

The disposition of calls which do not find a channel (trunk) immediately depends upon a number of things, including the equipment available and the habits of the caller. There are two extreme cases in the disposition of calls. On the one hand some calls upon not finding a channel immediately disappear from the system not to return again in the same period, say the busy hour. This assumption, *blocked calls cleared*, is the basis of a formula for obtaining the probability of loss known as the *Erlang B* formula, which is discussed later. On the other hand, some calls upon finding no channel available, wait in suspense until one becomes idle, at which time the channel is seized and then held for the full holding time. This assumption, *blocked calls delayed*, is the basis of another formula for obtaining probability of loss and is known as the *Erlang C* formula. Between these extremes is an intermediate assumption called, *blocked calls held*, upon which are based the binomial and Poisson formulas.[3] In this case an offered call upon finding no channel idle waits for an interval of time exactly equal to its holding time and then disappears from the system. If a trunk becomes idle while the call is waiting, the trunk is seized and occupied for the portion of the holding time remaining. This is known as the lost-calls-held assumption to distinguish it from the other assumptions used in the Erlang formulas.

Blocked-Calls-Cleared Discipline (Erlang B)

The blocked-calls-cleared (BCC) queueing discipline is a very important concept for telephone-traffic engineering. The most fundamental BCC system in traffic engineering is that consisting of a finite number, c, of channels and arrivals that occur according to a Poisson process with intensity λ. Arrivals which occur when a channel is idle are served immediately. An arrival that occurs when all channels are busy is blocked, leaves the system, and does not return. The call durations (holding times) are assumed to be independent and identically distributed according to a negative-exponential distribution and have mean $1/\mu$.

We assume the system to be in statistical equilibrium and define the state $\mathfrak{N}$ of the system to be the number of busy channels. Let $p_n = P\{\mathfrak{N} = n\}$. We then see that

$$\lambda p_0 = \mu p_1 \qquad (7.1\text{-}20)$$

and for $1 \leqslant n \leqslant c - 1$

$$(\lambda + \mu n)p_n = \lambda p_{n-1} + (n+1)\mu p_{n+1}. \qquad (7.1\text{-}21)$$

At the upper boundary

$$c\mu p_c = \lambda p_{c-1}. \qquad (7.1\text{-}22)$$

Of course

$$\sum_{n=0}^{c} p_n = 1. \qquad (7.1\text{-}23)$$

One can see that Eq. (7.1-20) through (7.1-22) will be satisfied if

$$\lambda p_n = (n+1)\mu p_{n+1} \qquad \text{for} \quad 0 \leqslant n \leqslant c - 1,$$

which implies that

$$p_n = \frac{(\lambda/\mu)^n}{n!} p_0. \qquad (7.1\text{-}24)$$

Equation (7.1-23) is then used to obtain

$$p_0 = \left[\sum_{n=0}^{c} \frac{(\lambda/\mu)^n}{n!} \right]^{-1}. \qquad (7.1\text{-}25)$$

The probability that all channels are busy (blocking probability) which is obtained by combining (7.1-24) and (7.1-25) is an important relation and is often referred to as the Erlang B blocking probability or the first Erlang loss function. Both $B(c,a)$ and $E_{1,c}(a)$ are used to denote the blocking probability, that is,

$$B(c,a) = E_{1,c}(a) = \frac{a^c/c!}{\displaystyle\sum_{n=0}^{c} a^n/n!}, \qquad a = \frac{\lambda}{\mu}. \qquad (7.1\text{-}26)$$

Figure 7.1-5 shows the number of channels required as a function of both offered traffic and probability of blocking for the Erlang B formula.

Blocked-Calls-Held Discipline

A very useful result can be obtained from the Erlang B relation by letting $c \to \infty$. Such a system is often described by assuming that arrivals that occur when all c channels are busy remain in the system for one holding time (blocked-calls held). If a channel becomes available during the holding time, a waiting customer seizes the channel for the remainder of the holding time. In any case, each customer leaves the system after one holding time. The assumptions concerning arrivals and service times are the same as in the Erlang B model.

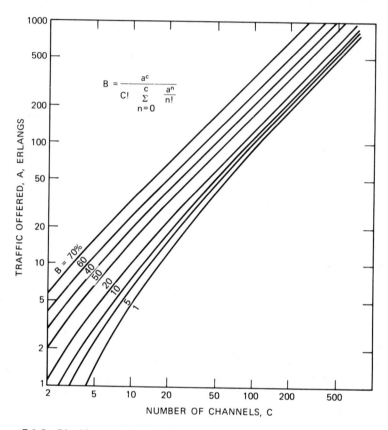

Figure 7.1-5 Blocking probability versus offered traffic for C radio channels from Erlang B formula.

Assume that the system is in statistical equilibrium. The state $\mathfrak{N}$ of the blocked-calls-held (BCH) model is defined to be the number of customers in the system at an arbitrary instant (either waiting or being served). Again we set $p_n = P\{\mathfrak{N} = n\}$ and we see

$$\lambda p_0 = \mu p_1,$$

and for $n \geqslant 1$,

$$(\lambda + n\mu)p_n = \lambda p_{n-1} + (n+1)p_{n+1}.$$

As before,

$$p_n = \frac{(\lambda/\mu)^n}{n!} p_0 \qquad \text{for} \quad n \geqslant 0,$$

so that

$$p_0^{-1} = \sum_{n=0}^{\infty} \frac{(\lambda/\mu)^n}{n!} = e^{\lambda/\mu}.$$

Hence,

$$p_n = \frac{(\lambda/\mu)^n}{n!} e^{-\lambda/\mu} \qquad \text{for} \quad n \geqslant 0; \qquad (7.1\text{-}27)$$

that is, the state probabilities are Poisson with the mean holding time $1/\mu$ times the length of the mean call duration.

When the preceding distribution is used for traffic engineering, the blocking probability is defined to be

$$P\left(c, \frac{\lambda}{\mu}\right) = \sum_{n=c}^{\infty} p_n$$

$$= \sum_{n=c}^{\infty} \frac{(\lambda/\mu)}{n!} e^{-\lambda/\mu}. \qquad (7.1\text{-}28)$$

Notice again that the blocking probability depends only on the load $a = \lambda/\mu$ and the number c of channels.

Erlang C Model (Blocked Calls Delayed)

The blocked-calls-delayed (BCD) or Erlang C model arises when blocked calls are allowed to queue up and wait for service in order of arrival. More precisely, the system is composed of c channels and an unlimited waiting queue. Customers arriving to find an idle channel are served immediately. Customers arriving when all channels are busy, queue up in order of arrival. If a channel becomes idle when customers are waiting, the channel serves the customer at the head of the queue. Customers do not defect from the queue.

Assume the system to be in statistical equilibrium and define the state $\mathfrak{N}$ to be number of customers in the system (either waiting or being served). Let $p_n = P\{\mathfrak{N} = n\}$ at an arbitrary instant. From the equilibrium state equations for p_n one can show that when $\lambda < c\mu$

$$p_n = \frac{(\lambda/\mu)^n}{n!} p_0 \qquad \text{for} \quad 0 \leqslant n \leqslant c,$$

$$p_{n+c} = \left(\frac{\lambda}{c\mu}\right) p_c \qquad \text{for} \quad n \geqslant 0,$$

$$p_0 = \left(\sum_{n=0}^{c-1} \frac{(\lambda/\mu)^n}{n!} + \frac{(\lambda/\mu)^c}{c!} \frac{c\mu}{c\mu - \lambda}\right)^{-1},$$

and the probability of delay is

$$C\left(c, \frac{\lambda}{\mu}\right) = p(0)\frac{(\lambda/\mu)^c}{c!} \frac{c}{c-a}. \tag{7.1-29}$$

Regarding what happens to blocked calls it is readily seen that the three assumptions (blocked calls cleared, held, and delayed) have differing influences on the theoretical probability in each case of calls being lost. Where lost calls are cleared from the system, they do not occupy a channel, and thus for a given load this probability of lost calls is the lowest among the three assumptions. Where lost calls are held they occupy some channel time. If the probability of loss is low, they occupy nearly the full holding time. Where lost calls are delayed, they occupy channels for the full holding time, and the probability of loss is thus the highest of the three assumptions. Mobile telephone service is more closely described by the blocked-calls-cleared assumption, while dispatch service is closer to the blocked-calls-delayed assumption.

7.1.3 An Example of Traffic in a Single Coverage Area

Consider the case of only one base station providing service to an entire area and assume that all of the significant traffic offered is contained within a circle of radius R. We further assume all channels are available both at the base station and all mobiles, and will show how the number of channels, blocking probability, and coverage radius are related.

We now assume some representative values to determine the offered traffic:

Car spacing	$S = 0.02$ miles
Street spacing	$w = 0.1$ mile
Attempt rate per vehicle	$A_v = 0.02$ attempts per minute
Average call length	$H = 3$ minutes

Standard deviation
  for mean signal $\sigma = 7.5$

Attenuation law $n = 3$

Fraction of cars with signal
  above threshold (see Chapter 2) $F_u = \frac{5}{6}$

The offered traffic, T, is the product of the call-attempt rate on the radio system and the mean call duration. Thus from Eqs. (7.1-3) and (7.1-4)

$$T = A_v F_u H \frac{4\pi R^2}{wS} = 100 \, \pi R^2. \tag{7.1-30}$$

(T is in Erlangs for R in miles.)

Using the curves of Figure 7.1-5, the number of channels required can be calculated versus R for various blocking probabilities, as shown in Figure 7.1-6. Alternatively, the number of channels required versus blocking probability can be plotted for various values of coverage radius, as shown in Figure 7.1-7.

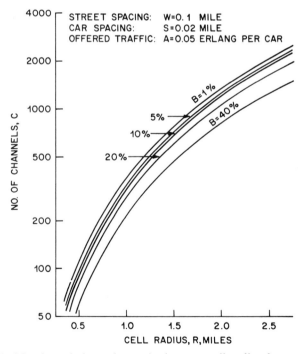

Figure 7.1-6 Number of channels required versus cell radius for several blocking probabilities, B (Erlang B).

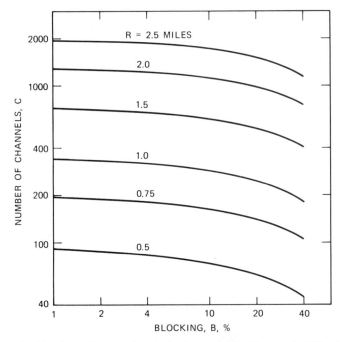

Figure 7.1-7 Number of channels required versus blocking probability for several values of cell radius, R. Traffic and spacing parameters same as Figure 7.1-6 (Erlang B).

If the available RF bandwidth sets a limit on the number of channels, we can estimate the coverage radius. For example, if $N = 150$ channels, the radius is about 1.5 miles or less for reasonable values of P, that is, less than 10%. Since a coverage cell of 3 miles diameter falls far short of covering a typical large metropolitan area, this approach is obviously inadequate, and we must turn to channel reuse plans with multiple cells.

7.2   COVERAGE LAYOUT OF SMALL CELL SYSTEMS

Often the total number of radio channels available to a mobile radio system will not provide satisfactory service within a metropolitan area on a large coverage area basis. The number of circuits available per unit area can be increased by simultaneously using radio channels in small radio coverage areas or cells within the metropolitan area but separated sufficiently to prevent excessive cochannel interference. This approach of increasing the system capacity by spatial reuse of radio channels is a form of spatial multiplexing. One distinguishing feature of different small cell systems is whether the coverage areas are initially laid out using statistical parameters of mobile radio propagation or by using propagation parameters for the actual region that have been determined by direct measurement or by prediction techniques. More emphasis has been on statistical layout rather than deterministic layout.[6-10]

7.2.1 Statistical Layout

For the purpose of illustrating some of the concepts involved in the statistical layout of a small radio coverage cell system, consider two base stations, A and B, separated by a distance D as illustrated in Figure 7.2-1.

Assume for illustrative purposes that they are equipped with omnidirectional transmitting and receiving antennas. The median signal power received at a point, P, from A is proportional to R^{-n}, where R is the distance from A to P and n is the propagation law discussed in Chapter 2. If we assume that P lies near the line connecting A and B, then the median signal power received at P from B is proportional to $(D - R)^{-n}$. Of course, the actual received power at P for a collection of actual base station layouts will be a random process consisting of a rapidly varying Rayleigh distributed component (Chapter 1) and a slowly varying log-normally distributed component (Chapter 2). If A and B are transmitting on the same frequency, then the average signal to interference ratio, $\overline{S}/\overline{I}$, taking into account both the Rayleigh and log-normal distributed fading and any diversity improvements as analyzed in Chapter 5, is equal to $[(D - R)/R]^n$. Thus the relative size R of the coverage area compared to the base-station separation distance D is determined for a given system when the allowable $\overline{S}/\overline{I}$ ratio is specified and n is selected as described in Chapter 2. If, in addition to B, there are other base stations located a distance D from A and all received signal statistics are independent at P, then $\overline{S}/\overline{I} = (1/M) [(D - R)/R]^n$, where M is the total number of base stations at distance D away. In general,

$$\frac{\overline{S}}{\overline{I}} = \frac{1}{R^n \sum_{i=1}^{M} 1/(R_i)^n} , \qquad (7.2\text{-}1)$$

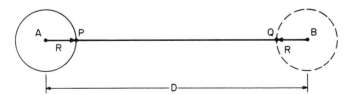

Figure 7.2-1 An illustration of frequency reuse by two base stations separated by a distance D.

where the R_i are the distances from the observation point to the M base stations. This discussion assumes that the power level of the signal radiated from A is sufficient to provide an adequate signal-to-noise ratio at points located a distance R or greater from A. That is, the system is interference limited, not signal power and thermal noise limited.

If total coverage of a region in space is to be achieved by reusing radio channels in small interference limited coverage areas, the total number of channels available to the entire mobile radio system must be distributed in some manner so that channels are available for use in all coverage areas. An example of channel distribution according to a fixed channel allocation plan with base stations distributed in one dimension to cover a long narrow region is illustrated in Figure 7.2-2. Such a layout might actually be used along a multilane highway. For the example shown, $(D - R)/R = 7$ or $D/R = 8$ for the adjacent base stations using the same channel and four radio channel subsets are required. Channel assignment methods are described in more detail later in this chapter.

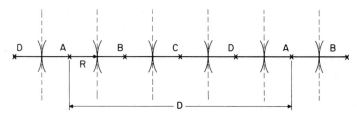

Figure 7.2-2   An example of a one-dimensional small cell mobile radio system with $D/R = 8$.

Similar cochannel interference situations exist in Figure 7.2-1 at base stations A and B if mobiles located at P and Q transmit on the same frequency.

Large planar regions can be covered by two-dimensional networks of small coverage cells. Covering such regions with equal regular polygons which do not overlap is known as tessellation.[11] Tessellations provide approximations to the interference limited small coverage cells defined on the basis of propagation statistics. The only regular polygons that tessellate a planar region are the equilateral triangle, the square, and the regular hexagon. These are illustrated in Figure 7.2-3. The parameters D/R and C are useful for comparing various two-dimensional small cell system layouts. For these comparisons D is the distance between centers of the closest coverage cells that simultaneously can use the same radio channel,

R is the maximum distance from the center of a cell to the cell boundary, and C is the minimum number of channel sets required to fully cover any planer area.

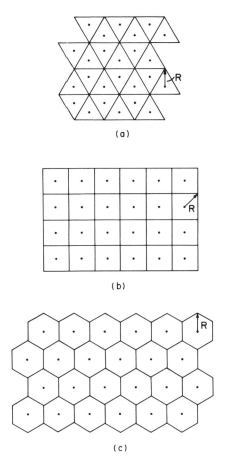

(a)

(b)

(c)

Figure 7.2-3 Tessellations of planar regions with small coverage cells in the shape of (*a*) equilateral triangles, (*b*) squares, and (*c*) regular hexagons.

Small cell systems based on squares have been used for determining performance of proposed systems because of the simplicity of the associated rectangular coordinate system.[12-14] For the square systems the relationship between D/R and the number of channel sets, C, required to completely cover a planar area with a fixed channel assignment plan is

$$C = \frac{1}{2}\left(\frac{D}{R}\right)^2 \quad \text{(squares)}. \tag{7.2-2}$$

Of course, for system design with a specified D/R, the C determined from the preceding equation is rounded to the next larger integer in the set: $C = 2, 4, 5, 8, 9, \ldots$ since the allowable values of C are given by $C = k^2 + l^2$, where k and l range over the positive integers.

Small cell systems based on regular hexagonal tessellation of a region have received considerable attention because a hexagon is a better approximation to the circular coverage area that results if a system is transmitter power and thermal noise limited. The actual average shape of equal $\bar{S}/\bar{I}$ contours for interference limited systems has not been investigated. For hexagonal systems with channels allocated on a fixed block assignment plan,

$$C = \frac{1}{3}\left(\frac{D}{R}\right)^2 \quad \text{(hexagons)}, \tag{7.2-3}$$

and C can take on only the selected values

$$C = 3, 4, 7, 9, 12, 13, \ldots,$$

determined from $C = (k + l)^2 - kl$, where again k and l range over the positive integers.

The ratio of D/R required by a given system is determined as described in earlier chapters from system S/I specifications and the characteristics of the system (the use of diversity, and so on). The actual size of the cells, that is, the actual values of D and R, are determined from the following specified parameters: density of traffic offered, the total number of radio channels available to the radio system, the allowable call blocking rate, and the channel assignment strategy implemented for system control. Some of these items already have been covered, while others, such as channel assignment, are covered later in this chapter. Often the density of traffic offered used for design is the hourly average for the busiest hour of the average business day.

If the density of offered traffic were uniform over the entire service area of the mobile radio system, the design of a grid of small coverage areas would be straight-forward. However, the traffic density usually decreases to much lower values at the outskirts of metropolitan areas than at the centers of cities. Also, base stations are expensive. Therefore, in order to minimize system costs, larger cells (larger R) are usually prescribed on the outskirts than around the city centers. Subdivision of regular coverage areas into smaller areas cannot be done exactly while still maintaining a constant specified D/R for all cells. Thus, some engineering judgment must be applied when changing cell size to fit significant changes in the density of offered traffic. Often smaller size cells are continued to a point where all channels are not required in the larger size cells. Channels used along the transition region then can be allocated in such a way that no channel is reused closer than the specified D/R. Figure 7.2-4 is an example[15] of such a tapered system designed based on a fixed channel assignment plan.

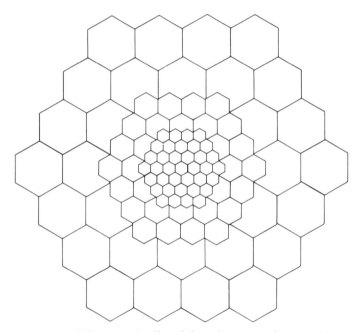

Figure 7.2-4   A small hexagonal cell mobile radio system layout with tapered cell sizes.

7.2.2 Deterministic Layout

Mobile radio systems with base-station coverage cells layed out as regular contiguous polygons (triangles, rectangles, or hexagons) must be based upon a statistical description of radio propagation effects. The number of coverage cells required between base stations using the same radio channel is then determined by statistical parameters and service requirements. In many cases the resulting reuse interval would be greater than necessary to prevent cochannel interference. On the other hand, excessive cochannel interference will result in some isolated cases and satisfactory radio coverage from the desired base station(s) may not be possible at some few locations in some cells.

In a deterministic coverage plan it should be possible to arrange the base-station positions in such a way that the average radio channel reuse interval is minimized and that adequate coverage is maintained essentially throughout the system. This optimizing of base-station positions would require knowledge of the average signal strength from a potential base station at many locations out to distances beyond which it no longer can cause interference. This required signal strength information could be determined by measurement and/or prediction. Service areas associated with different base stations would have different shapes and sizes. The optimizing of base-station positions would minimize the number of base stations required to produce a specified number of channels within a given area. At any base-station site more than one coverage area could be defined by using directional antennas at that location. This permits some additional tailoring of the shapes and sizes of service areas and the reduction of the channel reuse interval in some cases. Several topics require further study before an effective deterministic coverage plan can be outlined in detail. Special attention must be focused on base-station layout procedures and the extent and detail of measured signal strength data required.

In high base-station density areas where normal radio service areas from base stations are determined by cochannel interference considerations and not thermal noise limits, it should be possible because of known overlap in coverage areas to serve peaks in demand within the normal service area of one base station by using radio equipment located at an adjacent base station. This would provide more efficient utilization of radio equipment and also better service to areas during peaked loading conditions.

The complexity of such systems makes it difficult if not impossible to obtain quantitative results by any means but large-scale computer simulation.

7.3 BASE-STATION ASSIGNMENT IN SMALL CELL SYSTEMS

In small cell systems initial call setup must be made from information received from a mobile by several base stations. At some point in any call setup procedure, both mobile subscriber-originated and fixed base-originated new call attempts will receive similar processing. Mobiles for which calls have originated in the land communication network (fixed-base originated) must be paged either through specially reserved paging channels or through "idle channels" to which mobiles not making calls are automatically tuned. Mobiles recognizing their paging code probably will transmit on separate channels reserved for call startup or on a companion idle channel similar to the way they do for mobile-originated call initiations.

It should be possible to assign a coverage area and associated base station or stations to serve a new call initiation solely on the basis of the strength of the signal received at several base stations. Pattern recognition techniques and stored tables of measured or predicted signal strength (possibly from deterministic system layout) could undoubtedly be used to improve assignments made on this basis. There has been little analytical or experimental work done on this type of base-station assignment technique.

Most small cell mobile radio system analysis and design to date has been based on geometrically locating a vehicle requesting mobile radio service within a coordinate system on which the coverage areas and associated base stations are positioned. The base station (or stations) to be assigned is then the one associated with the coverage area that the vehicle is determined to be within. The accuracy of vehicle-locating systems for this application is not as stringent as that required by some other applications (such as police departments). Errors in locating vehicles well within a coverage area generally will not result in incorrect base-station assignment and thus have no detrimental effect on system performance. The coverage area boundaries in terms of signal-to-noise and interference are not abrupt but in reality are quite "fuzzy." Thus, errors made in base-station assignment to vehicles near boundaries will not have a serious effect on communication system performance. The study of this type of base-station assignment then reduces to investigating ways of locating vehicles by radio means and assessing locating errors due to the multipath propagation. Narrow-band locating techniques are preferred for use in conjunction with the narrow-band mobile radio systems discussed in the previous chapters. The detailed study of vehicle locating systems is beyond the scope of this book, but some of the applicable techniques will be summarized briefly.

The radio vehicle locating technique that has received the most attention is trilateration. Trilateration systems depend on measuring the distance between the vehicle desiring service and base stations receiving signals from it. An estimate of the vehicle location is made from the set of distance measurements that contain errors due to the the multipath propagation. Statistical filtering (to reject "bad" data) can be used to improve accuracy in many cases. The distance measurements can be made with either pulse systems or phase comparison systems. The pulse systems are less sensitive to multipath propagation effects and produce greater accuracy[16,17] since only the leading edge of a distorted pulse must be detected but the wide bandwidths required (several megahertz) appear incompatible with communication system objectives. According to Turin,[16] pulse system ranging errors greater than 700 feet will be made at fewer than 10% of the locations in areas with densely packed high-rising buildings. With different assumptions on the distribution of different types of buildings and different numbers of base stations involved, overall location errors should be less than 200–400 ft at 90% of the locations over an entire city when using pulse ranging and trilateration.

Phase ranging is directly compatible with mobile communication system bandwidths but is considerably more sensitive to multipath propagation effects. In these systems, a low-frequency sinusoid is modulated onto an RF carrier and its phase compared with a phase reference after propagating between the mobile and base stations. Considerable effort has been expended in determining the accuracy of such systems in the presence of severe multipath propagation.[16-19] Work by Engel[18] indicates that the mode of the distribution of phase ranging errors due to multipath occurs at the average delay (center of gravity) of the power versus delay profile for a given location. Phase ranging errors should be less than 2000–3000 ft at 90% of the locations in densely packed high-rise building areas.[16,17] With different assumptions on the distribution of different types of buildings and different numbers of base stations involved, overall location errors should be less than 800–1400 ft at 90% of the locations over an entire city.[16]

Radio location techniques based on triangulation have not received as much attention as have those based on trilateration. Triangulation systems depend on measuring the bearing angle to the vehicle desiring service from base stations receiving signal from it. A rapidly scanning narrow-beam antenna or several fixed narrow-beam antennas can be used to estimate bearing angles. These techniques generally involve finding the direction of maximum arriving signals and are also sensitive to multipath propagation characteristics. They represent one way of processing signal strength information to make base-station assignments. Few analytical or experimental data are available for assessing the performance of these systems. Accuracy adequate for coverage area determination and base-station assignment in small cell mobile systems has been indicated by preliminary investigation.[15,20]

The direct use of the strength of signals received at several base stations is a very attractive parameter for use in base-station assignment algorithms since no extra equipment is required for its implementation. In order for the signal strength measurements to be useful, the Rayleigh distributed small-scale fading must be averaged out either by diversity of some type or perhaps by time averaging for moving vehicles. Measured or predicted signal strength information should increase the accuracy of base-station assignment or vehicle location if that intermediary is used in the assignment procedure. Measurements in an unspecified environment with four base stations and no diversity indicates locating errors of less than 800–20,000 ft (depending on various conditions) for 90% of the locations investigated.[21] The use of diversity, statistical filtering, and more than four base stations should significantly improve the performance of systems based on signal strength measurements.

An "electronic fence" also has been proposed as a means of determining the region occupied by a mobile vehicle.[22] This system requires a "fence post" at all points where a vehicle can enter or exit a particular region. Vehicles identify the region they are crossing into when they pass a "post" and store this information. Region information could then be transmitted to base stations receiving a request for service from a mobile. This system is potentially very accurate as long as all electronic "fence posts" are operating and the mobile receiver is on when the vehicle crosses a region boundary. The requirement for a large quantity of special equipment distributed throughout a city is not desirable, however.

Base-station reassignment will be required when vehicles pass from one coverage area to antother. Thus, the base-station assignment system must be able to detect when this occurs and to determine the new base station to be assigned. Cochannel interference will have a detrimental effect on the ability of most systems to accomplish these functions, but neither analysis nor experiment has been done on this specific problem.

In a deterministic coverage plan one way to assign the base station or base stations to serve a call attempt would be to employ assignment algorithms that use the average signal strength received from the vehicle at surrounding base stations and the previously determined signal strength information stored in a computer. This call setup procedure would assume the use of a channel (or channel set) that is reserved for processing new call attempts. The reuse interval for this channel would be considerably greater than that of the normal calling channels, to permit interference-free measurement of signal strength at base stations surrounding the vehicle initiating the call. The base-station assignment algorithms would use methods similar to those used in pattern recognition and statistical decision theory to compare the measured signal strengths with information stored in the computer. Continuous or semicontinuous monitoring of the received signal strength from a calling vehicle at its assigned base station(s) and at surrounding base stations would permit the detection of situations where a moving vehicle requires a new base station(s) for its call to continue. In cases where it is not possible to determine which new base station should be assigned because of the presence of signals from more than one vehicle at the surrounding base stations, the vehicles's call would be switched back momentarily to a call start channel where a new base-station assignment can be made interference free. Obviously, in these cases a new channel assignment also would be required to avoid cochannel interference. Priority for available channels could be given to these callers.

Analysis to determine the accuracy and performance of the various base-station assignment algorithms and the size, speed, and complexity of the system controlling computer has not been done. Considerable further study is required on this approach to base-station assignment.

7.4. CHANNEL ASSIGNMENT IN SMALL CELL SYSTEMS

After a base station or base-station set has been assigned to serve a call in a multichannel small cell system, a channel assignment procedure must be followed to determine if a channel is available to serve the call. The channel selected will usually be a duplex radio channel. Channel selection must be done within the constraints of the channel reuse criterion discussed in Section 7.2.

7.4.1 Fixed Channel Assignment

In fixed channel assignment systems a subset of the channels available to the radio system is permanently reserved for use within each coverage cell. The channel subsets are reused in coverage cells separated by a reuse interval. Only channels from the reserved subset can be used to serve a call determined to be in a particular coverage cell. If all channels from the reserved subset are in use, service will not be provided to another caller even though there may be vacant channels among those which are reserved for use in adjacent coverage cells. Figure 7.2-2 illustrates a one-dimensional fixed channel assignment system with four reserved subsets A, B, C and D allocated in a system with $D/R = 8$. Calculation of the number of channel subsets required for different D/R ratios and different geometrical cell plans is discussed in Section 7.2.

Channel search for an available channel in a particular cell only involves searching for a channel from the reserved subset that is not in use. Thus, the channel subsets are equivalent to independent groups of trunks in telephone traffic theory, and the relationship between offered traffic, blocking (grade of service) and traffic carried in a cell can be determined from the traffic formulas derived in Section 7.1. If we assume that no mobiles cross coverage cell boundaries then the traffic offered to a channel subset is just the product of the new call attempt rate, A, and the average call duration, H (holding time). This condition is still approximated if the product of the average call duration and the average vehicle speed, S, is small compared to the cell size, that is if $HS \ll R$.

Vehicles crossing cell boundaries and requiring new base station and radio channel assignments produce the following effects:

1. The average call duration experienced by each cell is shortened since a call exiting a cell vacates a channel prematurely,

2. The effective call attempt rate increases by the boundary crossing rate since a call entering a cell requires a channel in the new cell just as a call attempt originating in the cell does.

3. Some calls are forced to terminate prematurely because channels are not available for their continuation after they enter new cells. This also reduces the average call duration.

Items (1) and (2) considered together do not affect the traffic offered product, $H \cdot A$. The effect of item (3) can be minimized by giving priority for channels to boundary-crossing calls in progress over new call attempts. If boundary-crossing calls are not given priority, the forced call terminations expressed as a percent of boundary crossings is equal to B, the probability of blocking a new call attempt. If original estimates of system loading are made in terms of offered traffic, these boundary-crossing effects do not enter directly. If, however, traffic offered is obtained from estimates of the new call attempt rate and call duration, the call duration must be adjusted for the boundary-crossing effect, which can be estimated from vehicle velocity, call durations, and cell size.

In most cases the average offered load per cell will not be uniform but will peak up in cells near city centers. If M, the total number of channels available to the mobile radio system, is large enough to meet the overall system channel requirements, then the number of channels required in the cells removed from the load peak will be less than M/C, where C is the number of channel subsets required to meet the cochannel interference criterion (Section 7.2), and the number of channels required in a few cells near the load peak will be greater than M/C. Some of the channels not needed in the outlying cells within a mutually interfering group of cells can be allocated on the fixed basis to the reserved channel subsets of the cells near the peak. This tapering of channel subset size to match the loading profile is sometimes referred to as fixed or static "borrowing" of channels from outlying cells.

7.4.2 Dynamic Channel Assignment Systems

The most general form of dynamic channel assignment assumes that any channel can be used in any coverage cell. Channels are assigned to serve calls based on the state of the system (channel usage at the time a call is set up) and some channel assignment strategy that attempts to optimize some system parameter within the channel reuse constraint.

Channel search for a channel to assign in a particular cell at a particular time involves searching through all channels allocated to the radio system to find an available channel or set of channels. Available channels are those not being used closer than the permitted minimum D/R from the cell. If no such channel is found, service cannot be provided in the cell at that time. The channel search may be done in some specified order through the channels or it may be done at random. If more than one channel is available for assignment, selection of the channel to use must be made by applying some optimizing strategy. These strategies may be based on criteria such as the order of search, the weighted distance to other cells using the same channel, the number of times the channel is used at the minimum D/R interval, or even on random selection. Some strategies and their relative performance under certain conditions are described later.

Control of a dynamic channel assignment system requires access to and processing of large quantities of data. A fast digital computer most certainly is required. The status (in-use or not) of each channel in each cell must be stored in a readily accessible form and must be updated with each status change. The system controller must identify call terminations, new call attempts, and cell boundary crossings and take appropriate action. The identity of vehicles making calls must be available both for determining customer busy status and for revenue purposes. Vehicle locating procedures may also be handled by this controller. Base-station radio equipment used with dynamic assignment must be switchable from channel to channel as channel assignments in the cell change. The added complexity of flexible system control yields increased channel occupancy over fixed channel assignment at low blocking rates, as will be illustrated later.

The interaction among cells and the large number of possible system states and state transitions make analysis of dynamic channel assignment systems appear intractable. So far the only available estimates of system performance have been obtained by simulation of large-scale systems on digital computers.[12, 23–27, 29]

7.4.3 Performance of One-Dimensional Systems

Performance characteristics of one-dimensional dynamic channel assignment systems of different size and under different loading conditions have been obtained from a large-scale computer simulation. Since the results are tightly coupled to the simulation, the simulation warrants describing here.[23]

This simulation is based on a one-dimensional cellular layout like the one illustrated in Figure 7.2-2.

In such a layout of equal size cells, the D/R ratio is equivalent to a reuse interval, D_R, of an integral number of cells. For example, as indicated in Figure 7.2-2 with $D/R = 8$, channels can be reused every fourth cell so the reuse interval is $D_R = 4$. In the simulation, vehicles making calls are identifiable entities whose locations and movements are stored in the computer. The data presented later were obtained by counting and storing the number of actual events that occurred during a simulation run. For each data point the simulation was run and events counted until the statistical fluctuation of the point decreased to about the size of the symbols plotted on the figures. The number of call attempts processed for each data point ranged between 10^3 and 10^4. More attempts were required at light loading of the system to ensure that a significant number (on the order of 100) of attempts were blocked, and so on. The fact that all the data points lie on smooth curves illustrates that the simulation runs were long enough to produce stable statistics.

Call initiation and vehicle movement in the simulation are handled in a single subroutine that is used to drive several different operating systems' subroutines. The flowchart in Figure 7.4-1 illustrates the major sequences in this driver subroutine. For each system cycle, the subroutine steps through each subscriber in sequence and checks his activity status. Depending on events which have occurred in previous system cycles, a subscriber will be either "on" in the system or "off."

If the subscriber is not "on" in the system, the right-hand branch of the chart is followed. The first step along this branch is to generate a random number that determines whether or not a call attempt will be made. If the result is not to make an attempt, that fact is noted and the subroutine steps to the next subscriber. If the result is to attempt a call, then another random number is generated from a uniformly distributed set to determine the subscriber's location in the one-dimensional universe of the simulation. This method of generating call attempts produces call arrivals that are Poisson processes in time at each base station and are uniformly distributed in space. This uniform distribution in space can be weighted to produce any desired spatial distribution by using counters that reject certain attempts at some locations. A velocity is assigned to the subscriber for each call attempt produced. This velocity is assigned from a random number set that has a prescribed density function. The velocity distribution was specified to be a Gaussian function truncated at velocities of ± 60 mi/hr. The mean (and median) velocity was 0 miles/hr and the standard deviation of the Gaussian before truncation was 30 miles/hr. If the call

attempt is located near an "edge" of the one-dimensional space, it is
checked against criteria used to compensate for the effects of the edges and
simulate an infinite system. This compensation then cancels some attempts
near the edges before they actually enter the system.

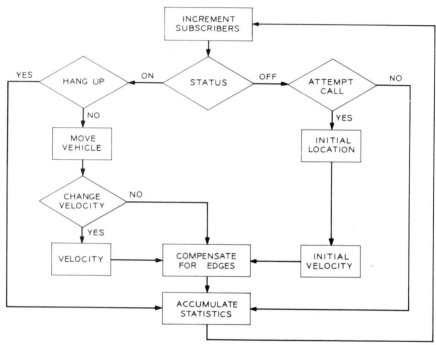

Figure 7.4-1 Flowchart for driver program in channel assignment simulation.

The left-hand branch of Figure 7.4-1, starting at the status block,
indicates the procedure followed for subscribers that are already "on" in
the system. The first check is whether or not the subscriber hangs up. Call
terminations are determined in a way that permits specifying the density
function for call durations. The call duration density function assumed for
the one-dimensional simulation was a truncated Gaussian with a mode of
90 sec, a minimum call duration of 30 sec and a maximum call duration of
600 sec. If the Gaussian had not been truncated it would have had a
standard deviation of 60 sec. The actual mean of the truncated distribution
is 103.5 sec.

Subscribers terminating calls are accounted for in the system statistics
and flagged for the system operating subroutine. A subscriber who does

not hang up is moved according to his predetermined velocity and the length of time between system cycles. His new location is then passed on to the system operating subroutine.

After a vehicle is moved, a random number is generated that determines if the velocity should be changed. If a velocity change is indicated, the new velocity is assigned at random, following the same procedure used to determine vehicle velocities for call attempts. After vehicles are moved, those near the "edges" are again checked against an edge compensation criterion and appropriate statistical values are accumulated.

In the simulation a single operating system subroutine determines the appropriate coverage cell to serve a call attempt, assigns a radio channel if one is available or refuses service otherwise, and updates channel assignments as vehicles move about. The flowchart in Figure 7.4-2 illustrates the major sequences in this operating system. For each system cycle, this subroutine steps through each subscriber who is either "on" in the system or is attempting a call and checks his activity status.

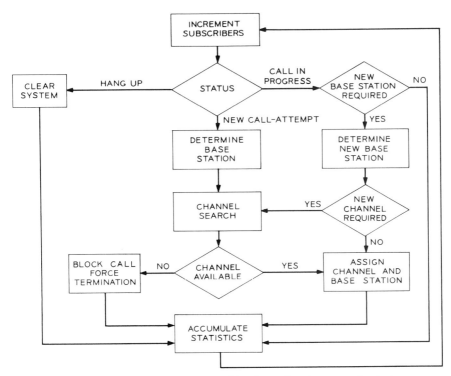

Figure 7.4-2   Flowchart for system control in channel assignment simulation.

New call attempts are processed along the center branch of the flowchart. The first step in processing an attempt is to determine which coverage cell the caller is in. This is readily done since the caller's geometrical location is part of the information available to the simulated operating system. If the cell is near an "edge" of the one-dimensional space, additional checks of the calls "on" in the cell are made. Some call attempts near the edges of the one-dimensional space are refused service at this point if the total calls "on" in the caller's cell exceed some value determined by the edge compensation criterion. Edge effects in a finite system come into play in the following way. As cells surrounding a given cell are being checked for channel availability, checks toward the center of a system will be made in more cells than checks toward an edge if the given cell is within a reuse interval of an edge. Channels will have a higher probability of being in use (not available) if there are more cells in which they may be used. Thus, cells on edges will have, on the average, more channels available to serve calls and will carry more traffic and block fewer attempts at a given average offered traffic than will cells in the center of a large system. These effects were observed in the simulations before edge compensation was implemented.

The next step in processing a new call attempt is to search for a radio channel that satisfies some channel assignment criteria. The simulation permits many different assignment criteria to be investigated. The dynamic channel assignment criteria studied assumed that all radio channels were available in all cells. It also assumed no limit on available radio equipment in the cells. This last assumption is somewhat unrealistic, but it permits the determination of upper bounds on actual radio equipment requirements. Channel search is done channel-by-channel starting with channel 1. If the channel being checked is in use in the caller's cell, then that channel is rejected and the next channel is checked. If a channel is not in use in that cell, a check is made cell-by-cell on both sides of that cell, If the channel is in use in a cell closer than a reuse interval away from the caller's cell, the channel is rejected and the next channel is checked starting again at the beginning of the channel check procedure. If the channel is not in use in any cell less than a reuse interval from the caller's cell, then further checks on that channel are made and it is compared with other channels, if any, that also have met this reuse interval criterion. If, at any time during the channel searching, a channel is found that is not in use in any other cell in the system, that channel is flagged for assignment and the channel search is terminated. Otherwise, channels that meet the reuse criteria are inter-compared to find the "most desirable one." The most desirable one will depend on the system operating parameter that is to be minimized, and thus the optimum channel assignment strategy depends on the overall system objectives.

The strategies studied certainly do not exhaust the multitude of possible channel assignment strategies. They were selected on the basis of relative simplicity and the fact that they intuitively attempt to optimize some aspect of system performance. Channel assignment strategies investigated in the simulation[25] were as follows.

First Available

Perhaps the simplest and cheapest strategy to implement is the assignment of the first available channel found during the channel search. Although this might appear to be far from an optimum assignment strategy, the data to be presented later indicate that this strategy performs as well as some of the more complicated schemes considered. This strategy assigns available channels randomly and minimizes the amount of data processing necessary to set up a call.

Mean Square (MSQ)

Channel activity is checked at cells in the interval between one and two reuse intervals (between D_R and $2D_R$) away from the assigned cell. This was done because the channel in use a distance greater than $2D_R$ away would allow the use of that channel in a cell within that distance. The mean square assignment strategy minimizes the quantity

$$\frac{1}{n} \sum_{j=1}^{n} D_j^2, \qquad D_R \leqslant D_j < 2D_R,$$

for the available channels. D_j is the distance between the assigned coverage cell and the cell using that channel within the specified interval and n is the number of cells using the channel within this interval. If $n=0$ for some channels, that is, if the channel is not in use in any cell between D_R and $2D_R$ on either side of the assigned cell, then the first such channel encountered is assigned to serve the call. In Figure 7.4-3 the mean square assignment strategy would choose channel $i+3$ to serve a call attempt in cell k.

Nearest Neighbor (NN)

This strategy chooses for assignment that channel which is "in use" at a cell nearest to the assigned coverage cell, but still at least a reuse interval, D_R, away. That is, it minimizes D over the available channels, where D is the distance to the first cell using that channel on either side of the assigned cell. If more than one channel has the same minimum D then the first channel encountered is assigned without regard to the distance to the first cell using that channel in the opposite direction. In Figure 7.4-3 the nearest-neighbor strategy would choose channel $i+1$ to serve a call attempt in cell k.

CELL NUMBER / CHANNEL NUMBER	1	2	...	K-7	K-6	K-5	K-4 = K-D_R	K-3	K-2	K-1	K	K+1	K+2	K+3	K+4 = K+D_R	K+5	K+6	K+7
1							X				X				X			
2									X						X			
.																		
.																		
.																		
I				X													X	
I+1				X											X			
I+2						X												X
I+3						X									X			

Figure 7.4-3 A channel-cell activity matrix. X indicates channel in use in cell. For text examples k is the assigned cell. *Note: X indicates channel in use at base station. For text examples κ is the assigned base station.*

Nearest Neighbor + 1(NN + 1)

This strategy is similar to the nearest-neighbor strategy except that it finds the minimum D for the available channels with

$$D \geqslant D_R + 1.$$

If no such channel exists then a channel with $D = D_R$ is assigned. This strategy tends to allow more callers to keep their assigned channel when they cross a radio coverage cell boundary. In Figure 7.4-3 the NN + 1 strategy would choose channel $i+2$ to serve a call attempt in cell k.

Fixed-Channel Assignment (FIXSYS)

A fixed-channel assignment system as described in Section 7.4-1 was also simulated.

We continue on with the description of the operating system subroutine at the branching block immediately following the channel search block on Figure 7.4-2. If, after checking all radio channels in the system, no available channel is found, the call attempt is refused service (the call is blocked), as indicated by the branching block on the flow chart. Blocked call attempts are completely removed from the system (cleared), and no other attempt to serve them is made. On the other hand, if an available channel is found, it is assigned to the vehicle and the base station to be assigned and the call setup is complete. Statistics are accumulated and the next subscriber checked.

The processing of calls in progress can be followed by returning to the status check block at the top of Figure 7.4-2 and following the right-hand branch. The first step along this branch checks the suitability of the current base-station assignment. This is done by checking to see if the vehicle has moved (during its last pass through the driving subroutine) to some other cell than the one currently serving it. If it has not moved into such a situation, then its current base station and channel assignment are acceptable and no further action is taken except to account for the call in the system statistics. If, however, the vehicle has moved into another cell, the base station(s) associated with the new cell must be used to serve the call. As indicated on the flowchart, the assigned channel is then checked at the new cell to be used and at all cells within a reuse interval of the new one. If the channel passes this check, that is, is not in use at those cells, the new base station and old channel are assigned to serve the call and the old base station is cleared of the call. If the old channel is in use within a reuse interval of the new cell, then a whole new channel search is initiated identical to the one used for new call attempts. In the case of a call in progress, the only difference is that, if a substitute channel is not available, the call is forced to terminate instead of being refused. Only one attempt is made in the simulation to switch channels for such a call in progress and no priority is given to these calls. In either case, the old base station is cleared of the call, statistics are accumulated, and the subroutine steps to the next subscriber.

The only actions required to process call terminations are to clear the channel at the assigned base station and accumulate the appropriate statistics. These actions are indicated on the left-hand branch of Figure 7.4-2.

The simulated operating systems were run for a 24-cell network with each cell 2 miles long. The cells were laid end to end along a line as in Figure 7.2-2. The simulation assumed 1000 subscribers and operated in a 10-sec cycle time. That is, within 0.10 sec in simulated time the status of each subscriber was examined. Statistics were taken from only the central 14 base stations to avoid any effects from the edges of the finite system.

Call loading at the edges of the finite system was tapered in a way which insured that the central 14 coverage areas were operating as though they were members of an infinite set of coverage areas.

The simulation was run using the different channel-assignment strategies for a 40-channel system with a reuse interval of every fourth cell. For these runs the average number of attempts per cell (the spatial demand profile) was uniform across the central coverage areas. The data from the simulation that show directly the relationship between traffic carried and blocking for these conditions are plotted in Figure 7.4-4. The curves associated with the data points for the various strategies are identified in the key. Blocking is defined as the ratio of new call attempts blocked to new call attempts and does not include channel changes or forced call terminations. It is evident that the dynamic channel-assignment strategies perform very similarly in terms of blocking and traffic carried. The behavior of the FIXSYS strategy is markedly different from the dynamic channel assignment strategies. As discussed in Section 7.4-1, the performance of fixed channel assignment systems can be determined from telephone traffic theory (Section 7.1.2). The blocking discipline for the FIXSYS is blocked calls cleared for which the Erlang B equation is applicable. The NN strategy appears to perform better below 20% blocking than any of the other strategies considered. Choosing the first available channel results in performance that is not appreciably different from the other dynamic channel assignment strategies.

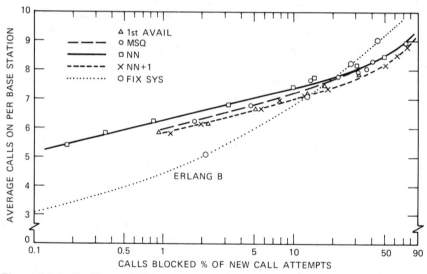

Figure 7.4-4 Traffic carried for some one-dimensional channel assignment strategies in a 40-channel system with a reuse interval of 4.

Forced call terminations that occur when vehicles cross coverage boundaries and do not find channels available for continuing their calls

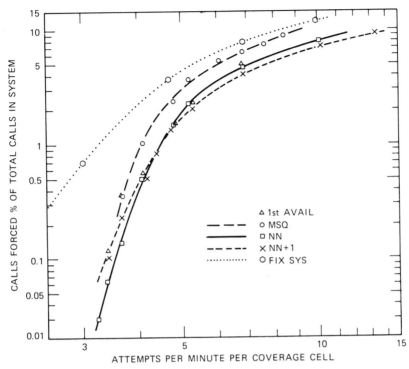

Figure 7.4-5 Forced call terminations relative to total calls in 40 channel one-dimensional systems with a reuse interval of 4.

will probably be more objectionable to customers than the blocking of new call attempts. In Figure 7.4-5 forced call terminations are expressed as the ratio of calls forced off in a time period to the total number of calls served in that time period. This ratio is plotted as a function of the call-attempt rate for the various channel assignment strategies. Since the Erlang B traffic equation does not include motion effects and thus does not include forced call terminations, the FIXSYS curve in Figure 7.4-5 is drawn through the simulation points without the aid of a theoretical curve. At low blocking rates the NN strategy forces off fewer calls than any of the other strategies. At higher attempt rates the NN+1 strategy forces the fewest calls. The cross-over of these curves is somewhat surprising considering that the NN+1 strategy produces fewer channel changes at boundary crossings than the NN strategy (the NN+1 strategy does not attempt to pack channel assignments as close together in space). At low attempt rates the NN strategy forces fewer calls to terminate because its blocking rate is significantly less and thus the chances of getting a new channel where one is required are considerably better. This characteristic apparently more than offsets the fact that fewer channel changes are required for the NN+1 strategy. The forced call termination ratio presented in Figure 7.4-5 is sensitive to vehicle velocities, cell sizes, and call durations. The ratio of the number of calls forced to terminate to the number of boundary crossings in a given time period is much less sensitive to these parameters.

Figure 7.4-6 shows forced call terminations presented as a percentage of boundary crossings and plotted as functions of the blocking of new call attempts. These curves show that for a given blocking rate the NN+1 strategy does not produce as many forced call terminations at boundary crossings as do the other strategies. This is because the NN+1 strategy sets up calls initially so that fewer channel changes are required at the boundary crossings. The termination rate produced at the boundary by the FIXSYS is the same as the blocking rate since all calls must either change channels or terminate at a boundary.

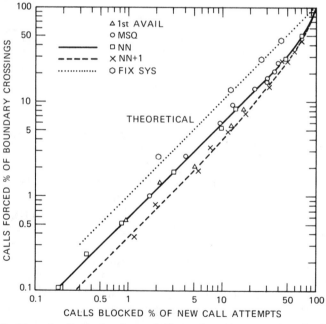

Figure 7.4-6 Forced call terminations relative to boundary crossings in 40 channel one-dimensional systems with a reuse interval of 4.

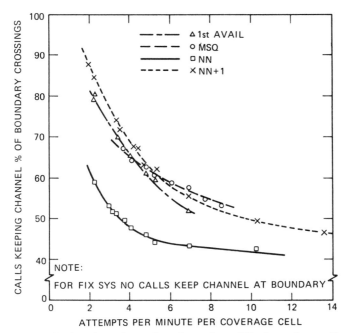

Figure 7.4-7 Calls keeping channel at boundary (40 channel one-dimensional systems with reuse interval of 4).

The ratio of the number of calls keeping the channel when crossing a coverage boundary to the total number of boundary crossings is plotted in Figure 7.4-7 as a function of the call attempt rate. These relationships indicate the extent to which the originally assigned radio channel floats with the vehicle and also the effort the system must expend in transferring calls from channel to channel at boundaries. At low attempt rates the NN + 1 strategy allows more callers to keep their originally assigned channel than any other strategy. The NN strategy, which packs calls closer together in space, allows significantly fewer callers to keep their channel. The other two dynamic strategies perform more nearly like the NN + 1 than the NN strategy. It is significant to note that for all of the dynamic channel assignment strategies considered, roughly half or more of the calls keep their channels when cell boundaries are crossed.

If callers crossing cell boundaries are given priority for channels or delay is permitted in channel switching at a boundary, then these boundary crossing results will require modification since the forced termination rates will be reduced by such procedures.

The dynamic channel assignment simulation allowed for the possibility that all channels could be used simultaneously in any cell. A histogram of the actual number of calls "on" for two different attempt rates and the NN+1 strategy is shown in Figure 7.4-8(a) and (b). It is immediately obvious that for a uniform spatial distribution of call attempts, halving the number of radio sets would not visibly affect the system performance, since the number of channels used in a cell seldom exceeds that value. The

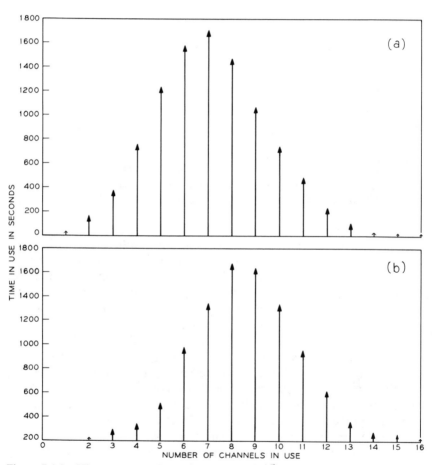

Figure 7.4-8 Histograms of channel usage per cell for 40 channel one-dimensional synamic channel assignment systems with a traffic carried of (a) 7.1 calls-on per cell and (b) 8.66 calls-on per cell. (NN+1 assignment strategy with channel reuse interval of 4).

design of the FIXSYS allows for a maximum of ten channels in any cell. Similar histograms of the calls-on distribution for the FIXSYS are presented in Figure 7.4-9(a) and (b).

For other simulation runs the number of system channels and the reuse interval were varied to determine the effects of these parameters on system performance. In one case the reuse interval was fixed at four and the total number of channels available to the dynamic channel assignment system

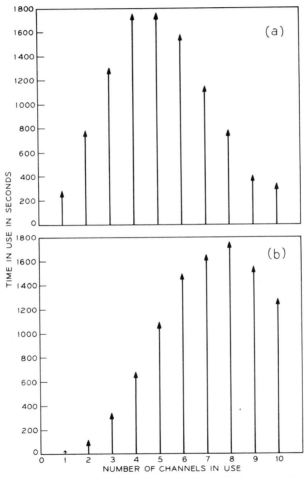

Figure 7.4-9 Histograms of channel usage per cell for 40 channel one-dimensional fixed channel assignement systems with a traffic carried of (a) 5.1 calls-on per cell and (b) 7.08 calls-on per cell.

was varied over the values 8, 20, 40, 80, and 120. With the reuse interval of every fourth base station, this results in a corresponding average number of channels per base station of 2, 5, 10, 20, and 30. All other system parameters remained as stated previously and the attempt rate was varied to produce data for performance curves.

The results of these runs are plotted in Figures 7.4-10 through 7.4-13. Figure 7.4-10 is a plot of the actual traffic carried per base station as a function of the number of new call attempts per minute per base station. The total number of channels in the system and the various channel assignment strategies are indicated on the figure. When new call attempts can be accommodated by a given system, the traffic carried is a linear function of the call attempts. As the systems fill up, new call attempts are blocked and the traffic carried tends to saturate as indicated. The blocking of new call attempts is illustrated in Figure 7.4-11. These two figures illustrate the result that systems with more channels perform better but they also provide a quantitative comparison of the behavior of the particular systems simulated. Not only do systems with more channels carry more actual traffic at a lower blocking, but they also utilize the channels available more efficiently. (This is true for a fixed channel assignment system also.) This fact is illustrated in Figures 7.4-12 and 7.4-13 for the NN assignment strategy. In Figures 7.4-12, the data from Figures 7.4-10 and 7.4-11 have been normalized on a per channel basis. This figure shows that the systems with large numbers of channels carry more traffic per channel and at a lower blocking rate for a given per channel call attempt rate than systems with fewer channels.

The data from the simulation that show directly the relationship between normalized traffic carried and blocking for the NN strategy are plotted in Figure 7.4-13. This figure again illustrates the increasing efficiency with numbers of channels. For example, Table 1 shows the calls on per channel at a blocking rate of 3%.

This increased efficiency results because there is a greater probability of having a circuit available at any given time from a set containing a large number of channels than from a set containing a small number of channels given that the offered traffic per channel is the same.

For another set of simulation runs both the reuse interval and the number of radio channels in the system were varied in such a way that their ratios remained constant. This resulted in the average number of radio channels available in a coverage cell remaining constant at a value of 10. Figure 7.4-14 is a plot of the relationship between the average traffic carried per coverage cell and the blocking of new call attempts for the NN channel assignment strategy and for a fixed channel assignment system

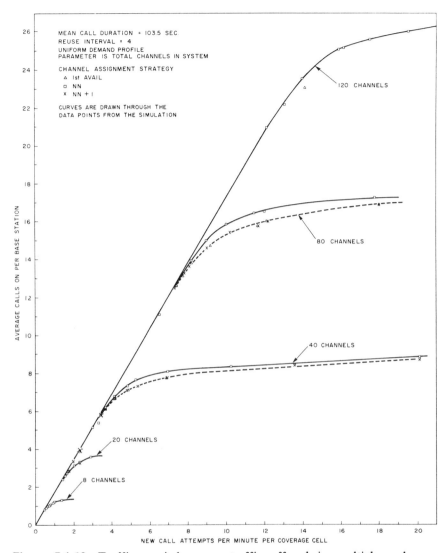

Figure 7.4-10 Traffic carried versus traffic offered in multichannel one-dimensional systems.

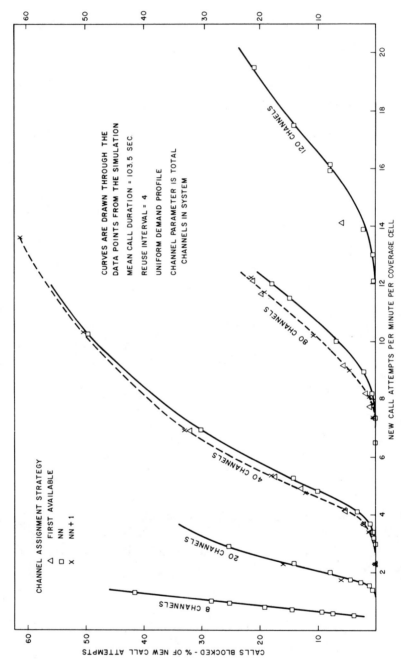

Figure 7.4-11 Blocking versus traffic offered in multichannel one-dimensional systems.

590

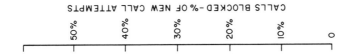

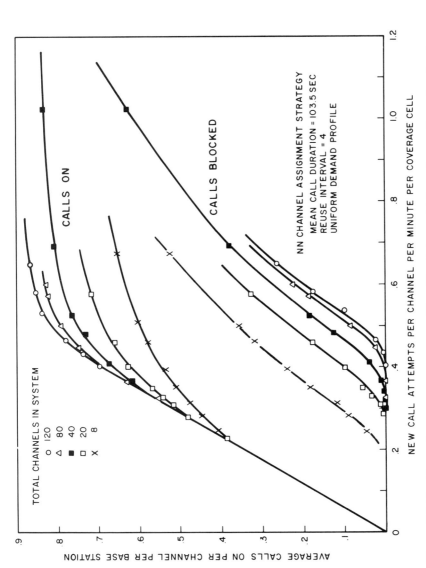

Figure 7.4-12 Channel occupancy versus traffic offered in multichannel one-dimensional systems.

591

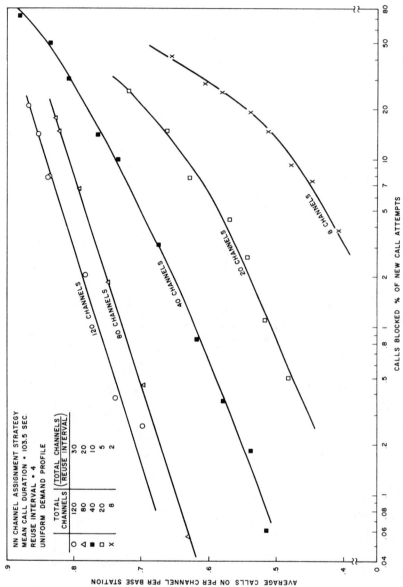

Figure 7.4-13 Channel occupancy versus blocking for multichannel one-dimensional systems.

592

Table 1 *Calls On Per
Channel Per Cell NN
Strategy; Reuse Interval of
4; 3% Blocking*

Number of Channels	Calls On Per Channel
8	0.40
20	0.55
40	0.67
80	0.765
120	0.80

with ten channels per cell. As indicated on the figure, the simulation was run with reuse intervals of 3, 4, and 5 for the dynamic channel assignment strategies. The FIXSYS performance does not depend on the reuse interval for the parameters considered.

The systems with more radio channels and correspondingly larger reuse intervals utilize their channels more efficiently, as indicated by the greater traffic carried for a given blocking. This increased efficiency is due to two factors. One factor is because of the larger number of channels alone as discussed in the previous section. This improvement, solely due to the increased number of radio channels, is also realized in the fixed channel assignment systems. The other factor comes about as follows: In meeting the Poisson distributed random load the dynamic channel assignment system operates with an actual average reuse interval that is greater than the minimum reuse interval specified for the system. The value of this excess actually decreases as the number of channels available increases. This is evidenced by the fact that fewer callers keep their channel when they are crossing coverage boundaries for the larger systems.[27] Fewer callers keeping their channel implies that the actual average reuse interval is smaller. This factor thus becomes less significant with larger reuse intervals because the ratio of the actual reuse interval to the specified reuse interval approaches unity. Note that this second factor does not affect system performance of fixed channel assignment systems and thus fixed channel assignment system performance depends only on the number of channels available in a particular cell.

The crossover points of the fixed channel assignment system curve with the dynamic channel assignment system curves occur at higher blocking for larger reuse intervals. This shows that systems for use in propagation environments that require large reuse intervals to achieve their cochannel

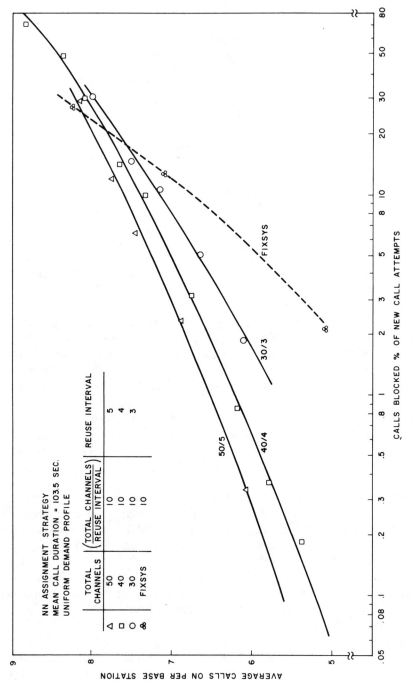

Figure 7.4-14 Channel occupancy versus blocking for multichannel one-dimensional systems.

interference design margins will benefit most from using dynamic channel assignment. For example, with a reuse interval of three this crossover occurs at a blocking of about 18%, whereas with a reuse interval of five it occurs at a blocking of about 25%.

The data from the computer simulation discussed in the previous paragraphs have been replotted in Figure 7.4-15 to illustrate the channels required to serve a given demand at two blocking rates. The curves marked Erlang B are theoretical curves obtained from the Erlang B telephone trunking formula that apply to the fixed channel assignment system. The percentage increase in channels required for the fixed channel assignment system over channels required for the dynamic channel assignment system for a given demand and blocking decreases as the demand increases and more channels are required. For example, at 1.6 attempts per minute per coverage cell and a blocking of 2%, the dynamic channel assignment system requires five channels per coverage whereas the fixed channel assignment system requires almost seven channels. At nine attempts per minute per cell and 2% blocking, the dynamic channel assignment system requires 20 channels per cell while the fixed channel assignment system requires 23 channels.

These results apply only to a fixed uniform spatial demand and do not take into account the advantages of dynamic channel assignment in meeting temporal variations in spatial demand.

For a few simulation runs the spatial demand profile was made periodic such that every other coverage cell operated with the same average call attempt rate. Several different average call attempt rates were used for each system configuration. The variations in call-attempt rates in alternate cells were chosen to be $\pm 11.1\%$, $\pm 20\%$, and $\pm 33\frac{1}{3}\%$ about the mean call-attempt rate. The variations that were simulated were chosen to span the range of potential day-to-day fluctuations about the long-term average. They were not intended to illustrate isolated demand peaks due to abnormal circumstances. It is assumed that a fixed assignment system would be engineered to serve the busy hour of the average business day distribution, which, for the purposes of the computer study, was assumed to be uniform. Examples of some of the simulated demand profiles for an average call-attempt rate of about four new call attempts per minute per base station are illustrated in Figure 7.4-16. The data points in this figure were obtained from actual simulation runs of 167 min in simulated system operating time. Data are included for a uniform demand profile to show the increase in capacity of a dynamic channel-assignment system, even over a perfectly engineered fixed assignment system. The data from the simulation that show directly the relationship between traffic carried and

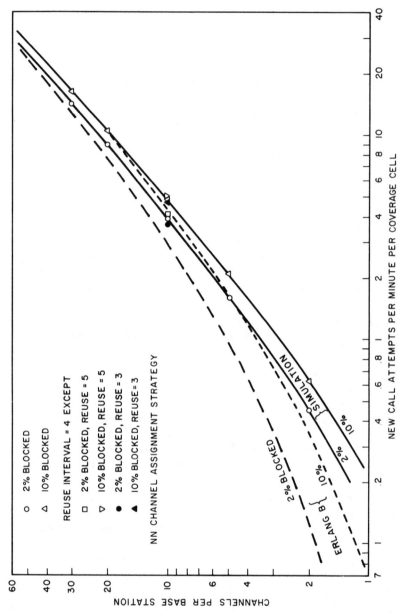

Figure 7.4-15 Channels required versus user demand for fixed blocking (one-dimensional systems).

596

blocking for the NN and FIXSYS channel assignment strategies are plotted in Figure 7.4-17. The solid curve for the NN strategy is drawn through the data points for the various loading profiles. The dashed curves are derived from the Erlang B telephone traffic formulas, which assume no vehicle motion. As can be seen from the figure, the periodic small-scale

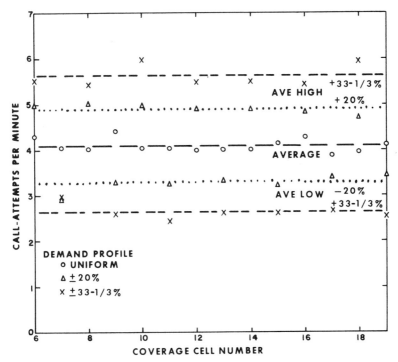

Figure 7.4-16 Average attempt rate in 14 cells for simulation of nonuniform spatial loading.

uneveness in the spatial distribution of call attempts does not significantly affect the performance of the NN strategy dynamic channel assignment system. The FIXSYS channel-assignment system, however, deteriorates with increasing amplitude of the periodic spatial distribution of call attempts.

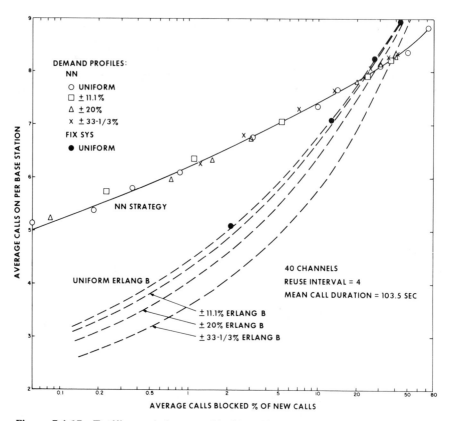

Figure 7.4-17 Traffic carried versus blocking for nonuniform spatial loading of one-dimensional systems.

7.4.4 Performance of Two-Dimensional Dynamic Channel Assignment Systems

The overall system performance of two-dimensional dynamic channel assignment mobile radio systems is similar to that of the one-dimensional systems. However, the channel reuse situation is more complex for two-dimensional systems, so there are some differences in performance between systems in one and two dimensions. Performance characteristics of two-dimensional dynamic channel assignment systems have been obtained by computer simulation also.[12]

The system simulated consists of a set of square radio coverage areas arranged to completely cover a square grid with no overlap as illustrated in Figure 7.4-18. For this model a channel used in a specified coverage area,

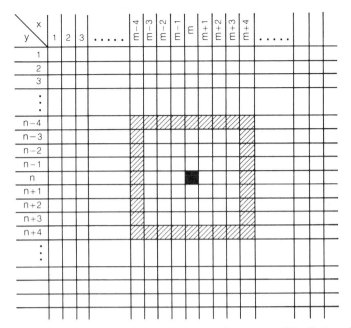

Figure 7.4-18 Coverage area configuration and a reuse "ring" for the two-dimensional simulation.

such as the shaded area at coordinates m, n may be reused anywhere on or outside of a specified square ring that surrounds that particular coverage area. This reuse ring is defined by those coverage areas that have either an X or Y coordinate that is the specified reuse interval of coverage areas separated from the center coverage area. In the example in Figure 7.4-18 the cross-hatched coverage areas on the reuse ring surrounding the coverage area at $X = M$, $Y = N$ are at a reuse interval of 4. Several coverage areas on the reuse ring may also use the same channel as long as all coverage areas are separated by at least the reuse interval. Thus, in the example, the maximum number that can use the same channel simultaneously on the reuse ring is eight. For the two-dimensional simulation the coverage squares were specified to be 2 miles on a side.

The initiation of call attempts and the movement of vehicles in this simulation are almost identical with that described previously for the one-dimensional simulation. Call attempts are generated in the simulation as a Poisson process in time that has a controllable average attempt rate in each coverage area. Initial call attempts from vehicles are uniformly distributed in both the X and Y coordinates and the X and Y locations are independent. Call attempts are tapered at the edges of the simulated universe and the performance data accumulated only from the central portion of the system (15 cells × 15 cells), so that again the results are representative of an infinite system.

Vehicle motions are primarily parallel to one of the coordinate axes with 45% of the vehicles having an X velocity component only and 45% a Y component only. Ten percent of the vehicles have both X and Y velocity components that are mutually independent. The velocity components (X and Y) have the same truncated Gaussian distribution with zero mean, standard deviation of 30 miles/hr, and maximum velocity magnitude of 60 miles/hr as used in the one-dimensional simulation.

Part of the data presented later is for a call duration distribution which is exponential with a mean of about 98 sec. Other data are for the truncated Gaussian call duration distribution used in the one-dimensional simulation that has a minimum call duration of 30 sec, a maximum call duration of 10 min, and a true mean of 103.5 sec.

The method of handling call attempts, calls in progress, call terminations, and cell-boundary crossings is similar to that described for the one-dimensional system. The blocking strategy is again the same as that used in deriving the Erlang B (blocked calls cleared) telephone traffic formula.

The procedures used in the channel assignment strategies are best described in terms of an example that can be referred to Figure 7.4-18. Assume that a call attempt located at coverage area $X = M$, $Y = N$ is awaiting channel assignment. The first step is to determine which channels (if any) are available to serve the call attempt. This is accomplished by compiling a list of the channels that are not being used currently within any coverage area surrounding the designated coverage area out to but not including the reuse ring defined previously. If more than one radio channel is available within this reuse interval, then some channel assignment strategy must be applied to determine the channel that should be assigned.

The first available strategy (random assignment of available channels) and two others were investigated in two dimensions.[12] Since two of the strategies perform similarly, only the simplest, the RING strategy, will be described here. In the RING strategy a search is made through the list of available channels to determine which of these channels is currently in use in the most coverage areas lying on the reuse ring. If more than one channel has this maximum usage, an arbitrary selection of one of the channels is made to serve the call attempt. If none of the available channels is in use on the reuse ring (an infrequent event as illustrated in Figure 7.4-21, below), then selection is made on a first-available basis. This strategy is described in detail in Ref. 24. Thus, this strategy attempts to maximize overall channel usage by maximizing usage on the reuse ring when initial channel assignments are made.

The simulated operating systems were run for a 27 by 27 grid of coverage areas that were 2 miles on a side. The systems had available 160 duplex radio channels. With the assumed reuse interval of four this results in 10 channels (160/16) available on the average at each base station. Of course, depending upon the system activity and the instantaneous demand, some coverage areas will have more than ten channels in use simultaneously and some will have fewer. The data from the simulation, which show directly the relationship between channel occupancy (traffic carried per channel per coverage area) and the blocking of new call attempts, are plotted in Figure 7.4-19. The key indicates the different conditions associated with the different data points.

In fixed channel assignment systems for which the Erlang B telephone traffic formula is applicable, it has been shown[28] that the shape of the call duration (holding time) distribution has no effect on the relationship between traffic carried and the blocking of new call attempts. The fact that the data points for the simulations with different call duration distributions but otherwise identical conditions lie on the same smooth curves indicate that also for these dynamic channel assignment strategies the performance parameters are not affected by the shape of the call duration distribution. Over a wide blocking range the traffic carried by the two-dimensional system is more nearly constant than for the one-dimensional system. We would expect that at very low blocking rates the two-dimensional system would perform better (carry more traffic at a given blocking rate) because it has more channels available to the system (160 channels) than the one-dimensional system (40 channels) even though the average number of channels available per coverage area is the same. However, as seen from Figure 7.4-19, the two-dimensional system does not perform as well as the one-dimensional system above a blocking rate of about 1%. This is probably due to the fact that the simultaneous use of the same channel in coverage areas separated by exactly a reuse interval is more difficult to achieve in the two-dimensional area situation than on a one-dimensional line.

The two-dimensional simulation using the RING strategy was also run for a reuse interval of five with 250 channels and for a reuse interval of six with 360 channels. The resulting data points are indicated by the $\otimes$ and $\otimes$ in Figure 7.4-19. As was found in the one-dimensional case there was a small increase in the traffic carried at a specified blocking as the reuse interval increased even though the average number of channels available per coverage area was held constant, (see Figure 7.4-14).

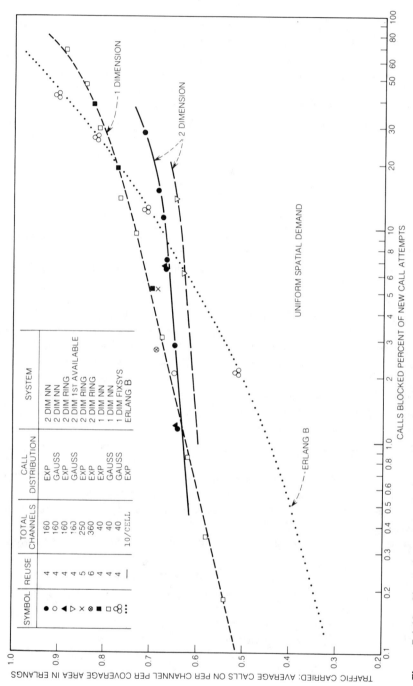

Figure 7.4-19 Channel occupancy for two-dimensional systems.

The figure contains the following legend table:

SYMBOL	REUSE	TOTAL CHANNELS	CALL DISTRIBUTION	SYSTEM
●	4	160	EXP	2 DIM NN
○	4	160	GAUSS	2 DIM NN
◀	4	160	EXP	2 DIM RING
▷	4	160	GAUSS	2 DIM 1ST AVAILABLE
×	5	250	EXP	2 DIM RING
⊗	6	360	EXP	2 DIM RING
■	4	40	EXP	1 DIM NN
□	4	40	GAUSS	1 DIM NN
⊗	4		GAUSS	1 DIM FIXSYS
⋮	—	10/CELL	EXP	ERLANG B

1 DIMENSION

2 DIMENSION

UNIFORM SPATIAL DEMAND

ERLANG B

CALLS BLOCKED PERCENT OF NEW CALL ATTEMPTS

TRAFFIC CARRIED: AVERAGE CALLS ON PER CHANNEL PER COVERAGE AREA IN ERLANGS

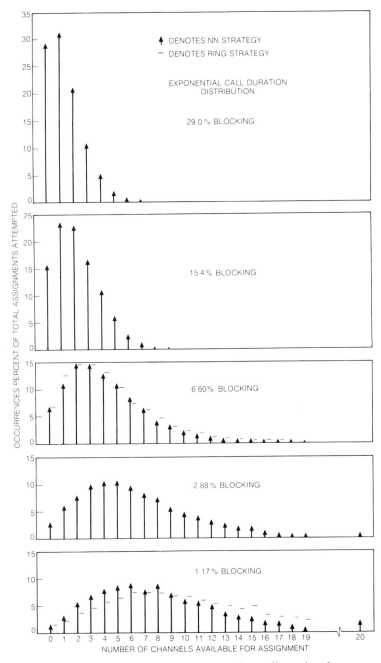

Figure 7.4-20 Channel availability in 160-channel two-dimensional systems with a reuse interval of 4.

603

More than one channel must be available for assignment in a particular coverage area if a channel assignment strategy is to improve system performance. Figure 7.4-20 shows the number of channels available for assignment expressed as the percentage of total channel assignments (new call attempts plus boundary crossing calls having channels reassigned) attempted for several blocking rates. At low blocking rates several channels are available to select from for most of the channel assignments attempted. As the loading on the system increases, fewer channels are available from which to choose. Thus, the effectiveness of the channel assignment strategies decreases as the system loading increases. Although slight differences exist in the histograms for the different channel assignment strategies, the overall availability of channels for each assignment strategy is about the same.

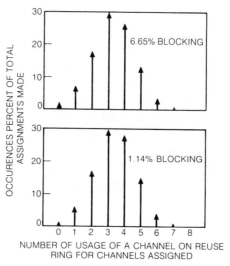

Figure 7.4-21 Channel useage on reuse ring for 160-channel two-dimensional ring strategy (reuse interval of 4).

The two-dimensional RING strategy attempts to maximize channel usage on the reuse ring. Figure 7.4-21 illustrates the degree to which this is achieved for two blocking rates. The percentage of total channel assignments made (new call attempts plus boundary crossings having channels reassigned) is given for each possible number of simultaneous channel usages on the reuse ring. The histograms of Figure 7.4-21 show that this dynamic channel-assignment strategy is not able to maximize channel reuse as it serves the randomly offered call attempts. The number of occurrences of channel assignments with each possible number of simultaneous channel usages on the reuse ring is expressed as the percentage of the total number of channel assignments made. For example, at a 6.67% blocking rate the channel assigned was in use in three coverage areas on the reuse ring for about 30% of the total channel assignments made. It is interesting to note that the assigned channel was being used in the maximum possible number (eight) of coverage areas on the reuse ring in fewer than 0.1% of the assignments made.

The results from the simulation of the two-dimensional dynamic channel assignment mobile radio systems show that these systems operate at very low blocking until the traffic offered reaches some critical value. Small increases in loading above this value produce a considerable increase in the blocking of new call attempts and result in very little increase in the traffic carried by the system. The loading at which blocking begins to occur in the two-dimensional systems is somewhat greater than the loading for which blocking begins to occur in one-dimensional dynamic channel assignment systems and is considerably greater than the loading for which a fixed channel assignment system begins to incur significant blocking. The reason that these dynamic channel assignment systems do not carry as much traffic at high blocking rates as fixed channel assignment systems is that they are not able to maximize channel reuse as they serve the randomly offered call attempts.

7.4.5 Increasing Channel Occupancy by Using Dynamic Channel Reassignment

If the spatial distribution of the traffic offered to a large-scale multiple-base-station mobile radio system can be estimated accurately and does not change from hour to hour, then the traffic handling capability of the system can be increased by utilizing dynamic channel reassignment.[13] Dynamic channel reassignment switches the channels assigned to some of the calls in progress so that the separation between coverage areas simultaneously using the same channel is more nearly optimum. Initial channel assignments are made using coordinated dynamic channel assignment and fixed channel assignment techniques. Again, because of the complexity of these systems, performance characteristics have been obtained by computer simulation.

The simulation used for the reassignment simulation was the same as the two-dimensional dynamic channel assignment simulation described in Section 7.4.4 except that provisions were added for including fixed assignment channels and for implementing channel reassignment.

X Y	1	2	3	4	5	6	•	•
1	A	B	C	D	A	B	C	•
2	E	F	G	H	E	F	G	•
3	I	J	K	L	I	J	•	
4	M	N	O	P	M	N	•	
5	A	B	C	D	A	B	•	
•	E	F	•	•	•	•	•	
•	•	•						

Figure 7.4-22 Coverage area grid with channel subsets as designated in Figure 7.4-23. Grid is applicable to 160-channel two-dimensional dynamic channel reassignment simulation.

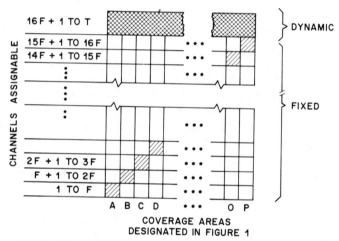

Figure 7.4-23 Channel allocations for coverage areas in Figure 7.4-22.

Channels allocated for possible use within coverage areas specified by A through P shown in Figure 7.4-22, are indicated in Figure 7.4-23. In Figure 7.4-23, duplex radio channels are numbered 1 through T, where T is the total number of channels available in the entire system. (For the simulation this number was specified to be 160.) In the example in Figure 7.4-23 the reuse interval is 4; F represents the number of fixed channels allocated to each coverage area. (F may vary from zero to T divided by reuse interval squared, in this case from zero to ten.) The remaining channels $[T-(\text{reuse interval})^2 \times F]$ are assumed to be available for assignment in any coverage area dynamically as discussed in Section 7.4.4.

In the simulated reassignment system a search is made first through the fixed channels allocated to the assigned coverage area. If there is a fixed channel not being used in the assigned coverage area, that channel is assigned to serve the call attempt. If all fixed channels are in use, a channel search is made through the dynamically assignable channels. The dynamic channel assignment procedure used was the RING strategy discussed previously. If no dynamic channel is available, the call is immediately blocked and cleared from the system (blocked calls cleared).

Efficient use of channels requires the simultaneous assignment (reuse) of a radio channel in radio coverage areas that are spaced as close together as possible without incurring excessive cochannel interference. Dynamic channel reassignment is initiated when a caller using a fixed channel terminates his call. A search is made within the coverage area where the fixed channel call termination occurred to determine whether a dynamically assignable channel is being used within that coverage area. If a dynamically assignable channel is being used to serve a call, that call is transferred to the just vacated fixed channel. This frees the dynamic channel for future assignment and ensures that a large number of calls are being served by the optimumly spaced fixed channels. Call terminations that occur on dynamically assigned channels are cleared from the system and no further action is taken.

The simulated system detects vehicles crossing coverage-area boundaries and initiates a new channel search when this occurs. If no channel is available, the call is immediately forced to terminate and is cleared from the system. Dynamic channel reassignment also is initiated when a caller using a fixed channel crosses a boundary and vacates the fixed channel.

The simulated operating systems were run for a 27 by 27 grid of coverage areas that were 2 miles on a side. The systems had available 160 duplex radio channels. With the assumed reuse interval of four, this results in ten channels $(160/16)$ available on the average at each base station. The number of fixed channels allocated per coverage area (F in Figure 7.4-23) was varied from five to nine and correspondingly the number of dynamically assignable channels available on the average per coverage area (total dynamic channels available divided by reuse interval squared) varied from five to one. Again, statistics were taken from the central 225 coverage areas (15 by 15) only to avoid any effects from the edges of the finite system.

System performance described in terms of traffic carried and the blocking of new call attempts as functions of traffic offered to the system is presented in Figure 7.4-24. The traffic offered to the system is defined as the product of the new call attempt rate and the mean call duration (see Section 7.1.2) and is expressed in the dimensionless quantity Erlangs (new calls per minute $\times$ minutes). The different curves are for the different mixtures of the number of fixed channels allocated per coverage area and the number of dynamic channels available on the average per coverage area as indicated in the figure key. The curves for the system with ten dynamic channels available per coverage area and no fixed channels are for the two-dimensional dynamic channel assignment system (DYNSYS) described in the previous section. The curves drawn through the system with ten fixed channels and no dynamic channels (FIXSYS) were obtained from the Erlang B telephone traffic formulas. The data points plotted along these FIXSYS curves were obtained by running the simulation with all of the vehicle velocities set equal to zero. The close fit of the simulation points to the theoretical curve provides a check on the accuracy of the simulation. All the other curves are drawn through the data points from the simulation using random velocities as described in Sections 7.4.3 and 7.4.4. Recall that the only effect of velocity is to shorten some of the call durations because of forced call terminations at coverage area boundaries. This effect will not change the curves, since call duration is included in the parameters plotted.

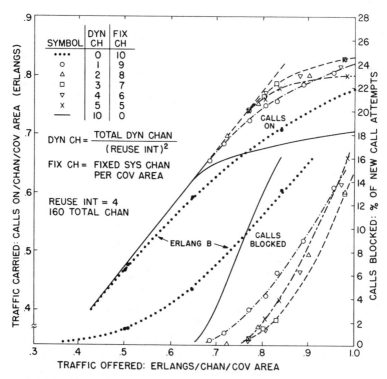

***Figure* 7.4-24**   Channel occupancy versus traffic offered for two-dimensional dynamic channel reassignment systems.

The effectiveness of dynamic channel reassignment is illustrated by the fact that the blocking does not becomes significant (greater than 1%) until the traffic offered is over 50% greater than the traffic offered that produces the same blocking for the FIXSYS. It is also evident from the curves that the better performing dynamic channel reassignment systems do not block significantly until the traffic offered is about 20% greater than the traffic offered for which the DYNSYS produces the same blocking. It is also obvious from the curves that all of the dynamic channel reassignment systems carry more traffic than either the FIXSYS or the DYNSYS within the region of good service (blocking rates less than 10%).

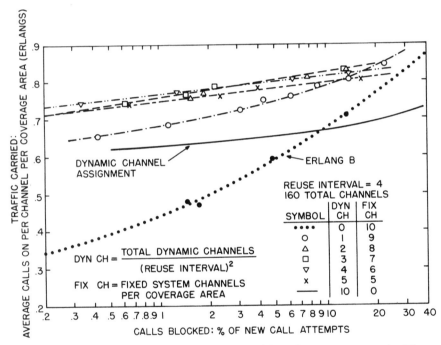

Figure 7.4-25 Channel occupancy versus blocking. Same system as in Figure 7.4-24.

The curves in Figure 7.4-25 show directly the relationship between the traffic carried and the blocking of new call attempts for the various systems. The performance of all of the dynamic channel reassignment systems is far superior to the performance of either the FIXSYS or the DYNSYS. The reasons that the dynamic channel reassignment systems perform better are that (1) they have channels available which can be moved around to serve the normal random fluctuations in the offered traffic and (2) most of the channels are reused at the specified minimum channel reuse distance by reassigning whenever possible calls using dynamic channels to fixed channels. The systems that have from two to five dynamic channels available on the average per coverage area perform similarly and have nearly an exponential relationship between the traffic carried and the blocking rate over the range of operation studied. The system with nine fixed channels per coverage area apparently does not have a sufficient number of dynamic channels to meet the normal fluctuations of the offered traffic. Since the performance of the systems with five and nine fixed channels is somewhat inferior to those with six, seven, and eight fixed channels, there appears to be a broad optimum mixture of the fixed and dynamic channels apportioned to a system. This fact is clearly illustrated in Figure 7.4-26 where the traffic carried is plotted versus the number of fixed channels per coverage area for three different blocking rates. Over a wide range of blocking, the system having only two dynamic

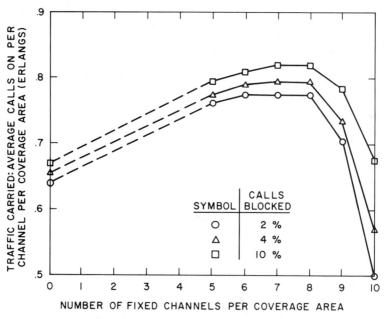

Figure 7.4-26 Channel occupancy at fixed blocking for different numbers of fixed channels in two-dimensional dynamic channel reassignment systems (160 channels total).

channels available on the average can provide nearly optimum performance. Of course, the actual values will change for different size systems (different numbers of channels, reuse interval, and so on) but similar improvement over pure fixed or dynamic channel assignment would be expected.

In addition to overall system performance characteristics just presented, more detailed aspects of system operation are needed to assess the relative operating complexity of the various systems. Systems that have many dynamic channels allocated and set up many calls on dynamic channels are more complicated and therefore will be more expensive to build and operate. Such systems are obviously more flexible and less dependent upon accurate matching to the spatial distribution of offered traffic for efficient use of available channels.

The percentage of calls that are set up on dynamic channels is an indication of the computation complexity required to operate a system. Assigning calls to dynamic channels requires many more steps in computer processing than assigning calls to fixed channels. Also, rapidly tunable radio equipment required for dynamic channel assignment will probably be more expensive than fixed tuned radios. Figure 7.4-27 shows the percentage of call attempts that are set up on dynamic channels as a function of the call blocking rate for the various channels indicated. In order to achieve the high system performance illustrated in the previous figures, a large fraction of the calls are initially assigned to dynamic channels. For example, a system which has only 32 dynamic channels out of its total of 160 channels, sets up 50% of its calls on dynamic channels at a blocking rate of 2%. Obviously, at high blocking rates ($>25\%$) most of the calls will be set up on dynamic channels, even for systems with few dynamic channels.

Efficient channel reuse is achieved in these dynamic channel reassignment systems by switching calls initially assigned dynamic channels to fixed channels as soon as possible. This channel switching dramatically improves system performance at the expense of operating complexity. Although these systems switch channels of many calls in progress, this function must already be provided in the radio equipment to handle the channel switching required when coverage boundaries are crossed. Figure 7.4-28 shows the percentage of total calls handled by the system that are switched from dynamic channels to fixed channels as a function of the blocking rate of new call attempts. As can be seen from the figure, this switching rate is not strongly dependent on the blocking rate. Again, to achieve the high overall system performance requires the switching of channels of a large fraction of the calls handled by the system. Since some

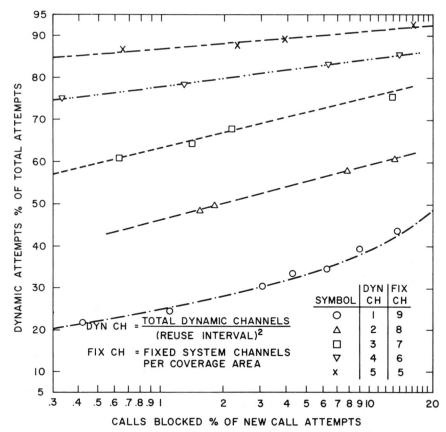

Figure 7.4-27 Activity of dynamically assignable channels versus blocking for different numbers of fixed and dynamic channels in two-dimensional dynamic channel reassignment systems (160 channels total).

of the calls cross coverage area boundaries and thus require more than one channel assignment, it is likely that a few calls will be switched from dynamic channels to fixed channels more than once while they are on in the system.

Another way of looking at the performance of these systems is to compare the channels required per coverage area for a FIXSYS to carry the same traffic as a 160-channel dynamic channel reassignment system using a specified mixture of fixed and dynamic channels. For example, Table 2 shows the number of fixed channels required to equal the performance of a dynamic channel reassignment system having eight fixed channels per coverage area and two dynamic channels available on the average per coverage area.

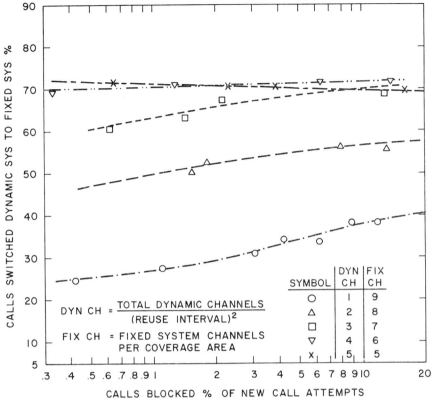

Figure 7.4-28 Channel switching activity versus blocking for different numbers of fixed and dynamic channels in two-dimensional dynamic channel reassignment systems (160 channels total).

Table 2 *Channels Required per Coverage Area for a FIXSYS to Carry the Same Traffice as a 160-Channel Dynamic Channel-Reassignment System Using Eight Fixed Channels per Square Coverage Area at a Reuse Interval of Four*

Blocking Probability	Number of Fixed Channels Required
1%	14+
2%	$13\frac{1}{2}$
5%	12

613

7.4.6   Channel Borrowing*

The dynamic channel assignment procedures covered in the preceding sections assumed that no fixed relationship existed between cells and channels, that is, no particular channel or channel set was preferred for use in a particular cell. Other dynamic channel assignment procedures that have been studied have a nominal relation between channels and cells like a fixed channel assignment system, but deviation from the nominal structure is permitted. Channels from the nominal channel set are used, if possible, to serve a call in a particular cell. If a channel from the nominal set is not available then a search is made through the nominal channel sets in the cells surrounding the caller's cell. If a channel is available in the nominal set of another cell, under certain conditions and subject to channel reuse constraints that channel may be temporarily used to serve the call. A channel being so used in a cell other than its nominal cell is then said to be "borrowed."[29,30] In many cases there is more than one channel that could be "borrowed" to serve a given call. Thus, channel assignment strategies are again needed to choose the "best" channel for borrowing. Such systems would be expected to perform similarly to the dynamic channel assignment systems treated in the previous sections since they both provide channel flexibility to meet load fluctuations. The major differences appear to be in point of view.

Channel borrowing systems also have been studied by computer simulation[29] because their complexity makes analysis intractable. The simulation run by Anderson was different from those described in previous sections in so many ways that direct comparison of the results is impossible. The simulated system was developed from estimated traffic requirements for the Philadelphia, Pennsylvania metropolitan area. It was a small system with 21 hexagonal cells and 360 channels total. The D/R ratio was specified as $D/R = 6$ and from Eq. (7.2-3) for fixed channel assignment and hexagonal cells the number of channel sets required was $C = \frac{1}{3}(D/R)^2 = 12$. The spatial distribution of offered traffic over the 21 cells was very nonuniform with a range of over 10 to 1 between the heaviest and lightest loaded cells. Fixed borrowing, that is, fixed reapportioning of channels to provide the closest "match" of available channels to load profile, was utilized in the system (see Section 7.4.1). After the channels were nominally assigned to cells, some cells had "excess" channel capacity to serve their offered traffic. The offered traffic was then adjusted in each cell so that all cells had a blocking rate of 2% on a fixed channel assignment (Erlang B)

---

*The material in this section is from L. G. Anderson[24] and from private communications from J. S. Engel and M. M. Peritsky.

basis. This loading adjustments was made to ensure that when channel borrowing was instituted in the simulation any improvement in system capacity (traffic carried at a given blocking) was due to the borrowing and not to residual capacity in the system. Three channel borrowing strategies or algorithms were investigated. Figure 7.4-29 illustrates the layout, loading, and channel reuse constraints used in the channel borrowing simulation.

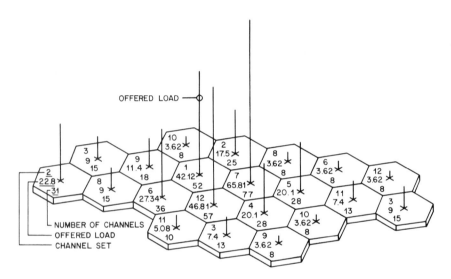

TOTAL OFFERED LOAD = 342.579 ERLANGS

Figure 7.4-29   System for simulation of channel borrowing.

The three channel borrowing strategies investigated in the channel borrowing system will be referred to as Strategies I, II, and III. In each case the nominal channels for a cell are checked before borrowing is attempted. If a nominal channel is available it is assigned.

Strategy I

If all nominal channels are busy, this strategy attempts to borrow a channel that is nominally assigned to an adjacent cell. This is accomplished by computing the number of channels that are available for borrowing in each adjacent cell. Then a channel is borrowed from the cell that has the largest number available for borrowing.

Strategy II

This is an improved version of Strategy I. It is also more complex. It is easiest explained with the help of the following definitions.

Nominal Channels—given a cell, those channels nominally assigned to it are its nominal channels.

Nominal Cells—given a channel, those cells to which it is nominally assigned are its nominal cells.

Interferable Cells—given a cell, any other cell with which it will interfere is one of its interferable cells.

This situation can be seen in Figure 7.4-29 where some channels from channel set 3 are nominal channels in three cells and conversely there are three nominal cells for some of the channels (13 of them) in channel set 3. These three cells meet the reuse requirement for $D/R = 6$ as discussed earlier. Note that because of the small size of the system, all 36 of the possible interferable cells for any given cell in a hexagonal grid with $D/R = 6$ or $C = 12$ do not exist in this system. This is a manifestation of the edge effect described earlier. In this strategy, candidates for borrowing are nominal channels assigned to adjacent cells. However, borrowing can have an effect on more than one nominal cell. It is the intent of Strategy II to take this fact into consideration and base the decision of which channel to borrow on the state of the worst case nominal cell of all nominal cells affected by the proposed borrowing. Worst case refers to the nominal cell that will have the fewest nominal channels available after the proposed borrowing. Stated in words, the objective of this algorithm is as follows: Choose the candidate channel that maximizes the available nominal channels in the worst case nominal cell that is also an interferable cell. Strategy I chooses the channel to be borrowed based on only the state of the adjacent cell. The general procedure is to compute a figure of merit for each candidate channel and choose the one for which this figure of merit is maximum. The figure of merit is the number of nominal channels available for use in the worst case nominal cell that would be affected by the proposed borrowing. (See Ref. 29 for details.)

Strategy III

This strategy is perhaps the simplest of the three and the point of view is slightly different. In the previous two strategies, nominal channels were assigned to cells based on the offered load in each cell. In this case we assign channels to channel sets. Then the nominal channel assignment for each cell is the entire channel set determined by the cell's location. This results in some cells having many more nominal channels than are needed. Service in these cells is limited by the number of radios provided, that is, by the number of channels permitted on simultaneously in a given cell. Borrowing is attempted when a radio is available, but there are no nominal channels available. The borrowing philosophy is also different. We search for an available channel by searching through channel sets in a prescribed sequence. The first available channel found is then used.

There is also the question of when a borrowed channel should be returned. Two strategies were tried. The easiest is to let the borrowed channel remain until the call it is serving is terminated. This is known as natural return. The other returns a borrowed channel as soon as a nominal channel becomes available. Whenever a nominal channel becomes idle, a call is transferred from one of the borrowed channels to it. No special effort was made to return channels to the cell most in need, analogous to the procedure for borrowing. The first strategy has been termed normal return, while the second has been called immediate return.

Reference 29 contains a large amount of quantitative performance data for several variations in the basic borrowing strategies, the channel return strategies, and the limitations placed on available radio equipment, that is, the number of simultaneous calls permitted in a cell. From these data were selected three sets of curves that are representative of the performance of this simulated channel borrowing system. For Strategy I, cell radios are provided at the level of 0.5% blocking. Strategy II uses the same number of cell radios. Strategy III however, was provided cell radios at the 0.5% level of blocking only in the highest loaded set of mutually interfering cells. Other cells were supplied at a blocking level of 2%. The procedure was to vary the offered load from 15% to 50% above the maximum fixed channel assignment case described earlier and in which fixed borrowing is used and loading adjusted per cell until all cells are driven at a 2% blocking level. For all cases run in the simulation, the natural return strategy produced better overall performance than did returning a borrowed channel as soon as a nominal channel became available. The data presented in the following figures are all for natural return, which is the return of a borrowed channel after the call it serves is terminated.

Average blocking is perhaps the statistic of most interest. It is computed by dividing the total number of blocked calls by the total number of call attempts throughout the system. It gives more weight to the high traffic cells since they receive more calls. Figure 7.4-30 illustrates the average blocking for the three channel borrowing strategies compared to the Erlang B blocking of a single large cell with 360 channels (a 360 server system). The figure shows that these particular small systems perform essentially as a single 360-channel system. This is to be expected since most of the traffic offered to the system is offered to the central set of mutually interfering cells that have access to the entire 360 channels. Increased blocking below

25% is due to radios being supplied at only the 0.005 level of blocking. Individually the strategies rank III, II, I, with III being the best by a narrow margin.

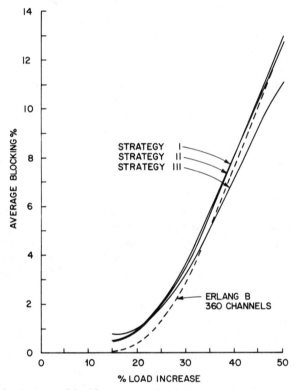

Figure 7.4-30 Average blocking versus load for simulated channel borrowing system.

The service deviation is a number computed from the fraction blocked, B_i, in each cell according to the following formula:

$$SD = \left(\sum_{i=1}^{21} \frac{\left(B_i - \bar{B} \right)^2}{20} \right)^{1/2} \tag{7.4-1}$$

where

$$\bar{B} = \frac{1}{21} \sum_{i=1}^{21} B_i. \tag{7.4-2}$$

Equation (7.4-1) indicates the relative amount of variation in blocking among cells. Figure 7.4-31 is a plot of the service deviation. Again the load increase is in percent of the load producing 2% blocking on the fixed channel-assignment system with fixed borrowing. Again strategies perform in order of III, II, I in the 25% region for which the average blocking is 2%.

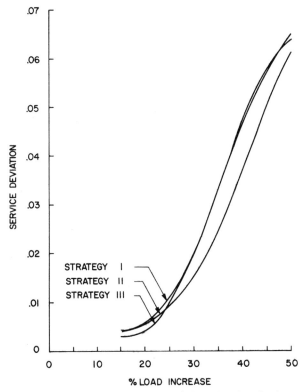

Figure 7.4-31 Service deviation versus load for simulated channel borrowing system.

The amount of borrowing is the fraction of completed calls that were set up on borrowed channels. Figure 7.4-32 plots the fraction of completed calls for which channels were borrowed. Strategy III requires substantially more borrowing than the other two.

This simulation was obviously different from the ones in the preceding sections in channel assignment strategies, in the spatial loading (traffic offered) profile, in the shape of coverage cells, and in overall system size. It allows a look at dynamic channel assignment in an entirely different

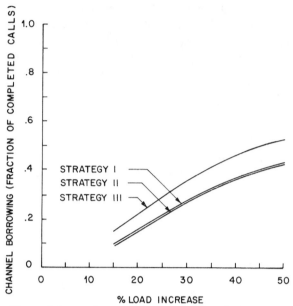

***Figure* 7.4-32** Channel borrowing activity for simulated channel borrowing system.

framework. A maybe not so obvious difference is a direct result of the differences in sizes of the systems and the way in which "edges" of the systems were treated. The large systems treated in the previous section were many reuse intervals across. The edge loading and system response were compensated (loading was tapered and number of simultaneous calls accommodated in edge cells limited) so that the interior cells, which still extended over several reuse intervals, operated as members of an infinite system. In contrast, cells in this relatively small system were all within a cell of an edge. Edge effects in this case would be substantial particularly in the narrow dimension of the system. These edge effects are described in Section 7.4.3.

Although the channel borrowing simulation was grossly different from the dynamic channel assignment simulations described in previous sections and the results are not directly comparable, the overall results from all the simulations indicate that over a wide range of system configurations, spatial loading profiles, numbers of channels, and assignment strategies, dynamic channel assignment produces an increased channel occupancy on the order of 25–50% over fixed channel assignment at blocking rates in the range of 1–5%.

REFERENCES

1. V. A. Douglas, "The MJ Mobile Radio Telephone System," *Bell Lab. Record*, December 1964, pp. 383–389.

2. R. E. Langseth, "Some Effects of Delay and Modulation-Index Mismatch on Cochannel FM Interference," *IEEE Trans. Veh. Tech.* **VT-20**, November 1971, pp. 124–132.

3. E. C. Molina, "Application of the Theory of Probability to Telephone Trunking Problems," *Bell System Tech. J.* **VI**, July 1927, pp. 461–94.

4. A. K. Erlang, "Solution of Some Problems in the Theory of Probabilities of Significance in Automatic Telephone Exchanges," *Post Office Electrical Engineers' Journal (London, England)*, **X**, January 1918, pp. 189–97.

5. A. W. Drake, *Fundamentals of Applied Probability Theory*, McGraw-Hill, New York, 1967, Chapter 5.

6. H. Schulte, Jr., and W. A. Cornell, "Multiarea Mobile Telephone System," *IRE Trans. Veh. Comm.*, **9,** May 1960, pp. 49–53.

7. W. D. Lewis, "Coordinated Broadband Mobile Telephone Systems," *IEEE Trans. Veh. Comm.*, **VC-9**, May 1960, pp. 43–48.

8. D. Araki, "Fundamental Problems of Nationwide Mobile Radio-Telephone System," *Rev. Elec. Comm. Lab. Japan*, **16**, May/June, 1968, pp. 357–373.

9. R. H. Frenkiel, "A High-Capacity Mobile Radiotelephone System Model Using a Coordinated Small-Zone Approach," *IEEE Trans. Veh. Tech.* **VT-19**, May 1970, pp. 173–177.

10. L. Schiff, "Traffic Capacity of Three Types of Common-User Mobile Radio Communication Systems," *IEEE Trans. Comm. Tech.*, **COM-18**, No. 1, February, 1970, pp. 12–21.

11. H. S. M. Coxeter, *Introduction to Geometry*, 2nd ed., Wiley, New York, 1969.

12. D. C. Cox and D. O. Reudink, "Dynamic Channel Assignment in Two-Dimensional Large-Scale Mobile Radio Systems," *Bell System Tech. J.* **51**, September 1972, pp. 1611–1630.

13. D. C. Cox and D. O. Reudink, "Increasing Channel Occupancy in Large Scale Mobile Radio Systems: Dynamic Channel Reassignment," *Joint IEEE Comm. Soc.–Veh. Tech. Group Special Trans. Mobile Radio Comm.*, November, 1973, pp. 1302–1306.

14. J. S. Engel, "The Effects of Cochannel Interference on the Parameters of a Small-Cell Mobile Telephone System," *IEEE Trans. Veh. Tech.* **VT-18**, November, 1969.

15. Bell Laboratories, "High-Capacity Mobile Telephone System Technical Report," Submitted to FCC December, 1971.

16. G. L. Turin, W. S. Jewell, and T. L. Johnston, "Simulation of Urban Vehicle-Monitoring Systems," *IEEE Trans. Veh. Tech.* **VT-21**, February, 1972.

17. H. Staras and S. Honickman, "The Accuracy of Vehicle Location by Trilateration in a Dense Urban Environment," *IEEE WESCON*, Session 27, August 24–27, 1971.

18. J. S. Engel, "The Effects of Multipath Transmission on the Measured Propagation Delay of an FM Signal," *IEEE Trans. Veh. Tech.* **VT-18**, May 1969, pp. 44–52.

19. S. G. Gustaffson, "Distribution of the Multipath Range Error in a Frequency Modulated Phase Ranging System Operating in an Urban Area," Spring Meeting of USNC/URSI, Washington, D. C., April 13–15, 1972.

20. P. T. Porter, "Supervision and Control Features of a Small-Zone Radiotelephone System," *IEEE Trans. Veh. Tech.* **20**, August 1971, p 75.

21. W. G. Figel, N. H. Sheperd, and W. F. Trammell, "Vehicle Location by a Signal Attenuation Method," *IEEE Trans. Veh. Tech.* **VT-18**, November, 1969.

22. J. Shefer and G. S. Kaplan, "X-Band Electronic Fence Automates Vehicle Location Without Spectrum Penalties," *Com. Equip. Systems Design*, February 1972, pp. 3–4.

23. D. C. Cox and D. O. Reudink, "Dynamic Channel Assignment in High Capacity Mobile Communications System," *Bell System Tech. J.* **50**, July–August 1971, pp. 1833–1857.

24. D. C. Cox and D. O. Reudink, "Dynamic Channel Assignment in Multidimensional Mobile Communications Systems," Patent application filed US patent office, December 1971.

25. D. C. Cox and D. O. Reudink, "A Comparison of Some Channel Assignment Strategies in Large-Scale Mobile Communications Systems," *IEEE Trans. Comm.* **COM-20**, April, 1972, pp. 190–195.

26. D. C. Cox and D. O. Reudink, "Effects of Some Nonuniform Spatial Demand Profiles on Mobile Radio System Performance," *IEEE Trans. Veh. Tech.* **VT-1**, May, 1972, pp. 62–67.

27. D. C. Cox and D. O. Reudink, "The Behavior of Dynamic-Channel-Assignment Mobile Communications Systems as a Function of Numbers of Radio Channels," *IEEE Trans. Comm.* **COM-20**, June, 1972, pp 471–479.

28. L. Kosten, "On the Validity of the Erlang and Engset Loss-Formulae," *P-T Bedrijs*, **2**, No. 1, 1948–49.

29. L. G. Anderson, "A Simulation Study of Some Dynamic Channel Assignment Algorithms in a High Capacity Mobile Telecommunication System," *Joint IEEE Comm. Soc.–Veh. Tech. Group Special Trans. on Mobile Radio Comm.*, November, 1973, pp. 1294–1302.

30. J. S. Engel and M. M. Peritsky, "Statistically Optimum Dynamic Server Assignment in Systems with Interfering Servers," *Joint IEEE Comm. Soc.–Veh. Tech. Group Special Trans. on Mobile Radio Comm.*, November 1973, pp. 1287–1293.

appendix a

computation of the spectra of phase-modulated waves by means of poisson's sum formula*

M. J. Gans

In this appendix, we describe a new method of computing the spectrum of a phase modulated wave as given in Equation (A-1)

$$v(t) = A \cos[\omega_c t + \theta(t) + \alpha], \tag{A-1}$$

where A is the amplitude of the wave and $f_c = \omega_c/2\pi$ is the carrier frequency. The phase angle, α, is a random variable uniformly distributed from 0 to 2π, and $\theta(t)$ is a stationary bandlimited (0 to W Hz) stochastic process. The autocorrelation of $v(t)$ is given by

$$R_v(\tau) = \langle v(t+\tau)v(t) \rangle$$

$$= \frac{A^2}{4}[e^{j\omega_c\tau}\Phi(\tau) + e^{-j\omega_c\tau}\Phi^*(\tau)], \tag{A-2}$$

where $\Phi(\tau)$ is the characteristic function of the joint probability density of $\theta(t+\tau)$ and $\theta(t)$, $p_{\theta(t+\tau),\theta(t)}(\beta_2,\beta_1)$, [10] with the transform variables evaluated at $+1$ and -1, respectively:

$$\Phi(\tau) \triangleq \int_{-\infty}^{\infty} d\beta_2 \int_{-\infty}^{\infty} d\beta_1 e^{j(\beta_2-\beta_1)} p_{\theta(t+\tau),\theta(t)}(\beta_2,\beta_1). \tag{A-3}$$

The two-sided power spectral density of $v(t)$ is then [6]

$$S_v(f) = \frac{A^2}{4}[F(f-f_c) + F^*(-f-f_c)], \tag{A-4}$$

*Reference numbers refer to references given in Chapter 4.

where we have defined

$$F(f) \overset{\Delta}{=} \int_{-\infty}^{\infty} e^{-j2\pi f\tau} \Phi(\tau)\, d\tau. \tag{A-5}$$

When $\Phi(\tau)$ is real [for example, when $-\theta(t)$ and $\theta(t)$ are equally probable sample functions], $F(f) = F^*(-f)$, and (A-4) becomes

$$S_v(f) = \frac{A^2}{4}[F(f-f_c) + F(f+f_c)], \qquad \text{if } \Phi(\tau) \text{ is real.} \tag{A-6}$$

We evaluate $(F(f)$ by Poisson's sum formula, [11] which states that for any constant f_0,

$$\sum_{n=-\infty}^{\infty} F(f + nf_0) = \frac{1}{f_0} \sum_{n=-\infty}^{\infty} \Phi\left(\frac{n}{f_0}\right) \exp\left[\frac{-j2n\pi f}{f_0}\right]. \tag{A-7}$$

For frequencies greater than W and greater than the rms frequency deviation, $(1/2\pi)\langle[\theta'(t)]^2\rangle^{1/2}$, the spectral density decreases very rapidly. [See Ref. 9, Equations (20) and (21), for example.] Thus if f_0 is chosen greater than

$$2\left\{ W + \frac{1}{2\pi}\langle[\theta'(t)]^2\rangle^{1/2} \right\}$$

and if $|f|$ is less than $f_0/2$, we may neglect all but the $n = 0$ term in the left-hand side of (A-7), giving

$$F(f) \doteq \frac{1}{f_0} \sum_{n=-\infty}^{\infty} \Phi\left(\frac{n}{f_0}\right) \exp\left[\frac{-j2\pi nf}{f_0}\right], \tag{A-8}$$

where the error, ϵ, is given by

$$\epsilon = \sum_{n=\pm 1, \pm 2, \ldots}^{\pm\infty} F(f + nf_0). \tag{A-9}$$

In the case where the modulation is a stationary process, [10]

$$\Phi(\tau) = \exp[-R_\theta(0) + R_\theta(\tau)], \tag{A-10}$$

where $R_\theta(\tau) \overset{\Delta}{=} \langle\theta(t+\tau)\theta(t)\rangle$ is the autocorrelation of the phase modulation, and $R(0)$ is the mean square modulation index in square radians. The angle-modulated-wave spectra for this case of Gaussian modulation have been computed in Ref. 9 with an exact formula [Equation (19), Ref. 9] for the case of phase modulation with a rectangular spectrum. Comparisons of Eq. (A-8), derived above, and the results of Ref. 9 indicate that Eq. (A-8) provides spectra accurate to within 10% if $|f| \leqslant 0.4 f_0$ and $f_0 \geqslant 2(\sqrt{R_\theta(0)} + 1)$ W, for rectangular spectrum Gaussian modulation. Figure 4.1-8 shows a comparison between Eq. (A-8) and the exact results for the case of rectangular spectrum Gaussian modulation with $R_\theta(0) = 6$ square radians and $f_0 = 20 W$.

Equation (A-8) represents still another addition to the myriad of formulas for the spectra of angle-modulated waves without accompanying formulas for bounds on the error. However, it is a convenient formula for our applications (indices on the order of unity or larger and baseband frequencies on the order of or less than W). Also, it is possible to check the fractional error by estimating ϵ in Eq. (A-9) from the shape of the spectrum out to $0.4 f_0$. Other formulas [Eq. (19) and (156) of Ref. 9] with better defined error bounds are restricted to rectangular baseband spectra, and either require double-precision computer calculations [Eq. (19), Ref. 9] or large indices [index $\geqslant \sqrt{10}$, Eq. (156), Ref. 9].

As an example of the use of Eq. (A-8) for computing the spectrum of an angle modulated wave, we will assume that the modulation, $\theta(t)$, is a stationary Gaussian process whose one-sided power spectrum is flat from 0 to W Hz. The characteristic function, $\Phi(\tau)$, for a Gaussian process is simply [10]

$$\Phi(\tau) = \exp[R_\theta(\tau) - R_\theta(0)]. \qquad \text{(A-11)}$$

and the rectangular baseband spectrum implies

$$R_\theta(\tau) = R_\theta(0) \frac{\sin 2\pi W \tau}{2\pi W \tau}. \qquad \text{(A-12)}$$

Substituting Eq. (A-11) and (A-12) into Eq. (A-8) gives

$$F(f) \doteq \frac{1}{f_0} \sum_{n=-\infty}^{\infty} \exp\left\{ -j(2\pi f)\frac{n}{f_0} \right.$$

$$\left. - R_\theta(0)\left[1 - \frac{\sin[2\pi W(n/f_0)]}{2\pi W(n/f_0)} \right] \right\}. \qquad \text{(A-13)}$$

The infinite sum in Eq. (A-13) does not lend itself to truncation since the terms do not approach zero. By adding and subtracting (Ref. 11, p. 44),

$$\frac{1}{f_0} \sum_{n=-\infty}^{\infty} \exp\left[-j(2\pi f)\frac{n}{f_0} - R_\theta(0) \right]$$

$$= e^{-R_\theta(0)} \sum_{n=-\infty}^{\infty} \delta(f+nf_0), \qquad (A-14)$$

and the sum in Eq. (A-13) is replaced by a convergent sum:

$$F(f) \doteq e^{-R_\theta(0)} \left\{ \sum_{n=-\infty}^{\infty} \delta(f+nf_0) + \frac{1}{f_0} \sum_{n=-\infty}^{\infty} e^{-j(2\pi f)n/f_0} \right.$$

$$\left. \times \left[\exp\left(R_\theta(0) \frac{\sin[2\pi W(n/f_0)]}{2\pi W(n/f_0)} \right) - 1 \right] \right\}. \qquad (A-15)$$

The step taken in Eq. (A-15) to improve the convergence is just one of a series of possible steps, as illustrated in the following equation:

$$\sum_{n=-\infty}^{\infty} e^{jnx} e^{a_n} = \sum_{n=-\infty}^{\infty} e^{jnx}\left[e^{a_n} - 1 - a_n - \frac{a_n^2}{2!} - \cdots \right]$$

$$+ \sum_{n=-\infty}^{\infty} e^{jnx} + \sum_{n=-\infty}^{\infty} a_n e^{jnx} + \sum_{n=-\infty}^{\infty} \frac{a_n^2}{2!} e^{jnx} + \cdots \qquad (A-16)$$

If the $\{a_n\}$ approach zero for large $|n|$, the first sum on the right-hand side of Eq. (A-16) converges more rapidly than the sum on the left-hand side. Often [as in (A-15)] the remaining sums on the right-hand side of Eq. (A-16) are known functions. Figure 4.1-8 was obtained with computer by using two terms in the series of Eq. (A-16) with $f_0 = 20W$, and the relation (Ref. 11, p. 45)

$$\sum_{n=-\infty}^{\infty} \exp\left[-j\left(\frac{\pi n}{10}\right)\frac{f}{W} \right]\left(\frac{\sin(\frac{1}{10}\pi n)}{(\frac{1}{10}\pi n)} \right) = \sum_{n=-\infty}^{\infty} p\left(\frac{f}{W} + 20n \right), \qquad (A-17)$$

where $p(t)$ is a rectangular pulse of amplitude, $\frac{1}{10}$, and width, 2, centered at $t = 0$. The remaining sum was truncated at 100 terms.

appendix b

click rate for a nonsymmetrical noise spectrum

M. J. Gans

Rice[3] defines a "click" as an increase or decrease of the quadrature component of the noise, $X_s(t)$, through zero, while the in-phase component, $X_c(t)$, plus the signal amplitude, Q, is negative. If the signal is stronger than the noise, the resultant phasor of signal plus noise experiences a rapid 2π phase change at this instant, producing a pulse ("click") in the output of the frequency demodulator.

In Ref. 3, Rice derives the formula for the "click" rate with a noise spectrum symmetrical about the signal carrier. The purpose of this appendix is to present a simple extension of his derivation to the case of asymmetrical noise spectra.

From Eq. (58) of Ref. 3, the "click" rate, N_c, is given by

$$N_c = \int_{-\infty}^{-Q} d\alpha \int_{-\infty}^{\infty} d\beta \, |\beta| \, p_{X_c, X_s, X_s'}(\alpha, 0, \beta), \tag{B-1}$$

where $p_{X_c, X_s, X_s'}(\alpha, \gamma, \beta)$ is the joint probability density of X_c, X_s, and X_s', the in-phase and quadrature components and time derivative of the quadrature component, respectively. [See Eq. (4.1-2), for example.]

From Eq. (4.4) of Ref. 1, the joint probability density is given by

$$p_{X_c, X_s, X_s'}(\alpha, 0, \beta) = \int_{-\infty}^{\infty} p_{X_c, X_s, X_c', X_s'}(\alpha, 0, \gamma, \beta) \, d\gamma$$

$$= \frac{1}{4\pi^2 \, \underline{B}} \int_{-\infty}^{\infty} \exp\left\{ -\frac{1}{2 \, \underline{B}} [\, \underline{b}_2 \alpha^2 + \underline{b}_0(\gamma^2 + \beta^2) \right.$$

$$\left. -2 \, \underline{b}_1 \alpha \beta \,] \right\} d\gamma$$

$$= (2\pi)^{-3/2} (\underline{Bb}_0)^{-1/2}$$

$$\times \exp\left\{ -\frac{1}{2 \, \underline{B}} [\, \underline{b}_2 \alpha^2 + \underline{b}_0 \beta^2 - 2 \, \underline{b}_1 \alpha \beta \,] \right\}, \tag{B-2}$$

where $\underline{B} \overset{\Delta}{=} \underline{b}_0 \underline{b}_2 - \underline{b}_1^2$ and $\underline{b}_n$ is defined in Eq. (4.1-11). We substitute Eq. (B-2) into (B-1) and integrate first with respect to β. The integration over β is accomplished by dividing the range of integration into two parts, $-\infty < \beta < 0$ and $0 < \beta < \infty$, and completing squares in the exponents. Thus,

$$N_c = \frac{\sqrt{\underline{B}/2}}{(\pi \underline{b}_0)^{3/2}} \int_{-\infty}^{Q} \left[1 + \sqrt{\frac{\pi}{2\underline{b}_0 \underline{B}}} \, \underline{b}_1 \alpha \right.$$

$$\times \exp\left(\frac{\underline{b}_1^2 \alpha^2}{2\underline{b}_0 \underline{B}} \right) \mathrm{erf}\left(\frac{\underline{b}_1 \alpha}{\sqrt{2\underline{b}_0 \underline{B}}} \right) \Bigg]$$

$$\times \exp\left[-\frac{\underline{b}_2 \alpha^2}{2\underline{B}} \right] d\alpha. \tag{B-3}$$

The remaining integration in (B-3) is performed by using integration by parts, giving

$$N_c = \frac{1}{2\pi \underline{b}_0} \left\{ \underline{b}_1 e^{-\rho} \mathrm{erf}\left(\frac{\underline{b}_1 \sqrt{\rho}}{\sqrt{\underline{B}}} \right) + \sqrt{\underline{b}_0 \underline{b}_2} \, \mathrm{erfc}\sqrt{\frac{\underline{b}_0 \underline{b}_2}{\underline{B}} \rho} \right\}. \tag{B-4}$$

The corresponding one-sided baseband output spectrum due to "clicks" is [Ref. 3, Eq. (20)]

$$\mathcal{W}_c(f) = 8\pi^2 N_c. \tag{B-5}$$

Let f_n be the center of the offset noise spectrum, $\eta G(f)$; that is, f_n is defined by the condition

$$\int_0^\infty (f - f_n) \eta G(f) \, df \overset{\Delta}{=} 0. \tag{B-6}$$

Then from (4.1-11)

$$\underline{b}_1 = 2\pi \int_0^\infty (f - f_c) \eta G(f) \, df$$

$$= 2\pi (f_n - f_c) \underline{b}_0 \tag{B-7}$$

and

$$\underline{b}_2 = 4\pi^2 \int_0^\infty (f - f_c)^2 \eta G(f) \, df$$

$$= \hat{\underline{b}}_2 + 4\pi^2 (f_n - f_c)^2 \underline{b}_0, \tag{B-8}$$

where we define

$$\hat{\underline{b}}_2 \overset{\Delta}{=} 4\pi^2 \int_0^\infty (f - f_n)^2 \eta G(f) \, df, \tag{B-9}$$

the second moment of the noise spectrum relative to its center frequency f_n. Usually, for calculation of adjacent channel interference, $[2\pi(f_n - f_c)]^2 \gg \hat{\underline{b}}_2 / \underline{b}_0$, which allows the following approximation to Eq. (B-4) valid when $\rho \gtrsim 1$:

$$N_c \doteq (f_n - f_c) e^{-\rho} \quad \text{(adjacent-channel case)}. \tag{B-10}$$

appendix c

median values of transmission coefficient variations

M. J. Gans

When computing the distortion due to frequency-selective fading we require the median value of the random function of time,

$$x(t) = A\left[\mathrm{Re}\left\{\frac{T_1(t)}{T_0(t)}\right\}\right]^2 + \left[B\,\mathrm{Re}\left\{\frac{T_2(t)}{T_0(t)}\right\}\right]^2$$

$$+ C\left[\mathrm{Re}\left\{\frac{T_1(t)}{T_0(t)}\right\}\mathrm{Im}\left\{\frac{T_1(t)}{T_0(t)}\right\} - \mathrm{Im}\left\{\frac{T_2(t)}{T_0(t)}\right\}\right]^2, \tag{C-1}$$

where A, B, and C are positive constants and $\{T_n(t)\}$ are parameters of the channel defined in Eqs. (4.1-114)–(4.1-116).

Bounds on the median value of expressions like (C-1) can be computed from the following inequalities[24]:

$$\mathrm{Pr}\{u \geqslant s_1\} \leqslant \mathrm{Pr}\{x \geqslant a_1\} + \mathrm{Pr}\{y \geqslant b_1\} + \mathrm{Pr}\{z \geqslant c_1\}, \tag{C-2}$$

and

$$\mathrm{Pr}\{u \leqslant s_2\} \leqslant \mathrm{Pr}\{x \leqslant a_2\} + \mathrm{Pr}\{y \leqslant b_2\} + \mathrm{Pr}\{z \leqslant c_2\}, \tag{C-3}$$

where the random variables satisfy

$$u = x + y + z, \tag{C-4}$$

and the constants satisfy

$$s = a + b + c. \tag{C-5}$$

631

If we choose a_i, b_i, and c_i $(i=1,2)$ such that

$$\Pr\{x \geqslant a_1\} + \Pr\{y \geqslant b_1\} + \Pr\{z \geqslant c_1\} = \Pr\{x \leqslant a_2\}$$

$$+ \Pr\{y \leqslant b_2\} + \Pr\{z \leqslant c_2\} = \tfrac{1}{2}, \qquad \text{(C-6)}$$

then s_1 and s_2 serve as bounds on the median,

$$s_2 \leqslant u_{\text{med}} \leqslant s_1. \qquad \text{(C-7)}$$

The probability distribution of

$$\left[\text{Re}\left\{ \frac{T_1(t)}{T_0(t)} \right\} \text{Im}\left\{ \frac{T_1(t)}{T_0(t)} \right\} - \text{Im}\left\{ \frac{T_2(t)}{T_0(t)} \right\} \right]^2$$

is plotted in Figure 4 of Ref. 23. From Eqs. (4.1-114)–(4.1-116) and the central limit theorem, it follows that the $\{T_n(t)\}$ are joint Gaussian variables. As given in Ref. 23, their correlations are

$$\langle T_0(t) T_0^*(t) \rangle \overset{\Delta}{=} 2b_0,$$

$$\langle T_0(t) T_1^*(t) \rangle = 0,$$

$$\langle T_1(t) T_1^*(t) \rangle = 2\sigma^2 b_0, \qquad \text{(C-8)}$$

$$\langle T_2(t) T_0^*(t) \rangle = \sigma^2 b_0,$$

$$\langle T_2(t) T_2^*(t) \rangle = \tfrac{1}{2}\mu_4 b_0,$$

where the central moments of the power delay distribution (cf. Section 1.4) are found from

$$T_0 \overset{\Delta}{=} \int_{-\infty}^{\infty} p(T) T \, dT,$$

$$\sigma^2 \overset{\Delta}{=} \int_{-\infty}^{\infty} p(T)(T - T_0)^2 \, dT, \qquad \text{(C-9)}$$

$$\mu_4 \overset{\Delta}{=} \int_{-\infty}^{\infty} p(T)(T - T_0)^4 \, dT.$$

It follows from the joint Gaussian statistics of the $\{T_n(t)\}$ that the probability distributions of

$$R_1 \overset{\Delta}{=} \text{Re}\left\{\frac{T_1(t)}{T_0(t)}\right\}, \qquad R_2 \overset{\Delta}{=} \text{Re}\left\{\frac{T_2(t)}{T_0(t)}\right\} \qquad \text{(C-10)}$$

are

$$\Pr\{R_1 \leqslant a\} = \frac{a}{\sqrt{a^2 + \sigma^2}}, \qquad \text{(C-11)}$$

$$\Pr\{R_2 \leqslant b\} = \frac{1}{4}\left\{\frac{2b - \frac{1}{2}\sigma^2}{\sqrt{b^2 - \frac{1}{2}\sigma^2 b + \frac{1}{4}\mu_4}} + \frac{2b + \frac{1}{2}\sigma^2}{\sqrt{b^2 + \frac{1}{2}\sigma^2 b + \frac{1}{4}\mu_4}}\right\}, \quad \text{(C-12)}$$

where σ is the spread of time delays as defined in Eq. (4.1-104). The probability distribution of

$$R_3 \overset{\Delta}{=} \text{Re}\left\{\frac{T_1(t)}{T_0(t)}\right\} \text{Im}\left\{\frac{T_1(t)}{T_0(t)}\right\} - \text{Im}\left\{\frac{T_2(t)}{T_0(t)}\right\} \qquad \text{(C-13)}$$

is plotted as $P(\gamma)$ in Figure 4 of Ref. 23, with $R_3^2 = (\sigma^2/2)^2\gamma$. The three probability distributions then allow the bounds to be computed as described above.

In general $R_1 \gg R_2$ and $R_1 \gg R_3$, so that a tight lower bound s_2, on $x(t)$ is obtained by using $\Pr\{R_2 \geqslant a_2\} = 0.5$ and $b_2 = c_2 = 0$. From Eq. (C-11) this implies that $a_2 = \sigma/\sqrt{3}$. In most cases, an appropriate upper bound, s_1, is found by using $\Pr\{R_1 \leqslant a_1\} = 0.4$ and $\Pr\{R_2 \leqslant b_1\} = \Pr\{R_3 \leqslant c_1\} = 0.05$. From Eq. (C-11) and (C-13) this implies $a_1 = 2\sigma/\sqrt{21}$ and $b_1 = \sigma^2/20$. From Ref. 23, assuming a Gaussian power delay distribution, $c_1 = \sqrt{50}\,\sigma^2$.

index

Adaptive diversity array, 489
Adaptive retransmission, 490
 continuous, 504
  pilot reception, 505
 continuous with pilot filter, 505
 division of diversity branches, 498
 double mixing, 510
 frequency division with feedback, 508
 probability distributions, 497
 transmitted signal as local
  oscillator, 508
Adaptive transmitting array, 490
Adjacent channel interference, 182, 199
 baseband spectrum, 183
 nonfading case, 182
 probability density, 199
 Rayleigh fading, 199
 signal-to-interference ratio, 201
AM multicarrier diversity, 512
 carrier separation, 512
 Rayleigh fading, 513
 signal-to-nose ratio, 515
 with filtering, 517
Amplitude fading, 162
Angle diversity, 311
Angular power distribution, 21
Antenna arrays, average CNR, 335
 circular, 152
 CNR probability distribution, 336
 element spacing, 329, 335
 equivalent circuit, 332
 gain, 335
 impedance matrix, 332
 in-line, 335
 level crossing rate, 341

linear, 329
 load impedance matrix, 333
 matching network, 333
 maximal ratio combining, 334
 mutual coupling, 329
 mutual impedance effects, 337
 mutual impedance with monopoles, 335
 planar, 335
 resistive loading, 333
 space diversity, 329
Antenna directivity, 138
Antenna pattern, effect of vehicle, 140
 horizontal plane, 139
 vertical plane, 139
Antenna power gain, 136
Antennas, 133
 field component, 148
 loop, 21
 mobile, 134
Atmospheric absorption, 88
Atmospheric index of refraction, 84
Attenuation by foliage, 107
Attenuation of median signal,
 correction factor, 105
 antenna height, 102
 distance, 98
 frequency, 101
Automatic gain control, 207
Average signal variations, 79
Azimuth pattern shaping, 152

Baseband phase noise spectrum, 175
Base station, antenna separation required, 65
 model of field, 61
 spatial correlation, 60
Base station antennas, 150
 azimuth directivity, 150
 bearing shift, 150

front-to-back ratio, 152
height effect, 103
improved horizontal directivi-
    ty, 151
Base station coverage, 126
Base station diversity, reduc-
    tion of power, 386
Base station height-gain fac-
    tor, 124
Base station reassignment, 571
Binary AM, DPSK detection, 233
Blocked calls, cleared, 556
delayed, 559
held, 557
Blocking, Erlang B, 617
probability, 549

Call attempt rate, actual, 549
model, 547
upper bound, 547
Capture effect, FM, 172, 178,
    383
Carson's rule, 172, 248, 453,
    499
in multiplexing, 245
peak frequency deviation, 246
Channel assignment, 572
dynamic, 573
fixed, 572
strategy, 579
  first available, 600
  fixed, 580
  mean square, 579
  nearest neighbor, 579
  nearest neighbor plus one,
    580
  ring, 600, 607
tapering, 573
Channel borrowing, 614
Channel borrowing strategies,
    615
average blocking, 617
immediate return, 617
normal return, 617
Channel reassignment simula-
    tion, 605
Channel reuse, 173
interval, 575
Chi-square distribution, 474
Circular antenna array, 152

Clicks, FM, 184, 452
Cochannel interference, 173,
    196, 278, 362, 382
averaged over fading, 198
baseband spectrum, 178
coherent combiner, advantage,
    480
 separate pilot, 475
 SNR distribution, 480
comparison of diversity com-
    biners, 377
multiple interferers, 198
multiplex, 278
nonfading case, 173
probability of, 384
probability density, 197
Rayleigh fading case, 196
reduction by diversity, 362
base station, 382
 perfect pilot combiner, 365
 separate pilot combiner, 371
signal-to-interference ratio,
    198
Coherence bandwidth, 47, 51, 54
definition, 45
Coherent combiners, 423, 464
baseband SNR, separate pilot,
    483
delayed pilot, index reduc-
    tion, 453
 phase error, 441, 444
delayed signal as pilot, 439
filtered signal as pilot, 426
four branch tests, 484
level crossing rates, 430
nonstationary noise, separate
    pilot, 483
 spectra, 484
power spectrum, 429
random FM, delayed pilot, 447
 filtered pilot, 430
 probability density of, 437
 spectrum, 434
receiver noise, delayed pilot,
    449
separate frequency pilot, 444
separate pilot, 464
 condition for in-phase addi-
    tion, 467
 design factors, 484

noise performance, 469
output SNR, 473
threshold improvement, 454
Complex conjugate retrans-
    mission, 490
array excitation, 494
average SNR, 495
Convolution theorem, 28
Correction factor, land-sea,
    124
rolling hilly terrain, 124
sloping terrain, 124
suburban terrain, 124
Correlation, field components,
    22
Cross-correlation, field com-
    ponents, 22, 36
Cross-covariance, two Gaussian
    signals, 324
Cross spectra, 24
Cumulative distribution func-
    tion, 18

Diffraction by trees, 107
Differential phase shift key-
    ing, 223
detection of, 223
Digital diversity, binary AM,
    518
error rate, coherent detec-
    tion, 519
differentially coherent de-
    tection, 519
FSK, 521
on-off AM, optimum threshold,
    528
postdetection DPSK, 525
examples, 529
frequency selective fading,
    528
irreducible error rate, 528
on-off AM, 518
performance, 528
postdetection DPSK, 522
FSK, 525
Digital modulation, 218
carrier recovery, 233
diversity, 517
DSB AM, 219
intersymbol interference, 220

moving threshold, 232
nonfading case, 219
Rayleigh fading, 230
Diversity branch weighting
    factors, 325
Diversity combiners, 311, 313
equal gain, 319
feedback, 321
maximal ratio, 319
pilot signal, 325
scanning, 321
selection, 313
Diversity effects, 341
Diversity impairments, branch
    correlation, 324
combining errors, 325
Diversity systems, against
    shadowing, 377
classifications, 310
cochannel interference, 362
fundamentals of, 309
combining methods, 313
multiple base stations, 378
receiver type, 316
transmitter type, 316
Diversity systems comparison,
    531
maximal ratio, receiver, 537
transmitter, 538
reference system, 531
reliability, 531
selection, receiver, 535
reliability, 536
transmitter, 535
transmitter power saving, 541
Diversity techniques, 389
Doppler, fading, 208
frequency, 216
shift, 14, 16, 30, 46, 135,
    429, 439, 448, 525, 531
maximum, 196
spectrum, 192, 350, 354
Ducting, 90
Duration of fades, 35
average, 36
Dynamic channel assignment,
    573, 574, 598
control, 574
Dynamic channel reassignment,
    605

effectiveness of, 609

Electromagnetic field components, 15
Electronic fence, 571
Elevation distribution of
  waves, 139
Ensemble average, 48
Envelope, amplitude measurements, 13
autocorrelation, 26
  of field components, 36
autocovariance at base station, 64
correlation, at base station,
  65
  at two frequencies, 50, 51
cross-covariance of field
  components, 38
cumulative distribution, 18
probability density, 17
ratio at two frequencies, 59
spectrum, electric field, 29
  magnetic field, 30
power, 24, 27
Equal gain combiner, 319, 349
mean SNR, 321
perfect pilot, 375
probability distribution, 321
reduction of random FM, 350
  of cochannel interference,
  375
Erlang, 550, 607
Error rate, nonfading, binary
  AM, 223
comparison of systems, 227
DPSK, 225
on-off AM, optimum threshold,
  226
Rayleigh fading, 231
  binary AM, 233
  DPSK, 234
  on-off AM, 231
Excess path loss, 120

Fading reduction, AM with pilot, 203
diversity, 319
Feedback combiner, 456
probability distribution, 457

Feedback diversity, 321, 401,
  422
baseband SNR, 423
principles, 322
Field component antennas, 148
Forced call terminations, 582
Fourier transform, 27, 42
Free-space transmission, 80,
  124
Frequency diversity, 312
Frequency division multiplex,
  241
Frequency modulation, 162
improvement over AM, 173
FM baseband noise spectrum,
  exact, 167
large signal approximation,
  167
FM baseband SNR improvement,
  162
FM capture effect, 172, 178,
  383
FM clicks, 184, 452
approximation, 169
nonsymmetrical noise, 627
FM noise performance, comparison with AM, 201
large signal approximation,
  166
nonfading case, 162
quasistatic approximation, 189
Rayleigh fading, 188
FM threshold, 163, 172
defined by clicks, 167
Frequency scaling, 113
Frequency selective fading distortion, 207, 270
digital modulation, 236
  irreducible error, 238
moving vehicle, large delay
  spread, 211
  small delay spread, 209
multiple echoes, 209
single echo, 209
stationary vehicle, 209
FSK, 228

Gaussian FM, 176
Gaussian noise, 471
Gaussian process, 16, 140, 481

cochannel interference, 376
complex, 136
joint density function, 34,
    49
with time delay, 47
Ground constants, 83
reflection coefficient, 81

Height-gain factor, 104
Hexagonal cells, 566
History of mobile telephony, 1
Horizontal magnetic dipole,
    148

Impulse response, 221
Instantaneous frequency auto-
    correlation, 193
Intermodulation with multi-
    plexing, 281
Intersymbol interference, 236
Irregular terrain, field pre-
    diction, 115
Isotropic antenna, 137

Knife edge diffraction, 87

Large area coverage, 546
Level crossing rate, 31
directive antenna, 140, 142
omnidirectional antenna, 142
reduction, 143
Local mean signal distribution,
    119
Log-normal distribution, 381
mean signal, 120, 125, 381,
    549

Man-made noise, 295
available data, 297
characterization, 295
measurement, 296
quasipeak meter, 296
Maximal ratio combiner, 316
effect of errors on SNR, 329
impairments, combining er-
    rors, 328
decorrelation, 325
mean SNR, 319
probability distribution,
    319

reduction of cochannel inter-
    ference, 365
reduction of random FM, 350
with FM, 343 ´
baseband SNR improvement, 344
Median signal, attenuation, 124
strength, 123
Minimum bandwidth FSK, 230
Mobile antennas, 134
average received power, 138
equivalent circuit, 135
height effect, 104, 124
Mobile frequencies, alloca-
    tions, 3
channel splitting, 3
domestic public land, 3
Mobile received signal, 134
complex envelope, 135
autocorrelation, 136
correlation of received vol-
    tage, 138
Mobile telephony classes, citi-
    zens band, 2
land transportation, 2
public safety, 2
Moments, 25
electric field, 26
in-phase, 24
magnetic field, 26
quadrature, 24
Monopole self-impedance, 335
Multipath interference, 11
Multipath simulation, 65
Multiplex bandwidth, FM/FM,
    257
FM/SSB, 260
SSB/FM, 253
Multiplex cochannel interfer-
    ence, FM/FM, 281
FM/SSB, 281
SSB/FM, 281
SSB/SSB, 278
Multiplex equal threshold con-
    dition, 258
Multiplex intermodulation, FM/
    FM, 285
FM/SSB, 287
SSB/FM, 285
Multiplex modulation index,
    FM/FM, 257

Multiplex required power, 244
  FM/FM, 259
  SSB/FM, 253
  with fading, FM/FM, 270
    FM/SSB, 266
    SSB/FM, 268
Multiplex selective fading
    distortion, FM/FM, 276
  FM/SSB, 271
  SSB/FM, 274
Multiplex SNR, FM/FM, 256
  FM/SSB, 260
  SSB/FM, 253
Multiplex systems comparison,
  261
  fading case, 263, 270
Multiplex threshold, FM/FM,
  256, 257
  SSB/FM, 253
Multiplexing, 240
  channel frequency require-
    ments, 251
  definitions, FM/SSB, 241
    FM/FM, 242
    SSB/FM, 241
    SSB/SSB, 241
  subcarrier frequency, 246
  subchannel, filter bandwidth,
    245
    spacing, 245

New Jersey, coherent combiner
    tests, 427
  postdetection combiner tests,
    395
New York City, excess path
    loss, 121
  field strength measurements,
    98
  irreducible error rate in,
    239
  level crossing rate in, 146
  street channeling effects,
    107
  time delay spread in, 46
Noise, above FM threshold, 190
  below FM threshold, 190
  components, 164
  Gaussian, 471
  man-made, 295

  nonstationary, 483
    FM detector performance, 484
  signal suppression, 190
  spectrum, 164
Number of channel sets, 565
N-unit, 85

One-dimensional traffic simu-
    lation, 574
  call duration function, 576
  edge effects, 578
  velocity, change, 577
    distribution, 575

PCM, 218
Phase conjugate retransmission,
  490
  covergence, 493
Phase difference, probability
  distribution, 54
Phase correlation with frequency, 52
Phase ranging, 570
Philadelphia, excess path loss,
  121
  field strength measurements,
    92, 98
  four-branch coherent combiner
    tests, 487
  mobile height-gain measure-
    ments, 140
  traffic requirements, 614
Poisson distribution, 553, 599
Polarization diversity, 311
Polarization effects, 133, 152
  colocated antennas, 155
  cross-coupling, 156
  potential for diversity, 153
Postdetection combining, 390
  maximal ratio, 391
    branch weighting factor, 394
    diversity advantage, 397
    receiver design, 395
Power spectra, fading signal,
  19
  field components, 20, 21
  different antenna patterns,
    141
  directive antenna, 141
  phase modulated waves, 623

Predetection combining, 390
 maximal ratio, separate pilot, 465
 switched diversity, 399
Prediction of field strength,
   123
Preemphasis, 287
 choice of functions, 258
 SNR, 291
Probability distribution of
   the field, 17
Propagation, in tunnels, 110
 over plane earth, 81
  direct wave, 81
  grazing angle, 83
  induction field, 81
  reflected wave, 81
  surface wave, 81
 over smooth spherical earth,
   85

Radio frequency interference,
   295
Rain attenuation, 88
Raised cosine filter, 221
Random FM, 39, 188, 211, 349
 irreducible error rate, 234
 limiting SNR, 191
 mean square value, 40
 power spectrum, 41, 354
 probability distribution, 39
 reduction by coherent combiner, 430, 447
Rayleigh criterion, 84
Rayleigh, density function,
   17, 231, 364
 distribution function, 13,
   18, 313
 fading, 189, 264, 472, 531
  effect on AM, 201
  simulation of, 68
 process, 401
Received power, 124
 corrected, 124
Reciprocity, adaptive retransmission, 491
Refraction by the atmosphere,
   84
Retransmission diversity, 489
 complex conjugate, 490

 phase conjugate, 490
Reuse interval, 575
Rough surface criterion, 84

Scanning diversity, 321
Secondary transmitter, 546
Selection diversity, 313
 base stations, 381
 impairments by correlation,
   324
 mean SNR, 316
  dependence on order, 316
 probability distribution, 313
 reduction of cochannel interference, 362
 reduction of random FM, 351
 with FM, 345
  baseband SNR improvement,
   348
 with squelch, 361
Separate pilot combiner, 371
Service area, 126
Service deviation, 618
Shadow losses, 79
Signal-suppression noise, 189
 reduction by audio filter,
   207
Simulation, of fading, 70
 of random FM, 72
Single coverage area traffic,
   560
Small area signal coverage,
   125
Small cell systems, 562
 base station assignment, 568
 deterministic layout, 567
 hexagonal cells, 566
 statistical layout, 563
 square cells, 565
Space diversity, 310
 antenna arrays, 329
 basic requirement, 311
Spatial field, distribution, 13
 model, 13
  with direct component, 31
  with time delays, 46
Speech coding, 218
Squelch, 196, 356
 effect on SNR, 356
 optimum muting point, 359

improvement, 360
Street orientation effects, 106
Switched diversity, 399
  allowable switching delay, 411
  baseband SNR, 418
  distortion with blanking, 417
  envelope probability distri-
    bution, 401
  optimum threshold, 420
  soft switching, 415
  switching rate, 416
  switching strategies, 401
  switching transient, 401
    noise, 411
System control, 545

Telephone traffic, 549
  birth-death process, 554
  call conversation time distri-
    bution, 551
  call duration, 550
  formulas, 556
    blocked calls cleared, 556
    blocked calls delayed, 556
    blocked calls held, 556
  negative exponential distribu-
    tion, 551
  offered load, 550
  Poisson distribution, 553
  simulation, 574
Terrain, irregular, 112
  open, 90
  quasismooth, 90
  suburban, 91
  undulating, sloping land-sea,
    117
  urban, 91
Tesselation, 564
Time-delayed pilot diversity
  combiner, 439

Time delay spread, 46
Time diversity, 312
  fundamental limitation, 313
Time-division retransmission,
    498
  maximum range, 500
  measured probability distri-
    bution, 503
  timing sequence, 498
Tokyo, field strength measure-
    ments, 98
  local mean signal in, 122
Transmission coefficient, me-
    dian values, 631
Transmission factors, 80
Triangulation, 570
Trilateration, 569
Trophospheric scattering,
    90
Tunnels, propagation in, 110
Two-dimensional traffic simu-
    lation, 598
  call attempts, 599
  call duration distribution,
    600
  velocity distribution, 600

Urban field strength variations,
    91

Vehicle locating, 569, 574
  electronic fence, 571
  phase ranging, 570
  pulse ranging, 570
  triangulation, 570
  trilateration, 569
Vertical dipole, 148
Voice signal, models, 289
  peaking factor, 246